Quasi-Geostrophic Theory of Oceans and Atmosphere

ATMOSPHERIC AND OCEANOGRAPHIC SCIENCES LIBRARY

VOLUME 45

For further volumes:
http://www.springer.com/series/5669

Fabio Cavallini • Fulvio Crisciani

Quasi-Geostrophic Theory of Oceans and Atmosphere

Topics in the Dynamics and Thermodynamics of the Fluid Earth

Springer

Fabio Cavallini
Ist. Naz. di Oceanografia e
di Geofisica Sperimentale – OGS
Geofisica della Litosfera (GDL)
Borgo Grotta Gigante 42/c
Sgonico
Italy

Fulvio Crisciani
Dipartimento di Fisica
Università di Trieste
Trieste
Italy

ISSN 1383-8601
ISBN 978-94-007-4690-9 ISBN 978-94-007-4691-6 (eBook)
DOI 10.1007/978-94-007-4691-6
Springer Dordrecht Heidelberg New York London

Library of Congress Control Number: 2012945304

Printed on acid-free paper

Springer is part of Springer Science+Business Media (www.springer.com)

Respectfully dedicated
to Professor Emeritus Paolo Budinich[1] *and*
to Professor Emeritus Giuseppe Furlan[2]

[1]Father of Trieste, "Town of Science".

[2]A Master of Theoretical Physics and the first to introduce Geophysical Fluid Dynamics as a curricular course for the degree in physics at the University of Trieste.

Preface

Large-scale flows, both in the atmosphere and in the ocean, satisfy to a large extent a balance between the Coriolis and the pressure gradient forces, that is, the geostrophic equilibrium. The time evolution of the fields involved in this equilibrium constitutes the subject of quasi-geostrophic theory.

This book has grown out of a course of advanced undergraduate and graduate lectures that one of the authors (Fu. Cr.) gave at the University of Trieste and at the "Abdus Salam International Center for Theoretical Physics" (ICTP) in the years 1997–2011.

There are excellent books on geophysical fluid dynamics (GFD) already available, but they are addressed either to undergraduate students (e.g. Cushman-Roisin 1994; Vallis 2006) or to advanced graduate courses and researchers (e.g. the "bible" by Pedlosky (1987)). The principal aim of this book is to fill the gap between these two levels of presentation by giving an overview of applied quasi-geostrophic theory suitable to advanced undergraduate and beginning graduate courses.

In this book, a detailed derivation of quasi-geostrophic fluid dynamics is given by means of an almost exclusively analytical approach, starting from the classical mechanics of fluids and having in mind the behaviour of large-scale currents and winds. Scale analysis, asymptotic expansions and boundary layer techniques are used wherever possible and repeatedly illustrated in different models; at the same time, the vorticity equations, derived at different scales by means of the above cited techniques, are systematically interpreted in terms of the fundamental Ertel's theorem.

Moreover, basic arguments, such as flow energetics or Rossby waves, are discussed in various contexts depending on the assumed stratification of each model. Thus, the application-oriented reader will be able to appreciate the conditions under which quasi-geostrophic models may be safely used.

The outline of this book contains the following; Part I (Chaps. 1 and 2) aims to fill the possible gaps in fluid dynamics the reader may have, while Part II (Chaps. 3–5) develops quasi-geostrophic theory in view of realistic geophysical applications. The different quasi-geostrophic models are presented in order of increasing complexity of the vertical structure, from the single-layer to the continuous stratification.

Chapter 1 presents basic ideas, computational rules and equations pertaining classical mechanics of fluids in an inertial system, although in the presence of gravity.

Chapter 2 consists, schematically, of four parts. In the first part, one deals with the concepts and the equations of thermodynamics that are necessary to close the set of governing equations of the fluids, and as an application, internal waves are investigated with full details. In the second part, the motion is reformulated with respect to a uniformly rotating frame of reference, just like we perceive the ocean and the atmosphere. Based on equations above, both Ertel's and Jeffrey's theorems are presented. The first application of the so obtained equations is considered in the third part, where the dynamics of various types of long gravity waves is explored. The last part of the chapter constitutes a comprehensive introduction to the quasi-geostrophic dynamics, which starts from the basic phenomenology and ends with the non-dimensional formulation of the governing equations of large-scale, geophysical flows. Chapters 1 and 2 contain the background material necessary to develop the quasi-geostrophic dynamics as presented, within different ambits, in the following three chapters (Chaps. 3–5).

Chapter 3 begins with the single-layer, shallow-water model and derives a set of related solutions, such as the steady motion over topography, the Fofonoff mode and many features of Rossby waves. Then, after the introduction of the surface and benthic Ekman layers, the homogeneous model of wind-driven circulation is expounded; in particular, full details are devoted to the boundary layer approach in the linear context (Stommel's and Munk's solutions). Finally, a classical application of the Ekman layer to the atmosphere is also described.

Chapter 4 is devoted to the two-layer model, which serves as a typical introduction to baroclinicity. Once the governing equations have been derived and solved in some simple case, the energetics is evaluated, and in particular, the basic concept of available potential energy is pointed out.

Chapter 5 deals with continuously stratified flows, both in the ocean and the atmosphere. In this chapter, the laws of the thermodynamics introduced in Chap. 2 are recovered; the mesoscale and the basin scale circulation in an adiabatic ocean are considered, while thermally forced stationary waves in the atmosphere are expounded in detail, and a class of solutions is presented.

Fabio Cavallini
Fulvio Crisciani
Trieste

Acknowledgements

Acknowledgements are due to many people for their advice and encouragement.

First of all, I wish to mention Joseph Pedlosky (WHOI) whose books, papers, seminars and personal communications represent an incomparable beacon to sail in the realm of Geophysical Fluid Dynamics. The same gratitude is also due to Giuseppe Furlan (ICTP, Trieste), who transformed my curiosity in Geophysical Fluid Dynamics into a serious and stimulating task culminated in teaching it at the University of Trieste.

I remind with emotion that the first quasi-geostrophic equation I read, many years ago, was written by my friend and colleague Roberto Purini (ISAC CNR, Rome) who first introduced me to this theory. It is a pleasure to mention several collaborations with George Carnevale (Scripps Institute of Oceanography) and with Renzo Mosetti (OGS, Trieste) in the context of relaxed but deep discussions which lead to several papers. Helpful discussions with Benoit Cushman-Roisin are also gratefully acknowledged.

For his skillful and fruitful collaboration, I wish to thank Gualtiero Badin (now researcher at Princeton University) who has been my best student at Trieste University. Moreover, Sara Bacer has collaborated in the proofreading of the draft. Finally, thanks are due to Emilio Caterini (ISMAR CNR, Trieste) for his careful drawing of the figures.

Fulvio Crisciani

In addition to the acknowledgements above, which I share, I would like to warmly thank my colleagues and friends José M. Carcione and Géza Seriani for many helpful conversations on physical, mathematical and computational related topics.

Fabio Cavallini

Contents

Part I Fundamentals

Part I
Fundamentals

Cependant tout ce que la Théorie des fluides renferme, est contenu dans les deux équations rapportées cy-dessus ($. XXXIV.),[1] *de sorte que se ne sont pas les principes de Méchanique qui nous manquent dans la poursuite de ces recherches, mais uniquement l'Analyse ...*[2]

Leonhard Euler (1757)

---

[1]Here are Euler's equations in original notation:

$$\left(\frac{du}{dt}\right)+u\left(\frac{du}{dx}\right)+v\left(\frac{du}{dy}\right)+w\left(\frac{du}{dz}\right)=X$$

$$\left(\frac{dv}{dt}\right)+u\left(\frac{dv}{dx}\right)+v\left(\frac{dv}{dy}\right)+w\left(\frac{dv}{dz}\right)=Y$$

$$\left(\frac{dw}{dt}\right)+u\left(\frac{dw}{dx}\right)+v\left(\frac{dw}{dy}\right)+w\left(\frac{dw}{dz}\right)=Z$$

which Euler further condenses into a single equation to be coupled with the continuity equation.

[2]English translation: *However, all that the Theory of fluids contains, is included within the two equations reported above ($. XXXIV.), so that what we miss in the development of these researches is not the principles of Mechanics, but only the Analysis ...*

Chapter 1
Basic Continuum Mechanics

Abstract At all scales, from galaxies to elementary particles, matter is discrete. Yet, the "continuum" paradigm is still of paramount importance, both in theory and in practice. This chapter briefly elucidates this paradox, with emphasis on fluid mechanics, and presents the basic conceptual tools needed to approach heuristically the physics of fluids.

Both the Lagrangian and the Eulerian descriptions of motion are introduced. The link between these two descriptions stems from the request that a property of an Eulerian field, at a given point and at a given time, coincides with the corresponding property of the Lagrangian particle that occupies that point at that time.

In the Lagrangian framework, the governing equations of fluid mechanics, the vorticity dynamics and the parameterization of turbulence are discussed. Illustrative applications of these results to simple models (e.g. flow in a channel and 2-D breezes) are also given.

1.1 Kinematics of Continua

The intuitive idea of "individual piece of fluid" is made precise in terms of "elementary material volume of fluid", or simply *parcel*, and the physico-mathematical approach to the physics of parcels is expounded.

Since, by definition of material volume, the total mass of fluid enclosed into a parcel is conserved, the concepts of velocity and acceleration can be attributed to each individual parcel; so, its kinematics can be established.

Hence, the behaviour of a parcel with respect to impermeable boundaries is also determined on general grounds.

F. Cavallini and F. Crisciani, *Quasi-Geostrophic Theory of Oceans and Atmosphere*,
Atmospheric and Oceanographic Sciences Library 45, DOI 10.1007/978-94-007-4691-6_1,

1.1.1 *The Continuum Hypothesis and the Concept of Material Element*

What Is a Continuum?

Fluid mechanics is concerned with the behaviour of matter on a macroscopic scale which is large with respect to the distance between molecules whose structure does not need to be taken into account explicitly. The behaviour of fluids is assumed to be the same as they were perfectly continuous in structure and the physical properties of the matter contained within a given small volume are regarded as being spread uniformly over that volume. This is the so-called *continuum hypothesis*, which is supposed to be hereafter valid without exceptions. From the observational viewpoint, the reason why the particle structure of the fluid is irrelevant is that the sensitive volume of a certain instrument embedded in the fluid itself is small enough for the measurement to be a local one relative to the macroscopic scale, even if it is large enough for the fluctuations arising from the molecular motion to have no effect on the observed average. If the volume of fluid to which the instrument responds were comparable with the volume in which variations due to molecular fluctuations take place, observations would fluctuate from one observation to another and the results would vary in an irregular way with the size of the sensitive volume of the instrument. On the opposite, if the volume of fluid to which the instrument responds were too large relatively to the variations associated to spatial distribution of physical quantities, the instrument would not be able to detect even macroscopic features of the fluid. Therefore, the fluid is regarded as a continuum when the measured fluid property is constant for sensitive volumes small on the macroscopic scale, but large on the microscopic (molecular) scale. This concept is illustrated in Fig. 1.1. In any case, there is observational evidence that the common real fluids, both gases and liquids, move as they are continuous both under normal conditions and also for considerable departures from them.

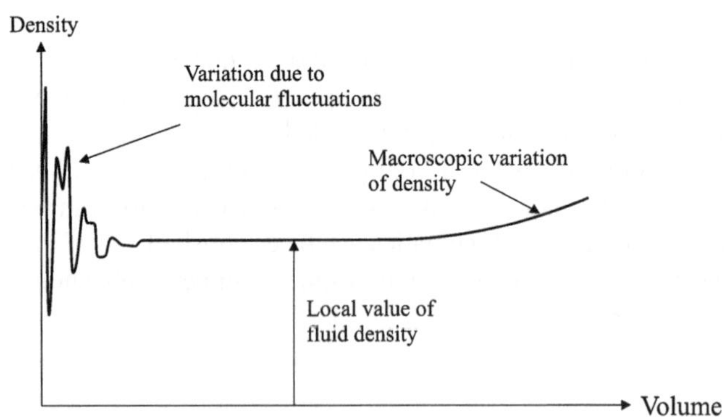

Fig. 1.1 The plot illustrates qualitatively the effect of size of sensitive volume of an instrument on the density measurement

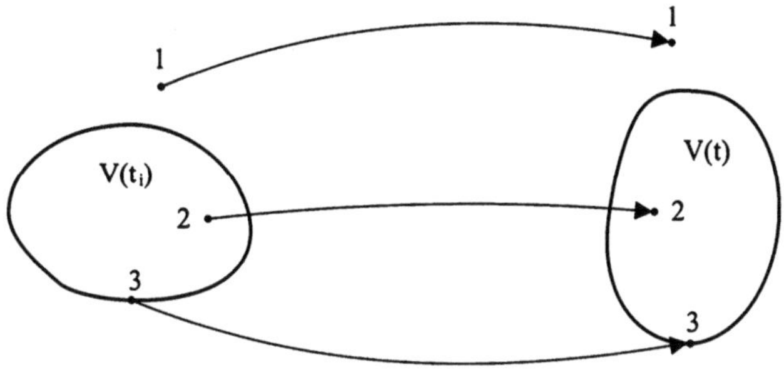

Fig. 1.2 Time evolution of a material volume V of fluid. Assume that, at a certain time t_i, point "1" lies outside $V(t_i)$, point "2" lies inside $V(t_i)$, while point "3" belongs to the surface enclosing $V(t_i)$. Then, at each subsequent time t, point "1" will be situated outside $V(t_i)$, point "2" will be situated inside $V(t_i)$ and point "3" will belong to the surface enclosing $V(t_i)$

The Concept of Material Element

Physical laws apply directly to a fixed collection of matter and for this reason the equations governing fluid mechanics and thermodynamics are developed most intuitively in a framework where the dynamical or physical quantities refer to identifiable pieces of matter (*Lagrangian description*) rather than to fields (*Eulerian description*). The former description relies on the concept of *material volume* of fluid (or *parcel*) which, by definition, always consists of the same fluid portion and moves with it, as Fig. 1.2 shows. In this perspective, the flow quantities are defined as functions of time and of the choice of a material element of fluid and describe the history of this selected fluid element.

Material elements of fluid change their shape as they move, so each selected element should be selected in such a way that its linear extension is not involved; therefore, the element is specified by the position of its centre of mass at some initial instant, on the understanding that its initial linear dimensions are so small as to guarantee smallness at all subsequent instants in spite of distortions and extensions of the element. The Eulerian description is related to the Lagrangian description by the following kinematic constraint: the field property at a given location and time must equal the property possessed by the material element occupying that position at that instant.

1.1.2 Kinematics of Material Elements

The Lagrangian Derivative

A material element of fluid is identified by the "initial" position of its centre of mass, say $\mathbf{x}_0 = (x_0, y_0, z_0)$ in a Cartesian coordinate frame, and by the trajectory

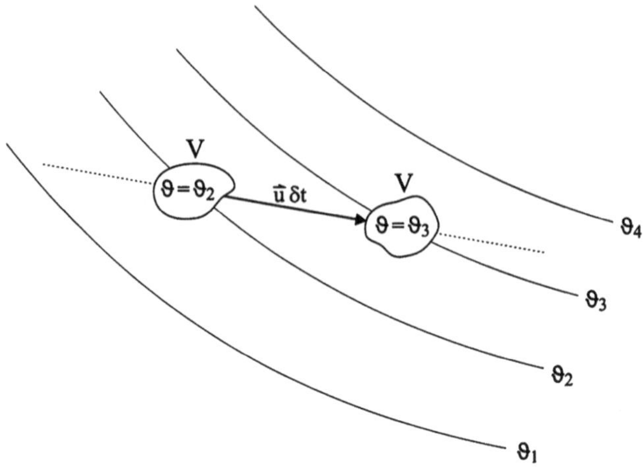

Fig. 1.3 Mr. Lagrange is situated inside the small material volume V, in motion with velocity $\mathbf{u}$ in a portion of space where the scalar field $\theta(\mathbf{x},t)$ is defined. He measures θ and finds the value θ_2 when the material volume that carries him crosses the isoline $\theta = \theta_2$. Then, after a time δt, the same volume displaces itself of the amount $\mathbf{u}\,\delta t$, thus crossing the isoline $\theta = \theta_3$. Mr. Lagrange repeats the measurement and now he finds the value θ_3, and so on. Finally, he states Eq. (1.8)

$\mathbf{x}(t) = (x(t), y(t), z(t))$ traced out in the course of its motion with velocity $\mathbf{u}(t) = (u, v, w)$, where

$$u = \frac{\mathrm{d}x}{\mathrm{d}t}, \qquad v = \frac{\mathrm{d}y}{\mathrm{d}t}, \qquad w = \frac{\mathrm{d}z}{\mathrm{d}t}. \tag{1.1}$$

The vector $\mathrm{d}\mathbf{x} = (\mathrm{d}x, \mathrm{d}y, \mathrm{d}z)$ is the incremental displacement of the given material element during the time interval $\mathrm{d}t$. By the kinematic constraint reported at the end of Sect. 1.1.1, velocity (1.1) is equal to the field value $\mathbf{u}(\mathbf{x}(t))$ located at the material element's position $\mathbf{x}(t)$ at time t. The time growth rate of a scalar, say θ, representing a physical quantity of a certain material element of fluid in motion depends both on the change of the position of the same fluid element in the course of time and on the explicit change in time of the scalar itself. The situation is playfully described in Fig. 1.3.

Therefore, if δt is a small time interval, the total variation of θ during this interval is

$$\theta(\mathbf{x}+\mathbf{u}\,\delta t, t+\delta t) - \theta(\mathbf{x}, t) \tag{1.2}$$

As $\delta t \to 0$, Taylor's expansion of (1.2) yields

$$\theta(\mathbf{x}+\mathbf{u}\,\delta t, t+\delta t) - \theta(\mathbf{x}, t) \tag{1.3}$$

$$= \frac{\partial\theta}{\partial x} u\,\delta t + \frac{\partial\theta}{\partial y} v\,\delta t + \frac{\partial\theta}{\partial z} w\,\delta t + \frac{\partial\theta}{\partial t}\delta t + O\left((\delta t)^2\right) \tag{1.4}$$

$$= \delta t\left(\frac{\partial}{\partial t} + \mathbf{u}\cdot\nabla\right)\theta + O\left((\delta t)^2\right) \tag{1.5}$$

where the components of $\mathbf{u}$ are given by (1.1). For $\delta t \to 0$, the time growth rate of $\theta(\mathbf{x}, t)$ is

$$\frac{\theta(\mathbf{x}+\mathbf{u}\,\delta t, t+\delta t) - \theta(\mathbf{x}, t)}{\delta t} = \left(\frac{\partial}{\partial t} + \mathbf{u}\cdot\nabla\right)\theta(\mathbf{x}, t) + O(\delta t) \tag{1.6}$$

and converges to the so-called *Lagrangian derivative* of θ. In other words,

$$\lim_{\delta t\to 0} \frac{\theta(\mathbf{x}+\mathbf{u}\,\delta t, t+\delta t) - \theta(\mathbf{x}, t)}{\delta t} = \frac{\mathrm{D}}{\mathrm{D}t}\theta(\mathbf{x}, t) \tag{1.7}$$

where, by definition,

$$\frac{\mathrm{D}}{\mathrm{D}t} := \frac{\partial}{\partial t} + \mathbf{u}\cdot\nabla$$

Thus, the Lagrangian derivative of θ, that is,

$$\frac{\mathrm{D}\theta}{\mathrm{D}t} := \frac{\partial\theta}{\partial t} + \mathbf{u}\cdot\nabla\theta \tag{1.8}$$

includes two contributions:

- The first, $\partial\theta/\partial t$, is introduced by temporal changes at the position $\mathbf{x}$ where the material volume is instantaneously located at time t. It is called the *Eulerian (or local) derivative.*
- The second, $\mathbf{u}\cdot\nabla\theta$, is due to the motion of the material volume to positions with different values of θ. It is the so-called *advective term.*

Note that the equation

$$\frac{\mathrm{D}\theta}{\mathrm{D}t} = 0 \tag{1.9}$$

does not mean that θ is a constant but, rather, that in each element of fluid in motion θ takes a value that does not change with time although, in general, it differs from element to element. In other words, the motion takes place along the isolines of θ. Moreover, Eq. (1.9) results to be the simplest covariant equation with respect to Galilean transformations (see Appendix at p. 38): this is a notable property, shared – under suitable conditions – by several physical quantities, such as seawater density, potential temperature of the atmosphere and potential vorticity.

The following differentiation rules immediately derive from (1.8):

$$\begin{aligned}
\frac{\mathrm{D}}{\mathrm{D}t}(\alpha\,\theta) &= \alpha\,\frac{\mathrm{D}\theta}{\mathrm{D}t} \qquad \text{if } \alpha \text{ is a constant}\\
\frac{\mathrm{D}}{\mathrm{D}t}(\theta_1+\theta_2) &= \frac{\mathrm{D}\theta_1}{\mathrm{D}t} + \frac{\mathrm{D}\theta_2}{\mathrm{D}t}\\
\frac{\mathrm{D}}{\mathrm{D}t}(\theta_1\,\theta_2) &= \theta_1\,\frac{\mathrm{D}\theta_2}{\mathrm{D}t} + \theta_2\,\frac{\mathrm{D}\theta_1}{\mathrm{D}t}
\end{aligned} \tag{1.10}$$

Example

Take, in particular, $\theta(\mathbf{x},t) = x$; then, from the definition of Lagrangian derivative, we obtain $\mathrm{D}x/\mathrm{D}t = u$. In the same way, we obtain the Lagrangian derivatives $Dy/Dt = v$ and $Dz/Dt = w$. On the whole, the vector

$$\mathbf{u} = \left(\frac{\mathrm{D}x}{\mathrm{D}t}, \frac{\mathrm{D}y}{\mathrm{D}t}, \frac{\mathrm{D}z}{\mathrm{D}t}\right)$$

is the Lagrangian velocity of the material element of fluid located at the position $\mathbf{x}$ and time t, which coincides – according to the kinematic constraint discussed in Sect. 1.1.1 – with Euler's velocity of the fluid at the same position and at the same time.

Reynolds' Transport Theorem

Let us consider the integral quantity

$$\int_{V(t)} \theta(\mathbf{x},t)\,\mathrm{d}V' \tag{1.11}$$

where $V(t)$ is a finite material volume of fluid. The time rate of change of (1.11) takes into account two terms:

- The first, that is,
$$\int_{V(t)} \frac{\partial\theta}{\partial t}\,\mathrm{d}V' \tag{1.12}$$
is generated by the unsteadiness of the field θ, and
- The second, that is,
$$\int_{S(t)} \hat{\mathbf{n}}\cdot(\theta\,\mathbf{u})\,\mathrm{d}S' \tag{1.13}$$
is the net transfer of θ across the boundary $S(t)$ of $V(t)$, where $\hat{\mathbf{n}}$ is the outward unit vector locally normal to $S(t)$.

Collecting (1.12) and (1.13) and using the divergence theorem in the latter integral, we have

$$\frac{\mathrm{d}}{\mathrm{d}t}\int_{V(t)} \theta\,\mathrm{d}V' = \int_{V(t)} \left(\frac{\partial\theta}{\partial t} + \mathrm{div}\,(\theta\,\mathbf{u})\right)\mathrm{d}V' \tag{1.14}$$

where div denotes the divergence operator. Because of (1.8), Eq. (1.14) can be rewritten as follows:

$$\frac{\mathrm{d}}{\mathrm{d}t}\int_{V(t)} \theta\,\mathrm{d}V' = \int_{V(t)} \left(\frac{\mathrm{D}\theta}{\mathrm{D}t} + \theta\,\mathrm{div}\,\mathbf{u}\right)\mathrm{d}V' \tag{1.15}$$

to yield *Reynolds' transport theorem*.

The Conservation of Mass

In classical mechanics, the physical space has neither sources nor sinks of matter. This fact implies that the mass of any material volume of fluid is conserved in time, whereas in a geometric volume the matter content may vary in time. Thus, if $\rho(\mathbf{x},t)$ is the density field of the fluid under consideration, we conclude that

$$\frac{\mathrm{d}}{\mathrm{d}t}\int_{V(t)} \rho(\mathbf{x},t)\,\mathrm{d}V' = 0 \tag{1.16}$$

Statement (1.16) can be developed by resorting to Reynolds' theorem (1.15) to obtain

$$\int_{V(t)} \left(\frac{\mathrm{D}\rho}{\mathrm{D}t} + \rho \,\mathrm{div}\, \mathbf{u}\right) \mathrm{d}V' = 0 \qquad\qquad \forall V(t)$$

whence the governing equation for the density field

$$\frac{\mathrm{D}\rho}{\mathrm{D}t} + \rho \,\mathrm{div}\, \mathbf{u} = 0 \tag{1.17}$$

follows, that is, the *principle of mass conservation* for a fluid.

An Important Application of Reynolds' Theorem

One relevant application of Reynolds' transport theorem, through the important result concerning the conservation of mass (1.17), involves a scalar field of the kind $\theta(\mathbf{x},t)\,\rho(\mathbf{x},t)$, where the density field $\rho(\mathbf{x},t)$ satisfies (1.17). In fact, the substitution in (1.15) yields

$$\frac{\mathrm{d}}{\mathrm{d}t}\int_{V(t)} \rho\,\theta\,\mathrm{d}V' = \int_{V(t)} \left[\rho\,\frac{\mathrm{D}\theta}{\mathrm{D}t} + \theta\,\frac{\mathrm{D}\rho}{\mathrm{D}t} + \rho\,\theta \,\mathrm{div}\, \mathbf{u}\right] \mathrm{d}V' \tag{1.18}$$

but, because of (1.17), Eq. (1.18) simplifies into

$$\frac{\mathrm{d}}{\mathrm{d}t}\int_{V(t)} \rho\,\theta\,\mathrm{d}V' = \int_{V(t)} \rho\,\frac{\mathrm{D}\theta}{\mathrm{D}t}\,\mathrm{d}V' \tag{1.19}$$

The usefulness of (1.19) will be evident in the following computation.

The Acceleration of an Element of Fluid

A material element of fluid may experience acceleration through moving to a position where $\mathbf{u}$ has a different value. Equation (1.7) can be used to evaluate each component of the acceleration; for instance, if $\theta = u$,

$$\lim_{\delta t \to 0} \frac{u(\mathbf{x}+\mathbf{u}\,\delta t, t+\delta t) - u(\mathbf{x},t)}{\delta t} = \frac{\mathrm{D}}{\mathrm{D}t} u(\mathbf{x},t) \tag{1.20}$$

Then, by adding together the equations for the three scalar components u, v, w, the vector acceleration turns out to be

$$\frac{\mathrm{D}}{\mathrm{D}t}\mathbf{u}(\mathbf{x},t) = \left(\frac{\partial}{\partial t} + \mathbf{u}(\mathbf{x},t)\cdot\nabla\right)\mathbf{u}(\mathbf{x},t) \tag{1.21}$$

Note that the advective term of (1.21) is nonlinear with respect to $\mathbf{u}$, and this is a typical aspect of Fluid mechanics. Equation (1.19) can be applied, in particular, to the current field $\mathbf{u}(\mathbf{x},t)$ to obtain the equation

$$\frac{\mathrm{d}}{\mathrm{d}t}\int_{V(t)} \rho\,\mathbf{u}\,\mathrm{d}V' = \int_{V(t)} \rho\,\frac{\mathrm{D}\mathbf{u}}{\mathrm{D}t}\,\mathrm{d}V' \tag{1.22}$$

that will be involved in the momentum equation (see next section).

Kinematic Boundary Conditions

The equations which express the kinematic boundary conditions, to be applied to a flow bounded somewhere, derive from the following considerations. Let $\mathbf{x}_0$ be the position, at the time t_0, of a point-like fluid element moving, with velocity $\mathbf{u}$, on the surface S whose equation is $F(\mathbf{x},t) = 0$. In the course of time, from t_0 to $t_0 + \delta t$, the fluid element moves on S from $\mathbf{x}_0$ to $\mathbf{x}_0 + \mathbf{u}\,\delta t$, leaving unaffected the shape of S, so

$$F(\mathbf{x_0} + \mathbf{u}\,\delta t, t_0 + \delta t) = F(\mathbf{x_0}, t_0) \tag{1.23}$$

Then, following the same procedure as for (1.3), one concludes that

$$\frac{\mathrm{D}F}{\mathrm{D}t} = 0 \tag{1.24}$$

Equation (1.24) is not an equation for the surface S but, rather, a constraint on the current $\mathbf{u}$ (included in D/Dt) of the fluid element that is constrained to move on S.

Consideration of three special cases of (1.24) will clarify how to use it.

1. If F represents a "wall" located in $x = x_0$, it takes the form

$$F = x - x_0 \tag{1.25}$$

and (1.24) gives

$$u(x_0, y, x, t) = 0 \tag{1.26}$$

Equation (1.26) shows that the flow cannot cross the wall; in fact, the unit vector normal to the wall is $\hat{\mathbf{i}}$ and so $\hat{\mathbf{i}}\cdot\mathbf{u} = u(x_0, y, x, t)$, that is, $\hat{\mathbf{i}}\cdot\mathbf{u} = 0$ because of (1.26).

2. If F represents the bathymetric or topographic profile $z = b(x, y)$, then

$$F(x, y, z) = z - b(x, y) \tag{1.27}$$

and (1.24) gives

$$w(x,y,b,t) = \mathbf{u} \cdot \nabla b(x,y) \tag{1.28}$$

If the bottom is flat, then b is a constant function and (1.28) implies

$$w(x,y,b,t) = 0 \tag{1.29}$$

3. If F represents the time-dependent interface between two fluids, located in $z = \eta(x,y,t)$, then

$$F = z - \eta(x,y,t) \tag{1.30}$$

whence the boundary condition

$$w(x,y,\eta,t) = \frac{\mathrm{D}\eta}{\mathrm{D}t} \tag{1.31}$$

follows.

In some circumstances, additional boundary conditions must be applied to the flow in order to obtain a unique model solution. However, in this case, no general equation of the kind (1.24) can be invoked; so, these conditions will be considered only in some special model of Sects. 1.2.3 and 3.2.3.

1.2 Dynamics of Fluids

Parcels are massive objects; so, in inertial frames, their momentum is governed by Newton's second law, the forces being mainly due to gravity and to pressure gradient fields.

From the momentum equation, one may derive the equation for the relative vorticity of the flow: as some examples will show, the latter yields a useful alternative approach, especially when dealing with two-dimensional fluids.

Finally, the consideration of turbulence allows us to take into account two further aspects of real flows: the influence of boundaries as source of forcing and the tendency of flows to decay because of internal dissipation.

1.2.1 Momentum Equation

Euler's Equation

For the time being, we refer to an inertial system. Then, the equation of motion of a fluid is Newton's second law applied to a material volume $V(t)$: it equates the time rate of change of momentum $\rho\,\mathbf{u}$ of the fluid included into $V(t)$, that is,

$$\frac{\mathrm{d}}{\mathrm{d}t}\int_{V(t)} \rho\,\mathbf{u}\,\mathrm{d}V' \tag{1.32}$$

and the sum of the body and surface forces acting on the same material volume, that is,

$$\int_{V(t)} \rho \, \mathbf{F} \, dV' + \int_{S(t)} \boldsymbol{\tau} \hat{\mathbf{n}} \, dS' \tag{1.33}$$

where $\mathbf{F}$ is body force per unit mass, $\boldsymbol{\tau}$ is stress tensor and $\hat{\mathbf{n}}$ is the outwards unit normal vector. The first term of (1.33) refers to body forces, among which the most relevant is due to gravity. For most of our purposes, we put

$$\mathbf{F} = -g\hat{\mathbf{k}} \tag{1.34}$$

where g is taken constant and the unit vector $\hat{\mathbf{k}}$ is antiparallel to the local gravity acceleration vector $\mathbf{g}$. In the present context, we take into account an unrealistic Earth with its own gravitational field, but moving in an inertial way; so, Newton's second law can be applied. In particular, Earth's rotation around its polar axis is here ignored. All this is the result of an approximation that will be clarified at the end of Sect. 2.3.2.

The second term of (1.33), in which $S(t)$ is the surface enclosing $V(t)$ and $\hat{\mathbf{n}}$ is the local unit vector orthogonal to $S(t)$ and pointing outside $S(t)$, represents the sum of the forces exerted on $V(t)$ by the surrounding matter. We disregard the effects of molecular viscosity which is quite negligible in the context of geophysical fluid dynamics (GFD), and thus one could prove that the stress tensor τ simplifies into the diagonal form

$$\tau_{ij} = -p\,\delta_{ij} \tag{1.35}$$

where $p = p(\mathbf{x}, t)$ is the pressure field.

Using components, the second term of (1.33) becomes (see Appendix A, p. 365)

$$\int_{S(t)} \boldsymbol{\tau} \hat{\mathbf{n}} \, dS' = \int_{S(t)} \sum_{i,j=1}^{3} \tau_{ij} \, (\hat{\mathbf{e}}_i \otimes \hat{\mathbf{e}}_j) \hat{\mathbf{n}} \, dS' = \int_{S(t)} \sum_{i,j=1}^{3} \tau_{ij} \, (\hat{\mathbf{e}}_j \cdot \hat{\mathbf{n}}) \hat{\mathbf{e}}_i \, dS'$$

where $\hat{\mathbf{e}}_1$, $\hat{\mathbf{e}}_2$, $\hat{\mathbf{e}}_3$ form an orthonormal basis, and symbol $\otimes$ denotes the tensor product.[1] Hence, using (1.35), we get

$$\int_{S(t)} \boldsymbol{\tau} \hat{\mathbf{n}} \, dS' = -\sum_{k=1}^{3} \left(\int_{S(t)} \hat{\mathbf{n}} \cdot (p\hat{\mathbf{e}}_k) \, dS' \right) \hat{\mathbf{e}}_k \tag{1.36}$$

Then, by the divergence theorem, we get

$$\int_{S(t)} \boldsymbol{\tau} \hat{\mathbf{n}} \, dS' = -\sum_{k=1}^{3} \left(\int_{V(t)} \operatorname{div}(p\hat{\mathbf{e}}_k) \, dV' \right) \hat{\mathbf{e}}_k \tag{1.37}$$

[1] Recall that the *tensor product* between any two vectors $\mathbf{a}$ and $\mathbf{b}$ is the linear operator defined by $\mathbf{a} \otimes \mathbf{b} : \mathbf{x} \mapsto (\mathbf{b} \cdot \mathbf{x})\,\mathbf{a}$.

and finally, noting that $\mathrm{div}\,(p\hat{\mathbf{e}}_k) = \partial p/\partial x_k$ since $\mathrm{div} = \Sigma_{i=1}^{3}\hat{\mathbf{e}}_i\partial/\partial x_i$, we have

$$\int_{S(t)} \boldsymbol{\tau}\hat{\mathbf{n}}\,\mathrm{d}S' = -\int_{V(t)} \nabla p\,\mathrm{d}V' \tag{1.38}$$

Therefore, the forces (1.33) can be written just in terms of volume integrals as

$$\int_{V(t)} (\rho\,\mathbf{F} - \nabla p)\,\mathrm{d}V' \tag{1.39}$$

Because of (1.32) and (1.39), the continuum version of Newton's second law takes the form

$$\frac{\mathrm{d}}{\mathrm{d}t}\int_{V(t)} \rho\,\mathbf{u}\,\mathrm{d}V' = \int_{V(t)} (\rho\,\mathbf{F} - \nabla p)\,\mathrm{d}V' \tag{1.40}$$

and, using (1.22) in the l.h.s. of (1.40), the latter becomes

$$\int_{V(t)} \left(\rho\,\frac{\mathrm{D}\mathbf{u}}{\mathrm{D}t} - \rho\,\mathbf{F} + \nabla p\right)\mathrm{d}V' = 0 \tag{1.41}$$

Since (1.41) holds for every material volume, the quantity in brackets is necessarily zero. Thus, in an inertial frame, Newton's second law for individual bodies of fluid is

$$\rho\,\frac{\mathrm{D}\mathbf{u}}{\mathrm{D}t} = \rho\,\mathbf{F} - \nabla p \tag{1.42}$$

This is known as *Euler's equation*. Recalling (1.21) and using (1.34), Euler's equation takes the form

$$\frac{\partial\mathbf{u}}{\partial t} + (\mathbf{u}\cdot\nabla)\,\mathbf{u} = -\frac{1}{\rho}\nabla p - g\hat{\mathbf{k}} \tag{1.43}$$

The scalar components of (1.43) are

$$\frac{\partial u}{\partial t} + \mathbf{u}\cdot\nabla u = -\frac{1}{\rho}\frac{\partial p}{\partial x} \tag{1.44}$$

$$\frac{\partial v}{\partial t} + \mathbf{u}\cdot\nabla v = -\frac{1}{\rho}\frac{\partial p}{\partial y} \tag{1.45}$$

$$\frac{\partial w}{\partial t} + \mathbf{u}\cdot\nabla w = -\frac{1}{\rho}\frac{\partial p}{\partial z} - g \tag{1.46}$$

Navier-Stokes Equations

A constitutive law more general than (1.35) is (e.g. Fletcher 1988, Sect. 11.2.3)

$$\boldsymbol{\tau} = \boldsymbol{\tau}^{(1)} + \boldsymbol{\tau}^{(2)} \tag{1.47}$$

where

$$\boldsymbol{\tau}^{(1)} := -p\mathbf{I} - \frac{2}{3}\mu\,(\mathrm{div}\,\mathbf{u})\,\mathbf{I} \tag{1.48}$$

with $\mathbf{I}$ the identity matrix, while the components of $\boldsymbol{\tau}^{(2)}$ are

$$\tau_{ij}^{(2)} := \mu\left(\frac{\partial u_j}{\partial x_i} + \frac{\partial u_i}{\partial x_j}\right) \tag{1.49}$$

with μ the *molecular viscosity* of the fluid. Equations (1.47)–(1.49) define *Newtonian fluids*.

Using assumption (1.47) in place of (1.35) yields[2]

$$\rho\,\frac{\mathrm{D}\mathbf{u}}{\mathrm{D}t} = \rho\,\mathbf{F} - \nabla p + \mu\nabla^2\mathbf{u} + \frac{\mu}{3}\,\nabla\mathrm{div}\,\mathbf{u} \tag{1.50}$$

in place of (1.42). Indeed, arguing as in (1.36)–(1.38), one obtains

$$\int_{S(t)} \boldsymbol{\tau}^{(1)}\,\hat{\mathbf{n}}\,\mathrm{d}S' = \int_{V(t)} \left(-\nabla p - \frac{2}{3}\,\mu\,\nabla\mathrm{div}\,\mathbf{u}\right)\mathrm{d}V' \tag{1.51}$$

while straightforward computations yield

$$\int_{S(t)} \boldsymbol{\tau}^{(2)}\,\hat{\mathbf{n}}\,\mathrm{d}S' = \int_{V(t)} (\mu\,\nabla\mathrm{div}\,\mathbf{u} + \mu\,\nabla^2\mathbf{u})\,\mathrm{d}V' \tag{1.52}$$

Summing (1.51) and (1.52), we get

$$\int_{S(t)} \boldsymbol{\tau}\,\hat{\mathbf{n}}\,\mathrm{d}S' = \int_{V(t)} \left(-\nabla p + \mu\,\nabla^2\mathbf{u} + \frac{\mu}{3}\,\nabla\mathrm{div}\,\mathbf{u}\right)\mathrm{d}V' \tag{1.53}$$

and hence, arguing as in (1.38)–(1.42), we finally obtain (1.50). With the additional assumption of incompressibility, Eq. (1.50) yields the celebrated *Navier-Stokes equations*:

$$\begin{cases} \rho\,\dfrac{\mathrm{D}\mathbf{u}}{\mathrm{D}t} = \rho\,\mathbf{F} - \nabla p + \mu\nabla^2\mathbf{u} \\ \mathrm{div}\,\mathbf{u} = 0 \end{cases} \tag{1.54}$$

Fluid at Rest

By definition, the condition of fluid at rest means

$$\mathbf{u} = 0 \qquad\qquad \forall\,(\mathbf{x}, t) \tag{1.55}$$

[2]Equation (1.50) corresponds to Eqs. (1.4.4) and (1.4.5) of Pedlosky (1987).

so (1.44) and (1.45) imply $\partial p/\partial x = 0$ and $\partial p/\partial y = 0$, respectively, while (1.46) yields the *hydrostatic equilibrium*, that is,

$$\frac{\partial p}{\partial z} + g\rho = 0 \tag{1.56}$$

Equation (1.56) can be integrated with respect to z to give

$$p(z) = p(\eta) + g \int_z^{\eta} \rho(z')\,\mathrm{d}z' \tag{1.57}$$

In (1.57), $p(\eta)$ is the hydrostatic pressure that the fluid takes at a certain height (or depth) $z = \eta$ and plays the role of boundary condition in the integration of (1.56). In the atmospheric case, (1.57) takes the form

$$p(z) = g \int_z^{+\infty} \rho(z')\,\mathrm{d}z' \qquad \text{(atmosphere)} \tag{1.58}$$

while in the oceanic case

$$p(z) = p_a + g \int_z^{\eta} \rho(z')\,\mathrm{d}z' \qquad \text{(ocean)} \tag{1.59}$$

where $z = \eta$ is the ocean-atmosphere interface and p_a represents the atmospheric pressure at the sea level. Note that, while (1.55) implies (1.56), the latter equation does not imply the former but, rather, only the equation

$$\frac{Dw}{Dt} = 0 \tag{1.60}$$

that comes from (1.46). More realistically, hydrostatic equilibrium (1.56) is valid whenever the vertical acceleration of the fluid can be disregarded. Hydrostatic equilibrium is consistent with the horizontal fluid motion (i.e. $u \neq 0$, $v \neq 0$), and this is just what happens in the large-scale circulation both of the atmosphere and of the ocean. On the other hand, Eq. (1.56) is not satisfied by convective motions, in which the full 3-dimensional current is involved.

A water body may be at rest also in the case in which the atmospheric pressure at the sea level p_a and the air-sea interface η are spatially modulated and anti-correlated, as follows. If p_a is split into a constant component, say $p_{\mathrm{a}0}$, plus a space-dependent anomaly $\delta p_\mathrm{a}(x, y)$ and, for simplicity, a constant-density fluid is considered, then total pressure (1.59) takes the form

$$p(x, y, z) = p_{\mathrm{a}0} + \delta p_\mathrm{a}(x, y) + g\rho\,[\eta(x, y) - z] \tag{1.61}$$

While (1.56) is still valid, whatever δp_a and η may be, the horizontal pressure gradient evaluated from (1.61) is

$$\nabla_\mathrm{H} p = \nabla_\mathrm{H} \delta p_\mathrm{a} + g\rho\,\nabla_\mathrm{H}\eta \tag{1.62}$$

so (1.55) is realized provided that

$$\delta p_{\mathrm{a}} + g\rho\,\eta = \text{constant}$$

Under the assumption that the absence of pressure perturbation ($\delta p_{\mathrm{a}} = 0$) yields a flat air-sea interface ($\eta = 0$), the constant above is zero and the anti-correlation

$$\delta p_{\mathrm{a}} + g\rho\,\eta = 0 \tag{1.63}$$

between δp_{a} and η is finally established. Relationship (1.63) is called *inverted barometer effect*. The "golden rule" inferred from (1.63) says that an atmospheric pressure increase/decrease of 1 hPa yields a related sea-level decrease/increase of 1 cm. Note that (1.63) is independent of vertical translations of the z-axis.

A Class of Solutions of Euler's Equation

The governing equations of a fluid in motion, derived up to now, are the conservation of mass (1.17) and Euler's equation in the presence of gravity, that is, (1.42). On the whole, four scalar equations but five unknown fields (u, v, w, p and ρ) are taken into account and the reason for this under-determination lies in the temporary lack of any thermodynamic equation giving a further link between u, v, w, p and ρ. In any case, the restrictive hypotheses of a constant density fluid (whose value is assumed to be assigned) simplifies (1.17) into

$$\operatorname{div}\mathbf{u} = 0 \tag{1.64}$$

so, now, four scalar equations, that is, (1.44)–(1.46) and (1.64) in place of (1.17), and four unknown fields (u, v, w, p) are considered.

For instance, each current field of the class

$$\mathbf{u} = u(y)\,\hat{\mathbf{i}} \tag{1.65}$$

where the profile $u(y)$ is arbitrary and satisfies the relationships $\operatorname{div}\mathbf{u} = \partial u/\partial x = 0$ and

$$(\mathbf{u}\cdot\nabla)\mathbf{u} = \left[u(y)\,\frac{\partial}{\partial x}\right]u(y)\,\hat{\mathbf{i}} = 0$$

Therefore, (1.64) is identically verified, while (1.43) becomes

$$\nabla p + \rho g\,\hat{\mathbf{k}} = 0 \tag{1.66}$$

The components of (1.66) are

$$\frac{\partial p}{\partial x} = 0 \qquad \frac{\partial p}{\partial y} = 0 \tag{1.67}$$

and

$$\frac{\partial p}{\partial z} + \rho g = 0 \tag{1.68}$$

whence

$$p = \rho g (\eta - z) + p(\eta) \tag{1.69}$$

where η is a constant depth or height. Equation (1.67) shows that the flow is not forced by any horizontal pressure gradient, consistently with the fact that the motion is steady. Currently, (1.65) can be conceived also as a rectilinear flow in a channel; but, in a more realistic model, the profile $u(y)$ should be somehow sensitive to the presence of the lateral boundaries of the channel itself. In fact, one expects a frictional retardation in the proximity of the boundaries, which can be realized only by introducing further terms in momentum Eq. (1.42), as a future analysis will show (cf. Sect. 1.2.3).

1.2.2 Basic Concepts of Vorticity Dynamics

Relative Vorticity and Its Basic Properties

By definition, the *relative vorticity* $\boldsymbol{\omega}$ is the curl of velocity, that is,

$$\boldsymbol{\omega}(\mathbf{x}, t) := \text{rot}\, \mathbf{u}(\mathbf{x}, t) \tag{1.70}$$

where rot $\mathbf{u} := \nabla \times \mathbf{u}$. From (1.70), it follows immediately

$$\text{div}\, \boldsymbol{\omega} = \text{div}\, (\text{rot}\, \mathbf{u}) = 0 \tag{1.71}$$

where div $\boldsymbol{\omega} := \nabla \cdot \boldsymbol{\omega}$; that is, the vorticity vector is nondivergent: neither sources nor sinks of relative vorticity exist. A *vortex line* or *vortex filament* is a line in the fluid that, at each point, is parallel to $\boldsymbol{\omega}$ at that point. A *vortex tube* is formed by the surface consisting of the vortex filaments that pass through a given close curve C; therefore, no vorticity vector can cross the tube surface. Consider now a volume of integration defined by any finite segment of the vortex tube singled out by the closed curve C and included between two plane surfaces with areas σ_1 and σ_2, according to Fig. 1.4. Equation (1.71) implies, trivially,

$$\int_V \nabla \cdot \boldsymbol{\omega}\, \mathrm{d}V' = 0 \tag{1.72}$$

and the divergence theorem applied to (1.72) yields the equation

$$\int_S \hat{\mathbf{n}} \cdot \boldsymbol{\omega}\, \mathrm{d}S' = 0 \tag{1.73}$$

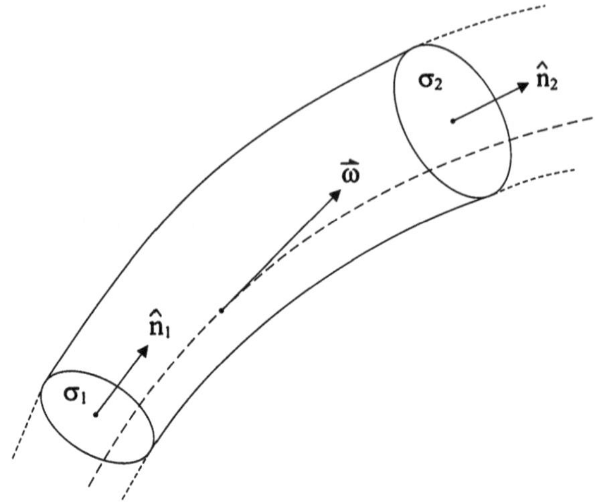

Fig. 1.4 Part of a vortex tube, included between the sections σ_1 and σ_2. A *vortex filament* (*dashed line*) is a line in the fluid that, at each point, is parallel to vector $\boldsymbol{\omega}$ at that point. Therefore, no vector $\boldsymbol{\omega}$ can penetrate into the vortex tube or exit from it

where S is the surface enclosing the vortex tube segment occupying the volume V and $\hat{\mathbf{n}}$ is the exterior normal to S. Since $\hat{\mathbf{n}} \cdot \boldsymbol{\omega} = 0$ on the surface of the tube formed by the filaments, one obtains from (1.73)

$$\int_{\sigma_1} (-\hat{\mathbf{n}}_1) \cdot \boldsymbol{\omega} \, \mathrm{d}\sigma' + \int_{\sigma_2} \hat{\mathbf{n}}_2 \cdot \boldsymbol{\omega} \, \mathrm{d}\sigma' = 0$$

and therefore the incoming flux of $\boldsymbol{\omega}$ across σ_1 equals the outgoing flux of $\boldsymbol{\omega}$ across σ_2. Thus, the *strength* Γ of the vortex tube, that is,

$$\Gamma := \int_{\sigma} \hat{\mathbf{n}} \cdot \boldsymbol{\omega} \, \mathrm{d}\sigma' \tag{1.74}$$

is independent from σ and is constant throughout the tube. Stokes' theorem applied to (1.74) gives

$$\Gamma = \oint_C \hat{\mathbf{t}} \cdot \mathbf{u} \, \mathrm{d}s \tag{1.75}$$

where $\hat{\mathbf{t}}$ is the tangent vector of curve C and shows that the strength of the vortex tube coincides with the circulation of the velocity $\mathbf{u}$ along the curve $C = \partial\sigma$ that encircles the generic section σ of the tube.

If the fluid is locally assimilated to a rigidly rotating disk with radius r_0 and angular frequency ω_0, then its tangential velocity is $\hat{\mathbf{t}} \cdot \mathbf{u} = \omega_0 \, r_0$ (see Fig. 1.5). By evaluating and equating the r.h.s. of (1.74) and (1.75), we get $\pi r_0^2 \hat{\mathbf{n}} \cdot \boldsymbol{\omega} = (2\pi r_0)(\omega_0 \, r_0)$ whence

$$\omega_{\mathrm{n}} := \hat{\mathbf{n}} \cdot \boldsymbol{\omega} = 2\,\omega_0 \tag{1.76}$$

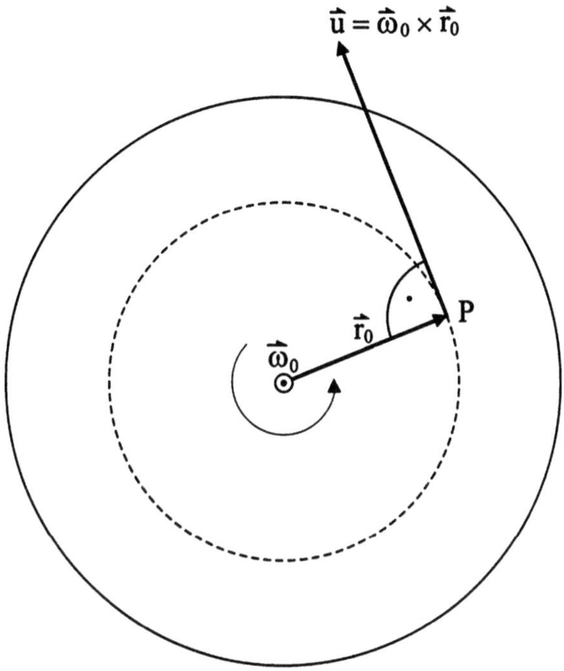

Fig. 1.5 Top view of a rigidly rotating fluid disk whose rotation vector points towards the reader. The component, along the rotation axis, of the vorticity of each material element *P* of the disk is twice the angular frequency ω_0

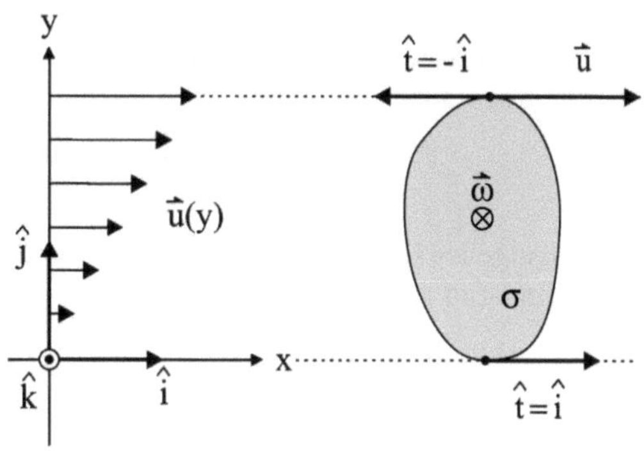

Fig. 1.6 *On the left*: velocity field of the kind $\mathbf{u}(y) = ay\hat{\mathbf{i}}$; *on the right*: cross section σ of a vortex tube determined by the related vorticity field $\boldsymbol{\omega} = -a\hat{\mathbf{k}}$. The shear of $\mathbf{u}(y)$, that is, the sign of a, determines the orientation of $\boldsymbol{\omega}$, incoming ($a > 0$, like in the figure) or outgoing ($a < 0$) the (x, y)-plane and, hence, also the sign of the strength $\Gamma = -a\sigma$ of the vortex tube

This equation shows that the normal component ω_n of relative vorticity is twice the angular frequency ω_0 of the fluid element. As (1.76) is independent of r_0, relation (1.76) is valid for each parcel of the rotating fluid body. Note, however, that relative vorticity is not an exclusive feature of rotating fluid bodies; for instance, if the rectilinear flow $\mathbf{u} = u(y)\hat{\mathbf{i}}$ has a cross stream shear $\mathrm{d}u/\mathrm{d}y \neq 0$, then (1.70) implies $\boldsymbol{\omega} = -(\mathrm{d}u/\mathrm{d}y)\hat{\mathbf{k}}$. The connection between velocity $\mathbf{u}$ and the relative vorticity ω, in the case of a constant shear, is depicted in Fig. 1.6.

The Vorticity Equation

The evolution equation of the relative vorticity (1.70) can be derived by taking the curl of each term of Euler's equation (1.43). First, with the aid of the identity

$$(\mathbf{u}\cdot\nabla)\,\mathbf{u} = \frac{1}{2}\nabla(\mathbf{u}\cdot\mathbf{u}) - \mathbf{u}\times\boldsymbol{\omega} \tag{1.77}$$

inserted into (1.43) and a rearrangement of the terms, the equation

$$\frac{\partial \mathbf{u}}{\partial t} - \mathbf{u}\times\boldsymbol{\omega} = -g\hat{\mathbf{k}} - \nabla B + p\nabla\frac{1}{\rho} \tag{1.78}$$

where

$$B = \frac{\mathbf{u}\cdot\mathbf{u}}{2} + \frac{p}{\rho} \tag{1.79}$$

is the *Bernoulli potential*. Then, the application of the curl operator to each term of (1.78) with the aid of the identities

$$\operatorname{rot}(\mathbf{u}\times\boldsymbol{\omega}) = (\boldsymbol{\omega}\cdot\nabla)\mathbf{u} - (\operatorname{div}\mathbf{u})\,\boldsymbol{\omega} - (\mathbf{u}\cdot\nabla)\,\boldsymbol{\omega}$$

$$\operatorname{rot}\left(p\nabla\frac{1}{\rho}\right) = -\frac{1}{\rho^2}\nabla p\times\nabla\rho$$

yields the desired *vorticity equation*

$$\frac{\mathrm{D}}{\mathrm{D}t}\boldsymbol{\omega} = (\boldsymbol{\omega}\cdot\nabla)\,\mathbf{u} - (\operatorname{div}\mathbf{u})\,\boldsymbol{\omega} - \frac{1}{\rho^2}\nabla p\times\nabla\rho \tag{1.80}$$

The divergence term can be eliminated from (1.17) and (1.80) in favour of density to obtain, after a slight rearrangement,

$$\frac{\mathrm{D}}{\mathrm{D}t}\frac{\boldsymbol{\omega}}{\rho} = \frac{1}{\rho}(\boldsymbol{\omega}\cdot\nabla)\,\mathbf{u} - \frac{1}{\rho^3}\nabla p\times\nabla\rho \tag{1.81}$$

In some cases, (1.80) is simplified by dropping the *distortion* term

$$(\boldsymbol{\omega}\cdot\nabla)\,\mathbf{u} = \left(\frac{\partial w}{\partial y} - \frac{\partial v}{\partial z}\right)\frac{\partial \mathbf{u}}{\partial x} + \left(\frac{\partial u}{\partial z} - \frac{\partial w}{\partial x}\right)\frac{\partial \mathbf{u}}{\partial y} + \left(\frac{\partial v}{\partial x} - \frac{\partial u}{\partial y}\right)\frac{\partial \mathbf{u}}{\partial z} \tag{1.82}$$

on geometrical grounds. This happens, for instance, for a two-dimensional motion or for a unidirectional flow.

In the former case (2-D motion), we can introduce a frame of reference such that

$$\mathbf{u} = u(x,\,y,\,t)\,\hat{\mathbf{i}} + v(x,\,y,\,t)\,\hat{\mathbf{j}} \tag{1.83}$$

and hence

$$\boldsymbol{\omega} = \left(\frac{\partial v}{\partial x} - \frac{\partial u}{\partial y}\right)\hat{\mathbf{k}} \tag{1.84}$$

Equation (1.84) implies, because of (1.82),

$$\boldsymbol{\omega}\cdot\nabla = \left(\frac{\partial v}{\partial x} - \frac{\partial u}{\partial y}\right)\frac{\partial}{\partial z}$$

and, owing to (1.83), $\partial\mathbf{u}/\partial z = 0$; so, $(\boldsymbol{\omega}\cdot\nabla)\,\mathbf{u} = 0$ follows.

In the latter case (unidirectional flow), the current can be taken as

$$\mathbf{u} = u(x, y, z, t)\hat{\mathbf{i}} \tag{1.85}$$

whence

$$\boldsymbol{\omega} = \frac{\partial u}{\partial z}\hat{\mathbf{j}} - \frac{\partial u}{\partial y}\hat{\mathbf{k}} \tag{1.86}$$

Then, from (1.82), (1.85) and (1.86), one easily checks that

$$(\boldsymbol{\omega}\cdot\nabla)\,\mathbf{u} = \left(\frac{\partial u}{\partial z}\frac{\partial}{\partial y} - \frac{\partial u}{\partial y}\frac{\partial}{\partial z}\right)u\hat{\mathbf{i}} = 0$$

2-D Motion

2-D motion of a constant-density fluid is governed by a special case of (1.80) in which $(\boldsymbol{\omega}\cdot\nabla)\mathbf{u} = 0$ and, because of (1.17), $\operatorname{div}\mathbf{u} = 0$. Thus, (1.80) simplifies into

$$\frac{\mathrm{D}\boldsymbol{\omega}}{\mathrm{D}t} = 0 \tag{1.87}$$

which states the conservation of the vorticity following the motion. Consider the special case of the motion on the (x, y)-plane, where condition $\operatorname{div}\mathbf{u} = 0$ takes the form

$$\frac{\partial u}{\partial x} + \frac{\partial v}{\partial y} = 0 \tag{1.88}$$

while relative vorticity is again given by (1.84). As the unit vector $\hat{\mathbf{k}}$ is constant, Eqs. (1.84) and (1.87) imply

$$\frac{\mathrm{D}}{\mathrm{D}t}\left(\frac{\partial v}{\partial x} - \frac{\partial u}{\partial y}\right) = 0 \tag{1.89}$$

that is, the conservation of the sole vertical component of vorticity.

Evolution Equation

The system of Eqs. (1.88) and (1.89) can be cast in a form in which only one scalar function, say $\chi = \chi(x, y, t)$ appears. In fact, if the latter is defined by

$$u = -\frac{\partial \chi}{\partial y} \qquad v = \frac{\partial \chi}{\partial x} \tag{1.90}$$

then (1.88) is identically satisfied, while (1.89) takes the form

$$\frac{\mathrm{D}}{\mathrm{D}t}\nabla^2\chi = 0 \tag{1.91}$$

and relative vorticity (1.84) becomes

$$\boldsymbol{\omega} = (\nabla^2\chi)\hat{\mathbf{k}} \tag{1.92}$$

where $\nabla^2 := \partial^2/\partial x^2 + \partial^2/\partial y^2$. Moreover, the Lagrangian derivative can be entirely expressed by means of χ; in fact, by using (1.90), we have

$$\frac{\mathrm{D}}{\mathrm{D}t} = \frac{\partial}{\partial t} + u\frac{\partial}{\partial x} + v\frac{\partial}{\partial y} = \frac{\partial}{\partial t} - \frac{\partial \chi}{\partial y}\frac{\partial}{\partial x} + \frac{\partial \chi}{\partial x}\frac{\partial}{\partial y} \tag{1.93}$$

whence (1.91) is written as

$$\frac{\partial}{\partial t}\nabla^2\chi + \frac{\partial \chi}{\partial x}\frac{\partial \nabla^2\chi}{\partial y} - \frac{\partial \chi}{\partial y}\frac{\partial \nabla^2\chi}{\partial x} = 0 \tag{1.94}$$

A concise form of (1.94) is

$$\frac{\partial}{\partial t}\nabla^2\chi + \mathscr{J}(\chi, \nabla^2\chi) = 0 \tag{1.95}$$

where

$$\mathscr{J}(a, b) := \frac{\partial a}{\partial x}\frac{\partial b}{\partial y} - \frac{\partial a}{\partial y}\frac{\partial b}{\partial x}$$

denotes the *Jacobian determinant* (see Appendix A, p. 369). If the current is given by (1.90), an advective term of the form $\mathbf{u} \cdot \nabla\theta$ can be written as $\mathscr{J}(\chi, \theta)$.

To summarize, Eq. (1.95) establishes the conservation of vorticity of a constant-density fluid moving on the (x, y)-plane, and at the same time, it also yields the evolution equation of the scalar function $\chi = \chi(x, y, t)$ from which both the current field $\mathbf{u}$ and the vorticity field $\boldsymbol{\omega}$ may be computed.

Boundary Condition

Equation (1.24), which constraints the current field by imposing the no–mass flux boundary condition across the surface S, takes the form

$$\frac{\partial F}{\partial t} + \mathscr{J}(\chi, F) = 0$$

In the steady case, this condition simplifies into (see Appendix A, p. 371)

$$\mathscr{J}(\chi, F) = 0 \tag{1.96}$$

where now $\chi = \chi(x, y)$ and $F = F(x, y)$. Equation (1.96) is equivalent to saying that χ and F are linked by a relationship of the kind

$$\chi = G(F) \tag{1.97}$$

where $G(\cdot)$ is an undetermined differentiable function of its argument (see Appendix A, p. 370). Hence, if $(x_0, y_0) \in S$, then $F(x_0, y_0) = 0$ and (1.97) implies

$$\chi(x_0, y_0) = G(0) \tag{1.98}$$

Therefore, χ takes the constant value $G(0)$ at each point of S. On the other hand, Eq. (1.94) shows that (1.95) is invariant under every transformation of the kind $\chi \mapsto \chi + \text{const.}$, so it is always possible to impose, without loss of generality, the boundary condition (see Appendix A, p. 368)

$$\chi(x, y) = 0 \qquad \forall (x, y) \in S \tag{1.99}$$

By analogy with the steady-state boundary condition (1.99), we shall couple the time dependent Eq. (1.95) with the boundary condition

$$\chi(x, y, t) = 0 \qquad \forall (x, y) \in S \quad \forall t \tag{1.100}$$

Example of Steady 2-D Flow

A simple example of steady circulation, governed by the steady version of (1.95), namely,

$$\mathscr{J}(\chi, \nabla^2 \chi) = 0 \tag{1.101}$$

in the square domain $D := [0 \leq x \leq L] \times [0 \leq y \leq L]$, is obtained through the stream function

$$\chi(x, y) = \chi_0 \sin\left(\pi \frac{x}{L}\right) \sin\left(\pi \frac{y}{L}\right) \tag{1.102}$$

In fact, (1.102) implies $\nabla^2\chi = -2\pi^2\chi/L^2 \propto \chi$, so (1.101) is identically satisfied because of the identity $\mathscr{J}(\chi, \chi) = 0$ (see Appendix A, p. 369). Moreover,

$$\begin{aligned}\chi(x,0) &= \chi(x,L) \qquad \forall x \in [0,L] \\ \chi(0,y) &= \chi(L,y) \qquad \forall y \in [0,L]\end{aligned} \tag{1.103}$$

so (1.99) is also verified. Thus, relative vorticity (1.92) is

$$\boldsymbol{\omega} = -2\pi^2 \frac{\chi}{L^2}\,\hat{\mathbf{k}} \tag{1.104}$$

while the current field is evaluated by substituting (1.102) into (1.90) to obtain

$$\begin{aligned}u(x,y) &= -\pi \frac{\chi_0}{L} \sin\left(\pi \frac{x}{L}\right) \cos\left(\pi \frac{y}{L}\right) \\ v(x,y) &= \pi \frac{\chi_0}{L} \cos\left(\pi \frac{x}{L}\right) \sin\left(\pi \frac{y}{L}\right)\end{aligned} \tag{1.105}$$

Solution (1.105) shows that the fluid parcels do not follow an inertial motion; so, the presence of a horizontal pressure gradient is expected to drive them while conserving steadiness. Indeed, the components of the pressure gradient are given by (1.44) and (1.45), which can be written here as

$$\mathscr{J}(\chi, u) = -\frac{1}{\rho}\frac{\partial p}{\partial x} \tag{1.106}$$

and

$$\mathscr{J}(\chi, v) = -\frac{1}{\rho}\frac{\partial p}{\partial y} \tag{1.107}$$

respectively. Equations (1.106) and (1.107) yield

$$\frac{\partial p}{\partial x} = -\frac{\pi^3}{2}\frac{\rho\chi_0^2}{L^3} \sin\left(2\pi \frac{x}{L}\right) \tag{1.108}$$

and

$$\frac{\partial p}{\partial y} = -\frac{\pi^3}{2}\frac{\rho\chi_0^2}{L^3} \sin\left(2\pi \frac{y}{L}\right) \tag{1.109}$$

whence, apart from an arbitrary additive constant, the pressure field is

$$p = \frac{\pi^2}{4}\frac{\rho\chi_0^2}{L^2}\left[\cos\left(2\pi \frac{x}{L}\right) + \cos\left(2\pi \frac{y}{L}\right)\right] \tag{1.110}$$

Note also that the average of each component of the velocity (1.105) on D is zero. (See also Appendix A, p. 373.)

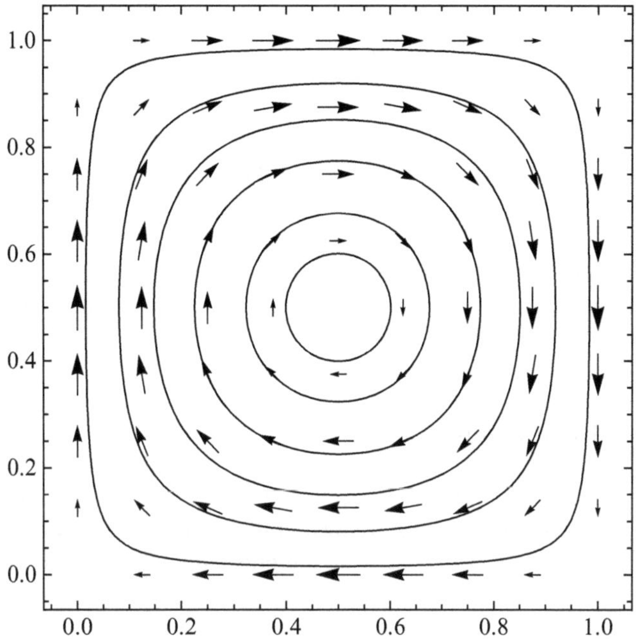

Fig. 1.7 Streamlines and velocity field of a simple vortex-like planar flow

For convenience, we introduce the non-dimensional streamfunction

$$\psi(\mathsf{x},\mathsf{y}) := \frac{\chi(L\mathsf{x},L\mathsf{y})}{\chi_0}$$

where

$$\mathsf{x} := \frac{x}{L} \qquad \mathsf{y} := \frac{y}{L}$$

are non-dimensional spatial coordinates running from 0 to 1. In terms of $\psi(\mathsf{x},\mathsf{y})$, Eq. (1.102) takes the form

$$\psi(\mathsf{x},\mathsf{y}) = \sin(\pi\mathsf{x})\sin(\pi\mathsf{y}) \tag{1.111}$$

whence the non-dimensional current components are given by

$$\mathsf{u} := \frac{L}{\chi_0}u \qquad \mathsf{v} := \frac{L}{\chi_0}v \tag{1.112}$$

Figure 1.7 shows the streamlines, which represent the trajectories of fluid parcels, as the contour lines of the scalar field (1.111). Also shown are the corresponding velocity vectors, whose components are given by (1.112).

Finally, we stress that, if the circular fluid domain $D = [0 \leq r \leq R]$ is considered in place of the square one, where r is the radial coordinate, equation $\mathscr{J}(\chi, \nabla^2\chi) = 0$ is identically satisfied by every differentiable function $\chi = \chi(r)$ (see Appendix A, p. 372). In this case, relative vorticity is

$$\boldsymbol{\omega} = \left(\frac{\mathrm{d}^2\chi}{\mathrm{d}r^2} + \frac{1}{r}\frac{\mathrm{d}\chi}{\mathrm{d}r}\right)\hat{\mathbf{k}}$$

Since the radial velocity in polar coordinates (r, θ) is given by $-(1/r)\partial\chi/\partial\theta$, here it is zero everywhere in D and, in particular, the no–mass flux condition across $r = R$ is verified tout court.

A Simple Model for Breezes

Equation (1.80) can be applied, for instance, to explain the evolution of sea-land-mountain breezes. To this purpose, we restrict the model to the vertical (x, z)-plane, so $\partial/\partial y = 0$ everywhere. Thus, the wind $\mathbf{u}$ representing the breeze is

$$\mathbf{u} = u(x, z, t)\hat{\mathbf{i}} + w(x, z, t)\hat{\mathbf{k}} \tag{1.113}$$

which yields the relative vorticity

$$\boldsymbol{\omega} = \mathrm{rot}\,\mathbf{u} = \left(\frac{\partial u}{\partial z} - \frac{\partial w}{\partial x}\right)\hat{\mathbf{j}} \tag{1.114}$$

In the meteorological context, it is quite reasonable to assume a nearly incompressible fluid whence, approximately,

$$\mathrm{div}\,\mathbf{u} = 0 \tag{1.115}$$

which here reads

$$\frac{\partial u}{\partial x} + \frac{\partial w}{\partial z} = 0 \tag{1.116}$$

Because of (1.113) and (1.115), Eq. (1.80) simplifies into

$$\frac{\mathrm{D}}{\mathrm{D}t}\boldsymbol{\omega} = -\frac{1}{\rho^2}\nabla p \times \nabla\rho$$

that is to say,

$$\frac{\mathrm{D}}{\mathrm{D}t}\boldsymbol{\omega} = \nabla p \times \nabla\frac{1}{\rho} \tag{1.117}$$

The baroclinic term $\nabla p \times \nabla\rho^{-1}$can become important, during day-time, when a thermal gradient ∇T is formed and maintained between the sea and land surfaces because of differential heating. In this situation, the pressure gradient ∇p is vertical, that is, $\nabla p = (\partial p/\partial z)\hat{\mathbf{k}}$, while the thermal gradient is tilted from the vertical direction because of its horizontal component due to the differential heating ($\partial T/\partial x \neq 0$). The interplay between pressure gradient and thermal gradient gives rise to a horizontal vorticity component (i.e. $\hat{\mathbf{j}}\cdot\boldsymbol{\omega} \neq 0$), and hence to a cell-like wind circulation. Then, during daytime, a net flux of air develops from the sea to the land. During the nighttime, the sea becomes warmer than the land and the vorticity changes its sign because of the change in the thermal gradient; thus, a net flux of air develops from the land towards the sea. This branch of the circulation pertains the side of the cell close to the ground and to the sea surface. The perfect gas law[3]

$$p = R\rho T \tag{1.118}$$

applied to the baroclinic term of (1.117) yields, by eliminating ρ in favour of p/T,

$$\nabla p \times \nabla \frac{1}{\rho} = R\nabla p \times \nabla \frac{T}{p} = \frac{R}{p}\nabla p \times \nabla T \tag{1.119}$$

and hence (1.117) becomes

$$\frac{\mathrm{D}}{\mathrm{D}t}\boldsymbol{\omega} = \frac{R}{p}\nabla p \times \nabla T \tag{1.120}$$

As anticipated,

$$\nabla p = \frac{\partial p}{\partial z}\hat{\mathbf{k}} \tag{1.121}$$

so the hydrostatic equilibrium $\partial p/\partial z + g\rho = 0$ allows us to write (1.121) as

$$\nabla p = -g\rho\hat{\mathbf{k}} \tag{1.122}$$

With the aid of (1.118) and (1.122), the baroclinic term at the r.h.s. of (1.120) becomes

$$\frac{R}{p}\nabla p \times \nabla T = -\frac{g}{T}\frac{\partial T}{\partial x}\hat{\mathbf{j}}$$

and hence Eq. (1.120) simplifies into

$$\frac{\mathrm{D}}{\mathrm{D}t}\boldsymbol{\omega} = -\frac{g}{T}\frac{\partial T}{\partial x}\hat{\mathbf{j}} \tag{1.123}$$

Equation (1.123) implies

$$\frac{\mathrm{D}}{\mathrm{D}t}\left(\hat{\mathbf{j}}\cdot\mathrm{rot}\,\mathbf{u}\right) = -\frac{g}{T}\frac{\partial T}{\partial x} \tag{1.124}$$

---

[3] Standard notation is understood.

Integration of (1.124) on the material surface S of the vertical (x, z)-plane enclosed by a certain path ∂S in the cell gives

$$\frac{\mathrm{d}}{\mathrm{d}t}\int_S \hat{\mathbf{j}}\cdot \mathrm{rot}\,\mathbf{u}\,\mathrm{d}S' = -g\int_S \frac{1}{T}\frac{\partial T}{\partial x}\,\mathrm{d}S' \tag{1.125}$$

according to Reynolds' theorem and (1.115). Note that $\hat{\mathbf{j}}$ enters into the (x, z)-plane of Fig. 1.8. By Stokes' theorem, Eq. (1.125) is equivalent to

$$\frac{\mathrm{d}}{\mathrm{d}t}\oint_{\partial S}\hat{\mathbf{t}}\cdot\mathbf{u}\,\mathrm{d}s = g\int_S \frac{1}{T}\frac{\partial T}{\partial x}\,\mathrm{d}S' \tag{1.126}$$

In (1.126), $\hat{\mathbf{t}}$ is the unit vector locally tangent to the given path (positive anticlockwise), and $\mathrm{d}s$ is the differential arclength along ∂S; so, $\oint_{\partial S}\hat{\mathbf{t}}\cdot\mathbf{u}\,\mathrm{d}s$ is the circuit integral of the wind along the considered path. Thus, values of $\partial T/\partial x$, alternatively positive and negative in the course of time, imply via (1.126) a related clockwise and anticlockwise rotation of the wind along its path, respectively. Close to the ground and the sea surface, the alternative rotation of the wind field is detected as a backwards and forwards air motion, that is, the *breeze*.

For time intervals during which the circulation along the boundary of the cell is always clockwise (or always counterclockwise), Eq. (1.126) yields

$$\frac{\mathrm{d}}{\mathrm{d}t}\left|\oint_{\partial S}\hat{\mathbf{t}}\cdot\mathbf{u}\,\mathrm{d}s\right| = \sigma g\int_S \frac{1}{T}\frac{\partial T}{\partial x}\,\mathrm{d}S' \tag{1.127}$$

where $\sigma := \mathrm{sgn}(\oint_{\partial S}\hat{\mathbf{t}}\cdot\mathbf{u}\,\mathrm{d}s)$. The correlation[4] between σ and $\partial T/\partial x$ that arises from (1.127) encompasses all the possible configurations of the breeze motion, according to Fig. 1.8. In particular, the breeze increases (i.e. l.h.s. (1.127) > 0) when σ and $\partial T/\partial x$ are positively correlated, while the wind field is damped by the thermal forcing in the case of negative correlation.

1.2.3 A Look at Turbulence

Mean and Turbulent Flows

With the application of fluid mechanics to geophysical flows in mind, the presence of turbulent fluctuations in fluid bodies in motion must be taken into account. The main hypothesis is that an "intermediate" timescale exists below which rapid fluctuations of velocity and pressure are fully developed in such bodies and above which a statistically significant mean flow can be singled out from them. In this

[4] Here, "correlation" means "sign of the product".

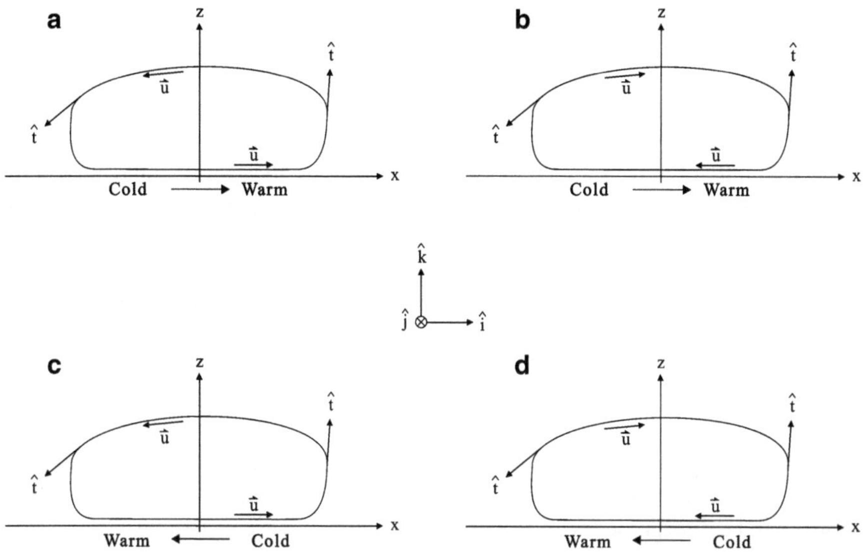

Fig. 1.8 Cell-like breeze circulation. Panel *a*. The anticlockwise circulation ($\sigma = 1$) is forced by $\partial T/\partial x > 0$, so the breeze is strengthened by the thermal gradient. Panel *b*. The circulation is clockwise ($\sigma = -1$), so the breeze is damped by the thermal gradient with $\partial T/\partial x > 0$. Panel *c*. The circulation is anticlockwise ($\sigma = 1$), so the breeze is damped by the thermal gradient with $\partial T/\partial x < 0$. Panel *d*. The clockwise circulation ($\sigma = -1$) is forced by $\partial T/\partial x < 0$, so the breeze is strengthened by the thermal gradient

perspective, the classical Reynolds approach resorts to a suitable, although not explicitly defined, time-averaging procedure by means of which any dynamical variable, say α, can be decomposed into a mean, denoted as $\overline{\alpha}$ or $\langle\alpha\rangle$, and a fluctuation α' so that

$$\alpha = \overline{\alpha} + \alpha' \tag{1.128}$$

where, by definition,

$$\langle\alpha'\rangle = 0 \tag{1.129}$$

Time averaging is linear, that is,

$$\langle c_1\,\alpha_1 + c_2\,\alpha_2\rangle = c_1\,\langle\alpha_1\rangle + c_2\,\langle\alpha_2\rangle \qquad \forall c_1, c_2 \in \mathbb{R} \tag{1.130}$$

From (1.128)–(1.130), the further rule

$$\langle\langle\alpha\rangle\rangle = \langle\alpha\rangle \tag{1.131}$$

immediately follows.

A good definition of the average is the following running mean:

$$\overline{\alpha(t)} := \frac{1}{T}\int_{t-T/2}^{t+T/2} \alpha(\tau)\,\mathrm{d}\tau$$

where the timescale T is selected long enough to filter out the high-frequency oscillations. However, no explicit formula for averaging is actually needed in the present discussion. Once the averaging technique has been established, we resolve to focus on the mechanical behaviour of the mean flow $\overline{\alpha}$, leaving aside all turbulent fluctuations α'. The point is that fluid mechanics is not linear, so the substitution of (1.128) into the governing equations does not allow to separate the equation for $\overline{\alpha(t)}$ from that for α'; instead, the final result is that the mean flow behaves as a fluid governed by a frictional law different from (1.48), parameterized by an isotropic molecular-viscosity term.

With reference to (1.128), the dynamic variables to take into account (velocity, density and pressure) are

$$\mathbf{u} = \overline{\mathbf{u}} + \mathbf{u}' \qquad \rho = \overline{\rho} + \rho' \qquad p = \overline{p} + p' \tag{1.132}$$

but a preliminary simplification can be introduced, based on the fact that the density fluctuation ρ' is usually negligible with respect to its averaged value $\overline{\rho}$. Hence, the former can be disregarded with respect to the latter; so we can assume

$$\rho = \overline{\rho} \tag{1.133}$$

The Averaged Continuity Equation

Consider first the continuity Eq. (1.17) written, for convenience, in the equivalent form

$$\frac{\partial \rho}{\partial t} + \operatorname{div}(\rho\,\mathbf{u}) = 0 \tag{1.134}$$

Substitution of the first equation of (1.132) together with (1.133) into (1.134) yields

$$\frac{\partial \rho}{\partial t} + \operatorname{div}(\rho\,\overline{\mathbf{u}} + \rho\,\mathbf{u}') = 0$$

whence, a fortiori,

$$\left\langle \frac{\partial \rho}{\partial t} + \operatorname{div}(\rho\,\overline{\mathbf{u}} + \rho\,\mathbf{u}') \right\rangle = 0 \tag{1.135}$$

and the rules (1.130), (1.129) and (1.131) can be applied to expand (1.135). One finds, after a little algebra, the following averaged continuity equation

$$\frac{\partial \rho}{\partial t} + \operatorname{div}(\rho\,\overline{\mathbf{u}}) = 0 \tag{1.136}$$

in which the effect of turbulence lies formally only in the presence of the averaged current field $\overline{\mathbf{u}}$ in place of the total current $\mathbf{u}$, as the comparison between (1.134) and (1.136) shows. We stress, however, that, unlike (1.134), the density ρ appearing in (1.136) refers to its averaged value $\overline{\rho}$.

The Averaged Momentum Equation

The momentum Eq. (1.43) written by using (1.132) and (1.133) takes the form

$$\frac{\partial}{\partial t}(\overline{\mathbf{u}}+\mathbf{u}') + \left[(\overline{\mathbf{u}}+\mathbf{u}')\cdot\nabla\right](\overline{\mathbf{u}}+\mathbf{u}') = -\frac{1}{\rho}\nabla(\overline{p}+p') - g\hat{\mathbf{k}}$$

and, a fortiori,

$$\left\langle \frac{\partial}{\partial t}(\overline{\mathbf{u}}+\mathbf{u}') + \left[(\overline{\mathbf{u}}+\mathbf{u}')\cdot\nabla\right](\overline{\mathbf{u}}+\mathbf{u}') \right\rangle = \left\langle -\frac{1}{\rho}\nabla(\overline{p}+p') - g\hat{\mathbf{k}} \right\rangle \tag{1.137}$$

Applying again the rules (1.130), (1.129) and (1.131)–(1.137) yields

$$\left(\frac{\partial}{\partial t}+\overline{\mathbf{u}}\cdot\nabla\right)\overline{\mathbf{u}} = -g\hat{\mathbf{k}} - \frac{1}{\rho}\nabla\overline{p} - \langle(\mathbf{u}'\cdot\nabla)\,\mathbf{u}'\rangle \tag{1.138}$$

and the formal difference between (1.43) and (1.138) is the presence, in the latter equation, of the new vector term

$$-\langle(\mathbf{u}'\cdot\nabla)\,\mathbf{u}'\rangle \tag{1.139}$$

which is originated by the nonlinear part of the Lagrangian derivative. Customarily, the notation of (1.136) and (1.138) is simplified by dropping the symbols of time averaging, under the convention that $\mathbf{u}$, p and ρ hereafter refer only to the statistically averaged velocity, pressure and density, respectively. Hence, Eqs. (1.136) and (1.138) are rewritten as

$$\frac{\partial\rho}{\partial t} + \mathrm{div}\,(\rho\,\mathbf{u}) = 0 \tag{1.140}$$

and

$$\left(\frac{\partial}{\partial t}+\mathbf{u}\cdot\nabla\right)\mathbf{u} = -g\hat{\mathbf{k}} - \frac{1}{\rho}\nabla p - \langle(\mathbf{u}'\cdot\nabla)\,\mathbf{u}'\rangle, \tag{1.141}$$

respectively.

Parameterization of Turbulence

Although the rapidly fluctuating field $\mathbf{u}'$ is present in (1.141) through the term (1.139), its governing equation is not known. This fact poses a closure problem in modelling flows governed by (1.141), and the way out from such impasse is to express (1.139) solely by means of the slowly varying field $\mathbf{u}$: this assumption leads to the so-called *parameterization of turbulence*. Unfortunately, no general principle exists to justify it, and its validity can be checked only a posteriori on an empirical basis.

In the large-scale atmospheric and oceanographic contexts, the typical horizontal and vertical length scales (say L and H, respectively) represent the distances at which the flow undergoes significant space modulations. These scales are markedly different in the sense that

$$\frac{H}{L} \ll 1 \tag{1.142}$$

Consistently, also the parameterization of (1.139) takes into account (1.142) as follows. The scalar components of (1.139) are

$$-\left\langle(\mathbf{u}'\cdot\nabla)\,u'\right\rangle \qquad -\left\langle(\mathbf{u}'\cdot\nabla)\,v'\right\rangle \qquad -\left\langle(\mathbf{u}'\cdot\nabla)\,w'\right\rangle$$

and each of them is expressed by means of the averaged current by postulating the positions

$$-\left\langle(\mathbf{u}'\cdot\nabla)\,u'\right\rangle = \mathrm{div}_{\mathrm{H}}\,(A_{\mathrm{H}}\,\nabla_{\mathrm{H}}u) + \frac{\partial}{\partial z}\left(A_{\mathrm{V}}\,\frac{\partial u}{\partial z}\right) \tag{1.143}$$

$$-\left\langle(\mathbf{u}'\cdot\nabla)\,v'\right\rangle = \mathrm{div}_{\mathrm{H}}\,(A_{\mathrm{H}}\,\nabla_{\mathrm{H}}v) + \frac{\partial}{\partial z}\left(A_{\mathrm{V}}\,\frac{\partial v}{\partial z}\right) \tag{1.144}$$

$$-\left\langle(\mathbf{u}'\cdot\nabla)\,w'\right\rangle = \mathrm{div}_{\mathrm{H}}\,(A_{\mathrm{H}}\,\nabla_{\mathrm{H}}w) + \frac{\partial}{\partial z}\left(A_{\mathrm{V}}\,\frac{\partial w}{\partial z}\right) \tag{1.145}$$

where $\nabla = \nabla_{\mathrm{H}} + \hat{\mathbf{k}}\partial/\partial z$, while A_{H} and A_{V} are the so-called *eddy viscosity coefficients*. If we assume that the horizontal and vertical turbulence mechanisms have similar importance, then the same order of magnitude holds for both the terms at the r.h.s. of (1.143)–(1.145), that is to say,

$$\mathrm{div}_{\mathrm{H}}\,(A_{\mathrm{H}}\,\nabla_{\mathrm{H}}u) \simeq \frac{\partial}{\partial z}\left(A_{\mathrm{V}}\,\frac{\partial u}{\partial z}\right)$$

(and so on for v and w), then $A_{\mathrm{H}}/L^2 \simeq A_{\mathrm{V}}/H^2$, that is, $A_{\mathrm{V}}/A_{\mathrm{H}} \simeq (H/L)^2$; hence, because of (1.142), the inequality

$$\frac{A_{\mathrm{V}}}{A_{\mathrm{H}}} \ll 1 \tag{1.146}$$

follows.

For "practical" computations, the eddy viscosity coefficients are assumed to be constant; so, (1.143)–(1.145) are simplified into

$$-\left\langle(\mathbf{u}'\cdot\nabla)\,u'\right\rangle = \mathscr{F}\,u \tag{1.147}$$

$$-\left\langle(\mathbf{u}'\cdot\nabla)\,v'\right\rangle = \mathscr{F}\,v \tag{1.148}$$

$$-\left\langle(\mathbf{u}'\cdot\nabla)\,w'\right\rangle = \mathscr{F}\,w \tag{1.149}$$

where the scalar differential operator

$$\mathscr{F} := A_{\mathrm{H}} \nabla_{\mathrm{H}}^2 + A_{\mathrm{V}} \frac{\partial^2}{\partial z^2} \tag{1.150}$$

and, on the whole, the final version of the momentum equation of a turbulent flow turns out to be

$$\left(\frac{\partial}{\partial t} + \mathbf{u} \cdot \nabla\right) \mathbf{u} = -g\hat{\mathbf{k}} - \frac{1}{\rho} \nabla p + \mathscr{F} \mathbf{u} \tag{1.151}$$

Hence, the scalar components of (1.151) read

$$\left(\frac{\partial}{\partial t} + \mathbf{u} \cdot \nabla\right) u = -\frac{1}{\rho} \frac{\partial p}{\partial x} + \mathscr{F} u \tag{1.152}$$

$$\left(\frac{\partial}{\partial t} + \mathbf{u} \cdot \nabla\right) v = -\frac{1}{\rho} \frac{\partial p}{\partial y} + \mathscr{F} v \tag{1.153}$$

$$\left(\frac{\partial}{\partial t} + \mathbf{u} \cdot \nabla\right) w = -\frac{1}{\rho} \frac{\partial p}{\partial z} + \mathscr{F} w - g \tag{1.154}$$

The vorticity equation for turbulent flows can be derived in the same manner as (1.80), but with two further terms coming from the application of the curl operator to the last two terms at the r.h.s. of (1.151). Since

$$\mathrm{rot} \mathscr{F} \mathbf{u} = \mathscr{F} \boldsymbol{\omega} \tag{1.155}$$

Equation (1.81) is modified simply by adding (1.155) to its r.h.s., whence the vorticity equation

$$\frac{\mathrm{D}}{\mathrm{D}t} \frac{\boldsymbol{\omega}}{\rho} = \frac{1}{\rho} (\boldsymbol{\omega} \cdot \nabla) \mathbf{u} - \frac{1}{\rho^3} \nabla p \times \nabla \rho + \frac{1}{\rho} \mathscr{F} \boldsymbol{\omega} \tag{1.156}$$

is established.

Due to their purely formal introduction, the eddy viscosity coefficients are not quantities susceptible of observational survey, and, for this reason, the values (in $\mathrm{m}^2 \cdot \mathrm{s}^{-1}$ in SI units) that should be ascribed to them are a priori quite uncertain. However, once that the governing equations have been cast in a non-dimensional form (fit for dealing with the mechanics of the fluid Earth, in the present context), these coefficients are usually transformed into ratios between length scales which, in turn, can be somehow inferred from observations. In this way, the uncertainty is appreciably reduced.

A Simple Application to a Flow in a Channel

Reconsider the model of a rectilinear current

$$\mathbf{u} = u(y)\hat{\mathbf{i}} \tag{1.157}$$

of a constant-density fluid, which flows in a channel included between $0 \leq y \leq L$ and governed by the equations

$$\operatorname{div} \mathbf{u} = 0 \tag{1.158}$$

and

$$-\frac{1}{\rho}\nabla p - g\hat{\mathbf{k}} + A_{\mathrm{H}}\frac{\mathrm{d}^2 u}{\mathrm{d}y^2}\hat{\mathbf{i}} = 0 \tag{1.159}$$

It is already known that (1.157) identically satisfies (1.158); on the other hand, (1.159) is nothing but (1.151) with the current given by (1.157). In components, (1.159) yields

$$-\frac{1}{\rho}\frac{\partial p}{\partial x} + A_{\mathrm{H}}\frac{\mathrm{d}^2 u}{\mathrm{d}y^2} = 0 \tag{1.160}$$

$$\frac{\partial p}{\partial y} = 0 \tag{1.161}$$

$$\frac{\partial p}{\partial z} + g\rho = 0 \tag{1.162}$$

Note that (1.161) coincides with the second equation of (1.67), and (1.162) coincides with (1.68). Thus, the effect of turbulent dissipation lies entirely into (1.160). Equations (1.160)–(1.161) imply $\mathrm{d}^3 u/\mathrm{d}y^3 = 0$, and hence u has a parabolic cross stream profile. The additional no-slip conditions

$$u(0) = u(L) = 0 \tag{1.163}$$

at the boundaries $y = 0$ and $y = L$ of the channel, together with the position

$$u\left(\frac{L}{2}\right) =: U \tag{1.164}$$

finally yield the parabolic velocity profile

$$u(y) = \frac{4U}{L^2}\left(Ly - y^2\right) \tag{1.165}$$

Note that a boundary condition is needed to achieve a unique solution, but the choice (1.163) is not dictated by turbulence and hence is somewhat arbitrary; yet the resulting solution agrees with the frictional retardation of the flow which is commonly observed at the walls.[5]

---

[5]Every generalization is a hypothesis; therefore hypothesis has a necessary role, which nobody has ever questioned. But hypothesis must always undergo verification, as soon as possible and as often as possible. (HENRI POINCARÉ, Science et hypothèse)

Given (1.165) and Eq. (1.160) becomes

$$\frac{\partial p}{\partial x} = -8\rho U \frac{A_H}{L^2} \tag{1.166}$$

so the horizontal gradient pressure is opposite to $\mathbf{u}$, according to (1.164). From (1.161), (1.162) and (1.166), the pressure field is evaluated to be

$$p = -g\rho \left[\frac{8UA_H}{gL^2}(x - x_0) + z \right] + p(x_0, 0) \tag{1.167}$$

where x_0 is a constant. Equation (1.166) shows that an along-channel forcing is requested in order to balance turbulent dissipation while preserving the steadiness of the motion.

Mechanical Energy and Turbulence

To elucidate the role of turbulence in fluid dynamics, we take into account a simplified version of (1.151), that is, we consider a fluid of constant density, say ρ, and assume, only for mathematical simplicity, $A_H = 0$. Hence, the momentum equation is

$$\frac{\mathrm{D}}{\mathrm{D}t}\mathbf{u} = -g\hat{\mathbf{k}} - \frac{1}{\rho}\nabla p + A_V \frac{\partial^2}{\partial z^2}\mathbf{u} \tag{1.168}$$

Note that a constant-density fluid implies, according to (1.17),

$$\operatorname{div}\mathbf{u} = 0 \tag{1.169}$$

The fluid is included into the volume

$$V = D \times [z_B, z_T] \tag{1.170}$$

where D represents its horizontal extent, while two rigid lids, placed in $z = z_B$ (bottom) and in $z = z_T$ (top), bound V vertically. For mathematical simplicity, we shall always assume that the domain D is a simply connected region of the plane. The no–mass flux condition across V takes the form

$$\hat{\mathbf{n}} \cdot \mathbf{u}(\mathbf{x}, t) = 0 \qquad \forall \mathbf{x} \in \partial D \qquad \forall t \tag{1.171}$$

$$w(z_B) = w(z_T) = 0 \tag{1.172}$$

where $\hat{\mathbf{n}}$ is the local unit vector normal to the boundary ∂D of D.

Now our aim is to compute the time rate of growth of the integrated mechanical energy of the flow included into V, and to point out the role of turbulence in the so obtained equation. To this effect, first we substitute

$$\rho w = \frac{\mathrm{D}}{\mathrm{D}t}(\rho z) \tag{1.173}$$

into the scalar product of (1.168) with $\rho\,\mathbf{u}$, that is,

$$\frac{\mathrm{D}}{\mathrm{D}t}\left(\frac{1}{2}\rho\,\mathbf{u}\cdot\mathbf{u}\right) = -\mathrm{div}\,(p\,\mathbf{u}) + \rho\,A_{\mathrm{V}}\,\mathbf{u}\cdot\frac{\partial^2\mathbf{u}}{\partial z^2} - g\,\rho\,w \tag{1.174}$$

where use has been made of (1.169) in dealing with the pressure term, to obtain

$$\frac{\mathrm{D}}{\mathrm{D}t}E = -\mathrm{div}\,(p\,\mathbf{u}) + \rho\,A_{\mathrm{V}}\,\mathbf{u}\cdot\frac{\partial^2\mathbf{u}}{\partial z^2} \tag{1.175}$$

with

$$E = \frac{1}{2}\rho\,\mathbf{u}\cdot\mathbf{u} + \rho\,g\,z \tag{1.176}$$

the total *mechanical energy density*. Second, volume integration of (1.175) yields the following results:

1. The no–mass flux conditions (1.171) and (1.172) imply that the fluid domain V is actually a finite material volume of fluid, so

$$\int_V \left(\frac{\mathrm{D}}{\mathrm{D}t}E\right)\mathrm{d}V' = \frac{\mathrm{d}}{\mathrm{d}t}\int_V E\,\mathrm{d}V' \tag{1.177}$$

2. Due to the divergence theorem, (1.171) and (1.172), we have

$$\int_V \mathrm{div}\,(p\,\mathbf{u})\,\mathrm{d}V' = \int_{\partial V} p\,\hat{\mathbf{n}}\cdot\mathbf{u}\,\mathrm{d}S' = 0 \tag{1.178}$$

3. Integration of the last term in (1.175) yields

$$\int_V \mathbf{u}\cdot\frac{\partial^2\mathbf{u}}{\partial z^2}\,\mathrm{d}V' = \int_D \left[\mathbf{u}\cdot\frac{\partial\mathbf{u}}{\partial z}\right]_{z=z_{\mathrm{B}}}^{z=z_{\mathrm{T}}}\mathrm{d}x\,\mathrm{d}y - \int_V \frac{\partial\mathbf{u}}{\partial z}\cdot\frac{\partial\mathbf{u}}{\partial z}\,\mathrm{d}V' \tag{1.179}$$

On the whole, (1.177), (1.178) and (1.179) state that

$$\frac{\mathrm{d}}{\mathrm{d}t}\int_V E\,\mathrm{d}V' = \rho\,A_{\mathrm{V}}\int_D \left[\mathbf{u}\cdot\frac{\partial\mathbf{u}}{\partial z}\right]_{z=z_{\mathrm{B}}}^{z=z_{\mathrm{T}}}\mathrm{d}x\,\mathrm{d}y - \rho\,A_{\mathrm{V}}\int_V \frac{\partial\mathbf{u}}{\partial z}\cdot\frac{\partial\mathbf{u}}{\partial z}\,\mathrm{d}V' \tag{1.180}$$

To explain the boundary term appearing in (1.180), we observe that the presence in (1.168) of the term that represents turbulence parameterization, that is, $A_{\mathrm{V}}\,\partial^2\mathbf{u}/\partial z^2$, rises the order of the differential Equation (1.168) with respect to (1.43); therefore, additional boundary conditions are required to single out a unique solution of Eq. (1.168). Such conditions physically reflect the effect of a shear stress generated by an external forcing given, in the marine case, by the wind stress at the air-sea interface (at $z = z_{\mathrm{T}}$) and/or the frictional retardation at the sea floor (at $z = z_{\mathrm{B}}$). On the other hand, the negative definite integral

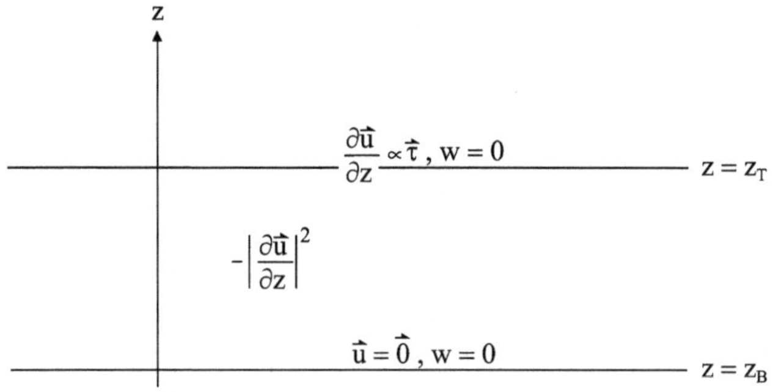

Fig. 1.9 Vertical section of a fluid layer which is forced by the shear stress (1.182) at $z = z_{\mathrm{T}}$ and subject to a frictional retardation, according to (1.181), at $z = z_{\mathrm{B}}$. In each point of the fluid interior, the dissipation of kinetic energy is proportional to $-|\partial \mathbf{u}/\partial z|$, and its global effect gives rise to the last term at the r.h.s. of (1.183)

$$-\rho A_{\mathrm{V}} \int_V \frac{\partial \mathbf{u}}{\partial z} \cdot \frac{\partial \mathbf{u}}{\partial z} \, \mathrm{d}V'$$

of (1.180) involves the material fluid volume as a whole and plays the role of an energy sink determined by eddy viscosity.

The additional boundary condition at the bottom is that of no-slip, that is,

$$\mathbf{u}(z = z_{\mathrm{B}}) = 0 \tag{1.181}$$

which demands that also the tangential components of the velocity be zero at $z = z_{\mathrm{B}}$ (Fig. 1.9). Condition (1.181) is analogous to (1.163).

At the air-sea interface, the shearing stress vector

$$\boldsymbol{\tau} = \rho A_{\mathrm{V}} \frac{\partial \mathbf{u}}{\partial z}(z_{\mathrm{T}}) \tag{1.182}$$

links the vertical derivative of the horizontal current at the surface with the applied wind stress (cf. Fig. 1.9). In general, a flow that shows direct proportionality between the applied shear stress and the resulting rate of deformation is a Newtonian fluid according to (1.49). With the aid of (1.181) and (1.182), Eq. (1.180) takes its final form, that is,

$$\frac{\mathrm{d}}{\mathrm{d}t} \int_V E \, \mathrm{d}V' = \int_D \mathbf{u}(z_{\mathrm{T}}) \cdot \boldsymbol{\tau} \, \mathrm{d}x \, \mathrm{d}y - \rho A_{\mathrm{V}} \int_V \frac{\partial \mathbf{u}}{\partial z} \cdot \frac{\partial \mathbf{u}}{\partial z} \, \mathrm{d}V' \tag{1.183}$$

Equation (1.183) means that the time rate of change of total mechanical energy is due to the imbalance between forcing and dissipation. It is worthy to point out that the r.h.s. of (1.183) comes entirely from the turbulent behaviour of the flow and from the additional boundary conditions (1.181) and (1.182).

1.2.4 Appendix: Galilean Covariance of the Lagrangian Derivative

Roughly speaking, a differential operator is said to be *covariant* under a coordinate transformation if it retains its form in both reference frames. In this Appendix, we shall see that two observers, that translate at a constant velocity with respect to each other, express the Lagrangian derivative with the same mathematical formula.

Two observers, say O and O', each fixed in its own inertial frame of reference, are in relative uniform motion with velocity $\mathbf{U}_{\text{rel}} := (U_{\text{rel}}, V_{\text{rel}}, W_{\text{rel}})$. They describe the same physical system S by means of the coordinates (x, y, z, t) and (x', y', z', t), which are linked by the well-known relations (Galilean transformation)

$$x = x' - U_{\text{rel}}t' \qquad y = y' - V_{\text{rel}}t' \qquad z = z' - W_{\text{rel}}t' \qquad t = t' \tag{1.184}$$

whose inverse ones are obvious. Therefore, if a certain material point has the velocity $\mathbf{u} := (u, v, w) := (\mathrm{d}x/\mathrm{d}t, \mathrm{d}y/\mathrm{d}t, \mathrm{d}z/\mathrm{d}t)$ relatively to O and the velocity $\mathbf{u}' := (u', v', w') := (\mathrm{d}x'/\mathrm{d}t', \mathrm{d}y'/\mathrm{d}t', \mathrm{d}z'/\mathrm{d}t')$ relatively to O', then relations (1.184) imply, in particular,

$$u = \frac{\mathrm{d}x}{\mathrm{d}t} = \frac{\mathrm{d}x'}{\mathrm{d}t} - U_{\text{rel}}\frac{\mathrm{d}t'}{\mathrm{d}t} = \frac{\mathrm{d}x'}{\mathrm{d}t'} - U_{\text{rel}} = u' - U_{\text{rel}}$$

and likewise for the other components, which yields

$$\mathbf{u} = \mathbf{u}' - \mathbf{U}_{\text{rel}} \tag{1.185}$$

Consider now a scalar field ψ that describes a certain property of S. Both the observers O and O' detect the same value of ψ referred to a given point of S and to a given time; so, inverting (1.184), we get

$$\psi(x', y', z', t') = \psi(x + U_{\text{rel}}t,\ y + V_{\text{rel}}t,\ z + W_{\text{rel}}t,\ t) \tag{1.186}$$

At this point, the transformation equations, between O and O', for the gradient operator and the local time derivative can be inferred.

Consider, first, the spatial derivative

$$\frac{\partial}{\partial x}\psi(x', y', z', t') = \frac{\partial \psi}{\partial x'}\frac{\partial x'}{\partial x} + \frac{\partial \psi}{\partial y'}\frac{\partial y'}{\partial x} + \frac{\partial \psi}{\partial z'}\frac{\partial z'}{\partial x} + \frac{\partial \psi}{\partial t'}\frac{\partial t'}{\partial x} \tag{1.187}$$

Equations (1.184) imply

$$\frac{\partial x'}{\partial x} = 1 \qquad \frac{\partial y'}{\partial x} = \frac{\partial z'}{\partial x} = 0 \qquad \frac{\partial z'}{\partial x} = 0 \qquad \frac{\partial t'}{\partial x} = 0$$

Thus, Eq. (1.187) is simply $\partial\psi/\partial x = \partial\psi/\partial x'$, and likewise for the other components. Therefore, the transformation equation for the gradient operator has the form

$$\nabla\psi = \nabla'\psi \tag{1.188}$$

where

$$\nabla := \left(\frac{\partial}{\partial x}, \frac{\partial}{\partial y}, \frac{\partial}{\partial z}\right) \qquad \nabla' := \left(\frac{\partial}{\partial x'}, \frac{\partial}{\partial y'}, \frac{\partial}{\partial z'}\right)$$

Consider, then, the time derivative

$$\frac{\partial}{\partial t}\psi(x', y', z', t') = \frac{\partial\psi}{\partial x'}\frac{\partial x'}{\partial t} + \frac{\partial\psi}{\partial y'}\frac{\partial y'}{\partial t} + \frac{\partial\psi}{\partial z'}\frac{\partial z'}{\partial t} + \frac{\partial\psi}{\partial t'}\frac{\partial t'}{\partial t} \tag{1.189}$$

Because of (1.184), we have

$$\frac{\partial x'}{\partial t} = U_{\text{rel}} \qquad \frac{\partial y'}{\partial t} = V_{\text{rel}} \qquad \frac{\partial z'}{\partial t} = W_{\text{rel}} \qquad \frac{\partial t'}{\partial t} = 1$$

so (1.189) becomes

$$\frac{\partial}{\partial t}\psi(x', y', z', t') = U_{\text{rel}}\frac{\partial\psi}{\partial x'} + V_{\text{rel}}\frac{\partial\psi}{\partial y'} + W_{\text{rel}}\frac{\partial\psi}{\partial z'} + \frac{\partial\psi}{\partial t'}$$

that is to say, in vector notation,

$$\frac{\partial}{\partial t}\psi = \left(\frac{\partial}{\partial t'} + \mathbf{U}_{\text{rel}}\cdot\nabla'\right)\psi \tag{1.190}$$

Equation (1.190) gives the transformation equation for the local time derivative.

The Lagrangian derivative of ψ is defined, according to observer O, by

$$\left(\frac{\partial}{\partial t} + \mathbf{u}\cdot\nabla\right)\psi \tag{1.191}$$

and we verify that (1.191) retains its form also when it is evaluated by observer O'. In fact, by substituting (1.185), (1.188) and (1.190) into (1.191), the latter becomes

$$\begin{aligned}\left(\frac{\partial}{\partial t} + \mathbf{u}\cdot\nabla\right)\psi &= \left(\frac{\partial}{\partial t'} + \mathbf{U}_{\text{rel}}\cdot\nabla' + (\mathbf{u}' - \mathbf{U}_{\text{rel}})\cdot\nabla'\right)\psi \\ &= \left(\frac{\partial}{\partial t'} + \mathbf{u}'\cdot\nabla'\right)\psi \end{aligned} \tag{1.192}$$

and the (Galilean) covariance is proved.

Finally, we point out that the transformation rules

$$\frac{\partial}{\partial t} = \frac{\partial}{\partial t'} + \mathbf{U}_{\mathrm{rel}} \cdot \nabla' \tag{1.193}$$

$$\mathbf{u} \cdot \nabla = (\mathbf{u}' - \mathbf{U}_{\mathrm{rel}}) \cdot \nabla' \tag{1.194}$$

ensure the covariance of the Lagrangian derivative, although neither of them is separately covariant.

Exercises

1. Given $\mathbf{u} = u_1(y)\hat{\mathbf{i}} + u_2(x)\hat{\mathbf{j}}$, evaluate $(\mathbf{u} \cdot \nabla)\mathbf{u}$.
What is the result if $u_1 = 0$ or $u_2 = 0$?
Comment this from the physical point of view, assuming that $\mathbf{u}$ is a current.

2. Reconsider Eq. (1.160) and boundary conditions (1.163) under the hypothesis $A_{\mathrm{H}} = A_{\mathrm{H}}(y)$, that is to say, consider problem

$$\begin{cases} \dfrac{\partial}{\partial y}\left(A_{\mathrm{H}}(y)\,\dfrac{\partial^2 u}{\partial y^2}\right) = 0 \\ \\ u(0) = u(L) = 0 \end{cases}$$

Then, assuming that $A_{\mathrm{H}}(y)$ is significantly different of zero only for $y \approx 0$ and for $y \approx L$ (i.e. in the proximity of the walls), solve the so posed problem and compare the solution with (1.165).

3. What does Galilean covariance of the Lagrangian derivative (1.192) state when it is applied to Eq. (1.24)?

4. Derive the mechanical energy equation starting from (1.168), but with $A_{\mathrm{H}}\nabla_{\mathrm{H}}^2\mathbf{u}$ in place of $A_{\mathrm{V}}\,\partial^2\mathbf{u}/\partial z^2$, and the boundary condition $\mathbf{u} = 0\ \forall (x, y) \in \partial D$ in place of (1.172). Compare the result with (1.183): is the new equation coupled with a forcing term like (1.182)?

5. By using (1.58) and (1.59), the full hydrostatic pressure p at the depth z below the sea surface (placed at $z = 0$) turns out to be

$$p(z) = g\left(\int_0^{+\infty} \rho_{\mathrm{a}}(\zeta)\,\mathrm{d}\zeta + \int_z^0 \rho_{\mathrm{w}}(\zeta)\,\mathrm{d}\zeta\right)$$

where ρ_{a} is air density, while ρ_{w} is sea water density. On the basis of approximate profiles of ρ_{a} (exponential) and ρ_{w} (constant), estimate at which depth atmosphere gives the same contribution as water to the full hydrostatic pressure.

Bibliographical Note

The topics of this chapter are found, variously distributed, in most of the books dealing with Fluid Dynamics and its geophysical applications. A classical reference is Batchelor (1967). In the geophysical ambit, a non-exhaustive list is Brown (1991), Gill (1982), Holton (1979), Massel (1999), Pedlosky (1987) and Salby (1996). We also mention Krauss (1973) and Fletcher (1988), which have a higher mathematical level.

Chapter 2
Basic Geophysical Fluid Dynamics

Abstract Geophysical Fluid Dynamics (GFD) governs the behaviour of large bodies of air in the atmosphere and of seawater in the ocean. Thus, the equations of mass and momentum balance must be supplemented with suitable information about the thermodynamics of these fluids. In this way, a complete set of governing equations, indeed hardly mathematically tractable, is obtained.

Moreover, the atmosphere and the ocean rotate as they were almost fixed with the Earth; so, large-scale winds and ocean currents, as detected by a terrestrial observer, are just the result of small departures of these fluids from the rest state. On the other hand, Earth's rotation heavily influences the motion of the winds and the ocean currents because of the Coriolis acceleration, and, therefore, the latter enters into the momentum equation in a way that deeply characterizes the dynamics of geophysical flows.

One of the most fascinating aspects of this dynamics lies in the common obedience of the atmospheric and the oceanic currents to two main constraints:

- The *geostrophic balance*, which is typical of rotating systems and involves the pressure gradient acting in the fluid interior
- The *hydrostatic equilibrium*, which holds in the absence of significant vertical accelerations of the fluid

All these ingredients contribute to the formulation of the so-called *quasi-geostrophic dynamics*, which, luckily, is (partially) amenable to an analytical treatment.

2.1 Constitutive Equations

Air and seawater satisfy different equations of state, but the first has a form close to that of the perfect gas (although water vapour is also taken into account), while the second may be hardly formulated by means of reasonably simple analytic expressions. However, for the purposes of the quasi-geostrophic (QG) dynamics, only a few features of seawater are involved, so many details of the actual equation of state are unnecessary.

F. Cavallini and F. Crisciani, *Quasi-Geostrophic Theory of Oceans and Atmosphere*, Atmospheric and Oceanographic Sciences Library 45, DOI 10.1007/978-94-007-4691-6_2,

Also from the thermodynamic point of view, air and seawater exhibit remarkable differences, mainly due to the compressibility of the former and the almost incompressibility of the latter. The result is that the thermodynamics of the atmosphere involves the potential temperature, while that of the ocean is expressed in terms of seawater density.

In any case, these differences are not dramatic, as the unified treatment of internal waves, in the atmosphere and the ocean, will show with full details.

2.1.1 Equations of State

Concept of "State"

Two samples of fluid have the same *state* if they can exist in contact with each other without a change in properties. The samples in contact that have the same state must have equal pressure (otherwise work will be done by one sample on the other), they must have equal temperatures (otherwise heat will be transferred from one sample to the other), and they must have the same concentration of each the constituents (otherwise there will be changes of concentration caused by diffusion). If pressure p, temperature T and the concentrations $C_1, \ldots, C_N$ of the constituents are taken as the set of variables defining the state, then the equation of state gives other state properties as functions of these variables.

In the case of air and seawater, the most important state property is the density of the fluid:

$$\rho = \rho(p, T, C_1, \ldots, C_N) \tag{2.1}$$

The Case of the Atmosphere

According to the heterogeneous nature of the Earth troposphere, the equation of state is derived by means of the application of Dalton's law to the mixture of gases composing the troposphere. The principal gases, molecular nitrogen (N_2) and molecular oxygen (O_2), account for almost the total density of an air volume. Furthermore, the water vapour present in the air contributes to the mixture, and although its mass fraction is very small in comparison with that of the principal two gases, its role in the thermodynamics and dynamics of the troposphere is very important.

It is useful to define some variables that express the water vapour content in the atmosphere, in particular the *specific humidity* q:

$$q := \frac{m_w}{m} = \frac{\rho_w}{\rho} \tag{2.2}$$

where m_w is the mass of water vapour of an air volume and m is the total mass of the same volume of air. Equation (2.2) also reports the equivalent specific humidity definition by means of the density of the water vapour ρ_{w} and the total air density ρ. Moreover, the *mixing ratio* r is defined as

$$r := \frac{m_{\mathrm{w}}}{m_{\mathrm{d}}} = \frac{\rho_{\mathrm{w}}}{\rho_{\mathrm{d}}} \tag{2.3}$$

where m_{d} and ρ_{d} are the mass and the density of the dry air, respectively, that is, the air mixture without any water vapour.

The masses and the densities appearing in (2.2) and (2.3) are related by the equations

$$m = m_{\mathrm{d}} + m_{\mathrm{w}} \qquad \rho = \rho_{\mathrm{d}} + \rho_{\mathrm{w}} \tag{2.4}$$

The specific humidity q and the mixing ratio r are tied by the relation

$$q = \frac{r}{1+r} \tag{2.5}$$

which simply follows from the definition of specific humidity with the use of (2.3) and (2.4).

Recalling Dalton's law, the total pressure p of a unit volume of the troposphere is the sum of the pressure contribution of the dry air mixture p_{d} and the water vapour p_{w}, that is,

$$p = p_{\mathrm{d}} + p_{\mathrm{w}} \tag{2.6}$$

The typical ranges of temperature and density of the troposphere for the dry air mixture allow the use of the ideal-gas equation of state with a very good approximation. The same is true for the water vapour, if phase change does not occur:

$$\begin{aligned} p_{\mathrm{d}} &= \rho_{\mathrm{d}} R_{\mathrm{d}} T && \text{(dry air)} \\ p_{\mathrm{w}} &= \rho_{\mathrm{w}} R_{\mathrm{w}} T && \text{(water vapour)} \end{aligned} \tag{2.7}$$

In relations (2.7), the temperature of dry air and water vapour is the same according to the fact that they coexist in thermal equilibrium. Parameters R_{d} and R_{w} are the gas constants of the dry air and the water vapour, respectively. They come from the universal molar gas constant $R = 8.31434\ \mathrm{J \cdot mol^{-1} \cdot {}^{\circ}K^{-1}}$ and the molar mass of the gas, $\mu_{\mathrm{d}} = 28.965\ \mathrm{g \cdot mol^{-1}}$ for the dry air and $\mu_{\mathrm{w}} = 18.002\ \mathrm{g \cdot mol^{-1}}$ for the water vapour. Then

$$\begin{aligned} R_{\mathrm{d}} &= \frac{R}{\mu_{\mathrm{d}}} && \text{(dry air)} \\ R_{\mathrm{w}} &= \frac{R}{\mu_{\mathrm{w}}} && \text{(water vapour)} \end{aligned} \tag{2.8}$$

By means of (2.6) and (2.7), the total air pressure is

$$p = (\rho_{\mathrm{d}} R_{\mathrm{d}} + \rho_{\mathrm{w}} R_{\mathrm{w}})\, T \tag{2.9}$$

It is useful to rearrange the equation of state (2.9) with the explicit use of the dry gas constant, to get a new equation formally similar to the ideal-gas equation. Through (2.9), substituting (2.8), we have

$$p = \frac{1 + r R_\mathrm{w}/R_\mathrm{d}}{1+r} \rho R_\mathrm{d} T \tag{2.10}$$

where $R_\mathrm{w}/R_\mathrm{d} \simeq 1.609$. Equation (2.10) shows that the equation of state for the mixture of gases composing the atmosphere is expressed in the same form as the ideal-gas equation (2.7), up to a numerical factor. This factor is a function of the mixing ratio r, which expresses the contribution of the water vapour. The equation of state (2.10) can be written as a function of the specific humidity q by using (2.5)

$$p = \left[1 + \left(\frac{R_\mathrm{w}}{R_\mathrm{d}} - 1 \right) q \right] \rho R_\mathrm{d} T \tag{2.11}$$

Virtual Temperature

In the troposphere, the mixing ratio r takes values up to 0.04, and it is usually about 0.01. So, in (2.11), the specific humidity q is even smaller because of (2.5). As a consequence, it is straightforward to define a new variable

$$T_\mathrm{v} := \left[1 + \left(\frac{R_\mathrm{w}}{R_\mathrm{d}} - 1 \right) q \right] T \tag{2.12}$$

called *virtual temperature*. The virtual temperature T_v is always greater than the real temperature T, and, due to the small value of the mixing ratio r, the difference between virtual and real temperature is, at most, of a few degrees. Anyway, in the thermodynamic analysis of the troposphere, that difference cannot be neglected.

By means of the definition of virtual temperature (2.12), the equation of state for the tropospheric fluid becomes

$$p = \rho R_\mathrm{d} T_\mathrm{v} \tag{2.13}$$

and hence we have $\rho = \rho(p, T_\mathrm{v})$, in accordance with (2.1). In the physics of the atmosphere and in meteorology, it is a common practice to use always the virtual temperature of the air, and thus to use the equation of state (2.13).

The Case of Seawater

As the concentrations of salts in seawater are very nearly in a constant proportion, the state may be defined very closely by giving only one concentration. The variable used to describe this concentration is *salinity* (S), which is equal to the mass of dissolved salt per unit mass of seawater. Note that salinity is non-dimensional. The equation of state of seawater is therefore of the kind

$$\rho = \rho(p, T, S) \tag{2.14}$$

Unlike for the atmosphere, an analytic form of the equation of state (2.14) has not been found for the seawater because of its complex chemical composition. In this case, the equation of state has to be established empirically using polynomial functions of p, T and S with the coefficients of thermal expansion, saline contraction and isothermal compressibility. Here we do not need the detailed expression of (2.14), but we wish only to point out the relation

$$\left(\frac{\partial \rho}{\partial p}\right)_{T,S} = \frac{1}{c_s^2} \tag{2.15}$$

where $c_s \simeq 1{,}500\,\mathrm{m/s}$ is sound speed in seawater. The smallness of (2.15) shows that seawater can be considered to be incompressible, as far as marine dynamics is concerned. Strictly speaking, incompressibility implies that sound speed is infinite. In other words, we set

$$\left(\frac{\partial \rho}{\partial p}\right)_{T,S} = 0$$

and therefore (2.14) simplifies into

$$\rho = \rho(T, S) \tag{2.16}$$

As a first approximation, (2.16) can be made linear around a certain state (T_0, S_0, ρ_0) to obtain the equation

$$\rho = \rho_0 \left[1 - \alpha_T (T - T_0) + \alpha_S (S - S_0)\right] \tag{2.17}$$

where $\alpha_T = O(10^{-4}\,{}^{\circ}\mathrm{K}^{-1})$ and $\alpha_S = O(5 \times 10^{-4})$. Note that cold saline waters have high density, so they tend to sink. This phenomenon arises, for instance, when strong dry winds induce evaporation and hence increase salinity and lower temperature, thus generating a positive density anomaly in the surface waters.

2.1.2 *Thermodynamic Equations of Dry Air and Seawater*

The First Law of Thermodynamics

According to the *first law of thermodynamics*, if a given uniform mass of fluid gains an amount Q of heat per unit mass, and an amount W of work per unit mass is performed on it, then its internal-energy change per unit mass is

$$\Delta E = Q + W \tag{2.18}$$

For an ideal gas, the internal energy E is a linear function of temperature alone; thus, the internal-energy change is given by

$$\Delta E = c_V T \tag{2.19}$$

where c_V is the specific heat capacity at constant volume, while T is the change of temperature from "absolute zero" (where $T = 0$). Moreover, c_p denotes the specific heat capacity at constant pressure. Over ranges of pressure and temperature relevant to the atmosphere, both c_V and c_p may be regarded as constant in the transition from one state to another.

The heat transferred into the fluid depends on one or more different mechanisms, such as temperature diffusion, internal heat sources and viscous dissipation, but, in any case, we represent them in (2.18) on the whole by means of the term Q (heat per unit mass).

The work done in deforming the material elements of the fluid, without changing its velocity but increasing its internal energy, is represented in (2.18) by the term W (work per unit mass).

Our aim is to derive the first law of thermodynamics for fluids in motion, starting from a hypothetical small Lagrangian control volume $\delta V = \delta x\, \delta y\, \delta z$ that includes a fluid of density ρ and velocity $\mathbf{u}$. We consider, at a certain time, a control volume of cubic shape with couples of faces parallel to the (x, y), (x, z) and (y, z) planes. The total energy in δV is

$$\left(\Delta E + \frac{\mathbf{u} \cdot \mathbf{u}}{2}\right) \rho\, \delta V \tag{2.20}$$

while the external forces that act on it are surface forces (such as pressure and viscosity) and body forces (such as gravity and the force due to Coriolis acceleration). Thus, total energy (2.20) is the sum of the thermal component $\Delta E\, \rho\, \delta V$ and the mechanical component $(\mathbf{u} \cdot \mathbf{u})\, \rho\, \delta V/2$.

Since pressure is a force per unit area, the rate at which a force does work is given by the dot product of the force and velocity vectors. Therefore, the rate at which the surrounding fluid does work on the control volume is due to the imbalance of the pressure force on couples of parallel faces. For instance, with reference to the (y, z)-plane, the imbalance is (Fig. 2.1)

$$[(pu)_x - (pu)_{x+\delta x}]\, \delta y\, \delta z \tag{2.21}$$

where u is the component of $\mathbf{u}$ normal to the (y, z)-plane. The assumed smallness of the Lagrangian control volume allows us to approximate the first factor in (2.21) as

$$(pu)_x - (pu)_{x+\delta x} \simeq \left[-\frac{\partial}{\partial x}(pu)\right]_x \delta x \tag{2.22}$$

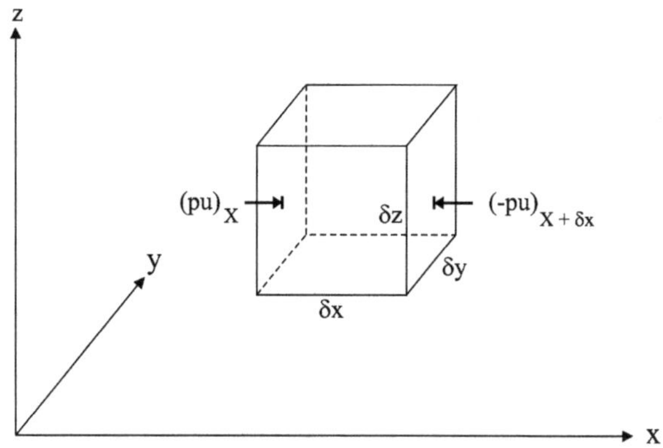

Fig. 2.1 Lagrangian control volume of a fluid in motion with velocity **u** and in the presence of an external pressure p. Only the rate of working of the pressure $(pu)_X$ at X and the rate of working of the pressure $-(pu)_{X+\delta x}$ at the opposite side $x = X + \delta x$ are reported for graphic simplicity

Hence, the net rate of work of the pressure force due to the x component of motion is obtained from (2.21) and (2.22), that is,

$$[(pu)_x - (pu)_{x+\delta x}]\, \delta y\, \delta z = \left[-\frac{\partial}{\partial x}(pu)\right]_x \delta V \tag{2.23}$$

Similarly to (2.23), the y and z components yield the contributions

$$\left[-\frac{\partial}{\partial x}(pv)\right]_y \delta V \qquad \text{and} \qquad \left[-\frac{\partial}{\partial z}(pw)\right]_z \delta V \tag{2.24}$$

respectively. The total rate of work done by the pressure force is the sum of (2.22)–(2.24), that is,

$$-\operatorname{div}(p\mathbf{u})\, \delta V \tag{2.25}$$

We disregard the molecular viscosity of air, which is zero in the perspective of GFD.

The rate at which gravity does work on the element of mass $\rho\, \delta V$ is

$$(\mathbf{g} \cdot \mathbf{u})\, \rho\, \delta V = -g w \rho\, \delta V \tag{2.26}$$

The time rate of change of (2.20) in the Lagrangian control volume is caused by (2.25) and (2.26), as well as by heating due to radiation, conduction and latent hear release, whose overall rate per unit mass is

$$\dot{Q} \rho\, \delta V \tag{2.27}$$

Note that "heat" is not a field variable, so that the heating rate $\dot{Q}$ is an ordinary, not a Lagrangian derivative. On the whole, the evolution equation of (2.20) takes into account (2.25)–(2.27) to give

$$\frac{\mathrm{D}}{\mathrm{D}t}\left[\left(\Delta E+\frac{\mathbf{u}\cdot\mathbf{u}}{2}\right)\rho\,\delta V\right]=\left[-\mathrm{div}\,(p\mathbf{u})-g w\rho+\dot{Q}\rho\right]\delta V \qquad (2.28)$$

Expanding the l.h.s. of (2.28), we get

$$\left[\frac{\mathrm{D}}{\mathrm{D}t}\left(\Delta E+\frac{\mathbf{u}\cdot\mathbf{u}}{2}\right)\right]\rho\,\delta V+\left(\Delta E+\frac{\mathbf{u}\cdot\mathbf{u}}{2}\right)\frac{\mathrm{D}}{\mathrm{D}t}(\rho\,\delta V)$$

but mass conservation in the Lagrangian control volume implies $\mathrm{D}(\rho\,\delta V)/\mathrm{D}t=0$, so (2.28) becomes

$$\frac{\mathrm{D}}{\mathrm{D}t}\Delta E+\frac{\mathrm{D}}{\mathrm{D}t}\frac{\mathbf{u}\cdot\mathbf{u}}{2}=-\frac{\mathrm{div}\,(p\mathbf{u})}{\rho}-g w+\dot{Q} \qquad (2.29)$$

To simplify (2.29), we evaluate preliminarily the dot product of the momentum equation (1.43) with **u** to obtain

$$\frac{\mathrm{D}}{\mathrm{D}t}\frac{\mathbf{u}\cdot\mathbf{u}}{2}=-\frac{\mathbf{u}\cdot\nabla p}{\rho}-g w \qquad (2.30)$$

Subtracting (2.30) from (2.29), the equation

$$\frac{\mathrm{D}}{\mathrm{D}t}\Delta E=\dot{Q}-\frac{p}{\rho}\,\mathrm{div}\,\mathbf{u}$$

follows. Using (2.19), the latter takes the final form

$$c_V\frac{\mathrm{D}T}{\mathrm{D}t}=\dot{Q}+\frac{\mathrm{D}W}{\mathrm{D}t} \qquad (2.31)$$

where

$$\frac{\mathrm{D}W}{\mathrm{D}t}=-\frac{p}{\rho}\,\mathrm{div}\,\mathbf{u} \qquad (2.32)$$

Thus, W is the deformation work done by the internal energy of a Lagrangian control volume without changing the volume itself.

The Thermodynamic Equation for the Dry Atmosphere

We start from the first law of thermodynamics and assume that the thermal behaviour of dry air is essentially the same as that of an ideal gas. This is consistent with (2.13), provided that the variable T now represents the virtual temperature (2.12).

Eliminating $\mathrm{div}\,\mathbf{u}$ from (2.32) and the equation for the conservation of mass $\mathrm{D}\rho/\mathrm{D}t+\rho\,\mathrm{div}\,\mathbf{u}=0$, we get

$$\frac{\mathrm{D}W}{\mathrm{D}t}=-p\frac{\mathrm{D}}{\mathrm{D}t}\frac{1}{\rho}$$

which, substituted into (2.31), yields

$$c_V \frac{\mathrm{D}T}{\mathrm{D}t} + p \frac{\mathrm{D}}{\mathrm{D}t} \frac{1}{\rho} = \dot{Q} \tag{2.33}$$

To eliminate, in (2.33), pressure in favour of temperature and density, we recall the state equation for the ideal gas, that is,

$$p = R \rho T \tag{2.34}$$

Substituting (2.34) into (2.33) and rearranging, we have

$$\frac{\mathrm{D}}{\mathrm{D}t} \ln \left[\frac{T}{T_0} \left(\frac{\rho_0}{\rho} \right)^{R/c_V} \right] = \frac{\dot{Q}}{c_V T} \tag{2.35}$$

where T_0 and ρ_0 are an arbitrary constant temperature and density, respectively, which we have introduced because the argument of the logarithm[1] must always be a pure number. At this point, it is useful to recall *Carnot's law* relating the specific heats c_p and c_V, that is,

$$c_p - c_V = R \tag{2.36}$$

and introduce the specific-heat ratio

$$\gamma := \frac{c_p}{c_V} \tag{2.37}$$

Equations (2.36) and (2.37) imply $R/c_V = \gamma - 1$; so, (2.35) can be written as

$$\frac{\mathrm{D}}{\mathrm{D}t} \ln \left[\frac{T}{T_0} \left(\frac{\rho_0}{\rho} \right)^{\gamma-1} \right] = \frac{\dot{Q}}{c_V T} \tag{2.38}$$

Noting that

$$\frac{T}{T_0} \left(\frac{\rho_0}{\rho} \right)^{\gamma-1} = \left(\frac{T}{T_0} \right)^{\gamma} \left(\frac{p_0}{p} \right)^{\gamma-1} = \left[\frac{T}{T_0} \left(\frac{p_0}{p} \right)^{\frac{\gamma-1}{\gamma}} \right]^{\gamma}$$

where $p_0 := R \rho_0 T_0$ is a constant reference pressure and using (2.37) with the state equation (2.34), we may rewrite (2.38) as

$$\frac{\mathrm{D}}{\mathrm{D}t} \ln \left[\frac{T}{T_0} \left(\frac{p_0}{p} \right)^{\frac{\gamma-1}{\gamma}} \right] = \frac{\dot{Q}}{c_p T} \tag{2.39}$$

[1] The same holds, of course, for any function that can be expanded as a power series, except monomials.

Potential Temperature

We define the *potential temperature* as

$$\theta := T\left(\frac{p_0}{p}\right)^{\frac{\gamma-1}{\gamma}} \tag{2.40}$$

Thus, in terms of potential temperature, (2.39) takes the form

$$\frac{1}{\theta}\frac{\mathrm{D}\theta}{\mathrm{D}t} = \frac{\dot{Q}}{c_p T} \tag{2.41}$$

Equation (2.41) is the desired thermodynamic equation for the dry atmosphere, for which we have the estimates

$$c_V = 717.5\,\mathrm{J\cdot kg^{-1}\cdot K^{-1}} \qquad c_p = 1{,}004.5\,\mathrm{J\cdot kg^{-1}\cdot K^{-1}}$$

whence, by (2.36) and (2.37),

$$R = 287.0\,\mathrm{J\cdot kg^{-1}\cdot K^{-1}} \qquad \gamma = 1.4$$

Under adiabatic conditions, the potential temperature θ behaves as a tracer of air motion, in the sense that particular values of θ track the movement of bodies of air.

From (2.40) and (2.34), an equivalent definition of potential temperature follows, namely,

$$\theta := C_0\frac{p^{1/\gamma}}{\rho} \tag{2.42}$$

where

$$C_0 := \frac{1}{R}p_0^{\frac{\gamma-1}{\gamma}}$$

is a constant.

The Dry Lapse Rate of the Atmosphere

The *dry lapse rate* is defined as the rate of decrease of air temperature with respect to height, that is, $\partial T/\partial z$.

Differentiating (2.40) with respect to z and using the identity

$$\frac{\gamma-1}{\gamma} = \frac{R}{c_p}$$

we obtain

$$\frac{1}{\theta}\frac{\partial\theta}{\partial z} = \frac{1}{T}\frac{\partial T}{\partial z} - \frac{R}{c_p}\frac{1}{p}\frac{\partial p}{\partial z} \tag{2.43}$$

Moreover, (1.56) and (2.34) imply

$$\frac{1}{p}\frac{\partial p}{\partial z} = -\frac{g}{RT} \tag{2.44}$$

Thus, (2.43) and (2.44) yield

$$\frac{1}{\theta}\frac{\partial \theta}{\partial z} = \frac{1}{T}\left(\frac{\partial T}{\partial z} + \frac{g}{c_p}\right) \tag{2.45}$$

from which the dry lapse rate follows:

$$\frac{\partial T}{\partial z} = \frac{T}{\theta}\frac{\partial \theta}{\partial z} - \frac{g}{c_p} \tag{2.46}$$

For an atmosphere in which the potential temperature is constant with respect to height, the lapse rate is

$$\left(\frac{\partial T}{\partial z}\right)_{\theta=\text{const.}} = -\frac{g}{c_p} \tag{2.47}$$

where $g/c_p = O(10^{-2}\ \text{K/m})$.

Introducing the notation

$$\Gamma_\text{d} := \frac{g}{c_p} \qquad (\textit{adiabatic lapse rate})$$

$$\Gamma := -\frac{\partial T}{\partial z} \qquad (\textit{actual lapse rate})$$

where subscript "d" stands for "dry", Eq. (2.45) becomes

$$\frac{T}{\theta}\frac{\partial \theta}{\partial z} = \Gamma_\text{d} - \Gamma \tag{2.48}$$

So, the adiabatic and the actual lapse rates coincide only if the iso-surfaces $\theta =$ const. are horizontal. Equation (2.48) will be useful in dealing with small vertical oscillations in an adiabatic and compressible atmosphere (cf. Sect. 2.1.3).

The Thermodynamic Equation for Seawater

In the case of seawater, the material volumes of fluid are nearly not deformable; so, $W = 0$ and $\gamma \simeq 1$ for water at temperatures and pressures in the proximity of the normal values. Therefore, (2.33) here becomes

$$c_p \frac{\mathrm{D}T}{\mathrm{D}t} = \dot{Q} \tag{2.49}$$

The equation of state for seawater (2.16) implies

$$\frac{\mathrm{D}\rho}{\mathrm{D}t} = \left(\frac{\partial\rho}{\partial T}\right)_S \frac{\mathrm{D}T}{\mathrm{D}t} + \left(\frac{\partial\rho}{\partial S}\right)_T \frac{\mathrm{D}S}{\mathrm{D}t}$$

that is to say, because of (2.49),

$$\frac{\mathrm{D}\rho}{\mathrm{D}t} = \left(\frac{\partial\rho}{\partial T}\right)_S \frac{\dot{Q}}{c_p} + \left(\frac{\partial\rho}{\partial S}\right)_T \frac{\mathrm{D}S}{\mathrm{D}t} \tag{2.50}$$

In the ocean interior, both the adiabatic condition $Q = 0$ and the conservation of salinity $\mathrm{D}S/\mathrm{D}t = 0$ are realized at a high degree of approximation, so (2.50) simplifies into

$$\frac{\mathrm{D}\rho}{\mathrm{D}t} = 0 \tag{2.51}$$

Equation (2.51) is the thermodynamic equation for the seawater mostly adopted in GFD, provided that the hypotheses $Q = 0$ and $\mathrm{D}S/\mathrm{D}t = 0$ are verified.

2.1.3 *Compressibility and Incompressibility*

Incompressible Fluids

An incompressible fluid has a constant density following the motion, that is, each element of material volume conserves its density:

$$\frac{\mathrm{D}\rho}{\mathrm{D}t} = 0 \tag{2.52}$$

Thus, by (1.17), in an incompressible fluid, the velocity divergence vanishes:

$$\mathrm{div}\,\mathbf{u} = 0 \tag{2.53}$$

The Case of Seawater

Equation (2.52) is the same as the thermodynamic equation for seawater (2.51), whence we immediately conclude that marine currents satisfy (2.53).

Vertical Oscillations in an Incompressible Seawater

Suppose now that seawater density is vertically stratified, that is,

$$\rho = \rho(z) \tag{2.54}$$

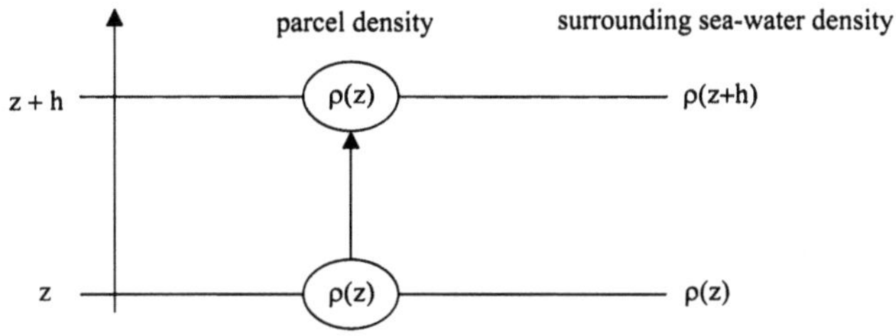

Fig. 2.2 A seawater parcel is initially in static equilibrium with the surrounding water. Then it is lifted from z to $z+h$ keeping constant its density $\rho(z)$, while the environmental density at the depth $z+h$ is $\rho(z+h) < \rho(z)$. Once the parcel is released, its oscillations are governed by (2.56)

and consider a parcel of fluid of volume δV, initially located at the depth z, so its density is $\rho(z)$. Then, the same parcel is lifted from the depth z to the depth $z+h$. Because of (2.52), the parcel conserves its density $\rho(z)$ but experiences the ambient density $\rho(z+h)$, as Fig. 2.2 shows. Therefore, the parcel undergoes two forces: (1) that of Archimedes, upward, given by $g\rho(z+h)\,\delta V$ and (2) the weight, downward, given by $-g\rho(z)\,\delta V$. According to *Newton's second law* (mass times acceleration equals force), the governing equation of the parcel's motion, expressed in terms of the time-dependent displacement $h(t)$, is

$$\rho(z)\,\delta V\,\frac{\mathrm{d}^2 h}{\mathrm{d}t^2} = g\,[\rho(z+h) - \rho(z)]\,\delta V \tag{2.55}$$

Under the assumption of a small-amplitude displacement h, the approximation

$$\rho(z+h) - \rho(z) \approx \frac{\partial \rho}{\partial z}\,h$$

can be substituted into (2.55) to give the evolution equation of the displacement h in the form

$$\frac{\mathrm{d}^2 h}{\mathrm{d}t^2} + N^2(z)\,h = 0 \tag{2.56}$$

where $N(z) > 0$ is the *buoyancy frequency* (or *Brunt-Väisälä frequency*) defined by

$$N^2(z) := -\frac{g}{\rho}\frac{\partial \rho}{\partial z} \tag{2.57}$$

under the assumption $\partial\rho/\partial z < 0$.

For a stably stratified fluid, for which $\partial\rho/\partial z < 0$ (as the vertical axis is positively oriented upward), the parcel exhibits a harmonic oscillation with angular frequency $N(z)$. Figure 2.3 shows the qualitative features of a typical buoyancy frequency profile in the ocean.

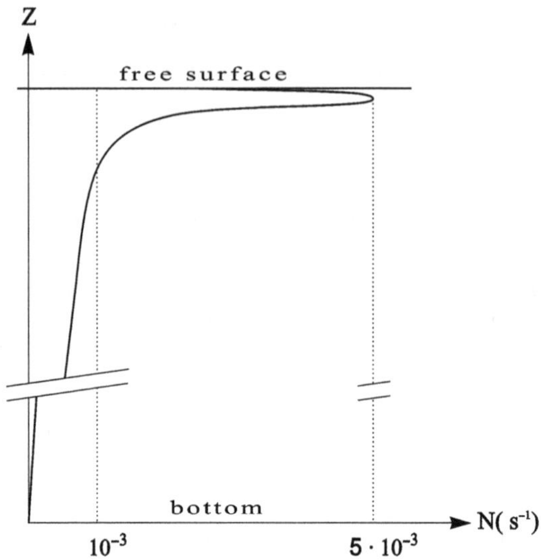

Fig. 2.3 Sketch of the profile of the global mean buoyancy frequency in the case of the ocean

Static Instability

In the presence of meteorological phenomena, such as the wind-induced evaporation shortly described at the end of Sect. 2.1.1, the density of the marine water included into a certain subsurface layer decreases with depth. In this case (i.e. for $\partial\rho/\partial z > 0$), the equation of motion (2.55) is still valid, but the r.h.s. of (2.57) is negative. Therefore, it is convenient to use the *escape rate* $\tilde{N}(z) > 0$ defined by

$$\tilde{N}^2(z) := \frac{g}{\rho}\frac{\partial\rho}{\partial z}$$

instead of buoyancy frequency $N(z)$ defined by (2.57). With this notation, a fluid parcel shifted of the quantity h from its "initial" position (say, at $t = 0$) follows the evolution equation

$$\frac{\mathrm{d}^2 h}{\mathrm{d}t^2} - \tilde{N}^2(z)\,h = 0 \qquad (2.58)$$

Note the "minus" sign in (2.58) in contrast with the "plus" sign in (2.56).

The general integral of (2.58) is

$$h(t) = A\cosh(\tilde{N}t) + B\sinh(\tilde{N}t) \qquad (2.59)$$

where A and B are arbitrary constants.

By assuming, for simplicity, that the time derivative of h is zero at $t = 0$, Eq. (2.59) gives

$$h(t) = h(0)\cosh(\tilde{N}t) \qquad (2.60)$$

and, therefore, the displacement of the parcel increases indefinitely, in the course of time, in the same direction as initially: upward if $h(0) > 0$, downward if $h(0) < 0$. In fact, in the upward motion, due to the increasing ambient density, the parcel (which retains its density) experiences a stronger and stronger buoyancy that pulls it always upward while, in the downward motion, the opposite happens. However, at the same time, the mixing of the water body involved in this process tends to homogenize the fluid and, as a consequence, the system evolves towards a neutral equilibrium in which Eq. (2.58) does not hold any longer.

The Case of the Atmosphere

Unlike seawater, the thermodynamic equation for the dry atmosphere (2.41) does not involve the air density; therefore, Eq. (2.53) cannot be derived. To obtain information about the divergence of large-scale atmospheric flows, we express the density field as the superposition of its horizontally averaged value $\rho_s(z)$ (the subscript stands for standard) plus a local departure $\tilde{\rho}(\mathbf{x},t)$:

$$\rho = \rho_s(z) + \tilde{\rho}(\mathbf{x},t) \tag{2.61}$$

For synoptic scale motions of the troposphere, the density anomaly is typically within one percent of the mean field:

$$\frac{\tilde{\rho}}{\rho_s} = O(10^{-2}) \tag{2.62}$$

Substitution of (2.61) into (1.17) yields, after a trivial rearrangement,

$$w\frac{\mathrm{d}\rho_s}{\mathrm{d}z} + \rho_s \,\mathrm{div}\,\mathbf{u} + \frac{\mathrm{D}\tilde{\rho}}{\mathrm{D}t} + \tilde{\rho}\,\mathrm{div}\,\mathbf{u} = 0 \tag{2.63}$$

Noting that the sum of the first two terms is div $(\rho_s\,\mathbf{u})$, (2.63) can be written as

$$\mathrm{div}\,(\rho_s\,\mathbf{u}) + \frac{\partial\tilde{\rho}}{\partial t} + \mathrm{div}\,(\tilde{\rho}\,\mathbf{u}) = 0 \tag{2.64}$$

Consider now the following estimates:

$$\mathrm{div}\,(\rho_s\,\mathbf{u}) = O\left(\rho_s\frac{U}{L}\right) \tag{2.65}$$

and

$$\frac{\partial\tilde{\rho}}{\partial t} + \mathrm{div}\,(\tilde{\rho}\,\mathbf{u}) = O\left(\tilde{\rho}\frac{U}{L}\right) \tag{2.66}$$

where the advective timescale L/U has been used in scaling (2.66). From (2.66), (2.65) and (2.62), we conclude that

$$\frac{\partial\tilde{\rho}/\partial t + \text{div}\,(\tilde{\rho}\,\mathbf{u})}{\text{div}\,(\rho_s\,\mathbf{u})} = O(10^{-2})$$

and, therefore, the leading balance of (2.64) turns out to be

$$\text{div}\,(\rho_s\,\mathbf{u}) = 0 \tag{2.67}$$

Equation (2.67) states that, for the purely horizontal flow, the atmosphere behaves as though it were an incompressible fluid; in fact, if $w = 0$, then (2.67) implies $\partial u/\partial x + \partial v/\partial y = 0$.

However, in the presence of vertical motion, the compressibility associated with the height dependence of $\rho_s(z)$ must be taken into account. For instance, if $u = v = 0$, (2.67) shows that the height-independent quantity is $\rho_s(z)\,w(z)$, rather than the sole vertical velocity $w(z)$.

Vertical Oscillations in an Adiabatic and Compressible Atmosphere

In the adiabatic case, Eq. (2.41) yields that potential temperature θ is conserved following the motion. Thus, if the transformation $(p_0, T_0) \mapsto (p, T)$ is adiabatic, we get from (2.40) $\theta(p, T) = \theta(p_0, T_0)$, whence

$$T\left(\frac{p_0}{p}\right)^{\frac{\gamma-1}{\gamma}} = T_0 \tag{2.68}$$

which is the first of the three adiabatic equations we are going to derive. Then, from (2.68), we obtain

$$\frac{p}{p_0} = \left(\frac{T}{T_0}\right)^{\frac{\gamma}{\gamma-1}} \tag{2.69}$$

Since $T = p/(R\rho)$ and $T_0 = p_0/(R\rho_0)$, Eq. (2.69) yields

$$\frac{p}{p_0} = \left(\frac{\rho}{\rho_0}\right)^{\gamma} \tag{2.70}$$

which is the second adiabatic state equation to be used in what follows. Moreover, elimination of pressure between (2.69) and (2.70) produces the third adiabatic state equation:

$$\frac{T}{T_0} = \left(\frac{\rho}{\rho_0}\right)^{\gamma-1} \tag{2.71}$$

Now we consider a stratified atmosphere in hydrostatic equilibrium and the adiabatic upward motion of an air parcel that does not conserve its density. Rather, the density

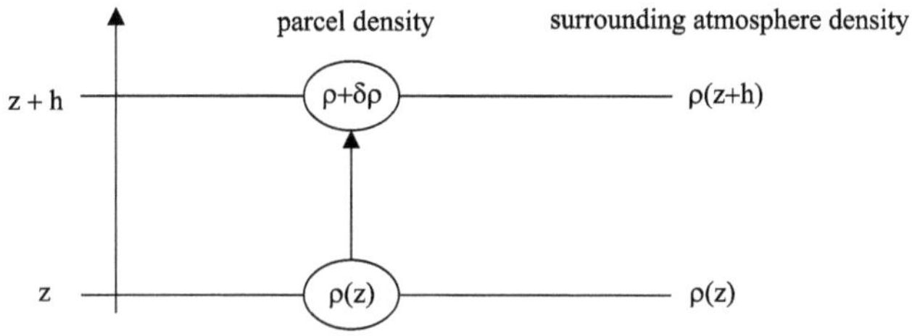

Fig. 2.4 An air parcel is initially in static equilibrium with the surrounding atmosphere. Then it is lifted adiabatically, according to (2.70), from z to $z+h$. Thus, its density decreases from ρ to $\rho+\delta\rho$, where $\delta\rho<0$ is given by (2.73), while the environmental density at the height $z+h$ is $\rho(z+h)<\rho(z)$. Once the parcel is released, its time evolution is governed by (2.74)

anomaly of the parcel is ascribed to the pressure decrease in moving from the starting height z to $z+h$. In the upward displacement (see Fig. 2.4), the ambient hydrostatic pressure changes from p to $p+\delta p = p-\rho\, g h$, thus influencing the parcel density according to (2.70). From the latter equation, we infer

$$\frac{\partial\rho}{\partial p}=\frac{1}{\gamma}\frac{\rho}{p} \tag{2.72}$$

and hence, using the law of ideal gases and the hydrostatic pressure anomaly $\delta p=-g\rho\, h$, we derive the density anomaly

$$\delta\rho := \frac{\partial\rho}{\partial p}\,\delta p = -\frac{\rho\, g h}{\gamma R T} \tag{2.73}$$

On the whole, at the height $z+h$, the ambient density is $\rho(z+h)$, while the parcel density is the sum of the value taken at z plus the anomaly, that is, $\rho+\delta\rho$. Unlike (2.55), here Newton's second law for the parcel's motion takes the form

$$\rho\,\delta V\,\frac{\mathrm{d}^2 h}{\mathrm{d}t^2} = g\left[\rho(z+h)-(\rho+\delta\rho)\right]\delta V$$

where $\rho(z+h)-(\rho+\delta\rho)\approx(\partial\rho/\partial z)\,h-\delta\rho$ with $\delta\rho$ given by (2.73). Thus, on the whole,

$$\rho\,\delta V\,\frac{\mathrm{d}^2 h}{\mathrm{d}t^2} = g\left[\frac{\partial\rho}{\partial z}+\frac{\rho\, g}{\gamma R T}\right] h\,\delta V$$

At this point, we can proceed as in the case of the incompressible fluid and write again Newton's law in the form of a harmonic oscillator:

$$\frac{\mathrm{d}^2 h}{\mathrm{d}t^2}+N^2 h = 0 \tag{2.74}$$

where now the *buoyancy frequency* $N = N(z)$ is defined by

$$N^2 := -\frac{g}{\rho}\left(\frac{g\rho}{\gamma RT} + \frac{\partial \rho}{\partial z}\right) \tag{2.75}$$

In order to relate the buoyancy frequency N with the adiabatic lapse rate g/c_p, the quantity in the parenthesis of the previous equation will be evaluated as a function of temperature. To do this, we start from

$$\frac{\partial \rho}{\partial z} = \frac{1}{R}\frac{\partial}{\partial z}\left(\frac{p}{T}\right) = \frac{1}{RT}\left(\frac{\partial p}{\partial z} - \frac{p}{T}\frac{\partial T}{\partial z}\right) \tag{2.76}$$

and make the substitutions

$$\frac{\partial p}{\partial z} \mapsto -g\rho \qquad\qquad p \mapsto R\rho T \tag{2.77}$$

at the r.h.s. of (2.76) to obtain

$$\frac{\partial \rho}{\partial z} = -\frac{\rho}{RT}\left(g + R\frac{\partial T}{\partial z}\right) \tag{2.78}$$

Hence,

$$\frac{\partial \rho}{\partial z} + \frac{g\rho}{\gamma RT} = -\frac{\rho}{RT}\left[g\left(1 - \frac{1}{\gamma}\right) + R\frac{\partial T}{\partial z}\right] \tag{2.79}$$

and, recalling that $1 - 1/\gamma = R/c_p$, Eq. (2.79) can be written as

$$\frac{\partial \rho}{\partial z} + \frac{g\rho}{\gamma RT} = -\frac{\rho}{T}\left(\frac{g}{c_p} + \frac{\partial T}{\partial z}\right) \tag{2.80}$$

Using (2.80), the buoyancy frequency (2.75) becomes

$$N^2 = \frac{g}{T}\left(\frac{g}{c_p} + \frac{\partial T}{\partial z}\right) \tag{2.81}$$

that is, recalling (2.45),

$$N^2 = \frac{g}{\theta}\frac{\partial \theta}{\partial z} \tag{2.82}$$

In the troposphere, $N = O(10^{-2}\,\mathrm{rad/s})$. Comparison of (2.48) with (2.82) shows that

$$N^2 = g\frac{\Gamma_\mathrm{d} - \Gamma}{T} \tag{2.83}$$

and substitution of (2.83) into (2.74) gives the evolution equation of a compressible air parcel embedded in a stratified atmosphere, that is,

$$\frac{\mathrm{d}^2 h}{\mathrm{d}t^2} + g \frac{\Gamma_\mathrm{d} - \Gamma}{T} h = 0 \tag{2.84}$$

The behaviour of a parcel displaced from its equilibrium position critically depends on the sign of $\Gamma_\mathrm{d} - \Gamma$ appearing in (2.84), as follows:

1. If $\Gamma_\mathrm{d} - \Gamma > 0$, the decrease of the environmental temperature with height is slower than the parcel's temperature. Thus, a positive restoring force brings the parcel back to its equilibrium position, whatever the sense of the displacement may be.
2. If $\Gamma_\mathrm{d} - \Gamma = 0$, the environmental temperature and the parcel's temperature decrease with the same rate. Thus, no restoring force acts on the displaced parcel.
3. If $\Gamma_\mathrm{d} - \Gamma < 0$, the decrease of the environmental temperature with height is faster than the parcel's temperature. Thus, a negative restoring force drives the parcel indefinitely away from its equilibrium position, whatever the sense of the displacement may be.

For future purposes, we stress that, based on (2.74) and (2.82), small air displacements from the rest space in an adiabatic and compressible atmosphere are governed by

$$\frac{\mathrm{d}^2 h}{\mathrm{d}t^2} + \frac{g}{\theta} \frac{\partial \theta}{\partial z} h = 0 \tag{2.85}$$

Potential Density

A noticeable link between compressible and incompressible fluids is found with the aid of the *potential density* σ, defined by

$$\sigma := \rho \left(\frac{p_0}{p} \right)^{1/\gamma} \tag{2.86}$$

Note, in passing, that multiplication of (2.40) by (2.86) gives

$$\theta \, \sigma = T \rho \frac{p_0}{p} = \frac{p_0}{R}$$

so, the equation of state of an ideal gas expressed as a function of potential temperature and potential density is simply

$$p_0 = R \sigma \theta \tag{2.87}$$

By using (2.86) to evaluate $\partial \rho / \partial z$ in terms of $\partial \sigma / \partial z$, we obtain with the aid of substitution (2.77),

$$\frac{\partial \rho}{\partial z} = -\frac{g \rho}{\gamma R T} + \frac{\rho}{\sigma} \frac{\partial \sigma}{\partial z} \tag{2.88}$$

and substitution of (2.88) into (2.75) eventually yields

$$N^2 = -\frac{g}{\sigma}\frac{\partial \sigma}{\partial z} \tag{2.89}$$

Comparison of (2.89) with (2.57) immediately shows that the substitution of potential density for density allows us to treat compressible fluids as incompressible.

2.2 Internal Gravity Waves in Adiabatic and Frictionless Fluids

In this section, we present a detailed account of internal waves, which constitute a special kind of motion that arises both in the atmosphere and in the ocean. Our choice is dictated by two main reasons: first, for oscillations suitably "fast" with respect to Earth's rotation speed, these waves can be investigated also in an inertial frame[2] and, second, we have the opportunity to deal with the important thermodynamical concepts involved in the derivation of the governing equation of these waves. Moreover, we resort to scaling arguments that will be systematically applied also in the geophysical context (i.e. when Earth's rotation is taken into account).

A disturbance on a fluid, initially at rest, causes the displacement of each parcel from its initial position: if the displacement is upward, the motion of the parcel is forced downward by gravity, while, if the displacement is downward, the motion is forced upward by buoyancy. Under adiabatic conditions, the parcel differs from the fluid ambient in which it is embedded after the displacement, and therefore, it oscillates around its initial position. In the absence of friction, these oscillations are undamped, and a wave-like motion takes place in the fluid body.

2.2.1 Definition of the Model

Governing Equations

A fluid at rest satisfies Eqs. (1.55) and (1.56). Thus, in this case,

$$\mathbf{u} = 0 \qquad \rho = \rho_s(z) \qquad p = p_s(z) \tag{2.90}$$

where the "standard" density ρ_s and the "standard" pressure p_s are linked by

$$\frac{\mathrm{d}p_s}{\mathrm{d}z} + g\rho_s = 0 \tag{2.91}$$

[2] Thus, the difficulties related to the treatment of the Coriolis acceleration are avoided altogether.

For the time being, the form of $\rho_s(z)$ is not explicitly established, apart from the request of static stability (i.e. lighter fluid lies above heavier fluid); so, recalling that z is increasing upward, we assume

$$\frac{\mathrm{d}\rho_s}{\mathrm{d}z} < 0 \tag{2.92}$$

Hence, a fluid at rest satisfies:

1. The incompressibility equations of seawater (2.53) and dry atmosphere (2.67)
2. The momentum equations (1.44)–(1.46)
3. The thermodynamic equation for adiabatic motions of the ocean (2.51), which here takes the form
$$\frac{\mathrm{D}\rho_s}{\mathrm{D}t} = 0 \tag{2.93}$$
4. The thermodynamic equation for adiabatic motions of the atmosphere (2.41) with $\dot{Q} = 0$, which here takes the form

$$\frac{\mathrm{D}\theta_s}{\mathrm{D}t} = 0 \tag{2.94}$$

where the standard potential temperature is

$$\theta_s = C_0 \frac{p_s^{1/\gamma}}{\rho_s} \tag{2.95}$$

with

$$C_0 := \frac{1}{R} p_0^{\frac{\gamma-1}{\gamma}}$$

It is useful to anticipate the definition of buoyancy frequency square N_s^2 of the standard density for liquids, namely,

$$N_s^2 := -\frac{g}{\rho_s} \frac{\mathrm{d}\rho_s}{\mathrm{d}z} \tag{2.96}$$

which is nothing but (2.57) with ρ_s in place of ρ. Moreover, the frequency square

$$N_s^2 := \frac{g}{\theta_s} \frac{\mathrm{d}\theta_s}{\mathrm{d}z} \tag{2.97}$$

is nothing but (2.82) with θ_s in place of θ, that is, the buoyancy frequency square for the standard potential temperature.

Consider now the fluid in motion, that is, $\mathbf{u} \neq 0$. Unless the fluid density is a constant, this configuration is not consistent with $\rho = \rho_s(z)$, but, rather, it demands $\rho = \rho(x, y, z, t)$. Thus, the actual density field is conceived as the superposition of the standard component $\rho_s(z)$ with a suitable density anomaly $\tilde{\rho}(x, y, z, t)$, namely,

$$\rho(x, y, z, t) = \rho_s(z) + \tilde{\rho}(x, y, z, t) \tag{2.98}$$

We stress that, while ρ and ρ_s are positive, the anomaly $\tilde{\rho}$ is not necessarily positive. In accordance with (2.62), ρ and ρ_s have the same order of magnitude; so, the approximation

$$\rho_s + \tilde{\rho} \approx \rho_s \tag{2.99}$$

can be applied, provided that density does not undergo differentiation.

In the same way, the pressure field of a fluid in motion can be expressed as

$$p(x, y, z, t) = p_s(z) + \tilde{p}(x, y, z, t) \tag{2.100}$$

where, unlike (2.91), pressure perturbation $\tilde{p}$ is not necessarily in hydrostatic equilibrium with the density anomaly $\tilde{\rho}$. We anticipate that just the imbalance between $\partial\tilde{p}/\partial z$ and $g\tilde{\rho}$ is responsible for the vertical acceleration of the fluid parcels involved in the wave motion.

The incompressibility equations of seawater (2.53) and dry atmosphere (2.67) are left unchanged, namely,

$$\operatorname{div}\mathbf{u} = 0 \qquad \text{(seawater)} \tag{2.101}$$

$$\operatorname{div}(\rho_s\,\mathbf{u}) = 0 \qquad \text{(atmosphere)} \tag{2.102}$$

while, recalling (2.98) and (2.100), the momentum equations (1.44)–(1.46) can be written as

$$(\rho_s + \tilde{\rho})\left(\frac{\partial}{\partial t} + \mathbf{u}\cdot\nabla\right)u = -\frac{\partial\tilde{p}}{\partial x} \tag{2.103}$$

$$(\rho_s + \tilde{\rho})\left(\frac{\partial}{\partial t} + \mathbf{u}\cdot\nabla\right)v = -\frac{\partial\tilde{p}}{\partial y} \tag{2.104}$$

$$(\rho_s + \tilde{\rho})\left(\frac{\partial}{\partial t} + \mathbf{u}\cdot\nabla\right)w = -\frac{\partial\tilde{p}}{\partial z} - \frac{\mathrm{d}p_s}{\mathrm{d}z} - g\,(\rho_s + \tilde{\rho}) \tag{2.105}$$

Because of (2.91), Eq. (2.105) becomes

$$(\rho_s + \tilde{\rho})\left(\frac{\partial}{\partial t} + \mathbf{u}\cdot\nabla\right)w = -\frac{\partial\tilde{p}}{\partial z} - g\tilde{\rho} \tag{2.106}$$

Finally, by using (2.99), Eqs. (2.103), (2.104) and (2.106) take the simpler form

$$\rho_s\left(\frac{\partial}{\partial t} + \mathbf{u}\cdot\nabla\right)u = -\frac{\partial\tilde{p}}{\partial x} \tag{2.107}$$

$$\rho_s \left(\frac{\partial}{\partial t} + \mathbf{u} \cdot \nabla \right) v = - \frac{\partial \tilde{p}}{\partial y} \tag{2.108}$$

$$\rho_s \left(\frac{\partial}{\partial t} + \mathbf{u} \cdot \nabla \right) w = - \frac{\partial \tilde{p}}{\partial z} - g \tilde{\rho} \tag{2.109}$$

respectively, usually known as the *Boussinesq approximation*.

Based on (2.98), density conservation in an adiabatic ocean (2.51) yields

$$w \frac{\mathrm{d}\rho_s}{\mathrm{d}z} + \left(\frac{\partial}{\partial t} + \mathbf{u} \cdot \nabla \right) \tilde{\rho} = 0 \tag{2.110}$$

The case of the atmosphere is less obvious. Here, we note preliminarily that, due to (2.95), (2.98) and (2.100), the potential temperature can be written as

$$\theta = C_0 \frac{(p_s + \tilde{p})^{1/\gamma}}{\rho_s + \tilde{\rho}} = \theta_s \frac{(1 + \tilde{p}/p_s)^{1/\gamma}}{1 + \tilde{\rho}/\rho_s} \tag{2.111}$$

where θ is given by (2.42). As we will see, the latter is not a mathematical identity, but, rather, it expresses a basic physical relation between the non-dimensional anomaly of potential temperature, say, θ', and the non-dimensional density anomaly, say, ρ'.

Approximations for Buoyancy Frequencies

In general, the buoyancy frequency $N_s(z)$ is depth dependent, but, for mathematical simplicity, we can reasonably approximate $N_s(z)$ with a constant value N_s, at least under suitable hypotheses about the rest state fluid.

In the case of the ocean, assumption

$$N_s^2 \approx \text{const.} \tag{2.112}$$

in the vertical range in which the wave motion takes place does not contradict any previous statement about the wave dynamics, so it can be taken into account, and its implications are pointed out below. Because of (2.112), the standard density $\rho_s(z)$ satisfies the ordinary differential equation

$$\frac{\mathrm{d}\rho_s}{\mathrm{d}z} + \frac{N_s^2}{g} \rho_s = 0$$

whence

$$\rho_s(z) = \rho_s(0) \exp\left(- \frac{N_s^2}{g} z \right) \tag{2.113}$$

The density-scale height

$$H_\rho := O\left[\left(-\frac{1}{\rho_s}\frac{d\rho_s}{dz}\right)^{-1}\right] \tag{2.114}$$

of (2.113) is the constant

$$H_\rho = \frac{g}{N_s^2} \tag{2.115}$$

If $N_s \approx 5\cdot 10^{-3}\,s^{-1}$, Eq. (2.115) yields $H_\rho \approx 4\cdot 10^5\,m$; so, one obtains the ratio $H/H_\rho \ll O(1)$ for realistic values of the depth H of the motion.

In the case of the atmosphere, in the vertical range in which the wave motion takes place, the rest state fluid is assumed to be isothermal, that is,

$$\frac{dT_s}{dz} = 0 \tag{2.116}$$

Equation (2.116) has two implications. First, Eq. (2.45) referred to the rest state has the form

$$\frac{1}{\theta_s}\frac{d\theta_s}{dz} = \frac{1}{T_s}\left(\frac{dT_s}{dz} + \frac{g}{c_p}\right) \tag{2.117}$$

so, because of (2.116), the buoyancy frequency square

$$N_s^2 = \frac{g}{\theta_s}\frac{d\theta_s}{dz} \tag{2.118}$$

takes the constant value

$$N_s^2 = \frac{g^2}{c_p T_s}$$

Second, the state equation $p_s = R\rho_s T_s$ and the hydrostatic equilibrium condition $dp_s/dz + g\rho_s = 0$ imply the ordinary differential equation

$$\frac{d\rho_s}{dz} + \frac{g}{RT_s}\rho_s = 0$$

whence

$$\rho_s(z) = \rho_s(0)\exp\left(-\frac{g}{RT_s}z\right) \tag{2.119}$$

Also in this case, the density-scale height is a constant, that is,

$$H_\rho = \frac{RT_s}{g} \tag{2.120}$$

The r.h.s. of (2.120) is equal to $p_s/(\rho_s g)$, and the estimate of the latter quantity, for instance, at the ground, yields $p_s/(\rho_s g) \approx 10^4\,m$. Thus, we get the ratio $H/H_\rho = O(1)$ for realistic values of the depth H of the motion in the troposphere.

To summarize, we have seen that approximation (2.112) is admissible, although under a restrictive hypothesis in the case of the atmosphere, and it leads to an exponentially decreasing standard-density profile. As a consequence, the density-height scales defined in (2.114) are also constant.

Scaling of the Governing Equations

Scaling analysis presupposes that each coordinate and field variable can be expressed in a unique way as the product of its "typical" value, which is assumed to characterize the considered scale, times the related non-dimensional coordinate or field variable, whose order of magnitude is 1.

Generalizing notation (1.112), we shall use the following typographic convention: non-dimensional quantities expressed by Latin characters will be written with sans-serif font, whereas non-dimensional quantities expressed by Greek or special characters will be written with a prime. For instance, if the original dimensional quantity is a and its typical magnitude (dimensional and positive) is A, then

$$a = A\,\mathsf{a} \tag{2.121}$$

where $\mathsf{a} = O(1)$ is the related non-dimensional quantity.

We recall that, in physics, the notation $O(x)$ usually indicates a typical value of variable x. For example, if x is a non-dimensional variable, then $O(\mathsf{x})$ may be thought of as the power of 10 that best approximates the average of x; moreover, the *order of magnitude* of x is $\log_{10}[O(\mathsf{x})]$. However, one should be aware that this notation and terminology is often abused in the literature; for example,

$$x = O(1)$$

is sometimes used in place of the more appropriate form

$$O(x) = 1$$

Once a certain equation is written only in non-dimensional variables, say, $\mathsf{a}_1, \mathsf{a}_2, \ldots$, the dominant terms of the equation (if any) can be singled out. For instance, consider the dimensional equation

$$a + b + c = 0 \tag{2.122}$$

which, according to (2.121), may be rewritten as

$$A\,\mathsf{a} + B\,\mathsf{b} + C\,\mathsf{c} = 0 \tag{2.123}$$

Assume $B = \max\{A, B, C\}$, and rewrite (2.123) as

$$\frac{A}{B}\mathsf{a} + \mathsf{b} + \frac{C}{B}\mathsf{c} = 0 \tag{2.124}$$

where $O(A/B) \leq 1$ and $O(C/B) \leq 1$. Then, one and only one of the following cases holds:

$$A < O(B) \text{ and } C < O(B) \Longrightarrow \mathsf{b} \approx 0 \tag{2.125}$$

$$A < O(B) \text{ and } C = O(B) \Longrightarrow \mathsf{b} + \mathsf{c} \approx 0 \tag{2.126}$$

$$A = O(B) \text{ and } C < O(B) \Longrightarrow \mathsf{a} + \mathsf{b} \approx 0 \tag{2.127}$$

$$A = O(B) \text{ and } C = O(B) \Longrightarrow \mathsf{a} + \mathsf{b} + \mathsf{c} \approx 0 \tag{2.128}$$

Thus, apart case (2.128), scaling analysis leads to a simplification of the original equation (2.122), as its non-dimensional counterpart contains, in general, a smaller number of terms.

The practical difficulty lies in the correct estimation of the orders of magnitude $A, B, C, \ldots$, which must be performed consistently with the phenomenology of the system under investigation. We stress that not every choice of $A, B, C, \ldots$ is expected to be referable to a physically realized system. According to Salmon (1998),

> Scaling analysis ... is the crudest sort of approximation theory

and

> Perhaps the best we can say about scaling analysis is that it forces us to impose our prejudices consistently.

while Cushman-Roisin (1994) states that

> ... the selection of scales for any given problem is more an art than a science.

After these preliminaries, we go back to our model and assume what follows:

1. All the space coordinates have the same length scale, say, L. Thus,

$$x = L\mathsf{x} \qquad y = L\mathsf{y} \qquad z = L\mathsf{z} \tag{2.129}$$

2. All the components of the fluid velocity have the same velocity scale, say, U. Thus,

$$u = U\mathsf{u} \qquad v = U\mathsf{v} \qquad w = U\mathsf{w} \tag{2.130}$$

3. The *local timescale*, say, T, such that

$$t = T\mathsf{t} \tag{2.131}$$

is at least one order of magnitude smaller than the *advective timescale* L/U, that is,

$$O\left(\frac{TU}{L}\right) < 1 \tag{2.132}$$

4. Let P and D be the scales of $\tilde{p}$ and $\tilde{\rho}$, respectively; so,

$$\tilde{p} = P\mathsf{p} \tag{2.133}$$

$$\tilde{\rho} = D\rho' \tag{2.134}$$

Then, P and D are estimated, assuming that the imbalance between $\partial\tilde{p}/\partial z$ and $g\tilde{\rho}$ be responsible for the vertical acceleration of the fluid parcels. Therefore, with reference to (2.109), we obtain

$$O\left(\rho_s \frac{\partial w}{\partial t}\right) = O\left(\frac{\partial \tilde{p}}{\partial z}\right) = O(g\tilde{\rho}) \tag{2.135}$$

and hence the estimates

$$P = O\left(\frac{\rho_s U L}{T}\right) \tag{2.136}$$

$$D = O\left(\frac{\rho_s U}{gT}\right) \tag{2.137}$$

Non-dimensional Governing Equations

We introduce the non-dimensional gradient operator

$$\nabla' := \hat{\mathbf{i}}\frac{\partial}{\partial \mathsf{x}} + \hat{\mathbf{j}}\frac{\partial}{\partial \mathsf{y}} + \hat{\mathbf{k}}\frac{\partial}{\partial \mathsf{z}}$$

and the non-dimensional divergence operator

$$\mathsf{div}\mathbf{f} := \nabla' \cdot \mathbf{f}$$

where $\mathbf{f}$ is any vector function of non-dimensional coordinates. As (2.129) and (2.130) imply

$$\frac{\partial u}{\partial x} = \frac{U}{L}\frac{\partial \mathsf{u}}{\partial \mathsf{x}}$$

and so on, Eq. (2.101) takes the form

$$\frac{U}{L}\mathsf{div}\mathbf{u} = 0 \tag{2.138}$$

where

$$\mathbf{u} := \mathsf{u}\hat{\mathbf{i}} + \mathsf{v}\hat{\mathbf{j}} + \mathsf{w}\hat{\mathbf{k}}$$

is the non-dimensional velocity vector. Therefore, the non-dimensional version of (2.101) is obtained by dividing (2.138) by U/L, whence

$$\mathsf{div}\,\mathbf{u} = 0 \tag{2.139}$$

In the case of internal waves, we have $L \ll H_\rho$, where H_ρ is defined in (2.114) and this inequality leads to a simplification of incompressibility equation (2.102), which is here rewritten as

$$\mathrm{div}\,\mathbf{u} + \frac{1}{\rho_\mathrm{s}}\frac{\mathrm{d}\rho_\mathrm{s}}{\mathrm{d}z}w = 0 \tag{2.140}$$

In fact, because of (2.114), Eq. (2.140)

$$\mathrm{div}\,\mathbf{u} = \frac{w}{H_\rho} \tag{2.141}$$

After little algebra, in non-dimensional variables, (2.141) takes the form

$$\mathsf{div}\,\mathbf{u} = \frac{L}{H_\rho}\mathsf{w} \tag{2.142}$$

where the divergence terms are, by definition, $O(1)$, while $O(L/H_\rho) \ll 1$. Therefore, the $O(1)$ non-dimensional equation of incompressibility valid for internal waves is

$$\mathsf{div}\,\mathbf{u} = 0 \tag{2.143}$$

Although (2.101) and (2.102) are formally different, their approximations do coincide once they are scaled in the context of internal waves (see (2.139) and (2.143)).

The non-dimensional version of the zonal momentum equation (2.107) is obtained as follows. First, by using (2.129), (2.130), (2.131) and (2.133), it can be restated as

$$\rho_\mathrm{s}\left(\frac{U}{T}\frac{\partial \mathsf{u}}{\partial \mathsf{t}} + \frac{U^2}{L}\mathbf{u}\cdot\nabla'\mathsf{u}\right) = -\frac{P}{L}\frac{\partial \mathsf{p}}{\partial \mathsf{x}} \tag{2.144}$$

Second, division of (2.144) by $\rho_\mathrm{s} U/T$ yields with the aid of (2.136) the non-dimensional equation

$$\frac{\partial \mathsf{u}}{\partial \mathsf{t}} + \frac{TU}{L}\mathbf{u}\cdot\nabla'\mathsf{u} = -\frac{\partial \mathsf{p}}{\partial \mathsf{x}} \tag{2.145}$$

Because of (2.132), the $O(1)$ balance of (2.145) gives

$$\frac{\partial \mathsf{u}}{\partial \mathsf{t}} = -\frac{\partial \mathsf{p}}{\partial \mathsf{x}} \tag{2.146}$$

Exactly in the same way, starting from (2.108), one obtains the $O(1)$ equation

$$\frac{\partial \mathsf{v}}{\partial \mathsf{t}} = -\frac{\partial \mathsf{p}}{\partial \mathsf{y}} \tag{2.147}$$

Consider now Eq. (2.109). By using (2.136) and (2.134), Eq. (2.109) is equivalent to

$$\frac{\rho_s U}{T}\frac{\partial \mathrm{w}}{\partial \mathrm{t}} + \frac{\rho_s U^2}{L}\mathbf{u}\cdot\nabla'\mathrm{w} = -\frac{\rho_s U}{T}\frac{\partial \mathrm{p}}{\partial \mathrm{z}} - Dg\rho' \tag{2.148}$$

Division of (2.148) by $\rho_s U/T$ yields, with the aid of (2.137), the non-dimensional equation

$$\frac{\partial \mathrm{w}}{\partial \mathrm{t}} + \frac{TU}{L}\mathbf{u}\cdot\nabla'\mathrm{w} = -\frac{\partial \mathrm{p}}{\partial \mathrm{z}} - \rho' \tag{2.149}$$

which, due to (2.132), can be approximated by

$$\frac{\partial \mathrm{w}}{\partial \mathrm{t}} = -\frac{\partial \mathrm{p}}{\partial \mathrm{z}} - \rho' \tag{2.150}$$

Thermodynamic Equation for the Ocean

With the aid of (2.96), after little algebra, the non-dimensional version of (2.110) turns out to be

$$\frac{\partial \rho'}{\partial \mathrm{t}} - N_s^2 T^2 \mathrm{w} + \frac{TU}{L}\mathbf{u}\cdot\nabla'\rho' = 0 \tag{2.151}$$

Because of (2.132), a balance is possible in (2.151), provided that

$$T = O(N_s^{-1}) \tag{2.152}$$

Equation (2.152) yields an estimate of the local timescale of the internal waves, which is the order of its oscillation period. From (2.151) and (2.152), we get the $O(1)$ non-dimensional equation

$$\frac{\partial \rho'}{\partial \mathrm{t}} = \mathrm{w} \tag{2.153}$$

Thermodynamic Equation for the Atmosphere

By using (2.133) and (2.136), pressure perturbation $\tilde{p}$ can be written as

$$\tilde{p} = \frac{U}{gT}\rho_s g L \mathrm{p} \tag{2.154}$$

The quantity U/gT in (2.154) is the ratio between two velocities. The numerator U is the typical velocity of a parcel involved in the wave motion. The denominator gT is the velocity reached after the time interval T by a massive object initially at rest and freely falling under gravity. Since T is of the order of the oscillation period of the wave, inequality

$$\frac{U}{gT} \ll 1 \tag{2.155}$$

reasonably holds. Introducing the non-dimensional parameter

$$a := \frac{U}{gT} \tag{2.156}$$

allows us to write (2.154) in the more compact form

$$\tilde{p} = a\rho_s g L \mathsf{p} \tag{2.157}$$

In particular,

$$\frac{\tilde{p}}{p_s} = a\frac{\rho_s g L}{p_s}\mathsf{p} \tag{2.158}$$

while the overall pressure field becomes

$$p = p_s\left(1 + a\frac{\rho_s g L}{p_s}\mathsf{p}\right) \tag{2.159}$$

By using (2.134) and (2.137), density anomaly $\tilde{\rho} = (U\rho_s/gT)\rho'$, that is to say, in terms of (2.156)

$$\tilde{\rho} = a\rho_s\rho' \tag{2.160}$$

and the overall density field takes the form

$$\rho = \rho_s(1 + a\rho') \tag{2.161}$$

To derive the non-dimensional anomaly of potential temperature, say, θ', we start from (2.111) in which (2.158) and (2.161) are used to obtain

$$\frac{\theta}{\theta_s} = \frac{1}{1 + a\rho'}\left(1 + a\frac{\rho_s g L}{p_s}\mathsf{p}\right)^{1/\gamma} \tag{2.162}$$

Due to the smallness of a, the r.h.s. of (2.162) can be expanded in a Maclaurin series of a (see the Appendix of this section), whose leading-order terms give

$$\theta \approx \theta_s\left[1 + a\left(\frac{\rho_s g L}{\gamma p_s}\mathsf{p} - \rho'\right)\right] \tag{2.163}$$

In (2.163), the quantity

$$\theta' = \frac{\rho_s g L}{\gamma p_s}\mathsf{p} - \rho' \tag{2.164}$$

is identified with the non-dimensional anomaly of potential temperature. Hence, substitution of (2.164) into (2.163) yields the overall potential temperature in the approximate form

$$\theta \approx \theta_s(1 + a\theta') \tag{2.165}$$

Equation (2.164) establishes the link between the non-dimensional anomaly of potential temperature θ', the non-dimensional pressure p and the non-dimensional density anomaly ρ'; in fact, (2.164) can be written as

$$\theta' = \frac{gL}{c_s^2}\,\mathsf{p} - \rho' \tag{2.166}$$

where $c_s := \sqrt{\gamma p_s/\rho_s}$ is the speed of sound. As $c_s \gg \sqrt{gL}$ in any case, the dominant balance in (2.166) is

$$\theta' = -\rho' \tag{2.167}$$

Moreover, the conservation of potential temperature θ yields, through (2.165), the time growth rate of its non-dimensional anomaly θ'. In fact, equation

$$\frac{\mathrm{D}}{\mathrm{D}t}\left[\theta_s\,(1+a\,\theta')\right] = 0$$

means

$$(1+a\,\theta')\,w\,\frac{\mathrm{d}\theta_s}{\mathrm{d}z} + a\,\theta_s\left(\frac{\partial\theta'}{\partial t} + \mathbf{u}\cdot\nabla\theta'\right) = 0 \tag{2.168}$$

and, because of the approximation $1+a\,\theta' \approx 1$, Eq. (2.168) becomes

$$w\,\frac{\mathrm{d}\theta_s}{\mathrm{d}z} + a\,\theta_s\left(\frac{\partial\theta'}{\partial t} + \mathbf{u}\cdot\nabla\theta'\right) = 0 \tag{2.169}$$

Recalling also (2.156), Eq. (2.169) is seen to be equivalent to

$$T^2\,\mathsf{w}\,\frac{g}{\theta_s}\frac{\mathrm{d}\theta_s}{\mathrm{d}z} + \frac{\partial\theta'}{\partial \mathsf{t}} + \frac{TU}{L}\,\mathsf{u}\cdot\nabla'\theta' = 0 \tag{2.170}$$

With this definition and (2.97) in mind, Eq. (2.170) becomes

$$\frac{\partial\theta'}{\partial \mathsf{t}} + T^2 N_s^2\,\mathsf{w} + \frac{TU}{L}\,\mathsf{u}\cdot\nabla'\theta' = 0 \tag{2.171}$$

where the full analogy with (2.151) is evident. As before, an $O(1)$ balance is possible, provided that

$$T = O(N_s^{-1}) \tag{2.172}$$

Equation (2.172) is formally identical to (2.152) and yields an estimate of the local timescale of the internal atmospheric waves. From (2.171) and (2.172), the $O(1)$ non-dimensional equation is

$$\frac{\partial\theta'}{\partial \mathsf{t}} + \mathsf{w} = 0 \tag{2.173}$$

that is to say, because of (2.167),

$$\frac{\partial \rho'}{\partial t} = w \tag{2.174}$$

which is the same as (2.153), but referred to the atmosphere.

2.2.2 *Evolution Equations for the Velocity, Pressure and Density Fields*

The previous analysis shows that the non-dimensional governing equations for the oceanic and atmospheric internal waves are exactly the same. This fact allows a unified derivation of the evolution equation for wave fields, as follows.

Vertical Velocity

Consider, first, vertical velocity w. From (2.146) and (2.147), one obtains

$$\frac{\partial}{\partial t}\left(\frac{\partial u}{\partial x} + \frac{\partial v}{\partial y}\right) = -\left(\frac{\partial^2 p}{\partial x^2} + \frac{\partial^2 p}{\partial y^2}\right) \tag{2.175}$$

while (2.139) or (2.143) yields

$$\frac{\partial u}{\partial x} + \frac{\partial v}{\partial y} = -\frac{\partial w}{\partial z} \tag{2.176}$$

so substitution of (2.176) into (2.175) gives

$$\frac{\partial^2 w}{\partial t \partial z} = \frac{\partial^2 p}{\partial x^2} + \frac{\partial^2 p}{\partial y^2} \tag{2.177}$$

On the other hand, (2.150) implies

$$\frac{\partial^2 w}{\partial t^2} = -\frac{\partial^2 p}{\partial t \partial z} - \frac{\partial \rho'}{\partial t} \tag{2.178}$$

that is to say, after the elimination of the density anomaly in favour of the vertical velocity by using (2.153) or (2.174),

$$\frac{\partial^2 w}{\partial t^2} + w = -\frac{\partial^2 p}{\partial t \partial z} \tag{2.179}$$

The pressure terms can be eliminated from (2.177) and (2.179) to obtain an equation for the sole vertical velocity

$$\left[\frac{\partial^4}{\partial t^2 \partial z^2} + \left(\frac{\partial^2}{\partial x^2} + \frac{\partial^2}{\partial y^2}\right)\left(\frac{\partial^2}{\partial t^2} + 1\right)\right] w = 0$$

whence

$$\left(\frac{\partial^2}{\partial t^2}\nabla'^2 + \frac{\partial^2}{\partial x^2} + \frac{\partial^2}{\partial y^2}\right) w = 0 \qquad (2.180)$$

Equation (2.180) is the desired evolution equation of w.

Pressure

Like in the case of the vertical velocity, we start from Eqs. (2.177) and (2.179), but now we eliminate vertical velocity w instead of p to obtain the evolution equation for the sole perturbation pressure:

$$\left[\frac{\partial^4}{\partial t^2 \partial z^2} + \left(\frac{\partial^2}{\partial x^2} + \frac{\partial^2}{\partial y^2}\right)\left(\frac{\partial^2}{\partial t^2} + 1\right)\right] p = 0$$

that is,

$$\left(\frac{\partial^2}{\partial t^2}\nabla'^2 + \frac{\partial^2}{\partial x^2} + \frac{\partial^2}{\partial y^2}\right) p = 0 \qquad (2.181)$$

Density

Once w is known, the density anomaly can be inferred from (2.153), which is equivalent to

$$\rho'(\mathbf{x}, t) = \rho'(\mathbf{x}, 0) + \int_0^t w(\mathbf{x}, \tau')\, d\tau' \qquad (2.182)$$

Horizontal velocities

Once p is known, the horizontal components of the velocity field can be inferred from (2.146) and (2.147), whence

$$u(\mathbf{x}, t) = u(\mathbf{x}, 0) - \int_0^t \frac{\partial}{\partial x} p(\mathbf{x}, \tau')\, d\tau' \qquad (2.183)$$

and

$$v(\mathbf{x}, t) = v(\mathbf{x}, 0) - \int_0^t \frac{\partial}{\partial y} p(\mathbf{x}, \tau')\, d\tau' \qquad (2.184)$$

respectively.

Dimensional Equations

The dimensional version of (2.180) is derived along the following lines. From (2.129), we get $\partial/\partial \mathrm{x} = L\,\partial/\partial x$, and so on, for the remaining space derivatives. Hence, $\partial^2/\partial \mathrm{x}^2 = L^2\,\partial^2/\partial x^2$, and $\nabla'^2 = L^2\nabla^2$. In the same way, from (2.131), we obtain $\partial/\partial \mathrm{t} = T\ \partial/\partial t$ and, because of (2.152) or (2.172), $\partial^2/\partial \mathrm{t}^2 = N_{\mathrm{s}}^{-2}\ \partial^2/\partial t^2$. Finally, from (2.130), we have $\mathrm{w} = w/U$. Thus, substitution of the above-listed non-dimensional variables with the related dimensional variables into (2.180) yields the dimensional governing equation of the vertical motion:

$$\left[\frac{\partial^2}{\partial t^2}\nabla^2 + N_{\mathrm{s}}^2\left(\frac{\partial^2}{\partial x^2} + \frac{\partial^2}{\partial y^2}\right)\right] w = 0 \tag{2.185}$$

Analogously, the remaining dimensional equations are

$$\left[\frac{\partial^2}{\partial t^2}\nabla^2 + N_{\mathrm{s}}^2\left(\frac{\partial^2}{\partial x^2} + \frac{\partial^2}{\partial y^2}\right)\right] \tilde{p} = 0 \tag{2.186}$$

$$\tilde{\rho}(\mathbf{x}, t) = \tilde{\rho}(\mathbf{x}, 0) + \frac{\rho_{\mathrm{s}} N_{\mathrm{s}}^2}{g}\int_0^t w(\mathbf{x}, \tau)\,\mathrm{d}\tau \tag{2.187}$$

$$u(\mathbf{x}, t) = u(\mathbf{x}, 0) - \frac{1}{\rho_{\mathrm{s}}}\int_0^t \frac{\partial}{\partial x}\tilde{p}(\mathbf{x}, \tau)\,\mathrm{d}\tau \tag{2.188}$$

$$v(\mathbf{x}, t) = v(\mathbf{x}, 0) - \frac{1}{\rho_{\mathrm{s}}}\int_0^t \frac{\partial}{\partial y}\tilde{p}(\mathbf{x}, \tau)\,\mathrm{d}\tau \tag{2.189}$$

Plane-Wave Solutions of the Governing Equations

Harmonic Waves

Consider the generic component

$$w_0 \cos(\mathbf{k}\cdot\mathbf{x} - \sigma t) \tag{2.190}$$

of a certain Fourier expansion of w in three-dimensional plane waves, where $\sigma > 0$ and

$$\mathbf{k} = k_1\hat{\mathbf{i}} + k_2\hat{\mathbf{j}} + k_3\hat{\mathbf{k}} \tag{2.191}$$

is the *wave number vector*, while

$$\mathbf{x} = x\hat{\mathbf{i}} + y\hat{\mathbf{j}} + z\hat{\mathbf{k}}$$

is the position vector. In the same way,

$$p_0 \cos(\mathbf{k}\cdot\mathbf{x} - \sigma t) \tag{2.192}$$

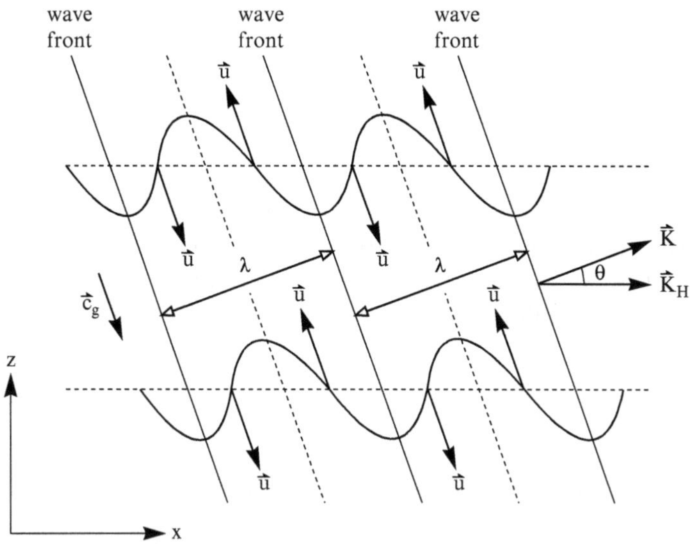

Fig. 2.5 Vertical structure of an internal wave

denotes the generic component of a certain Fourier expansion of $\tilde{p}$ in three-dimensional plane waves. About the meaning of (2.190) and (2.192), we recall what follows, with reference to Fig. 2.5.

Plane-wave fronts, at a given time, are represented by planes whose Cartesian equations are of the kind

$$\mathbf{k}\cdot\mathbf{x} = \Gamma \tag{2.193}$$

where Γ is any real constant. These planes propagate along the direction of the wave number vector with *phase speed*

$$c = \frac{\sigma}{|\mathbf{k}|} \tag{2.194}$$

where $|\mathbf{k}| = \sqrt{k_1^2 + k_2^2 + k_3^2}$, and *wavelength*

$$\lambda = \frac{2\pi}{|\mathbf{k}|} \tag{2.195}$$

Position (2.193) and the obvious identity $\cos(\Gamma - \sigma t) = \cos(\Gamma - \sigma t + 2\pi)$ show that the generic wavefront $\mathbf{k}\cdot\mathbf{x} = \Gamma + 2\pi$ at time t coincides with the wavefront $\mathbf{k}\cdot\mathbf{x} = \Gamma$ at time $t - 2\pi/\sigma$. Thus, during the wave *period* $T = 2\pi/\sigma$, each front moves along the direction (2.191) with speed (2.194) and covers the distance (2.195) between the parallel planes $\mathbf{k}\cdot\mathbf{x} = \Gamma$ and $\mathbf{k}\cdot\mathbf{x} = \Gamma + 2\pi$.

Substitution of (2.190) into (2.185), or (2.192) into (2.186), leads us to consider the equation

$$\left[\frac{\partial^2}{\partial t^2}\nabla^2 + N_s^2\left(\frac{\partial^2}{\partial x^2} + \frac{\partial^2}{\partial y^2}\right)\right]\cos(\mathbf{k}\cdot\mathbf{x} - \sigma t) = 0 \tag{2.196}$$

which is identically satisfied, provided that the dispersion relation

$$\sigma = N_s \frac{|\mathbf{k}_H|}{|\mathbf{k}|} \tag{2.197}$$

be satisfied with

$$\mathbf{k}_H := k_1 \hat{\mathbf{i}} + k_2 \hat{\mathbf{j}} \tag{2.198}$$

The form of propagation condition (2.197) is typical of the internal waves, and basic features of these waves can be inferred just from (2.197). Denoting by ϑ the angle between $\mathbf{k}$ and $\mathbf{k}_H$, so $|\mathbf{k}_H| = |\mathbf{k}| \cos\vartheta$ with $0 \le \vartheta < \pi/2$, then (2.197) gives

$$\sigma = N_s \cos\vartheta \tag{2.199}$$

Therefore:

- The *angular frequency* σ depends only on the orientation ϑ of $\mathbf{k}_H$ with respect to $\mathbf{k}$.
- If a wave can propagate along a direction, then it can also propagate along the opposite direction. Indeed, both $\mathbf{k}$ and $-\mathbf{k}$ yield the same magnitudes $|\mathbf{k}|$ and $|\mathbf{k}_H|$; hence, both $\mathbf{k}$ and $-\mathbf{k}$ (or none of them) fulfil the dispersion relation.
- The angular frequency is not higher than the buoyancy frequency: $\sigma \le N_s$.

The results above apply both to the vertical velocity w and to the perturbation pressure $\tilde{p}$. Consider a monochromatic wave

$$w = w_0 \cos(\mathbf{k} \cdot \mathbf{x} - \sigma t) \tag{2.200}$$

$$\tilde{p} = p_0 \cos(\mathbf{k} \cdot \mathbf{x} - \sigma t) \tag{2.201}$$

where σ and $\mathbf{k}$ satisfy propagation condition (2.197). What about the density anomaly (2.187) and the horizontal velocity (2.188)–(2.189)?

Substitution of (2.200) into (2.187) gives, after little algebra,

$$\tilde{\rho}(\mathbf{x}, t) = -\rho_s \frac{N_s^2 w_0}{g\sigma} \sin(\mathbf{k} \cdot \mathbf{x} - \sigma t) \tag{2.202}$$

Analogously, from (2.188) and (2.201), one obtains

$$u(\mathbf{x}, t) = \frac{p_0 k_1}{\rho_s \sigma} \cos(\mathbf{k} \cdot \mathbf{x} - \sigma t) \tag{2.203}$$

while, from (2.189) and (2.201),

$$v(\mathbf{x}, t) = \frac{p_0 k_2}{\rho_s \sigma} \cos(\mathbf{k} \cdot \mathbf{x} - \sigma t) \tag{2.204}$$

The incompressibility equation

$$\frac{\partial u}{\partial x}+\frac{\partial v}{\partial y}+\frac{\partial w}{\partial z}=0 \tag{2.205}$$

which, in the present context, constitutes the dimensional version of (2.139) and (2.143), links together the amplitudes w_0 and p_0 once (2.203), (2.204) and (2.200) are substituted into (2.205). The resulting equation is

$$p_0=-\frac{k_3\,\sigma\,\rho_s\,w_0}{|\mathbf{k}_H|^2} \tag{2.206}$$

Equation (2.206) can be substituted into (2.203) and (2.204) to give

$$u(\mathbf{x},t)=-w_0\frac{k_1\,k_3}{|\mathbf{k}_H|^2}\cos(\mathbf{k}\cdot\mathbf{x}-\sigma t)=-\frac{k_1\,k_3}{|\mathbf{k}_H|^2}w \tag{2.207}$$

$$v(\mathbf{x},t)=-w_0\frac{k_2\,k_3}{|\mathbf{k}_H|^2}\cos(\mathbf{k}\cdot\mathbf{x}-\sigma t)=-\frac{k_2\,k_3}{|\mathbf{k}_H|^2}w \tag{2.208}$$

because of (2.200). Thus, only the amplitude w_0 is left undetermined.

A consequence of (2.200), (2.207) and (2.208) is that each parcel of the wave belongs to a certain wavefront and oscillates in this plane; indeed,

$$\mathbf{k}\cdot\mathbf{u}=\left(-\frac{k_1^2\,k_3}{|\mathbf{k}_H|^2}-\frac{k_2^2\,k_3}{|\mathbf{k}_H|^2}+k_3\right)w=0 \tag{2.209}$$

Because of (2.206), the pressure wave (2.201) results to be proportional to the vertical velocity component w_0:

$$\tilde{p}=-w_0\frac{k_3\,\sigma\,\rho_s}{|\mathbf{k}_H|^2}\cos(\mathbf{k}\cdot\mathbf{x}-\sigma t)=-\frac{k_3\,\sigma\,\rho_s}{|\mathbf{k}_H|^2}w \tag{2.210}$$

just as the horizontal velocity components u and v given by (2.207) and (2.208).

Group Velocity

The linearity of the dynamics of internal waves allows us to take into account also wave packets, which consist in the superposition of monochromatic waves of the kind (2.200) or (2.201) whose wave numbers k_1, k_2, k_3 are peaked around "central" values (say, k_1^0, k_2^0, k_3^0, respectively). Wave packets are characterized by the so-called *group velocity* which, by definition, is the vector $\mathbf{c}_g$ of components

$$c_{gx}:=\frac{\partial\sigma}{\partial k_1}\qquad c_{gy}:=\frac{\partial\sigma}{\partial k_2}\qquad c_{gz}:=\frac{\partial\sigma}{\partial k_3} \tag{2.211}$$

In the case of internal waves, σ is determined by (2.197), whence

$$\mathbf{c}_g = \frac{N_s^2 k_3}{\sigma |\mathbf{k}|^4} (k_1 k_3,\, k_2 k_3,\, -|\mathbf{k}_H|^2) \tag{2.212}$$

Thus, for positive k_3, vertical component c_{gz} is negative and the group velocity points downward. Moreover,

$$\mathbf{k} \cdot \mathbf{c}_g = \frac{N_s^2 k_3}{\sigma |\mathbf{k}|^4} (k_1^2 k_3 + k_2^2 k_3 - k_3 |\mathbf{k}_H|^2) = 0 \tag{2.213}$$

that is to say, the group velocity $\mathbf{c}_g$ is always perpendicular to the wave number vector $\mathbf{k}$.

Finally, we argue that the group velocity $\mathbf{c}_g$ is parallel to the fluid velocity $\mathbf{u}$. To prove this last statement, we note that (2.207) and (2.208) yield

$$\mathbf{u} = \left(-\frac{k_1 k_3}{|\mathbf{k}_H|^2} w,\; -\frac{k_2 k_3}{|\mathbf{k}_H|^2} w,\; w \right) \propto (k_1 k_3,\, k_2 k_3,\, -|\mathbf{k}_H|^2) \tag{2.214}$$

while Eq. (2.212) implies

$$\mathbf{c}_g \propto (k_1 k_3,\, k_2 k_3,\, -|\mathbf{k}_H|^2) \tag{2.215}$$

whence we see that $\mathbf{u}$ and $\mathbf{c}_g$ are parallel, that is,

$$\mathbf{u} \times \mathbf{c}_g = 0 \tag{2.216}$$

Example: Horizontal Propagation

A noticeable special case of internal waves arises when the propagation vector is horizontal, namely,

$$k_3 = 0 \tag{2.217}$$

Because of (2.197), (2.207), (2.208), (2.210) and (2.212), Eq. (2.217) implies $\sigma = N_s$, $u = 0$, $v = 0$, $\tilde{p} = 0$ (while $\tilde{\rho} \neq 0$) and $\mathbf{c}_g = 0$, respectively. Thus,

$$\mathbf{u} = w_0 \cos(\mathbf{k}_H \cdot \mathbf{x} - N_s t)\,\hat{\mathbf{k}} \tag{2.218}$$

In this case, wavefronts are planes parallel to the z-axis, so the wave number vector (2.191) is parallel to the (x, y)-plane, and the wave parcels oscillate only vertically according to (2.218).

Appendix: The Approximation Inherent to (2.162)

For convenience, we write the r.h.s. of (2.162) as

$$f(a) := \frac{(1+Pa)^{1/\gamma}}{1+\rho' a}$$

where

$$P := \frac{\rho_s g L \mathrm{p}}{p_s}$$

The truncated Maclaurin expansion

$$f(a) \approx f(0) + f'(0)a = 1 + \left(\frac{P}{\gamma} - \rho'\right) a$$

for small a, gives (2.163).

2.3 Rotating Flows

In order to compare the implications of the physics of fluids with the data collected at fixed locations, the governing equations must be written relatively to a frame of reference fixed at the Earth's surface, and therefore rotating with respect to an inertial system. Thus, while the mass conservation and the thermodynamics are left unaffected by rotation, the momentum equation undergoes a fundamental change, which consists in the addition of further terms due to the Coriolis acceleration.

A partial simplification of the mathematics, which allows us to use Cartesian coordinates in place of spherical, is obtained by resorting to the so-called f- and β-planes. In this approximation, a relatively small spherical cap of fluid in motion is assimilated to a rotating planar layer. This enables us to consider different rotation rates along a special direction of the considered fluid layer, which mimics the local meridional direction on the real Earth.

2.3.1 Basic Parameters of the Earth and Local Cartesian Coordinate Systems

Basic Parameters of the Earth

A set of fundamental parameters, which characterizes our planet, enter into the governing equations of the atmosphere and the ocean. We refer to the nearly

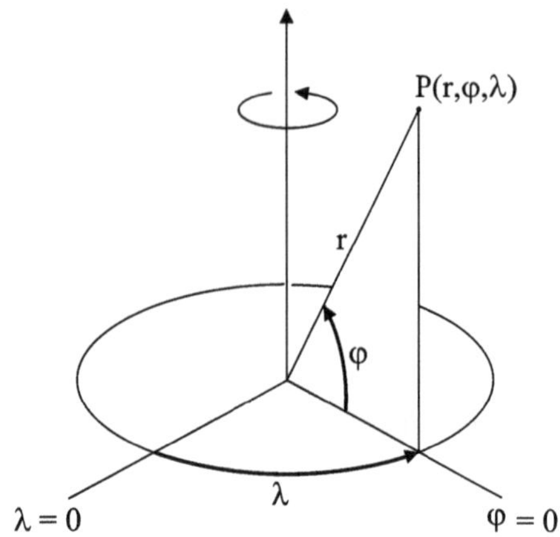

Fig. 2.6 Sketch of the natural coordinate system (r, ϕ, λ). The point P is determined by its linear distance r from the Earth's centre, its angular distance ϕ from the equatorial plane $\phi = 0$ and its angular distance λ from the meridian of Greenwich $\lambda = 0$

spherical shape of the Earth, its gravitational field and its almost uniform rotation rate around the polar axis.

For many purposes of Geophysical Fluid Dynamics (GFD), the Earth can be taken as a perfect sphere, to which we can associate the "natural" spherical coordinate system (r, ϕ, λ), where r is the radial coordinate (with $r = 0$ at the centre of the sphere), ϕ is the latitude and λ is the longitude, as Fig. 2.6 shows. In particular, $\phi = 0$ is the latitude of a point on the equatorial plane, while $\lambda = 0$ is the longitude on the prime meridian (say, Greenwich). Sphericity implies a unique value of the radius even if, because of the departure of the geopotential surfaces from the ideal spherical shape, we should distinguish, for instance, the equatorial radius $R_e = 6.378 \times 10^6$ m from the polar radius $R_p = 6.357 \times 10^6$ m. We take the value

$$R = 6.371 \times 10^6 \text{ m} \tag{2.219}$$

for the Earth's radius. It is the radius of a sphere having the volume of the actual geoid. The assumed sphericity of the Earth is justified by the smallness of the ratio $(R_e - R_p)/R \simeq 3.3 \times 10^{-3}$. Usually, the equation

$$r = R \tag{2.220}$$

represents the profile of the ground at sea level or the unperturbed air-sea interface.

Without Earth's rotation, the gravitational potential Φ would satisfy the equation

$$\nabla\Phi = g\hat{\mathbf{k}} \tag{2.221}$$

where the unit vector $\hat{\mathbf{k}}$ is parallel to r-axis and positively oriented outwards the sphere, while $g = O(10\,\text{m}\cdot\text{s}^{-2})$ in the proximity of Earth's surface. Should the centre of the Earth and the centre of gravity not coincide (because of spatial inhomogeneities in the mass structure of the Earth, or on account of the configuration of the Earth), this discrepancy would result in a correction of Eq. (2.221). But we will not take this effect into consideration.

Rotation around the polar axis is described by the constant vector $\boldsymbol{\Omega}$, whose magnitude Ω yields the Earth's angular frequency[3] given by

$$\Omega = 7.292 \times 10^{-5}\,\mathrm{s}^{-1} \tag{2.222}$$

The direction of Ω is that of the polar axis, and the orientation is such that Earth's rotation turns out to be counterclockwise for an observer placed above the North pole. This rotation induces the centrifugal acceleration

$$-\boldsymbol{\Omega} \times (\boldsymbol{\Omega} \times \mathbf{r}) \tag{2.223}$$

where $\mathbf{r} = r\hat{\mathbf{k}}$, so the body forces per unit of mass (2.221) and (2.223) are combined to become the effective force of gravity per unit of mass

$$\nabla \Phi + \boldsymbol{\Omega} \times (\boldsymbol{\Omega} \times \mathbf{r}) \tag{2.224}$$

We call "acceleration of gravity" the vector

$$\mathbf{g} = -[\nabla \Phi + \boldsymbol{\Omega} \times (\boldsymbol{\Omega} \times \mathbf{r})] \tag{2.225}$$

whose magnitude, for most of the purposes of GFD, can be taken as a constant, with the representative value

$$g = 9.80\,\mathrm{m} \cdot \mathrm{s}^{-2} \tag{2.226}$$

This approximation is acceptable, since the shape of the real Earth is such that the value of g at sea level varies depending on its latitude by $\pm 0.3\,\%$, and the inverse square law for gravitation results in a change in g of $0.3\,\%$ for a change in height of $10^4\,\mathrm{m}$.

The relative intensities of the body forces appearing into (2.224) can be estimated by means of the ratio $|\boldsymbol{\Omega} \times (\boldsymbol{\Omega} \times \mathbf{r})|\,/\,g$, which yields an upper bound of the order of 3.4×10^{-3}. As the centrifugal acceleration tends to take the fluid elements away from the Earth's surface, against gravity, the smallness of the ratio above denotes that little importance is given to centrifugal acceleration with respect to gravitation. Therefore, the approximation $\nabla \Phi + \boldsymbol{\Omega} \times (\boldsymbol{\Omega} \times \mathbf{r}) \approx \nabla \Phi$, that is to say,

$$\mathbf{g} \approx -\nabla \Phi = -g\hat{\mathbf{k}} \tag{2.227}$$

will be systematically applied to the momentum equation (Fig. 2.7). Finally, we anticipate that, unlike (2.223), another kind of acceleration due to rotation, that is, that of Coriolis, plays a central role in GFD as it moves the fluid elements along the Earth's surface, without doing work against gravity.

---

[3]We disregard the long-term variability of the Earth's angular frequency due to the Moon-Earth tidal interaction.

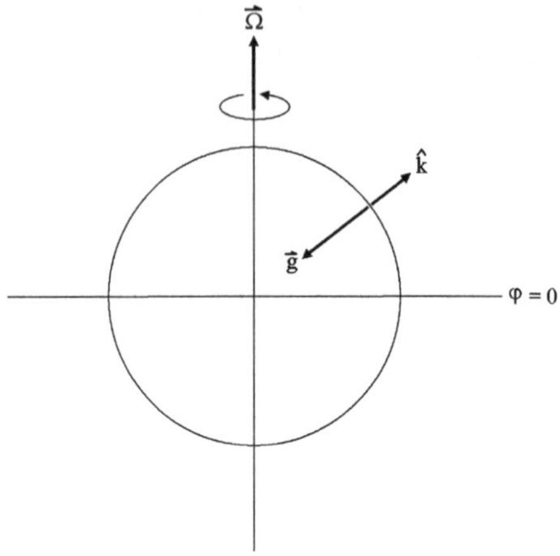

Fig. 2.7 Locations of the vectors $\hat{\mathbf{k}}$, $\hat{\boldsymbol{\Omega}}$ and $\mathbf{g}$ referred to a perfectly spherical Earth in which the centrifugal acceleration is negligibly small with respect to the gravity acceleration

Cartesian Coordinate Systems

In many relevant circumstances, GFD deals with phenomena whose space extent is small enough to resort to a local right-handed Cartesian coordinate system (x, y, z) in place of the spherical coordinates (r, ϕ, λ) quoted above.

The origin of the Cartesian frame is fixed at a given point (R, ϕ_0, λ_0) of the Earth's surface, and the x, y plane is located tangentially to the terrestrial sphere in that point. Among the ∞^1 orientations of the x, y axes around the z-axis, for future convenience, we choose that in which the y-axis points northward in the northern hemisphere and thus the x-axis points eastward. The orientation in the southern hemisphere is obtained by means of a suitable meridional translation of a given frame of the northern hemisphere, with the y-axis tangent to the considered meridian, and a subsequent suitable zonal translation, with the x-axis tangent to the considered circle of latitude.

As usual, $\hat{\mathbf{i}}, \hat{\mathbf{j}}$ and $\hat{\mathbf{k}}$ indicate the unit vectors associated to the Cartesian coordinate system. Obviously, $\hat{\mathbf{k}}$ coincides with the vector denoted in the same way as in (2.227).

In general, the governing equations of GFD should be expressed in terms of the coordinates (r, ϕ, λ), so the use of local Cartesian coordinates requires the transformation equations from (r, ϕ, λ) to (x, y, z):

$$\left.\begin{aligned} x &= R\cos(\phi_0)\tan(\lambda-\lambda_0) \\ y &= R\tan(\phi-\phi_0) \\ z &= \sqrt{r^2 - R^2\tan^2(\phi-\phi_0)} - R \end{aligned}\right\} \tag{2.228}$$

Equations (2.228) can be easily inferred from Fig. 2.8.

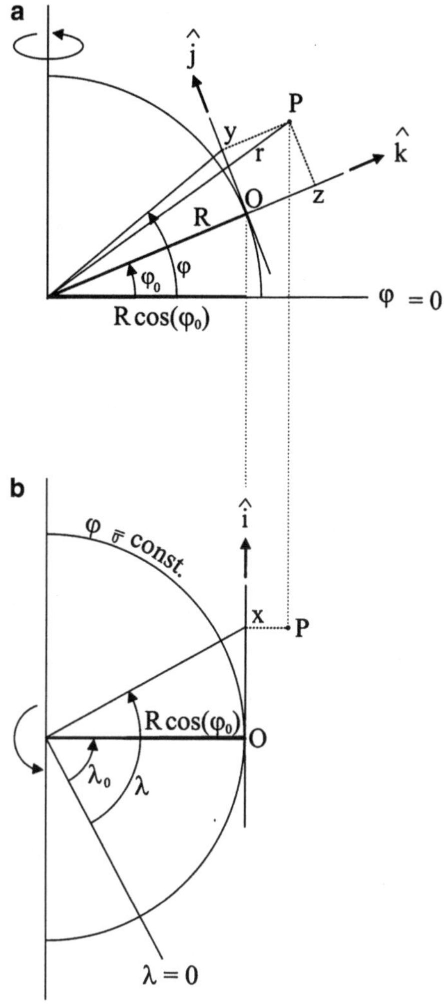

Fig. 2.8 Illustration of the relation between the natural coordinates (r, ϕ, λ) and the local Cartesian coordinates (x, y, z). Panel *a*: lateral view, panel *b*: top view

The details of the inversion of (2.228) are useful to clarify the conditions for the validity of the (local) Cartesian reference frame as an adequate substitute of the (global) spherical coordinates. To this purpose, consider the Maclaurin expansions

$$\tan(\lambda-\lambda_0)=\lambda-\lambda_0+\frac{1}{3}(\lambda-\lambda_0)^3+\cdots=(\lambda-\lambda_0)\left[1+\frac{1}{3}(\lambda-\lambda_0)^2+\dots\right] \tag{2.229}$$

and

$$\tan(\phi-\phi_0)=\phi-\phi_0+\frac{1}{3}(\phi-\phi_0)^3+\cdots=(\phi-\phi_0)\left[1+\frac{1}{3}(\phi-\phi_0)^2+\dots\right] \tag{2.230}$$

for

$$(\lambda - \lambda_0)^2 \ll 1 \qquad \text{and} \qquad (\phi - \phi_0)^2 \ll 1 \tag{2.231}$$

respectively. By neglecting the powers of $\lambda - \lambda_0$ and $\phi - \phi_0$ equal to or greater than 2, Eqs. (2.229) and (2.230) allow us to approximate the tangent functions with their arguments and, hence, to invert the first two equations of (2.228) thus, yielding

$$\lambda \approx \lambda_0 + \frac{x}{R\cos(\phi_0)} \tag{2.232}$$

$$\phi \approx \phi_0 + \frac{y}{R} \tag{2.233}$$

Moreover, the Maclaurin expansion of $\tan^2(\phi - \phi_0)$ shows that its smallest-order term is $(\phi - \phi_0)^2$ and therefore the same criterion as above justifies the approximation

$$\sqrt{r^2 - R^2\tan^2(\phi - \phi_0)} - R \approx r - R \tag{2.234}$$

of the third of Eqs. (2.228), whence

$$r \approx R + z \tag{2.235}$$

In order to get a deeper insight into the physical meaning of assumptions (2.231), we preliminarily introduce the horizontal length scale L of the motion, which we assume to be the same for both x and y, namely,

$$x = O(L) \qquad\qquad y = O(L) \tag{2.236}$$

This length scale represents the typical horizontal displacement of the flow under investigation. For instance, L might be the wavelength of an oscillating current, or the radius of a vortex. In terms of length scale L, the first of inequalities (2.231), together with (2.232), implies

$$\left(\frac{L}{R\cos(\phi_0)}\right)^2 < O(1) \tag{2.237}$$

while the second of (2.231) with (2.233) gives

$$\left(\frac{L}{R}\right)^2 < O(1) \tag{2.238}$$

Relationship (2.238) poses an upper bound on the horizontal length scale in the framework of the Cartesian approximation (2.232) and (2.233), and, customarily, it is satisfied by assuming

$$L \leq 1{,}000\,\text{km} \tag{2.239}$$

In fact, if $L = 1{,}000\,\text{km}$, then $L/R \approx 0.157$, and (2.238) is verified. On the other hand, relationship (2.237) is equivalent to

$$\left(\frac{L}{R}\right)^2 \frac{1}{\cos^2(\phi_0)} < O(1) \tag{2.240}$$

where, of necessity, $1/\cos^2(\phi_0) \geq 1$. Thus, to avoid that the order of magnitude of the product $[L/R\cos(\phi_0)]^2$ be equal to or greater than $O(1)$ in spite of (2.238), the central latitude ϕ_0 is constrained by the request

$$\frac{1}{\cos^2(\phi_0)} = O(1) \tag{2.241}$$

Equation (2.241) means that ϕ_0 is close enough to the equator, and the upper bound

$$|\phi_0| \leq \frac{\pi}{4} \tag{2.242}$$

is usually assumed in order to satisfy (2.241). For instance, $L/R = 0.157$ and $\phi_0 = \pi/4$ imply $[L/R\cos(\phi_0)]^2 \approx 5 \times 10^{-2}$.

To summarize, up to now, the validity of (2.232), (2.233) and (2.235) relies on conditions (2.239) and (2.242); however, in the applications of the Cartesian approximation to GFD, a further constraint – of dynamical origin – must be considered. Here, we simply anticipate that it takes the form

$$\frac{L}{R}|\cot(\phi_0)| < O(1) \tag{2.243}$$

while the explanation of (2.243) is postponed until the end of this chapter. Since $L/R < O(1)$, inequality (2.243) demands

$$|\cot(\phi_0)| \leq O(1) \tag{2.244}$$

Because of the pole at $\phi_0 = 0$ of $\cot(\phi_0)$, latitude ϕ_0 must be far enough from the equator; so, besides (2.242), one must also pose a lower bound, usually of the kind

$$|\phi_0| \geq \frac{\pi}{6} \tag{2.245}$$

On the whole, the central latitude belongs to one of the intervals

$$\begin{aligned} -\pi/4 \leq \phi_0 \leq -\pi/6 \qquad & \text{(southern hemisphere)} \\ \pi/6 \leq \phi_0 \leq \pi/4 \qquad & \text{(northern hemisphere)} \end{aligned} \tag{2.246}$$

For instance, if $L/R = 0.157$ and $\phi_0 = \pi/6$, then $L|\cot(\phi_0)|/R \approx 0.272$ and this value is within the limit of validity of (2.243).

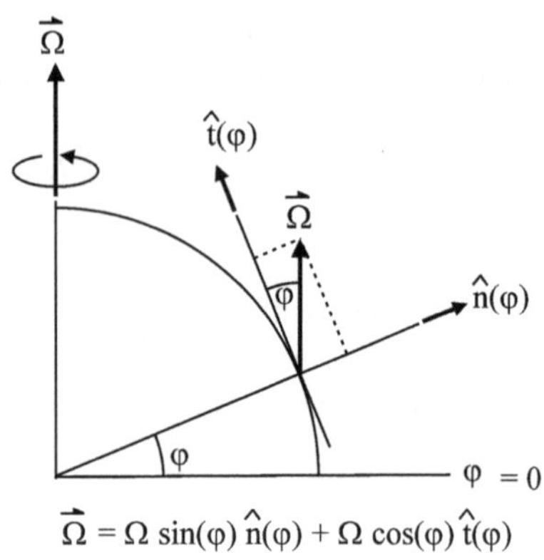

Fig. 2.9 The β-plane approximation consists in the linearization of the sinusoidal components of $\boldsymbol{\Omega}$ around a given latitude while keeping the unit vectors $\hat{\mathbf{n}}$ and $\hat{\mathbf{t}}$ fixed at this latitude

The f-Plane and the β-Plane Approximations

f-Plane

The components of $\boldsymbol{\Omega}$ normal and tangential to the terrestrial sphere play a basic role in the dynamics of the atmosphere and the ocean. Thus, we want to express these components in the above-introduced Cartesian coordinate system.

Let $\hat{\mathbf{n}}(\phi)$ and $\hat{\mathbf{t}}(\phi)$ be unit vectors at (R, ϕ, λ_0), normal and tangential to the sphere, respectively, and belonging to the longitudinal plane $\lambda = \lambda_0$. Then, vector $\boldsymbol{\Omega}$ can be written as (Fig. 2.9)

$$\boldsymbol{\Omega} = \Omega \sin(\phi)\hat{\mathbf{n}}(\phi) + \Omega \cos(\phi)\hat{\mathbf{t}}(\phi) \tag{2.247}$$

In the Cartesian coordinate system with the origin in (R, ϕ_0, λ_0), we have $\hat{\mathbf{n}}(\phi_0) = \hat{\mathbf{k}}$ and $\hat{\mathbf{t}}(\phi_0) = \hat{\mathbf{j}}$. If, moreover, the meridional displacement of the fluid elements is small enough to disregard latitudinal deviations from ϕ_0, then the r.h.s. of (2.247) can be approximated by setting $\phi \simeq \phi_0$, whence the approximation

$$\boldsymbol{\Omega} \simeq \Omega \sin(\phi_0)\hat{\mathbf{k}} + \Omega \cos(\phi_0)\hat{\mathbf{j}} \tag{2.248}$$

is obtained. In this reference frame, the normal component of $\boldsymbol{\Omega}$ is

$$\hat{\mathbf{k}} \cdot \boldsymbol{\Omega} \simeq \Omega \sin(\phi_0) \tag{2.249}$$

and the *Coriolis parameter* f_0 is, by definition,

$$f_0 := 2\Omega \sin(\phi_0) \tag{2.250}$$

Analogously, the tangential component of $\boldsymbol{\Omega}$ is

$$\hat{\mathbf{j}} \cdot \boldsymbol{\Omega} \simeq \Omega \cos(\phi_0) \tag{2.251}$$

and the *reciprocal Coriolis parameter* $\tilde{f}_0$ is, by definition,

$$\tilde{f}_0 := 2\Omega \cos(\phi_0) \tag{2.252}$$

In terms of (2.250) and (2.252), Eq. (2.248) can be written as

$$2\boldsymbol{\Omega} \simeq f_0 \hat{\mathbf{k}} + \tilde{f}_0 \hat{\mathbf{j}} \tag{2.253}$$

where

$$f_0 = O(10^{-4}\,\mathrm{s}^{-1}) \qquad \tilde{f}_0 = O(10^{-4}\,\mathrm{s}^{-1}) \tag{2.254}$$

Equation (2.253), by means of which $\boldsymbol{\Omega}$ can be referred to any Cartesian frame of reference, yields the so-called *f-plane approximation*. In the f-plane, each fluid element experiences the same components of $\boldsymbol{\Omega}$, that is, (2.249) and (2.251).

β-Plane

For meridional displacements larger than in the case of the f-plane, but consistent with (2.231), physics suggests that the latitudinal dependence of $\hat{\mathbf{k}} \cdot \boldsymbol{\Omega}$ and $\hat{\mathbf{j}} \cdot \boldsymbol{\Omega}$ should be preserved, also if a local Cartesian coordinate system is adopted. This request is satisfied by making, first, a compromise between (2.247) and (2.248) of the kind

$$\boldsymbol{\Omega} \simeq \Omega \sin(\phi) \hat{\mathbf{n}}(\phi_0) + \Omega \cos(\phi) \hat{\mathbf{t}}(\phi_0) \tag{2.255}$$

in which the unit vectors are kept constant, and, second, by adopting the subsequent linear approximation of $\sin(\phi)$ and $\cos(\phi)$ around the "central" latitude ϕ_0. On the whole, recalling that $\phi \simeq \phi_0 + y/R$, the result is the rotation vector in the form

$$2\boldsymbol{\Omega} \simeq \left(f_0 + \tilde{f}_0 \frac{y}{R} \right) \hat{\mathbf{k}} + \left(\tilde{f}_0 - f_0 \frac{y}{R} \right) \hat{\mathbf{j}} \tag{2.256}$$

whose components

$$2\hat{\mathbf{k}} \cdot \boldsymbol{\Omega} \simeq f_0 + \tilde{f}_0 \frac{y}{R}$$

$$2\hat{\mathbf{j}} \cdot \boldsymbol{\Omega} \simeq \tilde{f}_0 - f_0 \frac{y}{R}$$

generalize (2.249) and (2.251), respectively. The basic feature of (2.256) is that, unlike (2.248), its r.h.s. depends linearly on the *pseudo-latitude* y. In this way, (2.256) introduces the *latitude-dependent Coriolis parameter* $f(y)$ and the *latitude-dependent reciprocal Coriolis parameter* $\tilde{f}(y)$, that is,

$$f(y) := f_0 + \tilde{f}_0 \frac{y}{R} \tag{2.257}$$

$$\tilde{f}(y) := \tilde{f}_0 - f_0 \frac{y}{R} \tag{2.258}$$

respectively. In terms of (2.257) and (2.258), the Earth's rotation vector $\boldsymbol{\Omega}$ is given by

$$2\boldsymbol{\Omega} = f(y)\hat{\mathbf{k}} + \tilde{f}(y)\hat{\mathbf{j}} \tag{2.259}$$

which is an obvious generalization of (2.253). Conventionally, the shorthand notation

$$\beta_0 = \frac{\tilde{f}_0}{R} = \frac{2\Omega}{R}\cos(\phi_0) \tag{2.260}$$

is adopted, whence (2.257) takes the form

$$f(y) = f_0 + \beta_0 y \tag{2.261}$$

Because of (2.260), $\tilde{f}_0 = \beta_0 R$, and therefore, (2.258) can be written as

$$\tilde{f}(y) = \beta_0 R - f_0 \frac{y}{R} \tag{2.262}$$

Equations (2.261) and (2.262) yield the *β-plane approximation*, whose name derives from the parameter (2.260), which constitutes its distinctive aspect. Finally, from (2.219) and (2.222), one obtains, for low and midlatitudes, the estimate

$$\beta_0 = O(10^{-11}\,\mathrm{m}^{-1}\cdot\mathrm{s}^{-1}) \tag{2.263}$$

One can easily check that (2.261) and (2.262) have the same orders of magnitude, and thus, they should play equally important roles in GFD. On the contrary, we shall see in Sect. 2.3.7 that they are coupled with dynamic variables having different orders of magnitude (basically due to the fact that geophysical fluids are very thin with respect to their horizontal extent, and the related flows exhibit very small vertical velocities). For this reason, (2.261) is an ingredient of GFD much more fundamental than (2.262).

2.3.2 *Uniformly Rotating Coordinate Frames*

Connection Between Inertial and Uniformly Rotating Frames

Poisson's Formula

Let

$$\mathscr{I} := \{\hat{\mathbf{e}}_1, \hat{\mathbf{e}}_2, \hat{\mathbf{e}}_3\} \tag{2.264}$$

be a triple of mutually orthogonal unit vectors that constitute a right-handed inertial reference frame.

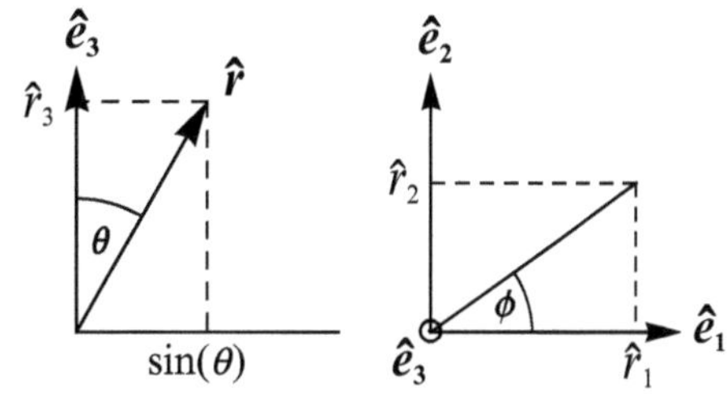

Fig. 2.10 Symbols related to Poisson's formula. Angle $\phi = \Omega t$ is longitude. *Left*: sagittal (i.e. meridional) plane. *Right*: basal (i.e. equatorial) plane

Let $\hat{\mathbf{r}}(t)$ be a unit vector that rotates about $\hat{\mathbf{e}}_3$ with constant angular speed Ω. Thus,

$$\hat{\mathbf{r}}(t) = \sin(\theta)\,\cos(\Omega t)\hat{\mathbf{e}}_1 + \sin(\theta)\,\sin(\Omega t)\hat{\mathbf{e}}_2 + \cos(\theta)\hat{\mathbf{e}}_3 \tag{2.265}$$

where $0 < \theta < \pi$ is co-latitude (see Fig. 2.10).

A straightforward computation using (2.265) yields *Poisson's formula*

$$\frac{\mathrm{d}\hat{\mathbf{r}}}{\mathrm{d}t} = \boldsymbol{\Omega} \times \hat{\mathbf{r}} \tag{2.266}$$

where $\boldsymbol{\Omega} := \Omega\hat{\mathbf{e}}_3$; indeed, both sides of (2.266) are equal to

$$\Omega\,\sin(\theta)\,[-\sin(\Omega t)\hat{\mathbf{e}}_1 + \cos(\Omega t)\hat{\mathbf{e}}_2]$$

Poisson's formula holds true if we substitute the unit vector $\hat{\mathbf{r}}$ with a vector of constant magnitude. In the next paragraph, we shall apply Poisson's formula (2.266) to the versors of a rotating reference system.

Velocity in a Rotating Frame

Let $\mathbf{r}(t)$ be a vector that describes the *absolute position* of a moving point in an inertial frame, say, (2.264). Moreover, we consider a reference frame

$$\mathscr{R} := \{\mathbf{R}(t), \hat{\mathbf{i}}_1(t), \hat{\mathbf{i}}_2(t), \hat{\mathbf{i}}_3(t)\} \tag{2.267}$$

that rotates about axis $\hat{\mathbf{e}}_3$ of frame (2.264) with constant speed Ω. In (2.267), point $\mathbf{R}(t)$ is the origin of frame $\mathscr{R}$ and coincides with a fixed point of Earth's surface, while $\hat{\mathbf{i}}_1, \hat{\mathbf{i}}_2, \hat{\mathbf{i}}_3$ constitute a right-handed triple of mutually orthogonal unit vectors. Therefore, by denoting the position vector in the rotating frame as in Fig. 2.11,

$$\mathbf{x}(t) := \mathbf{r}(t) - \mathbf{R}(t) \tag{2.268}$$

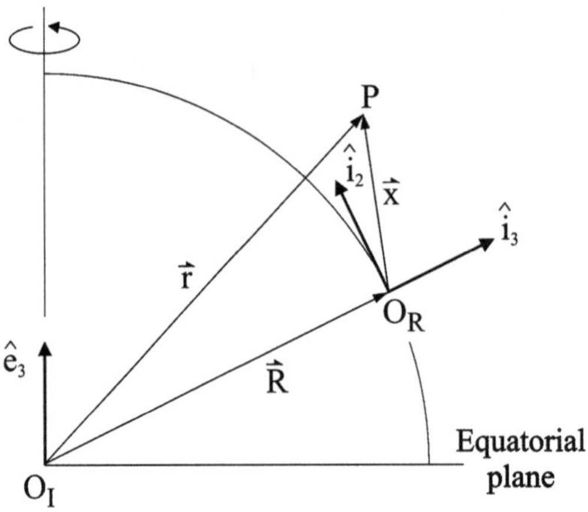

Fig. 2.11 Visualization of the vectors **r**, **R** and **x**. The frame of reference O_{I}, whose origin coincides with the *centre* of the (spherical) Earth, is inertial while O_{R} rotates with the Earth. For convenience, the unit vector $\hat{\mathbf{e}}_3$, which is fixed with O_{I}, is parallel to the Earth's rotation axis. Point P represents a point-like parcel of the atmosphere rotating with the Earth. The unit vectors $\hat{\mathbf{e}}_1$ and $\hat{\mathbf{e}}_2$, not reported in the figure, lie on the equatorial plane, while the unit vector $\hat{\mathbf{i}}_1$, not reported in the figure, is perpendicular to the plane of the figure and points inwards

we get

$$\mathbf{r}(t) = \mathbf{R}(t) + \sum_{k=1}^{3} x_k(t)\hat{\mathbf{i}}_k(t) \tag{2.269}$$

where $x_k := \hat{\mathbf{i}}_k \cdot \mathbf{x}$ with $k = 1, 2, 3$ are the components of the relative position in the rotating frame, and

$$\mathbf{x}(t) = \sum_{k=1}^{3} x_k(t)\hat{\mathbf{i}}_k(t) \tag{2.270}$$

By differentiating $\mathbf{r}(t)$ with respect to time, we get the velocity vector in the inertial frame

$$\mathbf{u}_{\mathrm{I}}(t) := \frac{\mathrm{d}\mathbf{r}}{\mathrm{d}t} \overset{(2.269)}{=} \frac{\mathrm{d}\mathbf{R}}{\mathrm{d}t} + \sum_{k=1}^{3} \frac{\mathrm{d}x_k}{\mathrm{d}t}\hat{\mathbf{i}}_k(t) + \sum_{k=1}^{3} x_k(t)\frac{\mathrm{d}\hat{\mathbf{i}}_k}{\mathrm{d}t} \tag{2.271}$$

Poisson's formula (2.266) applied to the first term in the r.h.s. of (2.271) yields

$$\frac{\mathrm{d}\mathbf{R}}{\mathrm{d}t} = \boldsymbol{\Omega} \times \mathbf{R}(t) \tag{2.272}$$

The second term in the r.h.s. of (2.271) is the velocity vector in the rotating frame,

$$\mathbf{u}_{\mathrm{R}}(t) := \sum_{k=1}^{3} \frac{\mathrm{d}x_k}{\mathrm{d}t}\hat{\mathbf{i}}_k(t) \tag{2.273}$$

The third term in the r.h.s. of (2.271) may be expanded using Poisson's formula to get

$$\sum_{k=1}^{3} x_k(t)\,\frac{\mathrm{d}\hat{\mathbf{i}}_k}{\mathrm{d}t} \overset{(2.266)}{=} \boldsymbol{\Omega} \times \sum_{k=1}^{3} x_k(t)\,\hat{\mathbf{i}}_k(t) \overset{(2.270)}{=} \boldsymbol{\Omega} \times \mathbf{x}(t) \tag{2.274}$$

Substituting (2.272)–(2.274) into (2.271), we get

$$\mathbf{u}_\mathrm{I}(t) = \mathbf{u}_\mathrm{R}(t) + \boldsymbol{\Omega} \times \mathbf{r}(t) \tag{2.275}$$

Through Eq. (2.275), we may express the absolute velocity in terms of relative quantities only (except for the origin $\mathbf{R}(t)$ and the angular velocity vector $\boldsymbol{\Omega}$, which are independent of the motion).

Acceleration in a Rotating Frame

Differentiating absolute velocity $\mathbf{u}_\mathrm{I}(t)$ with respect to time t, we get the *absolute acceleration* vector

$$\mathbf{a}_\mathrm{I}(t) := \frac{\mathrm{d}\mathbf{u}_\mathrm{I}}{\mathrm{d}t} \overset{(2.275)}{=} \frac{\mathrm{d}\mathbf{u}_\mathrm{R}}{\mathrm{d}t} + \boldsymbol{\Omega} \times \frac{\mathrm{d}\mathbf{r}}{\mathrm{d}t} \tag{2.276}$$

The first term in the r.h.s. of (2.276) may be expanded, using (2.273) and (2.266), to obtain

$$\begin{aligned}
\frac{\mathrm{d}\mathbf{u}_\mathrm{R}}{\mathrm{d}t} &\overset{(2.273)}{=} \frac{\mathrm{d}}{\mathrm{d}t}\sum_{k=1}^{3} \frac{\mathrm{d}x_k}{\mathrm{d}t}\hat{\mathbf{i}}_k(t) = \sum_{k=1}^{3} \frac{\mathrm{d}^2 x_k}{\mathrm{d}t^2}\hat{\mathbf{i}}_k(t) + \sum_{k=1}^{3} \frac{\mathrm{d}x_k}{\mathrm{d}t}\frac{\mathrm{d}}{\mathrm{d}t}\hat{\mathbf{i}}_k(t) \\
&\overset{(2.266)}{=} \mathbf{a}_\mathrm{R}(t) + \sum_{k=1}^{3} \frac{\mathrm{d}x_k}{\mathrm{d}t}\boldsymbol{\Omega} \times \hat{\mathbf{i}}_k(t) = \mathbf{a}_\mathrm{R}(t) + \boldsymbol{\Omega} \times \sum_{k=1}^{3} \frac{\mathrm{d}x_k}{\mathrm{d}t}\hat{\mathbf{i}}_k(t) \\
&\overset{(2.273)}{=} \mathbf{a}_\mathrm{R}(t) + \boldsymbol{\Omega} \times \mathbf{u}_\mathrm{R}(t)
\end{aligned} \tag{2.277}$$

where we have introduced the *relative acceleration* vector

$$\mathbf{a}_\mathrm{R}(t) := \sum_{k=1}^{3} \frac{\mathrm{d}^2 x_k}{\mathrm{d}t^2}\hat{\mathbf{i}}_k(t) \tag{2.278}$$

The second term in the r.h.s. of (2.276) may be expanded using, (2.275), to get

$$\begin{aligned}
\boldsymbol{\Omega} \times \frac{\mathrm{d}\mathbf{r}}{\mathrm{d}t} &\overset{(2.271)}{=} \boldsymbol{\Omega} \times \mathbf{u}_\mathrm{I}(t) \overset{(2.275)}{=} \boldsymbol{\Omega} \times [\mathbf{u}_\mathrm{R}(t) + \boldsymbol{\Omega} \times \mathbf{r}(t)] \\
&= \boldsymbol{\Omega} \times \mathbf{u}_\mathrm{R}(t) + \boldsymbol{\Omega} \times [\boldsymbol{\Omega} \times \mathbf{r}(t)]
\end{aligned} \tag{2.279}$$

Substituting (2.277) and (2.279) into (2.276), we finally get the kinematic identity

$$\underbrace{\mathbf{a}_\mathrm{I}(t)}_{\text{inertial}} = \underbrace{\mathbf{a}_\mathrm{R}(t)}_{\text{relative}} + \underbrace{2\boldsymbol{\Omega} \times \mathbf{u}_\mathrm{R}(t)}_{\text{Coriolis}} + \underbrace{\boldsymbol{\Omega} \times [\boldsymbol{\Omega} \times \mathbf{r}(t)]}_{\text{centripetal}} \tag{2.280}$$

Through Eq. (2.280), we may express the absolute acceleration in terms of relative quantities only (except for the origin $\mathbf{R}(t)$ and the angular velocity vector $\boldsymbol{\Omega}$, which are independent of the motion).

Remarks. The second term at the r.h.s. of (2.280) is the *Coriolis acceleration*

$$\mathbf{a}^{\mathrm{Cor}}(t) := 2\boldsymbol{\Omega} \times \mathbf{u}_{\mathrm{R}}(t) \tag{2.281}$$

which has been mentioned in Sect. 2.3.1.

The most striking feature of (2.281) comes from the identity

$$\mathbf{u}_{\mathrm{R}} \cdot \mathbf{a}^{\mathrm{Cor}} = 0 \tag{2.282}$$

which explains the observed tendency of large-scale flows to place them at right angles to the Coriolis acceleration: exactly this phenomenon constitutes the distinctive character of GFD.

The third term at the r.h.s. of (2.280) is the *centripetal acceleration*

$$\mathbf{a}^{\mathrm{cp}} := \boldsymbol{\Omega} \times (\boldsymbol{\Omega} \times \mathbf{r}) \tag{2.283}$$

The opposite of the centripetal acceleration is the *centrifugal acceleration*

$$\mathbf{a}^{\mathrm{cf}} := -\boldsymbol{\Omega} \times (\boldsymbol{\Omega} \times \mathbf{r}) \tag{2.284}$$

which has been anticipated in (2.223). To see the connection between (2.284) and the elementary definition of centrifugal acceleration, consider a point $\mathbf{r}$ fixed at the Earth's surface and the meridional plane (thus containing Earth's rotation axis) to which $\mathbf{r}$ belongs, and introduce in this plane the unit vector $\hat{\mathbf{n}}$ orthogonal to $\boldsymbol{\Omega}$. Then, $\mathbf{r}$ can be expanded as

$$\mathbf{r} = (\hat{\boldsymbol{\Omega}} \cdot \mathbf{r})\hat{\boldsymbol{\Omega}} + (\hat{\mathbf{n}} \cdot \mathbf{r})\hat{\mathbf{n}} \tag{2.285}$$

Substitution of (2.285) into (2.284) yields

$$\mathbf{a}^{\mathrm{cf}} = \Omega^2 (\hat{\mathbf{n}} \cdot \mathbf{r})\hat{\mathbf{n}} \tag{2.286}$$

The r.h.s. of (2.286) is just the well-known centrifugal acceleration of a material point that runs around $\hat{\boldsymbol{\Omega}}$ along a circle of radius $\hat{\mathbf{n}} \cdot \mathbf{r}$, with angular frequency Ω (see Fig. 2.12).

Connection Between Inertial and Uniformly Rotating Fields

The basic relationship (2.280) refers to a material point whose motion is described in terms of vectors (position, velocity and acceleration). As such, it is straightforward to extend its use to Lagrangian flow fields.

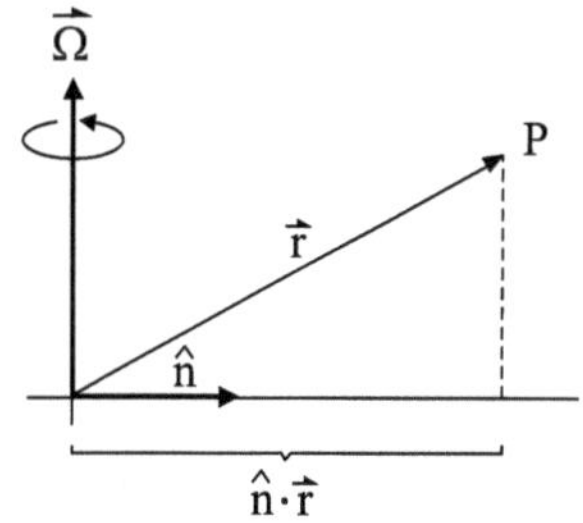

Fig. 2.12 The material point P runs around the Earth's rotation vector $\boldsymbol{\Omega}$ along a circle of radius $\hat{\mathbf{n}} \cdot \mathbf{r}$ with angular frequency $|\boldsymbol{\Omega}|$

On the other hand, in GFD, the natural approach to the motion of fluids is based on the concept of Eulerian field, and therefore, we need to derive the connection between inertial and uniformly rotating fields in the framework of the Eulerian description. This is the subject of the present paragraph.

Basic Results

We start from (2.275), namely,

$$\mathbf{u}_{\mathrm{I}}(t) = \mathbf{u}_{\mathrm{R}}(t) + \boldsymbol{\Omega} \times \mathbf{R}(t) + \boldsymbol{\Omega} \times \mathbf{x}(t) \tag{2.287}$$

and, only for convenience, we write in place of (2.273)

$$\mathbf{u}_{\mathrm{R}}(t) = \sum_{k=1}^{3} u_k(t)\,\hat{\mathbf{i}}_k(t) \qquad \left(u_k(t) := \frac{\mathrm{d}x_k}{\mathrm{d}t}\right) \tag{2.288}$$

Now, let $V(t)$ be an elementary material volume of fluid whose position and velocity in the rotating frame of reference are $\mathbf{x}(t)$ and $\mathbf{u}_{\mathrm{R}}(t)$, while in an inertial reference, they are $\mathbf{r}(t)$ and $\mathbf{u}_{\mathrm{I}}(t)$, respectively. Then we multiply each term of (2.287) by the fluid density ρ, integrate the products over $V(t)$ and apply the time derivative to each volume integral; in formulas, this yields

$$\frac{\mathrm{d}}{\mathrm{d}t}\int_{V(t)} \rho\,\mathbf{u}_{\mathrm{I}}\,\mathrm{d}V' = \frac{\mathrm{d}}{\mathrm{d}t}\int_{V(t)} \rho\,\mathbf{u}_{\mathrm{R}}(t)\,\mathrm{d}V' + \boldsymbol{\Omega} \times \frac{\mathrm{d}}{\mathrm{d}t}\left[\mathbf{R}(t)\int_{V(t)} \rho\,\mathrm{d}V'\right]$$
$$+ \boldsymbol{\Omega} \times \frac{\mathrm{d}}{\mathrm{d}t}\int_{V(t)} \rho\,\mathbf{x}(t)\,\mathrm{d}V' \tag{2.289}$$

where now ρ, $\mathbf{u}_{\mathrm{I}}$, $\mathbf{u}_{\mathrm{R}}$ and $\mathbf{x}$ may assume different values within $V(t)$.

Using corollary (1.22) to Reynolds' theorem, the l.h.s. of (2.289) becomes

$$\frac{\mathrm{d}}{\mathrm{d}t}\int_{V(t)} \rho\,\mathbf{u}_{\mathrm{I}}\,\mathrm{d}V' = \int_{V(t)} \rho\,\frac{\mathrm{D}\mathbf{u}_{\mathrm{I}}}{\mathrm{D}t}\,\mathrm{d}V' \tag{2.290}$$

The first term in the r.h.s. of (2.289) may be rearranged as

$$\frac{\mathrm{d}}{\mathrm{d}t}\int_{V(t)}\rho\,\mathbf{u}_{\mathrm{R}}(t)\,\mathrm{d}V' =$$

$$\overset{(2.288)}{=} \frac{\mathrm{d}}{\mathrm{d}t}\int_{V(t)}\rho\sum_{k=1}^{3}u_k(t)\hat{\mathbf{i}}_k(t)\,\mathrm{d}V' =$$

$$= \sum_{k=1}^{3}\left[\frac{\mathrm{d}\hat{\mathbf{i}}_k}{\mathrm{d}t}\int_{V(t)}\rho\,u_k(t)\,\mathrm{d}V' + \hat{\mathbf{i}}_k(t)\frac{\mathrm{d}}{\mathrm{d}t}\int_{V(t)}\rho\,u_k(t)\,\mathrm{d}V'\right] =$$

$$\overset{(2.266)}{=} \sum_{k=1}^{3}\left[\boldsymbol{\Omega}\times\hat{\mathbf{i}}_k(t)\int_{V(t)}\rho\,u_k(t)\,\mathrm{d}V' + \hat{\mathbf{i}}_k(t)\frac{\mathrm{d}}{\mathrm{d}t}\int_{V(t)}\rho\,u_k(t)\,\mathrm{d}V'\right] =$$

$$\overset{(1.19)}{=} \boldsymbol{\Omega}\times\int_{V(t)}\rho\sum_{k=1}^{3}u_k(t)\hat{\mathbf{i}}_k(t)\,\mathrm{d}V' + \sum_{k=1}^{3}\hat{\mathbf{i}}_k(t)\int_{V(t)}\rho\,\frac{\mathrm{D}(u_{\mathrm{R}})_k}{\mathrm{D}t}\,\mathrm{d}V' =$$

$$\overset{(2.288)}{=} \int_{V(t)}\rho\left[\boldsymbol{\Omega}\times\mathbf{u}_{\mathrm{R}}(t) + \left(\frac{\mathrm{D}\mathbf{u}_{\mathrm{R}}}{\mathrm{D}t}\right)_{\mathrm{R}}\right]\mathrm{d}V' \tag{2.291}$$

where, for convenience, we have put

$$\left(\frac{\mathrm{D}\mathbf{u}_{\mathrm{R}}}{\mathrm{D}t}\right)_{\mathrm{R}} := \sum_{k=1}^{3}\frac{\mathrm{D}(u_{\mathrm{R}})_k}{\mathrm{D}t}\,\hat{\mathbf{i}}_k(t) \tag{2.292}$$

The second term in the r.h.s. of (2.289) may be rearranged as

$$\boldsymbol{\Omega}\times\frac{\mathrm{d}}{\mathrm{d}t}\left[\mathbf{R}(t)\int_{V(t)}\rho\,\mathrm{d}V'\right] =$$

$$= \boldsymbol{\Omega}\times\left[\frac{\mathrm{d}\mathbf{R}}{\mathrm{d}t}\int_{V(t)}\rho\,\mathrm{d}V' + \mathbf{R}(t)\underbrace{\frac{\mathrm{d}}{\mathrm{d}t}\int_{V(t)}\rho\,\mathrm{d}V'}_{\overset{(1.16)}{=}0}\right] =$$

$$\overset{(2.272)}{=} \int_{V(t)}\rho\,\boldsymbol{\Omega}\times[\boldsymbol{\Omega}\times\mathbf{R}(t)]\,\mathrm{d}V' \tag{2.293}$$

The last term in the r.h.s. of (2.289) may be rearranged as

$$\boldsymbol{\Omega}\times\frac{\mathrm{d}}{\mathrm{d}t}\int_{V(t)}\rho\,\mathbf{x}(t)\,\mathrm{d}V'$$

$$\overset{(2.270)}{=} \boldsymbol{\Omega}\times\sum_{k=1}^{3}\frac{\mathrm{d}}{\mathrm{d}t}\left[\hat{\mathbf{i}}_k(t)\int_{V(t)}\rho\,x_k(t)\,\mathrm{d}V'\right]$$

$$= \boldsymbol{\Omega} \times \sum_{k=1}^{3} \Big[\underbrace{\frac{\mathrm{d}\hat{\mathbf{i}}_k}{\mathrm{d}t}}_{\overset{(2.266)}{=} \boldsymbol{\Omega}\times\hat{\mathbf{i}}_k(t)} \int_{V(t)} \rho\, x_k(t)\,\mathrm{d}V' + \hat{\mathbf{i}}_k(t) \underbrace{\frac{\mathrm{d}}{\mathrm{d}t}\int_{V(t)} \rho\, x_k(t)\,\mathrm{d}V'}_{\overset{(1.19)}{=} \int_{V(t)} \rho \frac{\mathrm{D}x_k}{\mathrm{D}t}\,\mathrm{d}V'} \Big]$$

$$= \boldsymbol{\Omega} \times \Big[\boldsymbol{\Omega} \times \int_{V(t)} \rho \underbrace{\sum_{k=1}^{3} x_k(t)\hat{\mathbf{i}}_k(t)}_{\overset{(2.270)}{=} \mathbf{x}(t)} \mathrm{d}V' \Big] + \boldsymbol{\Omega} \times \int_{V(t)} \rho \underbrace{\sum_{k=1}^{3} \frac{\mathrm{D}x_k}{\mathrm{D}t}\hat{\mathbf{i}}_k(t)}_{\overset{(2.288)}{=} \mathbf{u}_{\mathrm{R}}(t)} \mathrm{d}V'$$

$$= \int_{V(t)} \rho\, \{\boldsymbol{\Omega} \times [\boldsymbol{\Omega} \times \mathbf{x}(t)] + \boldsymbol{\Omega} \times \mathbf{u}_{\mathrm{R}}(t)\}\,\mathrm{d}V' \tag{2.294}$$

Substitution of (2.290)–(2.294) into (2.289) yields, after a trivial rearrangement,

$$\int_{V(t)} \rho \frac{\mathrm{D}\mathbf{u}_{\mathrm{I}}}{\mathrm{D}t}\,\mathrm{d}V' = \int_{V(t)} \rho \left\{ \left(\frac{\mathrm{D}\mathbf{u}_{\mathrm{R}}}{\mathrm{D}t}\right)_{\mathrm{R}} + 2\boldsymbol{\Omega} \times \mathbf{u}_{\mathrm{R}}(t) + \boldsymbol{\Omega} \times [\boldsymbol{\Omega} \times \mathbf{r}(t)] \right\} \mathrm{d}V' \tag{2.295}$$

Since (2.295) is valid whatever the elementary material volume $V(t)$ may be, the transformation law between the acceleration fields in an inertial and in the uniformly rotating frames takes its final form

$$\frac{\mathrm{D}\mathbf{u}_{\mathrm{I}}}{\mathrm{D}t} = \left(\frac{\mathrm{D}\mathbf{u}_{\mathrm{R}}}{\mathrm{D}t}\right)_{\mathrm{R}} + 2\boldsymbol{\Omega} \times \mathbf{u}_{\mathrm{R}}(t) + \boldsymbol{\Omega} \times [\boldsymbol{\Omega} \times \mathbf{r}(t)] \tag{2.296}$$

Remarks. Equation (2.296) gives the key tool to transform the Euler momentum equation (1.42), which is valid in inertial frames, into the version fit for the "terrestrial" point of view.

Note that the "almost inertial" behaviour of some systems fixed with the Earth (for instance, in hydraulics) does not contradict (2.296) but can be explained by means of a detailed analysis of the relative amplitudes of all the terms (Coriolis acceleration included!) appearing into the governing equations of the considered systems. For fluid motions whose timescale is much shorter than Ω^{-1}, the magnitude of Coriolis acceleration $2\boldsymbol{\Omega} \times \mathbf{u}_{\mathrm{R}}$ is much smaller than the magnitude of the local acceleration $\mathrm{D}\mathbf{u}_{\mathrm{R}}/\mathrm{D}t$; so, Coriolis acceleration plays a minor role within the momentum equation (cf. Appendix 2.3.8).

The opposite happens in GFD, where the order of magnitude, say, U, of $\mathbf{u}_{\mathrm{R}}$ is small enough to make much lesser than unity the ratio

$$\frac{|\mathrm{D}\mathbf{u}_{\mathrm{R}}/\mathrm{D}t|}{2\,|\boldsymbol{\Omega} \times \mathbf{u}_{\mathrm{R}}|} = O\left(\frac{U}{f_0 L}\right)$$

The non-dimensional quantity

$$\varepsilon := \frac{U}{f_0 L} \tag{2.297}$$

is the so-called *Rossby number*, which (in GFD) spans the interval $10^{-4} < \varepsilon < 10^{-1}$.

2.3.3 The Equations of Fluid Mechanics for Uniformly Rotating Systems

Governing Equations of Fluid Mechanics in Inertial Frames

Up to now, the governing equations for statistically averaged fields, in the sense explained in Sect. 1.2.3, have been established with respect to inertial frames. They are:

1. The mass-conservation principle (1.140), that is,

$$\frac{\mathrm{D}\rho}{\mathrm{D}t} + \rho \operatorname{div} \mathbf{u} = 0 \tag{2.298}$$

2. The thermodynamic equations (2.41), that is,

$$\frac{1}{\theta}\frac{\mathrm{D}\theta}{\mathrm{D}t} = \frac{\dot{Q}}{c_p T} \tag{2.299}$$

   for the atmosphere, and (2.51), that is,

$$\frac{\mathrm{D}\rho}{\mathrm{D}t} = 0 \tag{2.300}$$

   for the ocean.
3. The momentum equation (1.151), that is,

$$\frac{\mathrm{D}\mathbf{u}}{\mathrm{D}t} = -g\hat{\mathbf{k}} - \frac{1}{\rho}\nabla p + \left(A_{\mathrm{H}}\nabla_{\mathrm{H}}^2 + A_{\mathrm{V}}\frac{\partial^2}{\partial z^2}\right)\mathbf{u} \tag{2.301}$$

Here we prove that (2.298)–(2.300) preserve their forms when they are expressed in a uniformly rotating frame, while (2.301) undergoes a fundamental change due to (2.296). We anticipate that:

- Scalar quantities (such as density and pressure) and constant vectors (such as gravity acceleration) are left unaffected in passing from inertial to uniformly rotating frames.
- The gradient operator retains its form in the transformation from non-rotating to rotating coordinate frames: symbolically, $\nabla_{\mathrm{I}} = \nabla_{\mathrm{R}}$, as shown in Sect. 2.3.4 at the end of this subsection.

Invariance of the Lagrangian Derivative

Consider the Lagrangian derivative referred to an inertial frame

$$\left(\frac{\mathrm{D}}{\mathrm{D}t}\right)_{\mathrm{I}} = \left(\frac{\partial}{\partial t}\right)_{\mathrm{I}} + \mathbf{u}_{\mathrm{I}} \cdot \nabla_{\mathrm{I}} \tag{2.302}$$

and, separately, each of its terms: $(\partial/\partial t)_{\mathrm{I}}$ and $\mathbf{u}_{\mathrm{I}} \cdot \nabla_{\mathrm{I}}$.

Consider the first term, $(\partial/\partial t)_{\mathrm{I}}$. If a certain scalar field ϕ does not undergo advection relatively to the Earth, that is, if $\mathbf{u}_{\mathrm{R}} \cdot \nabla\phi = 0$, then

$$\left(\frac{\mathrm{D}\phi}{\mathrm{D}t}\right)_{\mathrm{R}} = \left(\frac{\partial\phi}{\partial t}\right)_{\mathrm{R}} \tag{2.303}$$

On the other hand, an inertial observer detects both its local variation

$$\left(\frac{\partial\phi}{\partial t}\right)_{\mathrm{I}} \tag{2.304}$$

and the variation due to Earth's rotation – which is generated by the tangential velocity $\boldsymbol{\Omega} \times \mathbf{r}$, where $\mathbf{r}$ is the position vector at which the variation of ϕ is evaluated. Thus, from the viewpoint of the inertial observer, the advective term is (see Fig. 2.13)

$$(\boldsymbol{\Omega} \times \mathbf{r}) \cdot \nabla_{\mathrm{I}}\phi \tag{2.305}$$

So, $(\partial\phi/\partial t)_{\mathrm{R}}$ is the sum of (2.304) and (2.305), whence

$$\left(\frac{\partial\phi}{\partial t}\right)_{\mathrm{I}} = \left(\frac{\partial\phi}{\partial t}\right)_{\mathrm{R}} - (\boldsymbol{\Omega} \times \mathbf{r}) \cdot \nabla_{\mathrm{I}}\phi \tag{2.306}$$

Consider the second term, $\mathbf{u}_{\mathrm{I}} \cdot \nabla_{\mathrm{I}}$, in the r.h.s. of (2.302). Because of (2.275), we have

$$\mathbf{u}_{\mathrm{I}} \cdot \nabla_{\mathrm{I}}\phi = (\mathbf{u}_{\mathrm{R}} + \boldsymbol{\Omega} \times \mathbf{r}) \cdot \nabla_{\mathrm{I}}\phi \tag{2.307}$$

Finally, the substitution of (2.306) and (2.307) into (2.302) gives

$$\left(\frac{\mathrm{D}\phi}{\mathrm{D}t}\right)_{\mathrm{I}} = \left(\frac{\mathrm{D}\phi}{\mathrm{D}t}\right)_{\mathrm{R}}$$

and the invariance of the Lagrangian derivative, that is,

$$\left(\frac{\mathrm{D}}{\mathrm{D}t}\right)_{\mathrm{I}} = \left(\frac{\mathrm{D}}{\mathrm{D}t}\right)_{\mathrm{R}} \tag{2.308}$$

is proved.

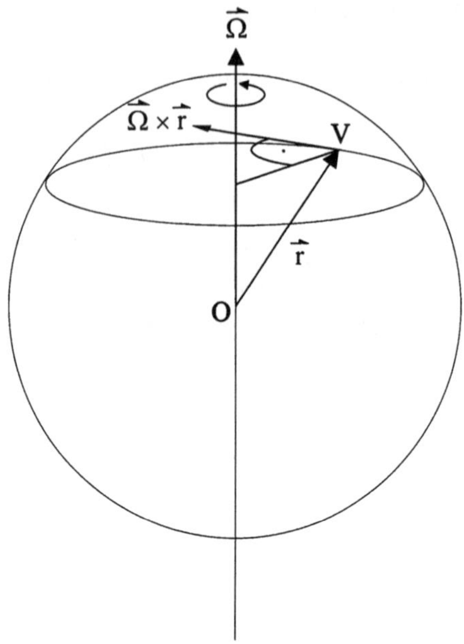

Fig. 2.13 Assume that a given scalar field ϕ, describing a property of a fluid, is fixed with the Earth; so it is, at most, a function of time only. Relatively to an inertial system, a material volume V of a fluid at rest with respect to the Earth rotates along a circle of latitude with orbital velocity $\boldsymbol{\Omega} \times \mathbf{r}$, where $\mathbf{r}$ ranges from Earth's centre O to V. Therefore, in this system, the scalar field ϕ is in motion and undergoes to the advection $(\boldsymbol{\Omega} \times \mathbf{r}) \cdot \nabla\phi$ due to Earth's rotation, where the gradient operator refers to the coordinates of the inertial frame

Note that, although the Lagrangian derivative of a scalar is the same in both frames, its individual components are not invariant.

Moreover, Eqs. (2.306) and (2.307) are formally analogous to (1.193) and (1.194), respectively, as can be seen through the correspondence $\mathbf{r} \times \boldsymbol{\Omega} \leftrightarrow \mathbf{U}_{\mathrm{rel}}$.

Invariance of the Divergence of the Velocity Vector

Consider the divergence of the velocity, $\nabla_{\mathrm{I}} \cdot \mathbf{u}_{\mathrm{I}}$. By using (2.275), we obtain

$$\nabla_{\mathrm{I}} \cdot \mathbf{u}_{\mathrm{I}} = \nabla_{\mathrm{R}} \cdot \mathbf{u}_{\mathrm{R}} + \nabla_{\mathrm{I}} \cdot (\boldsymbol{\Omega} \times \mathbf{r}) \tag{2.309}$$

The second term in the r.h.s. of (2.309) is

$$\nabla_{\mathrm{I}} \cdot (\boldsymbol{\Omega} \times \mathbf{r}) = \mathbf{r} \cdot (\nabla_{\mathrm{I}} \times \boldsymbol{\Omega}) - \boldsymbol{\Omega} \cdot (\nabla_{\mathrm{I}} \times \mathbf{r}) \tag{2.310}$$

Moreover, as $\nabla_{\mathrm{I}} \times \boldsymbol{\Omega} = 0$ and $\nabla_{\mathrm{I}} \times \mathbf{r} = 0$, Eq. (2.310) yields $\nabla_{\mathrm{I}} \cdot (\boldsymbol{\Omega} \times \mathbf{r}) = 0$. Hence, from (2.309), the invariance of the divergence follows:

$$\nabla_{\mathrm{I}} \cdot \mathbf{u}_{\mathrm{I}} = \nabla_{\mathrm{R}} \cdot \mathbf{u}_{\mathrm{R}} \tag{2.311}$$

Invariance of the Density and Thermodynamic Equations

First, we consider the mass balance (2.298) in an inertial frame:

$$\left(\frac{\mathrm{D}\rho}{\mathrm{D}t}\right)_{\mathrm{I}} + \rho\,\nabla_{\mathrm{I}}\cdot\mathbf{u}_{\mathrm{I}} = 0 \tag{2.312}$$

Second, we apply the transformation laws (2.308) and (2.311) to (2.312); the result is the following equation:

$$\left(\frac{\mathrm{D}\rho}{\mathrm{D}t}\right)_{\mathrm{R}} + \rho\,\nabla_{\mathrm{R}}\cdot\mathbf{u}_{\mathrm{R}} = 0 \tag{2.313}$$

which is valid in a uniformly rotating frame. The invariance of the density equation is so proved.

Likewise, because of the invariance of the Lagrangian derivative, expressed by Eq. (2.308), the thermodynamic equations (2.299) and (2.300) are also invariant.

The Momentum Equation in a Uniformly Rotating Frame

The momentum equation (2.301), when written in an inertial frame, takes the form

$$\left(\frac{\mathrm{D}\mathbf{u}_{\mathrm{I}}}{\mathrm{D}t}\right)_{\mathrm{I}} = -g\hat{\mathbf{k}} - \frac{1}{\rho}\nabla_{\mathrm{I}}p + \left[A_{\mathrm{H}}\left(\nabla_{\mathrm{H}}^{2}\right)_{\mathrm{I}} + A_{\mathrm{V}}\left(\frac{\partial^2}{\partial z^2}\right)_{\mathrm{I}}\right]\mathbf{u}_{\mathrm{I}} \tag{2.314}$$

The momentum equation in a frame fixed with the Earth comes from (2.314) using the fundamental transformation law (2.296), together with (2.275). Thus, the equation

$$\begin{aligned}\left(\frac{\mathrm{D}\mathbf{u}_{\mathrm{R}}}{\mathrm{D}t}\right)_{\mathrm{R}} &+ 2\boldsymbol{\Omega}\times\mathbf{u}_{\mathrm{R}} + \boldsymbol{\Omega}\times(\boldsymbol{\Omega}\times\mathbf{r})\\ &= -g\hat{\mathbf{k}} - \frac{1}{\rho}\nabla_{\mathrm{R}}p + \left[A_{\mathrm{H}}\left(\nabla_{\mathrm{H}}^{2}\right)_{\mathrm{R}} + A_{\mathrm{V}}\left(\frac{\partial^2}{\partial z^2}\right)_{\mathrm{R}}\right](\mathbf{u}_{\mathrm{R}} + \boldsymbol{\Omega}\times\mathbf{r})\end{aligned} \tag{2.315}$$

follows. As the vector $\boldsymbol{\Omega}\times\mathbf{r}$ is a linear function of its space variables, the second-order operators $\left(\nabla_{\mathrm{H}}^{2}\right)_{\mathrm{R}}$ and $\left(\partial^2/\partial z^2\right)_{\mathrm{R}}$ yield zero when applied to $\boldsymbol{\Omega}\times\mathbf{r}$. So, (2.315) simplifies into

$$\begin{aligned}\left(\frac{\mathrm{D}\mathbf{u}_{\mathrm{R}}}{\mathrm{D}t}\right)_{\mathrm{R}} &+ 2\boldsymbol{\Omega}\times\mathbf{u}_{\mathrm{R}} + \boldsymbol{\Omega}\times(\boldsymbol{\Omega}\times\mathbf{r})\\ &= -g\hat{\mathbf{k}} - \frac{1}{\rho}\nabla_{\mathrm{R}}p + \left[A_{\mathrm{H}}\left(\nabla_{\mathrm{H}}^{2}\right)_{\mathrm{R}} + A_{\mathrm{V}}\left(\frac{\partial^2}{\partial z^2}\right)_{\mathrm{R}}\right]\mathbf{u}_{\mathrm{R}}\end{aligned} \tag{2.316}$$

Consistently with (2.227), in (2.316), we disregard the centrifugal acceleration with respect to gravity acceleration $\mathbf{g} = -g\hat{\mathbf{k}}$, so that (2.316) further simplifies into

$$\left(\frac{\mathrm{D}\mathbf{u}_{\mathrm{R}}}{\mathrm{D}t}\right)_{\mathrm{R}} + 2\boldsymbol{\Omega} \times \mathbf{u}_{\mathrm{R}} = -g\hat{\mathbf{k}} - \frac{1}{\rho}\nabla_{\mathrm{R}}p + \left[A_{\mathrm{H}}\left(\nabla_{\mathrm{H}}^2\right)_{\mathrm{R}} + A_{\mathrm{V}}\left(\frac{\partial^2}{\partial z^2}\right)_{\mathrm{R}}\right]\mathbf{u}_{\mathrm{R}} \tag{2.317}$$

Due to the fact that GFD resorts only to frames of reference fixed with the Earth, the specification $(\,)_{\mathrm{R}}$ is hereafter dropped both from (2.313) and (2.317) to write simply

$$\frac{\mathrm{D}\rho}{\mathrm{D}t} + \rho\nabla\cdot\mathbf{u} = 0 \tag{2.318}$$

and

$$\frac{\mathrm{D}\mathbf{u}}{\mathrm{D}t} + 2\boldsymbol{\Omega} \times \mathbf{u} = -g\hat{\mathbf{k}} - \frac{1}{\rho}\nabla p + \left(A_{\mathrm{H}}\nabla_{\mathrm{H}}^2 + A_{\mathrm{V}}\frac{\partial^2}{\partial z^2}\right)\mathbf{u} \tag{2.319}$$

respectively. Because of the presence of Earth's rotation vector $\boldsymbol{\Omega}$, the momentum equation (2.319) deserves further developments. In the framework of the beta-plane approximation (2.259), the second term in the l.h.s. of (2.319) is given by

$$\begin{aligned} 2\boldsymbol{\Omega} \times \mathbf{u} &= (f\hat{\mathbf{k}} + \tilde{f}\hat{\mathbf{j}}) \times (u\hat{\mathbf{i}} + v\hat{\mathbf{j}} + w\hat{\mathbf{k}}) \\ &= (-fv + \tilde{f}w)\hat{\mathbf{i}} + fu\hat{\mathbf{j}} - \tilde{f}u\hat{\mathbf{k}} \end{aligned} \tag{2.320}$$

where $f = f(y)$ and $\tilde{f} = \tilde{f}(y)$ are given by (2.261) and (2.262), respectively. Because of the involved form of the cross product (2.320), it is convenient to expand the full momentum equation with respect to its Cartesian components. The result is

$$\frac{\mathrm{D}u}{\mathrm{D}t} - fv + \tilde{f}w = -\frac{1}{\rho}\frac{\partial p}{\partial x} + \left(A_{\mathrm{H}}\nabla_{\mathrm{H}}^2 + A_{\mathrm{V}}\frac{\partial^2}{\partial z^2}\right)u \tag{2.321}$$

$$\frac{\mathrm{D}v}{\mathrm{D}t} + fu = -\frac{1}{\rho}\frac{\partial p}{\partial y} + \left(A_{\mathrm{H}}\nabla_{\mathrm{H}}^2 + A_{\mathrm{V}}\frac{\partial^2}{\partial z^2}\right)v \tag{2.322}$$

$$\frac{\mathrm{D}w}{\mathrm{D}t} - \tilde{f}u = -\frac{1}{\rho}\frac{\partial p}{\partial z} + \left(A_{\mathrm{H}}\nabla_{\mathrm{H}}^2 + A_{\mathrm{V}}\frac{\partial^2}{\partial z^2}\right)w - g \tag{2.323}$$

The basic planetary parameters R, g and Ω, already introduced in Sect. 2.3.1, enter into (2.321)–(2.323) through f and $\tilde{f}$, or directly, as in the case of gravity acceleration g in (2.323).

2.3.4 Covariance of the Gradient Operator in Passing from an Inertial to a Rotating Frame of Reference

We denote the set of Cartesian coordinates in an inertial frame with

$$(X_1, X_2, X_3) \tag{2.324}$$

while the coordinates

$$(x_1, x_2, x_3) \tag{2.325}$$

refer to a rotating frame (fixed with the Earth, in the case under investigation). More precisely,

$$\mathbf{r} = \sum_{k=1}^{3} X_k \hat{\mathbf{e}}_k \qquad \text{and} \qquad \mathbf{x} = \sum_{k=1}^{3} x_k \hat{\mathbf{i}}_k$$

represent the position vector in the inertial and rotating systems, respectively (see Fig. 2.11). Since both (2.324) and (2.325) are Cartesian, the magnitude of a vector is given by the square root of the sum of the squares of the components. Further, since the actual vector remains unchanged no matter which coordinate system is used, the magnitude of the vector must be the same in both systems. On the basis of these assumptions, any linear transformation between (2.324) and (2.325), represented by

$$x_i = \sum_{i=1}^{3} a_{ij} X_j \tag{2.326}$$

must satisfy the orthogonality condition

$$\sum_{i=1}^{3} a_{ij} a_{ik} = \delta_{jk} \qquad \forall i, j = 1, 2, 3 \tag{2.327}$$

Consider now a scalar S which is a function of the coordinates and evaluate, in the inertial frame, the gradient

$$\nabla_{\mathrm{I}} S := \sum_{i=1}^{3} \frac{\partial S}{\partial X_i} \hat{\mathbf{e}}_i \tag{2.328}$$

where the unit vectors $\hat{\mathbf{e}}_i$ have the well-known properties

$$\hat{\mathbf{e}}_i \cdot \hat{\mathbf{e}}_j = \delta_{ij} \qquad \forall i, j = 1, 2, 3 \tag{2.329}$$

The r.h.s of (2.328) can be expanded to give

$$\nabla_{\mathrm{I}} S = \sum_{i=1}^{3} \sum_{j=1}^{3} \frac{\partial S}{\partial x_j} \frac{\partial x_j}{\partial X_i} \hat{\mathbf{e}}_i = \sum_{j=1}^{3} \frac{\partial S}{\partial x_j} \sum_{i=1}^{3} \frac{\partial x_j}{\partial X_i} \hat{\mathbf{e}}_i \tag{2.330}$$

Because of (2.326), we have

$$\frac{\partial x_j}{\partial X_i} = a_{ji} \tag{2.331}$$

and hence

$$\nabla_{\mathrm{I}} S = \sum_{j=1}^{3} \frac{\partial S}{\partial x_j} \hat{\mathbf{i}}_j \tag{2.332}$$

where

$$\hat{\mathbf{i}}_j := \sum_{i=1}^{3} a_{ji} \hat{\mathbf{e}}_i$$

are related by the orthogonality property

$$\hat{\mathbf{i}}_i \cdot \hat{\mathbf{i}}_j = \delta_{ij} \qquad \forall i, j = 1, 2, 3 \tag{2.333}$$

due to (2.327) and (2.333). In other words, the vectors $\hat{\mathbf{i}}_j$ are actually three mutually orthogonal unit vectors. More precisely, they are the versors of the rotating frame in which the coordinates are given by (2.325). Hence, we may consistently define the gradient operator in the rotating system as

$$\nabla_{\mathrm{R}} := \sum_{j=1}^{3} \hat{\mathbf{i}}_j \frac{\partial}{\partial x_j} \tag{2.334}$$

and the substitution of (2.334) into (2.332) allows us to derive the transformation rule

$$\nabla_{\mathrm{I}} S = \nabla_{\mathrm{R}} S \tag{2.335}$$

formally written as $\nabla_{\mathrm{I}} = \nabla_{\mathrm{R}}$, between the gradient operators in the transition from reference (2.324) and (2.325) and vice versa.

All the arguments of this subsection carry over to n-dimensional space; essentially, one has just to change the upper limit of summations from 3 to n.

2.3.5 *Governing Equations of Long Gravity Waves*

Long gravity waves consist in the non-dissipative motion of a thin fluid layer, whose density is constant. The surface of the layer oscillates under the restoring force of gravity and the influence of Earth's rotation (f-plane).

Governing Equations

In a constant-density fluid with

$$\rho = \rho_s = \text{const.} \tag{2.336}$$

there is no density anomaly.

An important consequence of (2.336) is that the thermodynamic equations (2.41) and (2.51), the former for dry air and the latter for seawater, are identically satisfied. The case of seawater is trivial, and (2.51) immediately implies (2.53), that is,

$$\operatorname{div}\mathbf{u} = 0 \tag{2.337}$$

In the case of dry air, assumption (2.336) again leads, via (2.67), to (2.337). Because of (2.337), Eq. (2.32) becomes $\mathrm{D}W/\mathrm{D}t = 0$, and hence, for $Q = 0$, Eq. (2.31) gives

$$\frac{\mathrm{D}T}{\mathrm{D}t} = 0 \tag{2.338}$$

Moreover, due to (2.336), Eq. (2.34) yields

$$p = C_s T \tag{2.339}$$

where $C_s := R\rho_s$ is a constant. Substitution of (2.339) into (2.40) allows potential temperature θ to be expressed in the form

$$\theta = C_{s0} T^{1/\gamma} \tag{2.340}$$

where

$$C_{s0} := \left(\frac{p_0}{C_s}\right)^{(\gamma-1)/\gamma}$$

is another constant. From (2.338) and (2.340), one concludes

$$\frac{\mathrm{D}\theta}{\mathrm{D}t} = 0 \tag{2.341}$$

so also (2.41) is identically verified.

Hereafter we are faced with a fluid layer included between the flat bottom, at

$$z = 0 \tag{2.342}$$

and the free surface, at

$$z = H + \eta(x, y, t) \tag{2.343}$$

where the constant H is the typical thickness of the layer, which is the depth of the motion, while η is the oscillating part, say, of the summit. The dynamics of this system, referred to the f-plane, is governed by the incompressibility and momentum equations listed below:

$$\frac{\partial u}{\partial x}+\frac{\partial v}{\partial y}+\frac{\partial w}{\partial z}=0 \tag{2.344}$$

$$\left(\frac{\partial}{\partial t}+\mathbf{u}\cdot\nabla\right)u-f_0\,v+\tilde{f}_0\,w=-\frac{1}{\rho_s}\frac{\partial p}{\partial x} \tag{2.345}$$

$$\left(\frac{\partial}{\partial t}+\mathbf{u}\cdot\nabla\right)v+f_0\,u=-\frac{1}{\rho_s}\frac{\partial p}{\partial y} \tag{2.346}$$

$$\left(\frac{\partial}{\partial t}+\mathbf{u}\cdot\nabla\right)w-\tilde{f}_0\,u=-\frac{1}{\rho_s}\frac{\partial p}{\partial z}-g \tag{2.347}$$

Like for internal waves (cf. Sect. 2.2), the pressure field is expressed by the superposition

$$p=p_s(z)+\tilde{p}(x,y,z,t) \tag{2.348}$$

where

$$\frac{\mathrm{d}p_s}{\mathrm{d}z}+\rho_s\,g=0 \tag{2.349}$$

while $\tilde{p}$ is still unspecified. Thus, (2.345)–(2.347) can be rewritten as

$$\left(\frac{\partial}{\partial t}+\mathbf{u}\cdot\nabla\right)u-f_0\,v+\tilde{f}_0\,w=-\frac{1}{\rho_s}\frac{\partial \tilde{p}}{\partial x} \tag{2.350}$$

$$\left(\frac{\partial}{\partial t}+\mathbf{u}\cdot\nabla\right)v+f_0\,u=-\frac{1}{\rho_s}\frac{\partial \tilde{p}}{\partial y} \tag{2.351}$$

$$\left(\frac{\partial}{\partial t}+\mathbf{u}\cdot\nabla\right)w-\tilde{f}_0\,u=-\frac{1}{\rho_s}\frac{\partial \tilde{p}}{\partial z} \tag{2.352}$$

Scaling Analysis

To restate (2.344) and (2.350)–(2.352) in non-dimensional form, the fact that the fluid layer is somehow thin must be taken into account explicitly. The thinness of the layer means that the horizontal scale of the motion, say, L, is much larger than that vertical, H, introduced in (2.343). Therefore, we set

$$x=L\mathsf{x} \qquad y=L\mathsf{y} \qquad z=H\mathsf{z} \tag{2.353}$$

and consider a certain timescale T such that

$$t=T\mathsf{t} \tag{2.354}$$

The vertical velocity scale W is significantly lesser than the horizontal velocity scale U, as can be seen from the following chain of inequalities:

$$W = O\left(\frac{\eta}{T}\right) < O\left(\frac{H}{T}\right) \ll O\left(\frac{L}{T}\right) = U$$

Thus, different scales must be ascribed also to the horizontal velocity U and to the vertical velocity W:

$$u = U\,\mathsf{u} \qquad v = U\,\mathsf{v} \qquad w = W\,\mathsf{w} \tag{2.355}$$

Note that, unlike the case of internal waves (Eqs. (2.129) and (2.130)), the motion depth H in (2.353) and the typical vertical velocity W in (2.355) differ from the corresponding horizontal scales. The relation between U and W follows by considering the non-dimensional version of (2.344), that is,

$$\frac{\partial \mathsf{u}}{\partial \mathsf{x}} + \frac{\partial \mathsf{v}}{\partial \mathsf{y}} + \frac{W L}{U H}\frac{\partial \mathsf{w}}{\partial \mathsf{z}} = 0 \tag{2.356}$$

In fact, in order that the horizontal convergence of the flow $\partial \mathsf{u}/\partial \mathsf{x} + \partial \mathsf{v}/\partial \mathsf{y}$ may give rise to an $O(1)$ vertical divergence $\partial \mathsf{w}/\partial \mathsf{z}$ (or vice versa), the estimate

$$W = O\left(\frac{U H}{L}\right) \tag{2.357}$$

immediately comes from (2.356). Hence, the latter equation becomes

$$\frac{\partial \mathsf{u}}{\partial \mathsf{x}} + \frac{\partial \mathsf{v}}{\partial \mathsf{y}} + \frac{\partial \mathsf{w}}{\partial \mathsf{z}} = 0 \tag{2.358}$$

In terms of the so-called *aspect ratio*

$$\delta := \frac{H}{L} \tag{2.359}$$

Equation (2.357) takes the concise form

$$W = O(\delta U) \tag{2.360}$$

We stress that $\delta \ll 1$ implies $W \ll U$.

To scale momentum equations (2.350)–(2.352), we also introduce the typical intensity P of the perturbation pressure through the position

$$\tilde{p} = P\,\mathsf{p} \tag{2.361}$$

Moreover, to shorten the notation, we use two non-dimensional parameters that naturally appear in scaling (2.350)–(2.352), that is, the *temporal Rossby number*

$$\varepsilon_T := \frac{1}{f_0 T} \tag{2.362}$$

and the advective Rossby number

$$\varepsilon = \frac{U}{f_0 L} \tag{2.363}$$

which has been already introduced in Eq. (2.297).

Consider (2.350) written as

$$\frac{U}{T}\frac{\partial \mathsf{u}}{\partial \mathsf{t}} + \frac{U^2}{L}\mathbf{u}\cdot\nabla'\mathsf{u} - f_0 U \mathsf{v} + \tilde{f}_0\,\delta\, U \mathsf{w} = -\frac{P}{\rho_s L}\frac{\partial \mathsf{p}}{\partial \mathsf{x}} \tag{2.364}$$

Division of (2.364) by $f_0 U$ and the use of (2.362) and (2.363) yield the non-dimensional version of (2.350) in the form

$$\varepsilon_T \frac{\partial \mathsf{u}}{\partial \mathsf{t}} + \varepsilon \mathbf{u}\cdot\nabla'\mathsf{u} - \mathsf{v} + \frac{\tilde{f}_0}{f_0}\delta\,\mathsf{w} = -\frac{P}{\rho_s f_0 U L}\frac{\partial \mathsf{p}}{\partial \mathsf{x}} \tag{2.365}$$

Long gravity waves arise from the dynamic balance among the Coriolis acceleration, the pressure gradient and the local acceleration. Therefore, with reference to (2.365), we demand

$$\frac{P}{\rho_s f_0 U L} = O(1) \tag{2.366}$$

and

$$\varepsilon_T = O(1) \tag{2.367}$$

whence

$$P = O(\rho_s f_0 U L) \tag{2.368}$$

and, recalling (2.362), equation (2.367) yields

$$T = O(f_0^{-1}) \tag{2.369}$$

while

$$\frac{\tilde{f}_0\,\delta}{f_0} < O(1) \tag{2.370}$$

In the case of wave-like motions, the timescale (2.369) is usually much shorter than the advective timescale L/U (i.e., $f_0^{-1} \ll L/U$); so (2.363) implies

$$\varepsilon < O(1) \tag{2.371}$$

while $\varepsilon_T = O(1)$. Hence, the $O(1)$ terms of (2.365) constitute the momentum equation

$$\frac{\partial \mathsf{u}}{\partial \mathsf{t}} - \mathsf{v} = -\frac{\partial \mathsf{p}}{\partial \mathsf{x}} \tag{2.372}$$

whose dimensional version is

$$\frac{\partial u}{\partial t} - f_0\, v = -\frac{1}{\rho_s}\frac{\partial \tilde{p}}{\partial x} \tag{2.373}$$

Analogously, the $O(1)$ equation

$$\frac{\partial \mathsf{v}}{\partial \mathsf{t}} + \mathsf{u} = -\frac{\partial \mathsf{p}}{\partial \mathsf{y}} \tag{2.374}$$

follows from (2.351), and the dimensional version of (2.374) is

$$\frac{\partial v}{\partial t} + f_0\, u = -\frac{1}{\rho_s}\frac{\partial \tilde{p}}{\partial y} \tag{2.375}$$

Consider now the vertical component (2.352). Once this equation is written with the aid of (2.360) and (2.368) as

$$\frac{\delta U}{T}\frac{\partial \mathsf{w}}{\partial \mathsf{t}} + \frac{\delta U^2}{L}\mathbf{u}\cdot\nabla'\mathsf{w} - \tilde{f}_0 U\,\mathsf{u} = -\frac{f_0 U}{\delta}\frac{\partial \mathsf{p}}{\partial \mathsf{z}} \tag{2.376}$$

multiplication of each term of (2.376) by $\delta/f_0 U$ leads to the non-dimensional version

$$\varepsilon_T\,\delta^2\frac{\partial \mathsf{w}}{\partial \mathsf{t}} + \varepsilon\,\delta^2\,\mathbf{u}\cdot\nabla'\mathsf{w} - \frac{\tilde{f}_0\,\delta}{f_0}\,\mathsf{u} = -\frac{\partial}{\partial \mathsf{z}} \tag{2.377}$$

of (2.352), where $\delta^2 < O(1)$, $\varepsilon\,\delta^2 < O(1)$ and $\tilde{f}_0\,\delta/f_0 < O(1)$. Therefore, the $O(1)$ term at the r.h.s. of (2.377) is left unbalanced unless

$$\frac{\partial \mathsf{p}}{\partial \mathsf{z}} = 0 \tag{2.378}$$

that is to say,

$$\frac{\partial \tilde{p}}{\partial z} = 0 \tag{2.379}$$

This result authorizes us to express the overall pressure field p at every depth (or height) $z \le H + \eta$ in the hydrostatic form

$$p(x,y,z,t) = \rho_s\, g\,[H + \eta(x,y,t) - z] \tag{2.380}$$

Comparison of (2.380) with (2.348) shows that

$$p_s(z) = -\rho_s\, g\, z \tag{2.381}$$

and

$$\tilde{p}(x,y,t) = \rho_s\, g\,[H + \eta(x,y,t)] \tag{2.382}$$

which is openly independent of z, in accordance with (2.379).

Using (2.382), the pressure gradient components in (2.373) and (2.375) are expressed through the free-surface elevation, which yields

$$\frac{\partial u}{\partial t} - f_0\, v = -g\,\frac{\partial \eta}{\partial x} \tag{2.383}$$

and

$$\frac{\partial v}{\partial t} + f_0\, u = -g\,\frac{\partial \eta}{\partial y} \tag{2.384}$$

respectively.

Taking the horizontal gradient of (2.382) and using (2.368), the estimate

$$\eta = O\left(\frac{f_0\, U\, L}{g}\right) \tag{2.385}$$

is derived, and the non-dimensional free-surface elevation η' defined through the position

$$\eta = \frac{f_0\, U\, L}{g}\,\eta' \tag{2.386}$$

can be introduced. Note that (2.386) is consistent with the choice of H as the vertical scale of the motion. In fact,

$$H + \eta \approx H + \frac{f_0\, U\, L}{g} = H\left(1 + \frac{f_0\, U\, L}{g\, H}\right) \tag{2.387}$$

where, in the present context (but not only), $f_0\, U\, L/g\, H < O(1)$. Thus, $H + \eta \approx H$. The quantity $f_0\, U\, L/g\, H$ appearing in (2.387) can be factorized as

$$\frac{f_0\, U\, L}{g\, H} = \frac{U}{f_0\, L}\,\frac{f_0^2\, L^2}{g\, H} = \varepsilon\, \mathsf{F} \tag{2.388}$$

where the *rotational Froude number*

$$\mathsf{F} := \frac{f_0^2\, L^2}{g\, H} = \left(\frac{L}{L_{\text{ext}}}\right)^2 \tag{2.389}$$

is the square ratio between the horizontal length scale L and the *external deformation radius* L_{ext} defined by

$$L_{\text{ext}} := \frac{\sqrt{g\, H}}{|f_0|} \tag{2.390}$$

Equation (2.379) is not an evolution equation, and so it cannot constitute a system of evolution equations in the unknowns u, v, η, when considered together with (2.383)and (2.384). Indeed, (2.379) barely means that the free-surface elevation is depth independent, since $\partial\eta/\partial z = 0$ because of (2.382). Rather, the desired evolution equation is obtained through the vertical integration of (2.344) from $z = 0$ up to $z = H + \eta$, that is,

$$\int_0^{H+\eta} \left(\frac{\partial u}{\partial x} + \frac{\partial v}{\partial y}\right) \mathrm{d}z + \left(\frac{\partial}{\partial t} + \mathbf{u}\cdot\nabla\right)\eta = 0 \tag{2.391}$$

because $w(z = H + \eta) = \mathrm{D}\eta/\mathrm{D}t$. Equation (2.391) involves the fields u, v, η and has the form of the desired evolution equation for the free-surface elevation η, although the integral term does not look to be easily manageable. However, if we go back to (2.383) and (2.384), we immediately realize that their r.h.s.'s are depth independent. Thus, assuming that also the velocity components u and v, at the l.h.s.'s of the same equations, are independent of z, (2.391) becomes

$$(H+\eta)\left(\frac{\partial u}{\partial x} + \frac{\partial v}{\partial y}\right) + \left(\frac{\partial}{\partial t} + \mathbf{u}\cdot\nabla\right)\eta = 0 \tag{2.392}$$

Using non-dimensional variables and in particular (2.386), one may recast Eq. (2.392) as

$$\frac{UH}{L}\left(1+\frac{f_0 UL}{gH}\eta'\right)\left(\frac{\partial \mathsf{u}}{\partial \mathsf{x}} + \frac{\partial \mathsf{v}}{\partial \mathsf{y}}\right) + \frac{f_0 UL}{gT}\frac{\partial \eta'}{\partial \mathsf{t}} + \frac{U}{L}\frac{f_0 UL}{g}\mathsf{u}\cdot\nabla'\eta' = 0 \tag{2.393}$$

and division of (2.393) by UH/L gives the non-dimensional version of (2.392)

$$(1+\varepsilon\mathsf{F}\eta')\left(\frac{\partial \mathsf{u}}{\partial \mathsf{x}} + \frac{\partial \mathsf{v}}{\partial \mathsf{y}}\right) + \left(\mathsf{F}\frac{\partial}{\partial \mathsf{t}} + \varepsilon\mathsf{F}\,\mathsf{u}\cdot\nabla'\right)\eta' = 0 \tag{2.394}$$

where use has been made of (2.389). For motions such that $\mathsf{F} = O(1)$, that is, $L = O(L_{\mathrm{ext}})$, and $\varepsilon < O(1)$, the $O(1)$ balance in (2.394) is

$$\frac{\partial \mathsf{u}}{\partial \mathsf{x}} + \frac{\partial \mathsf{v}}{\partial \mathsf{y}} + \mathsf{F}\frac{\partial \eta'}{\partial \mathsf{t}} = 0 \tag{2.395}$$

whose dimensional version is

$$H\left(\frac{\partial u}{\partial x} + \frac{\partial v}{\partial y}\right) + \frac{\partial \eta}{\partial t} = 0 \tag{2.396}$$

Summary of Governing Equations for Long Gravity Waves

To summarize, if the flow, of a constant-density fluid, satisfies the conditions

$$\delta < O(1) \qquad \varepsilon_T = O(1) \qquad \varepsilon < O(1) \qquad L = O(L_{\text{ext}}) \tag{2.397}$$

that is to say,

$$H \ll L \approx \frac{\sqrt{gH}}{f_0} \qquad T \approx \frac{1}{f_0} \ll \frac{L}{U} \tag{2.398}$$

then this flow is governed, on the f-plane, by the system of coupled equations (2.383), (2.384) and (2.396) which, for future convenience, are reported below:

$$\frac{\partial u}{\partial t} - f_0 v = -g\frac{\partial \eta}{\partial x} \tag{2.399}$$

$$\frac{\partial v}{\partial t} + f_0 u = -g\frac{\partial \eta}{\partial y} \tag{2.400}$$

$$H\left(\frac{\partial u}{\partial x} + \frac{\partial v}{\partial y}\right) + \frac{\partial \eta}{\partial t} = 0 \tag{2.401}$$

2.3.6 *Poincaré and Kelvin Waves*

This section is focussed on special solutions of the governing equations for long gravity waves (see Sect. 2.3.5). In the case of *Poincaré waves*, the motion takes place in an unbounded fluid, while *Kelvin waves* arise in a fluid domain laterally bounded by a rectilinear coastline.

Poincaré Waves

Because of their linearity, Eqs. (2.399), (2.400) and (2.401) possess wave-like solutions of the form $u = u_0 \exp[\mathrm{i}(kx + ny - \sigma t)]$, and likewise for v and η. The fields

$$u(x, y, t) = \mathrm{Re}\{u_0 \exp[\mathrm{i}(kx + ny - \sigma t)]\} \tag{2.402}$$

$$v(x, y, t) = \mathrm{Re}\{v_0 \exp[\mathrm{i}(kx + ny - \sigma t)]\} \tag{2.403}$$

$$\eta(x, y, t) = \mathrm{Re}\{\eta_0 \exp[\mathrm{i}(kx + ny - \sigma t)]\} \tag{2.404}$$

are taken as putative Fourier components of the waves governed by (2.399)–(2.401). The symbol "Re{...}" means "the real part of {...}", and the quantities u_0, v_0

and η_0 are complex constants such that $|u_0| = O(U)$, $|v_0| = O(U)$ and $|\eta_0| = O(f_0 U L/g)$. Moreover, the *angular frequency* σ is assumed to be positive.

Wavefronts are represented by planes

$$kx + ny = \Gamma \tag{2.405}$$

where k, n and Γ are real constants. Thus, wavefronts are parallel to the z-axis, and the *wave number vector*

$$\mathbf{k} := k\mathbf{i} + n\mathbf{j} \tag{2.406}$$

is parallel to the f-plane. Hence, each wavefront (2.405) propagates in the direction of the wave number (2.406) with *phase speed*

$$c := \frac{\sigma}{|\mathbf{k}|} \tag{2.407}$$

and, during the *wave period* $T := 2\pi/\sigma$, it covers a distance equal to the *wavelength*

$$\lambda := \frac{2\pi}{|\mathbf{k}|} = 2\pi \frac{c}{\sigma} \tag{2.408}$$

where $|\mathbf{k}| = \sqrt{k^2 + n^2}$.

The above-outlined procedure leads to the linear system

$$\begin{cases} -\mathrm{i}\sigma u_0 - f_0 v_0 + \mathrm{i} g k \eta_0 = 0 \\ f_0 u_0 - \mathrm{i}\sigma v_0 + \mathrm{i} g n \eta_0 = 0 \\ kH u_0 + nH v_0 - \sigma \eta_0 = 0 \end{cases} \tag{2.409}$$

in the unknowns u_0, v_0 and η_0. Nontrivial solutions of (2.409) exist, provided that

$$\det \begin{pmatrix} -\mathrm{i}\sigma & -f_0 & \mathrm{i}gk \\ f_0 & -\mathrm{i}\sigma & \mathrm{i}gn \\ kH & nH & -\sigma \end{pmatrix} = 0 \tag{2.410}$$

whence the *dispersion relation*

$$\sigma^2 = f_0^2 + gH|\mathbf{k}|^2 \tag{2.411}$$

follows. From (2.411), we immediately ascertain that:

- Phase speed (2.407) depends on the magnitude $|\mathbf{k}|$ of the wave number vector $\mathbf{k}$, but not on its direction $\hat{\mathbf{k}} := \mathbf{k}/|\mathbf{k}|$. In other words, the medium is *isotropic*.
- The wave frequency is bounded from below:

$$\sigma > |f_0| \tag{2.412}$$

For angular frequencies σ satisfying (2.412), both u_0 and v_0 can be expressed in terms of η_0 to obtain

$$u_0 = g\,\eta_0 \frac{\sigma k + \mathrm{i}\, f_0 n}{\sigma^2 - f_0^2} \overset{(2.411)}{=} \frac{\eta_0}{H} \frac{\sigma k + \mathrm{i}\, f_0 n}{|\mathbf{k}|^2}$$

$$v_0 = g\,\eta_0 \frac{\sigma n - \mathrm{i}\, f_0 k}{\sigma^2 - f_0^2} \overset{(2.411)}{=} \frac{\eta_0}{H} \frac{\sigma n - \mathrm{i}\, f_0 k}{|\mathbf{k}|^2} \tag{2.413}$$

Equations (2.413) show that, due to ambient rotation (i.e. for $f_0 \neq 0$), amplitudes u_0 and v_0 are complex quantities if η_0 is real.

In the special case in which the x-axis is aligned with the wave vector $\mathbf{k}$ (i.e. $n = 0$) and η_0 is real, Eq. (2.413) give

$$u_0 = U_0 \qquad\qquad v_0 = -\mathrm{i}\, V_0 \tag{2.414}$$

where

$$U_0 := \frac{\eta_0}{H} \frac{\sigma}{k} \qquad\qquad V_0 := \frac{\eta_0}{H} \frac{f_0}{k} \tag{2.415}$$

Hence, (2.402)–(2.404) yield

$$u(x,t) = U_0 \cos(kx - \sigma t) \tag{2.416}$$

$$v(x,t) = V_0 \sin(kx - \sigma t) \tag{2.417}$$

$$\eta(x,t) = \eta_0 \cos(kx - \sigma t) \tag{2.418}$$

respectively. From (2.416) and (2.417), the equation

$$\frac{u^2}{U_0^2} + \frac{v^2}{V_0^2} = 1 \tag{2.419}$$

can be easily derived. Thus, the velocity vector traces out an ellipse in the f-plane. Because of (2.412) and (2.415), we have $U_0^2 > V_0^2$; so, both the major axis and the wave vector are in the x–direction.

The group velocity

$$\mathbf{c}_\mathrm{g} := \left(\frac{\partial \sigma}{\partial k}, \frac{\partial \sigma}{\partial n} \right)$$

evaluated from (2.411) is

$$\mathbf{c}_\mathrm{g} = \frac{gH}{\sigma} \mathbf{k} \tag{2.420}$$

and hence $\mathbf{c}_\mathrm{g}$ is in the direction of the wave vector (2.406).

Two extreme cases of (2.411) are of interest:

Case 1: $gH|\mathbf{k}|^2 \gg f_0^2$.] Equation (2.408) shows that very short wavelengths are obtained for diverging values of $|\mathbf{k}|$. Hence, if $gH|\mathbf{k}|^2 \gg f_0^2$, then (2.411) approaches

$$\sigma^2 = gH|\mathbf{k}|^2 \tag{2.421}$$

that is, the dispersion relation for non-rotating gravity waves. This result is ascribed to the weak sensitivity to Earth's rotation exhibited by systems having a very short horizontal length scale, in conflict with some of hypotheses (2.398).

Case 2: $gH|\mathbf{k}|^2 \ll f_0^2$.] At the opposite extreme, very large wavelengths imply $gH|\mathbf{k}|^2 \ll f_0^2$, and (2.411) approaches

$$\sigma^2 = f_0^2 \tag{2.422}$$

In the limit case in which both (2.411) and (2.422) hold true, we have

$$k = n = 0 \tag{2.423}$$

Because of (2.423), the fields u, v and η depend on time only, and the governing equations (2.399)–(2.401) take the form

$$\frac{\partial u}{\partial t} - f_0 v = 0 \tag{2.424}$$

$$\frac{\partial v}{\partial t} + f_0 u = 0 \tag{2.425}$$

$$\frac{\partial \eta}{\partial t} = 0 \tag{2.426}$$

Equation (2.426) implies that η is a constant, which we can set to zero. On the other hand, (2.424) and (2.425) have the general integral

$$u = A\sin(f_0 t + \varphi) \qquad v = A\cos(f_0 t + \varphi) \tag{2.427}$$

that is,

$$\frac{\mathrm{d}x}{\mathrm{d}t} = A\sin(f_0 t + \varphi) \qquad \frac{\mathrm{d}y}{\mathrm{d}t} = A\cos(f_0 t + \varphi) \tag{2.428}$$

in which the amplitude A and the phase φ are arbitrary. Equations (2.428) can be integrated to give

$$x(t) = x_0 - \frac{A}{f_0}\cos(f_0 t + \varphi) \qquad y(t) = y_0 + \frac{A}{f_0}\sin(f_0 t + \varphi) \tag{2.429}$$

The parametric curve (2.429) represents the circular trajectory

$$(x - x_0)^2 + (y - y_0)^2 = \frac{A^2}{f_0^2} \tag{2.430}$$

followed by each parcel in the course of the inertial motion on the f-plane. The very existence of the horizontal current $\big(u(t), v(t)\big)$ for $k \to 0$ and $\eta_0 \to 0$ presupposes, according to (2.415), that η_0/k converges to a finite nonzero limit, say, Λ. In this case, ellipse (2.419) transforms into the circle

$$u^2 + v^2 = \left(\frac{\Lambda f_0}{H}\right)^2 \tag{2.431}$$

Hence, the amplitude A appearing in (2.427) coincides with the radius $\Lambda f_0/H$ of (2.431).

Steady State

Equation (2.410) has a further solution besides those given by (2.411), that is,

$$\sigma = 0 \tag{2.432}$$

Solution (2.432) implies that the related fields are steady and therefore they satisfy the steady version of (2.399)–(2.401):

$$f_0\, v = g\frac{\partial \eta}{\partial x} \tag{2.433}$$

$$f_0\, u = -g\frac{\partial \eta}{\partial y} \tag{2.434}$$

$$\frac{\partial u}{\partial x} + \frac{\partial v}{\partial y} = 0 \tag{2.435}$$

Equations (2.433) and (2.434) are the components of the vector equation

$$f_0\, \mathbf{u} = g\hat{\mathbf{k}} \times \nabla \eta \tag{2.436}$$

which is a special form of the so-called *geostrophic balance*. In general, this balance links the current field $\mathbf{u}$ and the pressure gradient field ∇p through the equation

$$\rho_s f_0\, \mathbf{u} = \hat{\mathbf{k}} \times \nabla p \tag{2.437}$$

which takes the form (2.436) in case (2.382). Geostrophic balance is a keystone of GFD and one of its distinguishing features: indeed, Eq. (2.437) represents, as a matter of fact, the basic diagnostic equation for large-scale atmospheric and oceanic flows. Note that the very existence of the geostrophic balance (2.437) presupposes $f_0 \neq 0$, that is, geostrophic balance can take place only in the presence of ambient rotation. As a consequence, the geostrophic current $\mathbf{u}$ and the pressure gradient ∇p are mutually orthogonal, as one can check immediately from (2.437).

Geostrophic flows (i.e., flows satisfying (2.433) and (2.434)) can be interpreted as *arrested waves*. Mathematically, Eq. (2.435) simply derives from (2.433) and (2.434), but the relevant physical consequence is that geostrophic flows are planar; so, whenever they are a good approximation of real flows, the possible vertical velocity component is a higher-order perturbation to the planar motion. Geostrophic motions will be explored with wide details in the following sections.

Response of the Sea to Travelling Disturbances of Atmospheric Pressure

Long gravity waves in large water bodies can be induced mechanically by the overhanging atmospheric pressure field p_{a}; the case in which pressure is the sum of a constant term $p_{\mathrm{a}0}$ plus a wave-like anomaly Δp_{a}, of the kind

$$\Delta p_{\mathrm{a}} = \cos(kx + ny - \sigma t)\,\Delta p_{\mathrm{a}0} \tag{2.438}$$

bears also a remarkable practical interest because of its connection with coastal flooding. We already know that if the water body is at rest, the time-independent pressure anomaly and the free-surface elevation η are linked by the so-called *inverted-barometer effect* (see (1.63)); the present model takes into account the dynamic counterpart of the same effect when it is caused by (2.438). In the latter case, not only η adjusts itself to (2.438), but also the marine current (u, v) undergoes accelerations which are requested to balance the nonzero horizontal pressure gradient.

The local timescale T of the system under investigation is given by the period $2\pi/\sigma$ of the forcing (2.438), and the special case

$$T f_0 = O(1) \qquad \text{that is,} \qquad \sigma = 2\pi f_0 \tag{2.439}$$

is preliminarily taken into account. Hypothesis (2.439) will be subsequently released in favour of timescales either larger or smaller than f_0^{-1}.

Consider the momentum equation (2.350) with the perturbation pressure

$$\tilde{p} = p_{\mathrm{a}0} + \Delta p_{\mathrm{a}} + \rho_{\mathrm{s}} g \eta \tag{2.440}$$

that is,

$$\frac{\mathrm{D}u}{\mathrm{D}t} - f_0 v + \tilde{f}_0 w = -\frac{1}{\rho_{\mathrm{s}}}\frac{\partial}{\partial x}(\Delta p_{\mathrm{a}}) - g\frac{\partial \eta}{\partial x} \tag{2.441}$$

Setting

$$\Delta p_{\mathrm{a}} = P_{\mathrm{a}} \Delta \mathsf{p}_{\mathrm{a}} \qquad \text{and} \qquad \eta = E\eta'$$

the non-dimensional version of (2.441) is again (2.365) in the form

$$\varepsilon_{\mathrm{T}}\frac{\partial \mathsf{u}}{\partial \mathsf{t}} + \varepsilon \mathbf{u}\cdot\nabla'\mathsf{u} - \mathsf{v} + \frac{\tilde{f}_0}{f_0}\delta \mathsf{w} = -\frac{P_{\mathrm{a}}}{\rho_{\mathrm{s}} f_0 U L}\frac{\partial}{\partial \mathsf{x}}(\Delta \mathsf{p}_{\mathrm{a}}) - \frac{gE}{f_0 U L}\frac{\partial \eta'}{\partial \mathsf{x}} \tag{2.442}$$

In order that both the gradients of atmospheric pressure and the free-surface elevation enter into the $O(1)$ equation that follows from (2.442), the orders of magnitude

$$P_{\mathrm{a}} = O(\rho_{\mathrm{s}} f_0 U L) \qquad \text{and} \qquad E = O\left(\frac{f_0 U L}{g}\right) \tag{2.443}$$

must be fixed. Note that conditions (2.443) imply $P_{\mathrm{a}} = O(\rho_{\mathrm{s}} g E)$ in accordance with the inverted-barometer effect.

The first relationship of (2.443) is verified for $UL = O(10^4\,\mathrm{m^2\,s^{-1}})$. In fact, this estimate yields $P_{\mathrm{a}} = O(10^3\,\mathrm{Pa})$, which is the typical horizontal pressure fluctuation of the atmosphere. Note that L represents an "atmospheric" length scale, so $UL = O(10^4\,\mathrm{m^2\,s^{-1}})$ implies marine currents of at most a few centimetres per second.

With these values of UL, the second relationship of (2.443) gives $E = O(10\,\mathrm{cm})$. Owing to (2.443), (2.371) and (2.370), the $O(1)$ equation derived from (2.442) is

$$\varepsilon_{\mathrm{T}} \frac{\partial \mathsf{u}}{\partial \mathsf{t}} - \mathsf{v} + \frac{\partial \eta'}{\partial \mathsf{x}} = -\frac{\partial}{\partial \mathsf{x}}(\Delta \mathsf{p}_{\mathrm{a}}) \tag{2.444}$$

and its dimensional version, evaluated with resort to (2.443), is

$$\frac{\partial u}{\partial t} - f_0 v + g\frac{\partial \eta}{\partial x} = -\frac{1}{\rho_{\mathrm{s}}}\frac{\partial}{\partial x}(\Delta p_{\mathrm{a}}) \tag{2.445}$$

Then, substituting (2.438) into the r.h.s. of (2.445), and likewise for the momentum equation (2.351), gives the horizontal momentum equations in the final form

$$\frac{\partial u}{\partial t} - f_0 v + g\frac{\partial \eta}{\partial x} = \frac{\Delta p_{\mathrm{a0}}}{\rho_{\mathrm{s}}} k \sin(kx + ny - \sigma t) \tag{2.446}$$

$$\frac{\partial v}{\partial t} + f_0 u + g\frac{\partial \eta}{\partial y} = \frac{\Delta p_{\mathrm{a0}}}{\rho_{\mathrm{s}}} n \sin(kx + ny - \sigma t) \tag{2.447}$$

The set of governing equations is completed by the vertically integrated continuity equation (2.401), rearranged here in the form

$$\frac{\partial u}{\partial x} + \frac{\partial v}{\partial y} + \frac{1}{H}\frac{\partial \eta}{\partial t} = 0 \tag{2.448}$$

The full set of governing equations is given by (2.446)–(2.448).

A special wave-like solution of (2.446)–(2.448) is the following. Define preliminarily the length

$$\bar{E} := E\left[1 - \frac{\lambda^2(\sigma^2 - f_0^2)}{4\pi^2 g H}\right]^{-1} \tag{2.449}$$

where $E := -\Delta p_{\mathrm{a0}}/\rho_{\mathrm{s}} g$ is the static response of the inverted-barometer effect (1.63) and $\lambda^2 := 4\pi^2/(k^2 + n^2)$ is the wavelength square of (2.438). Then, the free-surface elevation is

$$\eta = \bar{E}\cos(kx+ny-\sigma t) \tag{2.450}$$

and the components of the marine current are

$$u = \frac{g\bar{E}+\Delta p_{a0}/\rho_s}{\sigma^2 - f_0^2}[\sigma k\cos(kx+ny-\sigma t) - f_0 n\sin(kx+ny-\sigma t)] \tag{2.451}$$

$$v = \frac{g\bar{E}+\Delta p_{a0}/\rho_s}{\sigma^2 - f_0^2}[f_0 k\sin(kx+ny-\sigma t) + \sigma n\cos(kx+ny-\sigma t)] \tag{2.452}$$

Equation (2.449) shows that solution (2.450)–(2.452) is singular for

$$\sigma^2 = f_0^2 + (k^2+n^2)\,gH \tag{2.453}$$

that is, if the frequency of the pressure forcing satisfies the dispersion relation of Poincaré waves (2.411). This means that if (2.438) excites a long gravity wave which is close to a Poincaré wave of the considered marine basin, then the latter wave may exhibit a tendentially divergent amplitude $\bar{E}$, far greater than the static response E due to the inverted-barometer effect. When the resonance takes place in shallow seas and in the proximity of coastal areas, it is frequently associated with flooding phenomena.

Hypothesis (2.439) can be released in favour of assumption

$$T > O(f_0^{-1}) \tag{2.454}$$

or, alternatively, of assumption

$$T < O(f_0^{-1}) \tag{2.455}$$

Case (2.454) is further constrained by inequality $UT/L < O(1)$, which is required to neglect the advective terms in the governing equations; so, (2.454) should be completed to give

$$O(f_0^{-1}) < T < O\left(\frac{L}{U}\right) \tag{2.456}$$

In case (2.456), Eqs. (2.449)–(2.452) are reconsidered in the limit $\sigma/f_0 \to 0$, whence

$$\bar{E} = E\left(1+\frac{\lambda^2 f_0^2}{4\pi^2 gH}\right)^{-1} \tag{2.457}$$

$$\eta = \bar{E}\cos(kx+ny-\sigma t) \tag{2.458}$$

$$u = n\frac{g\bar{E}+\Delta p_{a0}/\rho_s}{f_0}\sin(kx+ny-\sigma t) \tag{2.459}$$

$$v = -k\frac{g\bar{E}+\Delta p_{a0}/\rho_s}{f_0}\sin(kx+ny-\sigma t) \tag{2.460}$$

Owing to (2.454), local acceleration is much smaller than Coriolis acceleration, and hence, the current given by (2.459) and (2.460) is in geostrophic balance according to (2.437), that is to say, we have

$$f_0 u = -\frac{1}{\rho_s}\frac{\partial}{\partial y}(\Delta p_a + \rho_s g\eta) \tag{2.461}$$

$$f_0 v = \frac{1}{\rho_s}\frac{\partial}{\partial x}(\Delta p_a + \rho_s g\eta) \tag{2.462}$$

as a direct calculation shows. Using vector notation, Eqs. (2.461) and (2.462) may be condensed into

$$\mathbf{u} = \frac{1}{\rho_s f_0}\hat{\mathbf{k}}\times\nabla(\Delta p_a + \rho_s g\eta)$$

Note that resonance cannot take place, and then $\bar{E} < E$.

In case (2.455), the limit $f_0/\sigma \to 0$ applies to (2.449)–(2.452), whence

$$\bar{E} = E\left(1 - \frac{\lambda^2\sigma^2}{4\pi^2 gH}\right)^{-1} \tag{2.463}$$

$$\eta = \bar{E}\cos(kx+ny-\sigma t) \tag{2.464}$$

$$u = k\frac{g\bar{E}+\Delta p_{a0}/\rho_s}{\sigma}\cos(kx+ny-\sigma t) \tag{2.465}$$

$$v = n\frac{g\bar{E}+\Delta p_{a0}/\rho_s}{\sigma}\cos(kx+ny-\sigma t)] \tag{2.466}$$

Thus, local acceleration prevails on Coriolis acceleration; so, the current given by (2.465) and (2.466) satisfies Euler's equations (1.44) and (1.45) in their linear version, that is to say,

$$\frac{\partial u}{\partial t} = -\frac{1}{\rho_s}\frac{\partial}{\partial x}(\Delta p_a + \rho_s g\eta) \tag{2.467}$$

$$\frac{\partial v}{\partial t} = -\frac{1}{\rho_s}\frac{\partial}{\partial y}(\Delta p_a + \rho_s g\eta) \tag{2.468}$$

as a direct calculation shows. Using vector notation, Eqs. (2.467) and (2.468) are conveniently written in the compact form

$$\frac{\partial \mathbf{u}}{\partial t} = -\frac{1}{\rho_s}\nabla(\Delta p_a + \rho_s g\eta)$$

Resonance is possible when the phase velocity $c := \lambda\,\sigma/2\pi$ of the pressure anomaly equals the velocity of the long gravity wave $\sqrt{gH}$ in the non-rotating limit (see (2.421)).

Kelvin Waves

Another remarkable solution for long gravity waves is obtained under the following assumptions. First, the fluid domain is the half-plane $x \geq 0$ with a rigid boundary at $x = 0$ and, second, the velocity component u of each parcel along the whole y-axis is zero. Thus, the no–mass flux condition across the boundary at $x = 0$ is implicitly satisfied.

The governing equations

$$f_0\,v = g\,\frac{\partial \eta}{\partial x} \tag{2.469}$$

$$\frac{\partial v}{\partial t} + g\frac{\partial \eta}{\partial y} = 0 \tag{2.470}$$

$$\frac{\partial \eta}{\partial t} + H\,\frac{\partial v}{\partial y} = 0 \tag{2.471}$$

are derived from (2.399) to (2.401). Equation (2.469) means that the alongshore velocity v is in geostrophic balance with the free-surface elevation η of the wave. On the other hand, (2.470) and (2.471) can be decoupled by cross differentiation to give

$$\frac{\partial^2 v}{\partial t^2} = c^2\frac{\partial^2 v}{\partial y^2} \tag{2.472}$$

and

$$\frac{\partial^2 \eta}{\partial t^2} = c^2\frac{\partial^2 \eta}{\partial y^2} \tag{2.473}$$

where $c := \sqrt{gH}$ is *phase speed*. The general integrals of (2.472) and (2.473) have the forms

$$v(x,y,t) = v_1(x, y+ct) + v_2(x, y-ct) \tag{2.474}$$

and

$$\eta(x,y,t) = \eta_1(x, y+ct) + \eta_2(x, y-ct) \tag{2.475}$$

respectively. In (2.474) and (2.475), v_n and η_n (with $n = 1, 2$) are, for the time being, any differentiable functions of two real variables. The link between v_n and η_n is inferred by substituting (2.474) and (2.475) into (2.470). As $v_1(x, y+ct)$ and $v_2(x, y-ct)$ satisfy the one-way wave equations

$$\frac{\partial v_1}{\partial t} = c\frac{\partial v_1}{\partial y} \quad \text{and} \quad \frac{\partial v_2}{\partial t} = -c\frac{\partial v_2}{\partial y}$$

respectively, such substitution yields

$$c\left(\frac{\partial v_1}{\partial y} - \frac{\partial v_2}{\partial y}\right) + g\left(\frac{\partial \eta_1}{\partial y} + \frac{\partial \eta_2}{\partial y}\right) = 0 \tag{2.476}$$

Since nothing prevents us from taking $v_2 = \eta_2 = 0$, Eq. (2.476) yields

$$c\frac{\partial v_1}{\partial y} + g\frac{\partial \eta_1}{\partial y} = 0 \tag{2.477}$$

In the same way, Eq. (2.476) also implies

$$c\frac{\partial v_2}{\partial y} - g\frac{\partial \eta_2}{\partial y} = 0 \tag{2.478}$$

The wave-like nature of the solution we are looking for suggests to integrate (2.477) and (2.478) without adding any further function of t and x at the r.h.s. of

$$c v_1 + g \eta_1 = 0 \tag{2.479}$$

and at the r.h.s. of

$$c v_2 - g \eta_2 = 0 \tag{2.480}$$

Due to (2.479) and (2.480), Eq. (2.475) takes the form

$$\eta(x, y, t) = \frac{c}{g}\left[v_2(x, y - ct) - v_1(x, y + ct)\right] \tag{2.481}$$

Consider now (2.469). Substitution of (2.474) and (2.481) into (2.469) gives

$$f_0 (v_1 + v_2) = c\frac{\partial}{\partial x}(v_2 - v_1) \tag{2.482}$$

For the same reason as above, Eq. (2.482) implies

$$f_0 v_1 = -c\frac{\partial v_1}{\partial x} \tag{2.483}$$

and

$$f_0 v_2 = c\frac{\partial v_2}{\partial x} \tag{2.484}$$

In the northern hemisphere, f_0 is positive; so, by resorting to position (2.390), the solution of (2.483) is

$$v_1(x, y+ct) = v_1(0, y+ct)\exp\left(-\frac{x}{L_{\text{ext}}}\right) \tag{2.485}$$

since (2.483) may be thought of as an ordinary differential equation with respect to x. In the same way, we obtain

$$v_2(x, y-ct) = v_2(0, y-ct)\exp\left(\frac{x}{L_{\text{ext}}}\right) \tag{2.486}$$

Now, (2.355) implies that v takes finite values in the whole fluid domain, but this requirement is not satisfied by (2.486) unless $v_2 = 0$. Therefore, Eqs. (2.474) and (2.485) with condition $v_2 = 0$ yield

$$v(x, y, t) = v_1(0, y+ct)\exp\left(-\frac{x}{L_{\text{ext}}}\right) \tag{2.487}$$

and likewise (2.481) gives

$$\begin{aligned}\eta(x, y, t) &= -\frac{c}{g}v_1(0, y+ct)\exp\left(-\frac{x}{L_{\text{ext}}}\right)\\ &= -\frac{c}{g}v(x, y, t)\end{aligned} \tag{2.488}$$

A typical graph of the free-surface elevation (2.488) is plotted in Fig. 2.14. In the southern hemisphere, f_0 is negative, whence

$$v(x, y, t) = v_2(0, y-ct)\exp\left(-\frac{x}{L_{\text{ext}}}\right) \tag{2.489}$$

and

$$\begin{aligned}\eta(x, y, t) &= \frac{c}{g}v_2(0, y-ct)\exp\left(-\frac{x}{L_{\text{ext}}}\right)\\ &= \frac{c}{g}v(x, y, t)\end{aligned} \tag{2.490}$$

From (2.487) to (2.490), we see that:

- Because of the exponential decay away from the boundary, the Kelvin wave is *trapped*.
- In the limit of no rotation, the trapping distance increases without bound and the wave reduces to a gravity wave.
- In the northern hemisphere, the wave travels with the coast at its right; in the southern hemisphere, with the coast at its left. In other words, when the coast is the western boundary of an ocean basin, Kelvin waves propagate towards the equator.

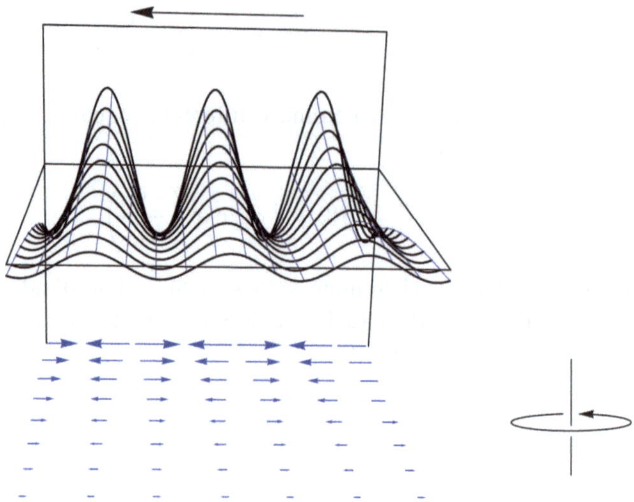

Fig. 2.14 Qualitative picture of the space structure of the Kelvin wave (2.488) in the proximity of the coastal wall, represented by a thick vertical rectangle. The arrow above the coast shows the direction of progress of the wave, while the small arrows at the bottom represent the convergence and the divergence (at nodes of elevation) of the horizontal current generated by the modulation of the free surface. The thick horizontal rectangular line shows the plane that corresponds to the mean elevation of the Kelvin wave. The curved arrow in the bottom-right panel depicts the rotation of the Earth as seen in the Northern Hemisphere

- If we assume that v_1 is a Fourier component, and we recall that the argument of a trigonometric function must be a non-dimensional quantity, then we get

$$v_1(0, y+ct) \propto \cos(ny+nc\,t)$$

Hence, we obtain the angular frequency $\sigma = cn$ and ascertain that the alongshore phase velocity $c_\mathrm{p} := \sigma/n = c$ coincides with group velocity $c_\mathrm{g} := \partial\sigma/\partial n = c$. Thus, in the alongshore direction, a Kelvin wave travels without dispersion at the speed of surface gravity waves.

2.3.7 Vorticity and Potential Vorticity of a Uniformly Rotating Flow

Governing Equation of Absolute Vorticity

In Sect. 1.2.2, the vorticity dynamics has been introduced on the basis of the momentum equation (1.43), which is valid in inertial systems. The vorticity equation of the averaged flow is (1.156), and the obvious question arises about the generalization of the latter equation starting from (2.319) rather than from (1.43), that is to say, when Earth's rotation is taken into account.

With the aid of (1.77) and after some trivial rearrangement, Eq. (2.319) takes the convenient form

$$\frac{\partial \mathbf{u}}{\partial t} - \mathbf{u} \times \boldsymbol{\omega}_{\mathbf{a}} = -g\hat{\mathbf{k}} - \nabla B + \mathscr{F}\mathbf{u} + p\nabla\frac{1}{\rho} \tag{2.491}$$

where

$$\boldsymbol{\omega}_{\mathrm{a}} := \boldsymbol{\omega} + 2\boldsymbol{\Omega} \tag{2.492}$$

is the *absolute vorticity* and

$$\mathscr{F} := A_{\mathrm{H}} \nabla_{\mathrm{H}}^2 + A_{\mathrm{V}} \frac{\partial^2}{\partial z^2}$$

is the *eddy-viscosity operator* introduced in (1.150), while B is the Bernoulli potential (1.79).

The application of the rot operator to (2.491) yields the equation

$$\frac{\partial \boldsymbol{\omega}}{\partial t} - \mathrm{rot}\,(\mathbf{u} \times \boldsymbol{\omega}_{\mathrm{a}}) = \mathrm{rot}\,\left(\mathscr{F}\mathbf{u} + p\nabla\frac{1}{\rho}\right) \tag{2.493}$$

that is to say, since $\boldsymbol{\Omega}$ is a constant vector,

$$\frac{\partial \boldsymbol{\omega}_{\mathrm{a}}}{\partial t} - \mathrm{rot}\,(\mathbf{u} \times \boldsymbol{\omega}_{\mathrm{a}}) = \mathrm{rot}\,\left(\mathscr{F}\mathbf{u} + p\nabla\frac{1}{\rho}\right) \tag{2.494}$$

In (2.494), the absolute vorticity appears as a whole: it constitutes the sole difference between the inertial and the rotating cases. The reason for (2.492) is that the Earth is a rigidly rotating body for which Eq. (1.76) holds.

Using a well-known vector identity, the second term in the l.h.s. of (2.494) may be written as

$$\mathrm{rot}\,(\mathbf{u} \times \boldsymbol{\omega}_{\mathrm{a}}) = (\boldsymbol{\omega}_{\mathrm{a}} \cdot \nabla)\,\mathbf{u} - \boldsymbol{\omega}_{\mathrm{a}}\,\mathrm{div}\,\mathbf{u} - (\mathbf{u} \cdot \nabla)\,\boldsymbol{\omega}_{\mathrm{a}}$$

and Eq. (2.494) eventually becomes

$$\frac{\mathrm{D}\boldsymbol{\omega}_{\mathrm{a}}}{\mathrm{D}t} = (\boldsymbol{\omega}_{\mathrm{a}} \cdot \nabla)\,\mathbf{u} - \boldsymbol{\omega}_{\mathrm{a}}\,\mathrm{div}\,\mathbf{u} - \frac{1}{\rho^2}\nabla p \times \nabla \rho + \mathrm{rot}\,(\mathscr{F}\mathbf{u}) \tag{2.495}$$

This equation shows that the "ambient" gives its own contribution to vorticity, besides that of the flow, represented by the vector $2\boldsymbol{\Omega}$ appearing in $\boldsymbol{\omega}_{\mathrm{a}}$ (see (2.492)), which is named *ambient vorticity*. The same definition is used also in the context of the β-plane approximation (2.259).

Potential Vorticity and Ertel's Theorem

Equation (2.495) is the starting point to infer the evolution law of an important quantity, the "potential vorticity", whose definition will be explained in the course of the derivation of the evolution law itself.

We reconsider (2.495) and substitute (1.17) into it, to obtain

$$\frac{\mathrm{D}\boldsymbol{\omega}_{\mathrm{a}}}{\mathrm{D}t} - \frac{\boldsymbol{\omega}_{\mathrm{a}}}{\rho}\frac{\mathrm{D}\rho}{\mathrm{D}t} = (\boldsymbol{\omega}_{\mathrm{a}} \cdot \nabla)\mathbf{u} - \frac{1}{\rho^2}\nabla p \times \nabla\rho + \mathrm{rot}\,(\mathscr{F}\mathbf{u})$$

and, after trivial rearrangements,

$$\frac{\mathrm{D}}{\mathrm{D}t}\frac{\boldsymbol{\omega}_{\mathrm{a}}}{\rho} = \left(\frac{\boldsymbol{\omega}_{\mathrm{a}}}{\rho} \cdot \nabla\right)\mathbf{u} - \frac{\nabla p \times \nabla\rho}{\rho^3} + \frac{1}{\rho}\mathrm{rot}\,(\mathscr{F}\mathbf{u}) \tag{2.496}$$

Now we suppose that q is a certain dynamic variable which is conserved following the motion, that is,

$$\frac{\mathrm{D}q}{\mathrm{D}t} = 0 \tag{2.497}$$

and take the scalar product of ∇q with (2.496), that is,

$$\nabla q \cdot \frac{\mathrm{D}}{\mathrm{D}t}\frac{\boldsymbol{\omega}_{\mathrm{a}}}{\rho} = \nabla q \cdot \left[\left(\frac{\boldsymbol{\omega}_{\mathrm{a}}}{\rho} \cdot \nabla\right)\mathbf{u}\right] - \nabla q \cdot \frac{\nabla p \times \nabla\rho}{\rho^3} + \frac{1}{\rho}\nabla q \cdot \mathrm{rot}\,(\mathscr{F}\mathbf{u}) \tag{2.498}$$

Because of (2.497), the identity

$$\frac{\boldsymbol{\omega}_{\mathrm{a}}}{\rho} \cdot \frac{\mathrm{D}}{\mathrm{D}t}\nabla q = -\nabla q \cdot \left[\left(\frac{\boldsymbol{\omega}_{\mathrm{a}}}{\rho} \cdot \nabla\right)\mathbf{u}\right] \tag{2.499}$$

holds independently of (2.498) (see the Appendix below). Thus, addition of (2.498) with (2.499) yields

$$\frac{\mathrm{D}\Pi}{\mathrm{D}t} = -\nabla q \cdot \frac{\nabla p \times \nabla\rho}{\rho^3} + \frac{1}{\rho}\nabla q \cdot \mathrm{rot}\,(\mathscr{F}\mathbf{u}) \tag{2.500}$$

where the quantity

$$\Pi := \frac{\boldsymbol{\omega}_{\mathrm{a}}}{\rho} \cdot \nabla q \tag{2.501}$$

appearing in (2.500) is called *potential vorticity*, and (2.500) is the evolution equation of Π.

What is q? This variable is not universally determined, but it usually represents a peculiar feature of the specific physical system under study, as some classical models will show in the following.

Ertel's Theorem

Equation (2.500) leads to the fundamental *Ertel's theorem*, which states that:

$$\text{if}\begin{cases}\text{friction is negligible, that is, } \mathscr{F}\mathbf{u}=0 & \textit{(H-1)}\\ \text{and} & \\ \begin{cases}\text{the fluid is barotropic, that is, } \nabla p\times\nabla\rho=0 & \textit{(H-2a)}\\ \text{or} & \\ q \text{ is a function only of } p \text{ and } \rho\text{, that is, } q=q(p,\rho) & \textit{(H-2b)}\end{cases}\end{cases}$$

then potential vorticity (2.501) is conserved following the motion of each element of fluid, that is,

$$\frac{\mathrm{D}\Pi}{\mathrm{D}t}=0 \tag{2.502}$$

In fact, because of *(H-1)* and (2.500), we have

$$\frac{\mathrm{D}\Pi}{\mathrm{D}t}=-\nabla q\cdot\frac{\nabla p\times\nabla\rho}{\rho^3}$$

and, in turn, *(H-2a)* immediately implies the conservation of Π. Alternatively, *(H-2b)* yields

$$\nabla q=\frac{\partial q}{\partial p}\nabla p+\frac{\partial q}{\partial\rho}\nabla\rho$$

and hence

$$\nabla q\cdot(\nabla p\times\nabla\rho)=\frac{\partial q}{\partial p}\nabla p\cdot(\nabla p\times\nabla\rho)+\frac{\partial q}{\partial\rho}\nabla\rho\cdot(\nabla p\times\nabla\rho)=0$$

Thus, (2.502) follows again from (2.500).

An Approximate Version of Ertel's Theorem

In most of the models taken into account by GFD, the approximation

$$\nabla q\approx\frac{\partial q}{\partial z}\hat{\mathbf{k}} \tag{2.503}$$

holds true. Hence, in the framework of the beta-plane approximation, (2.501) yields

$$\Pi\approx\frac{\boldsymbol{\omega}+2\boldsymbol{\Omega}}{\rho}\cdot\frac{\partial q}{\partial z}\hat{\mathbf{k}}=\frac{\zeta+f}{\rho}\frac{\partial q}{\partial z} \tag{2.504}$$

where, conventionally, ζ denotes the vertical component or relative vorticity, that is,

$$\zeta := \hat{\mathbf{k}} \cdot \boldsymbol{\omega} = \frac{\partial v}{\partial x} - \frac{\partial u}{\partial y} \tag{2.505}$$

Within the validity of (2.503) and recalling (1.155), Eq. (2.500) becomes

$$\frac{\mathrm{D}}{\mathrm{D}t}\left(\frac{\zeta + f}{\rho}\frac{\partial q}{\partial z}\right) = \frac{1}{\rho}\frac{\partial q}{\partial z}\left[\mathscr{F}\zeta - \frac{1}{\rho^2}\hat{\mathbf{k}} \cdot (\nabla p \times \nabla \rho)\right] \tag{2.506}$$

Note that, while the Coriolis parameter f is involved in (2.504) and in (2.506), the reciprocal $\tilde{f}$ is not, in accordance with the anticipation at the end of Sect. 2.3.1. Under the assumptions of Ertel's theorem, Eq. (2.506) yields

$$\frac{\mathrm{D}}{\mathrm{D}t}\left(\frac{\zeta + f}{\rho}\frac{\partial q}{\partial z}\right) = 0 \tag{2.507}$$

which involves only the vertical component of the absolute vorticity.

Jeffrey's Theorem

Equation (2.319) is equivalent to

$$\rho\left(\frac{\mathbf{Du}}{\mathrm{D}t} + 2\boldsymbol{\Omega} \times \mathbf{u}\right) = -g\rho\hat{\mathbf{k}} - \nabla p + \rho\mathscr{F}\mathbf{u} \tag{2.508}$$

Because of the identity $\mathrm{rot}\,(\rho\hat{\mathbf{k}}) = \nabla\rho \times \hat{\mathbf{k}}$, the equation

$$\mathrm{rot}\left[\rho\left(\frac{\mathbf{Du}}{\mathrm{D}t} + 2\boldsymbol{\Omega} \times \mathbf{u} - \mathscr{F}\mathbf{u}\right)\right] = g\hat{\mathbf{k}} \times \nabla\rho \tag{2.509}$$

immediately follows from (2.508). Assuming $\mathbf{u} = 0$ implies that the l.h.s. of (2.509) is zero and hence $\hat{\mathbf{k}} \times \nabla\rho = 0$. Thus, conversely,

$$\hat{\mathbf{k}} \times \nabla\rho \neq 0 \Longrightarrow \mathbf{u} \neq 0 \tag{2.510}$$

Implication (2.510) is the content of

Jeffrey's theorem: the state of rest $\mathbf{u} = 0$ is impossible if density variations occur on level surfaces, that is, on surfaces which are orthogonal to local gravity.

This theorem demonstrates in a direct way that the atmosphere is forced to circulate because of solar heating, which maintains a planetary north-south density gradient on level surfaces.

Appendix: Derivation of Eq. (2.499)

We expand the l.h.s. of (2.499), that is,

$$\frac{\boldsymbol{\omega}_\mathrm{a}}{\rho} \cdot \frac{\mathrm{D}}{\mathrm{D}t} \nabla q = \left(\frac{\boldsymbol{\omega}_\mathrm{a}}{\rho} \cdot \nabla \right) \frac{\partial q}{\partial t} + \frac{\boldsymbol{\omega}_\mathrm{a}}{\rho} \cdot [(\mathbf{u} \cdot \nabla) \nabla q] \tag{2.511}$$

to derive the r.h.s. of the same Eq. (2.499).

Making the following substitution:

$$\mathbf{a} \mapsto \frac{\boldsymbol{\omega}_\mathrm{a}}{\rho} \qquad \mathbf{b} \mapsto \mathbf{u} \qquad \phi \mapsto q$$

in the vector identity[4]

$$\mathbf{a} \cdot [(\mathbf{b} \cdot \nabla) \nabla \phi] = (\mathbf{a} \cdot \nabla) (\mathbf{b} \cdot \nabla \phi) - [(\mathbf{a} \cdot \nabla) \mathbf{b}] \cdot \nabla \phi$$

we obtain that the second term at the r.h.s. of (2.511) is

$$\frac{\boldsymbol{\omega}_\mathrm{a}}{\rho} \cdot [(\mathbf{u} \cdot \nabla) \nabla q] = \left(\frac{\boldsymbol{\omega}_\mathrm{a}}{\rho} \cdot \nabla \right) (\mathbf{u} \cdot \nabla q) - \left[\left(\frac{\boldsymbol{\omega}_\mathrm{a}}{\rho} \cdot \nabla \right) \mathbf{u} \right] \cdot \nabla q \tag{2.512}$$

Substituting (2.512) into (2.511) yields

$$\frac{\boldsymbol{\omega}_\mathrm{a}}{\rho} \cdot \frac{\mathrm{D}}{\mathrm{D}t} \nabla q = \left(\frac{\boldsymbol{\omega}_\mathrm{a}}{\rho} \cdot \nabla \right) \frac{\mathrm{D}q}{\mathrm{D}t} - \left[\left(\frac{\boldsymbol{\omega}_\mathrm{a}}{\rho} \cdot \nabla \right) \mathbf{u} \right] \cdot \nabla q \tag{2.513}$$

but, because of (9.8), we can drop the first term in the r.h.s. of (2.513) to obtain (2.499).

2.3.8 Appendix: Hydraulics and Earth's Rotation

We have learnt that, according to an observer fixed with the Earth, the motion of a terrestrial flow is governed by the momentum equations (2.321)–(2.323). Since there is no way to exclude Earth's rotation from the dynamics, one could wonder why in many problems of Fluid Dynamics, which do not involve oceanic or atmospheric motions, the resort to the Navier-Stokes equations for inertial systems (1.54) successfully leads to solutions that look "correct" also for an observer fixed with the Earth. Typical examples are those of Hydraulic and Environmental Engineering, in which Earth's rotation is usually neglected under suitable assumptions and approximations.

[4] Equation (2.512) is readily verified by using components.

Scaling Arguments

When the horizontal extent of the liquid body is far smaller than that of marine basins or great lakes, the first obvious approximation is the f-plane, that is,

$$f \approx \tilde{f} \approx f_0 \tag{2.514}$$

Well-mixed liquids are mostly concerned; so, the resulting density ρ is a constant, and no density anomaly exists. Moreover, at least a priori, fully isotropic flows can be conceived, and, therefore, a single length scale L and a single velocity scale U are sufficient to characterize the 3-D displacements and the 3-D velocity components, respectively. Hence,

$$\{x, y, z\} = L\{\mathsf{x}, \mathsf{y}, \mathsf{z}\} \qquad \{u, v, w\} = U\{\mathsf{u}, \mathsf{v}, \mathsf{w}\} \tag{2.515}$$

Consistently with the isotropy assumption, turbulence can be parameterized by

$$A\nabla^2 u \qquad A\nabla^2 v \qquad A\nabla^2 w \tag{2.516}$$

where ∇^2 is the 3-D Laplacian operator, while the constant A represents both the eddy-viscosity coefficients A_{H} and A_{V}. For most hydraulic systems, the timescale T is the advective one and much shorter than the period of Earth's rotation:

$$T := \frac{L}{U} \ll \frac{1}{f_0} \tag{2.517}$$

The scale of the local acceleration is $O(U/T) = O(U^2/L)$ and, due to the relative smallness of T, can reach values enormously higher than the Coriolis acceleration (which is $O(f_0 U)$ and, hence, independent of T). In such a context, the gradient of the perturbation pressure, of the order of P/L, balances the local acceleration, of the order of U^2/L, rather than Coriolis acceleration considered in the geostrophic case. The consequent relationship $U^2/L = P/\rho L$ yields the scale of the perturbation pressure

$$P = \rho U^2 \tag{2.518}$$

The total pressure p, the perturbation pressure $\tilde{p}$ and the hydrostatic pressure p_{s} are linked by the well-known equation

$$p = \tilde{p} + p_{\mathrm{s}} \tag{2.519}$$

Then, using (2.518), we get

$$\tilde{p} = p - p_{\mathrm{s}} = P\mathsf{p} \tag{2.520}$$

where p is the non-dimensional perturbation pressure and p_{s} satisfies the hydrostatic equation

$$\frac{\mathrm{d}p_{\mathrm{s}}}{\mathrm{d}z} = -\rho g \tag{2.521}$$

Non-dimensional Equations

By resorting to (2.514)–(2.521), Eqs. (2.321)–(2.323) can be written in non-dimensional form. After little algebra, one obtains the following set of non-dimensional momentum equations:

$$\frac{\mathsf{Du}}{\mathsf{Dt}} - \frac{1}{\varepsilon}(\mathsf{v}-\mathsf{w}) = -\frac{\partial \mathsf{p}}{\partial \mathsf{x}} + \frac{1}{\mathsf{Re}}\nabla'^2\mathsf{u} \tag{2.522}$$

$$\frac{\mathsf{Dv}}{\mathsf{Dt}} + \frac{1}{\varepsilon}\mathsf{u} = -\frac{\partial \mathsf{p}}{\partial \mathsf{y}} + \frac{1}{\mathsf{Re}}\nabla'^2\mathsf{v} \tag{2.523}$$

$$\frac{\mathsf{Dw}}{\mathsf{Dt}} - \frac{1}{\varepsilon}\mathsf{u} = -\frac{\partial \mathsf{p}}{\partial \mathsf{z}} + \frac{1}{\mathsf{Re}}\nabla'^2\mathsf{w} \tag{2.524}$$

where $\mathsf{D}/\mathsf{Dt} := \partial/\partial \mathsf{t} + \mathbf{u}\cdot\nabla'$, while ε is the advective Rossby number introduced in (2.550), and $\mathsf{Re} := UL/A$ is *Reynolds' number*. As $\varepsilon = 1/f_0 T \gg 1$, the limit $\varepsilon \to \infty$ can be considered in (2.522)–(2.524) to give

$$\frac{\mathsf{Du}}{\mathsf{Dt}} = -\frac{\partial \mathsf{p}}{\partial \mathsf{x}} + \frac{1}{\mathsf{Re}}\nabla'^2\mathsf{u} \tag{2.525}$$

$$\frac{\mathsf{Dv}}{\mathsf{Dt}} = -\frac{\partial \mathsf{p}}{\partial \mathsf{y}} + \frac{1}{\mathsf{Re}}\nabla'^2\mathsf{v} \tag{2.526}$$

$$\frac{\mathsf{Dw}}{\mathsf{Dt}} = -\frac{\partial \mathsf{p}}{\partial \mathsf{z}} + \frac{1}{\mathsf{Re}}\nabla'^2\mathsf{w} \tag{2.527}$$

that is, in vector form,

$$\frac{\mathsf{D}\mathbf{u}}{\mathsf{Dt}} = -\nabla'\mathsf{p} + \frac{1}{\mathsf{Re}}\nabla'^2\mathbf{u} \tag{2.528}$$

Dimensional Equation for High Rossby Number

Noting that

$$\nabla'\mathsf{p} = \frac{L}{P}\nabla(p - p_s) = \frac{L}{P}(\nabla p + \rho g\hat{\mathbf{k}})$$

the dimensional version of (2.528) turns out to be

$$\rho\frac{\mathrm{D}\mathbf{u}}{\mathrm{D}t} = -\nabla p - \rho g\hat{\mathbf{k}} + \rho A\nabla^2\mathbf{u} \tag{2.529}$$

Equation (2.529) has the same mathematical form as the first of Navier-Stokes equations (1.54), as soon as $-\rho g\hat{\mathbf{k}}$ is identified with body force $\rho\,\mathbf{F}$. Moreover, these equations also have the same physical meaning if (1.54) are referred to the mean motion, as determined by the Reynolds' averaging method expounded in Sect. 1.2.3, which produces the further term $\rho A \nabla^2 \mathbf{u}$ appearing in (2.529). In this hypothesis, the term of molecular viscosity $\mu \nabla^2 \mathbf{u}$ in the first of Eqs. (1.54) is quite negligible with respect to $\rho A \nabla^2 \mathbf{u}$.

2.4 Large-Scale Flows

Large-scale flows, both in the atmosphere and in the ocean, satisfy to a large extent the geostrophic balance and the hydrostatic equilibrium. A direct consequence is described by the so-called *thermal wind equation*, which shows a peculiar feature of stratified and rotating flows: they conserve steadily sloped isopycnals.

Based on the geostrophic balance and the hydrostatic equilibrium, a scale analysis of the governing equations leads to a special set of non-dimensional equations, which include several parameters, each of them depending on the scale of the motion and on basic planetary quantities, as gravity acceleration and Earth's rotation rate. This is a midway result in view of the derivation of the so-called quasi-geostrophic dynamics.

2.4.1 Phenomenology of Large-Scale Circulation of Geophysical Flows

Diagnostic Equations

The fluid motion can be conceived as the effect of a perturbation (of mechanical and/or thermal kind) superimposed to the *state of rest*, the latter being defined as

$$\mathbf{u} = 0 \qquad p = p_s(z) \qquad \rho = \rho_s(z) \tag{2.530}$$

where the so-called *standard pressure* p_s and *standard density* ρ_s satisfy (1.56), that is,

$$\frac{\mathrm{d}p_s}{\mathrm{d}z} + g\,\rho_s = 0 \tag{2.531}$$

Note that the state of rest (2.530) together with (2.531) verifies exactly both the continuity equation (2.318) and the momentum equation (2.319).

Unlike (2.530), in the *state of motion*, we observe

$$\mathbf{u} \neq 0 \qquad p = p_s(z) + \tilde{p}(\mathbf{x}, t) \qquad \rho = \rho_s(z) + \tilde{\rho}(\mathbf{x}, t) \tag{2.532}$$

where $\mathbf{u}$, $\tilde{p}$ and $\tilde{\rho}$ represent the departure from the state of rest induced by the above-cited disturbance. We have already met the third of Eqs. (2.532) in (2.61).

We presuppose that the motion takes place on a single well-defined scale such that the magnitude of the terms in the equations of motion can be systematically estimated in terms of these scales. On the basis of this assumption, we introduce:

- The non-dimensional $O(1)$ coordinates according to the following positions:

$$x = L\mathsf{x} \qquad y = L\mathsf{y} \qquad z = H\mathsf{z} \qquad t = T\mathsf{t} \tag{2.533}$$

- The analogous non-dimensional $O(1)$ fields:

$$u = U\mathsf{u} \qquad v = U\mathsf{v} \qquad w = W\mathsf{w} \qquad \tilde{p} = P\mathsf{p} \qquad \tilde{\rho} = D\rho' \tag{2.534}$$

We are following the typographical convention introduced with Eq. (2.121), that is, non-dimensional quantities are mostly identified by sans-serif font if represented through Latin characters, and by a prime if represented through Greek characters or special symbols. In (2.533), L is the horizontal length scale, H is the vertical length scale and T is the timescale of the motion. In (2.534), U is the horizontal velocity scale, W is the vertical velocity scale, P is the scale of the perturbation pressure $\tilde{p}$ and D is the scale of the density anomaly $\tilde{\rho}$. We anticipate that the phenomenology of large-scale circulation establishes definite dependences of the kind

$$D = D(\rho_s, L, U, H) \qquad P = P(\rho_s, L, U, H) \tag{2.535}$$

which are indispensable to derive the related governing equations by means of a suitable scaling of (2.319), (2.41) and (2.51). The derivation of $D(\rho_s, L, U, H)$ and $P(\rho_s, L, U, H)$ is the main aim of this section.

Phenomenology shows that, far from the boundaries, both the atmosphere and the ocean tend to arrange themselves according to two *diagnostic equations*:

- The *geostrophic balance* (2.437)

$$f_0 \rho \mathbf{u} = \hat{\mathbf{k}} \times \nabla p \tag{2.536}$$

- The *hydrostatic equilibrium* (1.56)

$$\frac{\partial p}{\partial z} + g\rho = 0 \tag{2.537}$$

Note that only the horizontal pressure gradient is involved in (2.536). In other words, the geostrophic balance and the hydrostatic equilibrium constitute the dominant horizontal and vertical momentum equations, respectively.

The geostrophic balance (2.536) demands a uniform ambient rotation rate ($f_0 \neq 0$) and implies that the current is orthogonal to the horizontal pressure gradient:

$$\mathbf{u} \cdot \nabla p = 0 \tag{2.538}$$

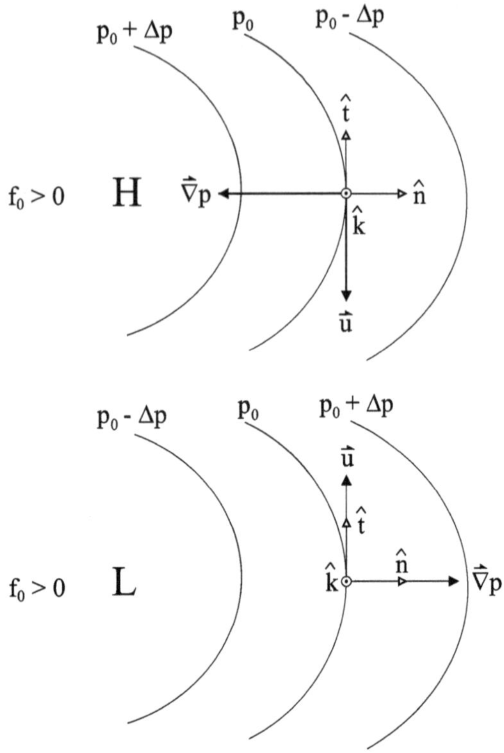

Fig. 2.15 In terms of the right-handed triple of unit vectors $(\hat{\mathbf{n}}, \hat{\mathbf{t}}, \hat{\mathbf{k}})$, the geostrophic balance (2.536) states that $\nabla p = (\partial p/\partial n)\hat{\mathbf{n}}$ implies $\mathbf{u} = (\rho f_0)^{-1}(\partial p/\partial n)\hat{\mathbf{t}}$. In the *upper panel*, $\partial p/\partial n < 0$ yields $\mathbf{u}$ clockwise; in the *lower panel*, $\partial p/\partial n < 0$ yields $\mathbf{u}$ anticlockwise

Relation (2.538) constitutes the most striking difference with respect to the dynamics of inertial flows. For instance, as $\nabla p = (\partial p/\partial n)\hat{\mathbf{n}}$, where $\hat{\mathbf{n}}$ is the unit vector locally normal to a certain isobar and pointing outside, then $f_0 \rho \mathbf{u} = (\partial p/\partial n)\hat{\mathbf{t}}$, where $\hat{\mathbf{t}} := \hat{\mathbf{k}} \times \hat{\mathbf{n}}$ is a unit vector locally tangent to the same isobar, so that $\hat{\mathbf{k}} = \hat{\mathbf{n}} \times \hat{\mathbf{t}}$. Thus, fluid parcels in geostrophic balance with pressure move along the isobars of the pressure field.

More precisely, since $\partial p/\partial n = f_0 \rho \hat{\mathbf{t}} \cdot \mathbf{u}$, in the northern hemisphere (where $f_0 > 0$), the directional derivative $\partial p/\partial n$ and the direction cosine $\hat{\mathbf{t}} \cdot \mathbf{u}$ have the same sign; therefore, the geostrophic current $\mathbf{u}$ flows:

- Clockwise (i.e. $\hat{\mathbf{t}} \cdot \mathbf{u} < 0$) around high-pressure areas (where $\partial p/\partial n < 0$)
- Anticlockwise (i.e. $\hat{\mathbf{t}} \cdot \mathbf{u} > 0$) around low-pressure areas (where $\partial p/\partial n > 0$)

These configurations are reported in Fig. 2.15. In the southern hemisphere (where $f_0 < 0$), the directional derivative $\partial p/\partial n$ and the direction cosine $\hat{\mathbf{t}} \cdot \mathbf{u}$ have opposite signs; therefore, the geostrophic current $\mathbf{u}$ flows:

- Anticlockwise (i.e. $\hat{\mathbf{t}} \cdot \mathbf{u} > 0$) around high-pressure areas (where $\partial p/\partial n < 0$)
- Clockwise (i.e. $\hat{\mathbf{t}} \cdot \mathbf{u} < 0$) around low-pressure areas (where $\partial p/\partial n > 0$)

The order of magnitude of geostrophic wind at mid-latitudes is quickly computed. Indeed, meteorological charts show that pressure typically changes of 10 hPa in 1,000 km; hence, $\nabla p = O(10^{-3}\,\text{Pa/m})$. Moreover, $f = O(10^{-4}\,\text{s}^{-1})$, and $\rho = O(1\,\text{kg/m}^3)$ for air. Therefore, wind speed is $u = O(10\,\text{m/s})$.

In the case of the ocean, a typical slope $\Delta\eta/L = O(10^{-7})$ of the free surface yields $u = O(10^{-2}\,\text{m/s})$ because of the geostrophic balance (2.436).

The hydrostatic equilibrium (2.537) does not hold only for the standard pressure p_s and density ρ_s, as (2.531) shows, but also for the perturbation pressure $\tilde{p}$ and density anomaly $\tilde{\rho}$, as one easily verifies by substituting $p = p_s + \tilde{p}$ and $\rho = \rho_s + \tilde{\rho}$ into (2.537) to find

$$\frac{\partial \tilde{p}}{\partial z} + g\tilde{\rho} = 0 \tag{2.539}$$

Scaling of Perturbation Pressure and Density and of Potential Temperature Anomaly

From the geostrophic balance (2.536) and the hydrostatic equation (2.539), a basic information can be derived about the relative amplitude of the perturbation pressure and density and, hence, of the total pressure and density.

Scaling of Perturbation Pressure and Density

Substitution of $\tilde{p} = P\mathrm{p}$ and $\tilde{\rho} = D\rho'$ into (2.536), recalling also that $\mathbf{u} = O(1)$ and $\hat{\mathbf{k}} \times \nabla'\mathrm{p} = O(1)$, yields the following relation among the typical values D, U and P:

$$f_0\,(\rho_s + D)\,U = \frac{P}{L} \tag{2.540}$$

where the symbol = means that the quantities at both sides of = have the same order of magnitude. In the same way, recalling that $\partial\mathrm{p}/\partial\mathrm{z} = O(1)$, substitution of $\tilde{p} = P\mathrm{p}$ and $\tilde{\rho} = D\rho'$ into (2.539) yields

$$P = gHD \tag{2.541}$$

where the meaning of "=" is the same as above.

From (2.540) and (2.541), D and P can be singled out to give

$$D = \rho_s \frac{L f_0 U}{gH - L f_0 U} \tag{2.542}$$

$$P = \rho_s L f_0 U \frac{gH}{gH - L f_0 U} \tag{2.543}$$

As (2.542) is equivalent to

$$D = \rho_s \frac{L f_0 U}{g H} \left(1 - \frac{L f_0 U}{g H}\right)^{-1} \tag{2.544}$$

and (2.543) can be written as

$$P = \rho_s L f_0 U \left(1 - \frac{L f_0 U}{g H}\right)^{-1} \tag{2.545}$$

comparison of the pure number

$$\mathsf{r} := \frac{L f_0 U}{g H} \tag{2.546}$$

with unity is in order. To achieve this, r is estimated as

$$\mathsf{r} = O\left(10^{-5} \frac{U L}{H}\right)$$

In the case of the ocean, $U L/H \leq O(10^2\,\mathrm{m/s})$, while, in the case of the troposphere, $U L/H \leq O(10^3\,\mathrm{m/s})$. Thus, in both cases, we get $\mathsf{r} \leq O(10^{-2}) \ll O(1)$. The latter inequality allows us to approximate (2.544) with

$$D = \frac{\rho_s L f_0 U}{g H} \tag{2.547}$$

and (2.545) with

$$P = \rho_s L f_0 U \tag{2.548}$$

Number r can be expressed in terms of the *rotational Froude number* (2.389)

$$\mathsf{F} := \frac{f_0^2 L^2}{g H} \tag{2.549}$$

and the advective Rossby number (2.363)

$$\varepsilon = \frac{U}{f_0 L} \tag{2.550}$$

to obtain

$$\mathsf{r} = \varepsilon \mathsf{F} \tag{2.551}$$

By using (2.546), Eqs. (2.548) yield

$$D = \mathsf{r} \rho_s \qquad P = \mathsf{r} \rho_s g H \tag{2.552}$$

respectively. Hence, the density anomaly $\tilde{\rho}$ takes the form

$$\tilde{\rho} = \mathsf{r}\,\rho_s\,\rho' \tag{2.553}$$

and the total density field is

$$\rho = \rho_s\,(1+\mathsf{r}\rho') = O(\rho_s) \tag{2.554}$$

Analogously, the perturbation pressure is

$$\tilde{p} = \rho_s L f_0 U \,\mathsf{p} = \mathsf{r}\,\rho_s\, g H\, \mathsf{p} \tag{2.555}$$

and, from (2.552), the total pressure field can be written as

$$p = p_s + \mathsf{r}\,\rho_s\, g H \mathsf{p} \tag{2.556}$$

Scaling of Potential Temperature Anomaly

To derive the non-dimensional anomaly θ' of potential temperature, the ratios $\tilde{\rho}/\rho_s = \mathsf{r}\rho'$ and $\tilde{p}/p_s = \mathsf{r}\,(\rho_s\, g H/p_s)\,\mathsf{p}$ inferred from (2.553) and (2.555), respectively, are substituted into (2.111) to obtain

$$\theta = \frac{\theta_s}{1+\mathsf{r}\rho'}\left(1+\mathsf{r}\frac{\rho_s\, g H}{p_s}\mathsf{p}\right)^{1/\gamma} \tag{2.557}$$

Due to the smallness of r, the r.h.s. of (2.557) can be approximated by its truncated Maclaurin series:

$$\theta \approx \theta_s\left[1+\mathsf{r}\left(\frac{\rho_s\, g H}{\gamma p_s}\mathsf{p}-\rho'\right)\right] \tag{2.558}$$

In (2.558), the quantity $(\rho_s\, g H/\gamma p_s)\,\mathsf{p}-\rho'$ is identified with the non-dimensional anomaly of potential temperature, that is,

$$\theta' = \frac{\rho_s\, g H}{\gamma p_s}\mathsf{p}-\rho' \tag{2.559}$$

Thus, substitution of (2.559) into (2.558) yields the overall potential temperature θ in the approximate form

$$\theta = \theta_s\,(1+\mathsf{r}\,\theta') \tag{2.560}$$

Note that (2.559) coincides with (2.164), although the former has been derived in the context of non-rotating internal waves, where the small parameter in the Maclaurin expansion of θ/θ_s was $a = U/gT$ in place of $\mathsf{r} = \varepsilon\,\mathsf{F}$ used here.

Non-dimensional Hydrostatic Equation and Geostrophic Balance

For future reference, we report the non-dimensional version of the hydrostatic equation (2.539), which is obtained with the aid of (2.553) and (2.555). A straightforward computation gives (note that $H\,\partial_z = \partial_{\mathsf{z}}$)

$$\frac{1}{\rho_s}\frac{\partial}{\partial \mathsf{z}}(\rho_s\,\mathsf{p}) + \rho' = 0 \tag{2.561}$$

The non-dimensional version of the geostrophic balance comes from (2.536), once (2.554) and (2.555) are substituted into it to obtain (note that $\hat{\mathbf{k}} \times \nabla = L^{-1}\hat{\mathbf{k}} \times \nabla'$)

$$(1 + \mathsf{r}\rho')\,\mathsf{u} = \hat{\mathsf{k}} \times \nabla'\mathsf{p} \tag{2.562}$$

Because $O(r) < 1$, Eq. (2.562) can be legitimately approximated with

$$\mathsf{u} = \hat{\mathsf{k}} \times \nabla'\mathsf{p} \tag{2.563}$$

which is the non-dimensional version of (2.536). Starting from (2.563), we can go back to its dimensional form by means of the obvious substitutions

$$\mathsf{u} \mapsto \frac{1}{U}\mathbf{u} \qquad \nabla' \mapsto L\nabla \qquad \mathsf{p} \mapsto \frac{\tilde{p}}{\rho_s f_0 U L}$$

to obtain

$$\mathbf{u} = \frac{1}{\rho_s f_0}\hat{\mathbf{k}} \times \nabla p \tag{2.564}$$

since ∇ is the horizontal gradient operator, so $\nabla\tilde{p} = \nabla p$. The Cartesian components of (2.564) are

$$u = -\frac{1}{\rho_s f_0}\frac{\partial p}{\partial y} \qquad v = \frac{1}{\rho_s f_0}\frac{\partial p}{\partial x} \qquad w = 0 \tag{2.565}$$

Strictly speaking, Eq. (2.564) selects only the first two equations of (2.565), while the reason for the third equation, that is,

$$w = 0 \tag{2.566}$$

will be discussed in Sect. 2.4.2, in the context of the incompressibility equations for air and seawater. The geostrophic current (2.564) is horizontally non-divergent, that is,

$$\operatorname{div}\mathbf{u} = 0 \tag{2.567}$$

as shown by a straightforward computation, based on (2.565). We stress that (2.567) relies on the approximation $1 + \varepsilon\,\mathsf{F}\rho' \approx 1$, which has been used to derive (2.563) from (2.562). In fact, if we start from

$$\mathbf{u} = \frac{1}{\rho f_0} \hat{\mathbf{k}} \times \nabla p \tag{2.568}$$

in place of (2.564), then

$$\operatorname{div} \mathbf{u} = -\frac{1}{f_0 \rho^2} \hat{\mathbf{k}} \cdot \nabla \rho \times \nabla p \tag{2.569}$$

but, unlike (2.564), $\mathbf{u}$ is no more divergence free, because of the baroclinic part $\nabla_{\mathrm{H}} \rho$ of the density gradient.

The Thermal Wind Equation

The thermal wind equation is a consequence of the geostrophic balance and of the hydrostatic equilibrium and points out a distinctive feature of uniformly rotating flows. It holds for both the ocean and the atmosphere, but it is preferable to follow separate derivations in each case because of the lack of homogeneity in the orders of magnitude that are involved and because of the use of the perfect gas law in the sole atmospheric case.

The Thermal Wind Equation for the Ocean

Application of the vertical derivative to the geostrophic balance in the form (2.564) yields

$$\frac{\partial \mathbf{u}}{\partial z} + \frac{1}{f_0 \rho_{\mathrm{s}}^2} \frac{\mathrm{d}\rho_{\mathrm{s}}}{\mathrm{d}z} \hat{\mathbf{k}} \times \nabla p = \frac{1}{f_0 \rho_{\mathrm{s}}} \hat{\mathbf{k}} \times \nabla \frac{\partial p}{\partial z} \tag{2.570}$$

Hence, using (2.537) into the r.h.s. of (2.570), one obtains

$$\frac{\partial \mathbf{u}}{\partial z} + \frac{1}{f_0 \rho_{\mathrm{s}}^2} \frac{\mathrm{d}\rho_{\mathrm{s}}}{\mathrm{d}z} \hat{\mathbf{k}} \times \nabla p = -\frac{g}{f_0 \rho_{\mathrm{s}}} \hat{\mathbf{k}} \times \nabla \rho \tag{2.571}$$

By using again (2.564) in the second term at the l.h.s. of (2.571), the *thermal wind equation*

$$\frac{\partial \mathbf{u}}{\partial z} + \frac{1}{\rho_{\mathrm{s}}} \frac{\mathrm{d}\rho_{\mathrm{s}}}{\mathrm{d}z} \mathbf{u} = \frac{g}{f_0 \rho_{\mathrm{s}}} \nabla \rho \times \hat{\mathbf{k}} \tag{2.572}$$

follows.

Equation (2.572) can be simplified on scaling arguments. In fact, the second term at the l.h.s. of (2.572) is

$$\frac{1}{\rho_{\mathrm{s}}} \frac{\mathrm{d}\rho_{\mathrm{s}}}{\mathrm{d}z} \mathbf{u} = O\left(\frac{U}{H_\rho}\right) \tag{2.573}$$

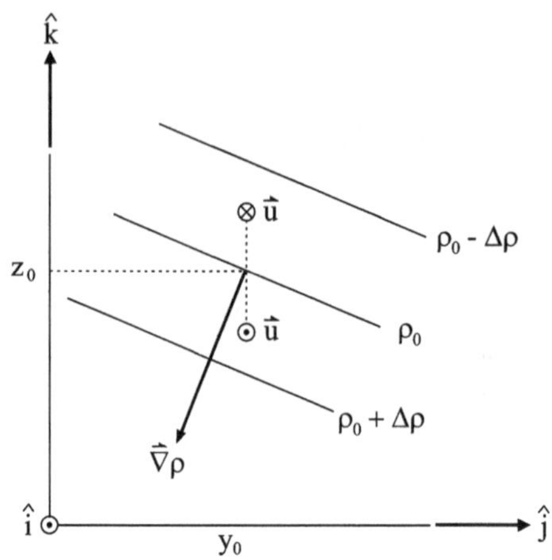

Fig. 2.16 The gradient of the isopycnals at $z = z_0$ has a component in the direction $\hat{\mathbf{j}}$ and opposite to it. Hence, the cross product of the gradient with $\hat{\mathbf{k}}$ yields $\partial u/\partial z = (g/f_0\,\rho)\,\partial\rho/\partial y$. As $\partial\rho/\partial y < 0$, the component u of the velocity decreases with height: for instance, it is *positive* below z_0 and *negative* above z_0

where H_ρ is the density height scale, already met in dealing with internal waves (see Sect. 2.2 and especially Eq. (2.114)). On the other hand, recalling (2.554) and (2.546), the estimate

$$\frac{g}{f_0\,\rho_s}\nabla\rho\times\hat{\mathbf{k}} = O\left(\frac{g}{f_0\,\rho_s}\frac{\mathrm{r}\rho_s}{L}\right) = O\left(\frac{U}{H}\right) \tag{2.574}$$

can be established. Equations (2.573) and (2.574) imply

$$\frac{1}{\rho_s}\frac{\mathrm{d}\rho_s}{\mathrm{d}z}\,|\mathbf{u}|\,\left(\frac{g}{f_0\,\rho_s}\,|\hat{\mathbf{k}}\times\nabla\rho|\right)^{-1} = O\left(\frac{H}{H_\rho}\right) \tag{2.575}$$

For the ocean, H/H_ρ is very small; in fact, $H_\rho = O(g/N_s^2) = O(10^6\,\mathrm{m})$ and hence $H/H_\rho = O(10^{-3})$. For this reason, Eq. (2.572) can be reasonably approximated by

$$\frac{\partial\mathbf{u}}{\partial z} = -\frac{g}{f_0\,\rho_s}\hat{\mathbf{k}}\times\nabla\rho \qquad \text{(ocean)} \tag{2.576}$$

which is the thermal wind equation for the ocean in its most common version. The main feature of (2.576) consists in that the isopycnals are not orthogonal to the local gravity vector (i.e. $\hat{\mathbf{k}}\times\nabla\rho \neq 0$), in the presence of a geostrophic current with a vertical shear (i.e. if $\partial\mathbf{u}/\partial z \neq 0$). This situation cannot be realized in inertial frames (i.e. for $f_0 = 0$) because steadiness in these frames demands minimum potential energy, which is reached only if the isopycnals are orthogonal to the local gravity vector. An example of flow geometry related to (2.576) is reported in Fig. 2.16.

Density data can be obtained with high accuracy from hydrology, and vertical integration of the thermal wind equation (2.576) allows one to evaluate the geostrophic current by integrating the density field from the *depth of no motion* z_0, namely,

$$\mathbf{u} = -\frac{g}{f_0\,\rho_s}\hat{\mathbf{k}} \times \int_{z_0}^{z} \nabla\rho \, dz' \tag{2.577}$$

At the depth of no motion, we have $u = v = 0$, and, below it, the (almost) quiescent abyss should be situated. To integrate (2.577), it is necessary to know this depth, and its location constitutes a nontrivial difficulty in the practical application of this equation. The method also demands that the density field does not depend explicitly on time, and this requisite is hardly met in the upper layer of the ocean. In any case, the method based on (2.577) was the sole at the disposal of the oceanographers, in the early days of Dynamical Oceanography, to compute large-scale currents.

The Thermal Wind Equation for the Atmosphere

Application of the vertical derivative to the geostrophic balance in the form (2.536) and the subsequent use of (2.537) yields

$$\frac{\partial \mathbf{u}}{\partial z} + \frac{1}{\rho}\frac{\partial \rho}{\partial z}\mathbf{u} = \frac{g}{f_0\,\rho}\nabla\rho \times \hat{\mathbf{k}} \tag{2.578}$$

From the perfect gas law $p = R\rho T$, one obtains

$$\frac{1}{\rho}\frac{\partial \rho}{\partial z} = \frac{1}{p}\frac{\partial p}{\partial z} - \frac{1}{T}\frac{\partial T}{\partial z} \qquad \frac{1}{\rho}\nabla\rho = \frac{1}{p}\nabla p - \frac{1}{T}\nabla T \tag{2.579}$$

With the aid of (2.579), Eq. (2.578) becomes, after a trivial rearrangement,

$$\begin{aligned}\frac{\partial \mathbf{u}}{\partial z} &= -\frac{g}{f_0}\hat{\mathbf{k}} \times \left(\frac{1}{p}\nabla p - \frac{1}{T}\nabla T\right) - \left(\frac{1}{p}\frac{\partial p}{\partial z} - \frac{1}{T}\frac{\partial T}{\partial z}\right)\mathbf{u} \\ &= -\frac{g}{f_0\,p}\hat{\mathbf{k}} \times \nabla p - \frac{1}{p}\frac{\partial p}{\partial z}\mathbf{u} + \frac{g}{f_0\,T}\hat{\mathbf{k}} \times \nabla T + \frac{1}{T}\frac{\partial T}{\partial z}\mathbf{u}\end{aligned} \tag{2.580}$$

In particular, because of (2.536) and (2.537), we have

$$-\frac{g}{f_0\,p}\hat{\mathbf{k}} \times \nabla p - \frac{1}{p}\frac{\partial p}{\partial z}\mathbf{u} = -\frac{1}{p}\left(g\rho + \frac{\partial p}{\partial z}\right)\mathbf{u} = 0$$

and hence (2.580) becomes

$$\frac{\partial \mathbf{u}}{\partial z} = \frac{g}{f_0\,T}\hat{\mathbf{k}} \times \nabla T + \frac{1}{T}\frac{\partial T}{\partial z}\mathbf{u} \tag{2.581}$$

Equation (2.581) can be simplified on scaling arguments. Indeed, in the troposphere, we have $\partial\mathbf{u}/\partial z = O(10^{-3}\,\mathrm{s}^{-1})$ and $(g/f_0\,T)\hat{\mathbf{k}} \times \nabla T = O(10^{-3}\,\mathrm{s}^{-1})$ while $(1/T)\,(\partial T/\partial z)\,\mathbf{u} = O(10^{-4}\,\mathrm{s}^{-1})$. Thus, (2.581) can be reasonably approximated by

$$\frac{\partial \mathbf{u}}{\partial z} = \frac{g}{f_0\,T}\hat{\mathbf{k}} \times \nabla T \qquad \text{(atmosphere)} \tag{2.582}$$

Like (2.576), Eq. (2.582) also shows the foundation of a steady equilibrium in which the isothermals are not orthogonal to the local gravity vector (i.e. $\hat{\mathbf{k}} \times \nabla T \neq 0$), in the presence of a geostrophic current with a vertical shear ($\partial \mathbf{u}/\partial z \neq 0$).

2.4.2 *Governing Equations of Large-Scale, Geophysical Flows*

Summary of the Dimensional Governing Equations

We list preliminarily the set of the governing equations that have been derived in the preceding Sections.

Incompressibility Condition

We start, on simplicity grounds, from the incompressibility equation of seawater (2.53), that is, div $\mathbf{u} = 0$ or, using components

$$\frac{\partial u}{\partial x} + \frac{\partial v}{\partial y} + \frac{\partial w}{\partial z} = 0 \qquad \text{(sea)} \tag{2.583}$$

while the incompressibility of the atmosphere is described by (2.67), that is,

$$\rho_s \left(\frac{\partial u}{\partial x} + \frac{\partial v}{\partial y} \right) + \frac{\partial}{\partial z}(\rho_s w) = 0 \qquad \text{(air)} \tag{2.584}$$

where $\rho_s = \rho_s(z)$.

Momentum Balance

The scalar components (2.321)–(2.323) of the momentum equation are

$$\frac{\mathrm{D}}{\mathrm{D}t}u - fv + \tilde{f}w = -\frac{1}{\rho}\frac{\partial p}{\partial x} + \mathscr{F}u \tag{2.585}$$

$$\frac{\mathrm{D}}{\mathrm{D}t}v + fu = -\frac{1}{\rho}\frac{\partial p}{\partial y} + \mathscr{F}v \tag{2.586}$$

$$\frac{\mathrm{D}}{\mathrm{D}t}w - \tilde{f}u = -\frac{1}{\rho}\frac{\partial p}{\partial z} + \mathscr{F}w - g \tag{2.587}$$

where $\mathrm{D}/\mathrm{D}t := \partial/\partial t + \mathbf{u} \cdot \nabla$ is the Lagrangian derivative operator (1.8), $f := f_0 + \beta_0 y$ is the Coriolis parameter (2.257) $\tilde{f} := \beta_0 R - f_0 y/R$ is the reciprocal Coriolis parameter (2.262), and $\mathscr{F} := A_H \nabla_H^2 + A_V \partial^2/\partial z^2$ is the eddy-viscosity operator (1.150).

Momentum equations (2.585)–(2.587) may be condensed in the compact form

$$\frac{\mathrm{D}\mathbf{u}}{\mathrm{D}t} + \mathbf{F}\,\mathbf{u} = -\frac{1}{\rho}\nabla p + \mathscr{F}\,\mathbf{u} - g\hat{\mathbf{k}}$$

where

$$\mathbf{F} = \begin{pmatrix} 0 & -f & \tilde{f} \\ f & 0 & 0 \\ -\tilde{f} & 0 & 0 \end{pmatrix}$$

but, as we shall see in the following, it is convenient to use components for scaling purposes: horizontal and vertical scales are much different.

Thermodynamic Equation

The thermodynamic equation which is valid for an adiabatic and salinity-conserving seawater body is (2.51), namely, $\mathrm{D}\rho/\mathrm{D}t = 0$ or

$$\frac{\partial \rho}{\partial t} + \mathbf{u}\cdot\nabla\rho = 0 \qquad \text{(sea)} \qquad (2.588)$$

Finally, the thermodynamic equation for the atmosphere in the presence of thermal forcing is expressed by (2.41), namely, $\mathrm{D}\theta/\mathrm{D}t = \theta\,\dot{Q}/c_p T_\mathrm{a}$ or

$$\frac{\partial \theta}{\partial t} + \mathbf{u}\cdot\nabla\theta = \frac{\theta\,\dot{Q}}{c_p T_\mathrm{a}} \qquad \text{(air)} \qquad (2.589)$$

Here, we denote by T_a the absolute temperature, whereas T has been used in Sect. 2.4.1 for the local timescale.

From this equation, we obtain the adiabatic case by putting $\dot{Q} = 0$, which yields

$$\frac{\partial \theta}{\partial t} + \mathbf{u}\cdot\nabla\theta = 0 \qquad (2.590)$$

Non-dimensional Governing Equations

The non-dimensional version of (2.583)–(2.590), which will be carried out consistently with the phenomenology analysed in Sect. 2.4.1, includes the fundamental physics of the large-scale circulation of the atmosphere and the ocean. Their formulation is the main aim of the present section.

Incompressibility Equation for the Sea

With (2.533) and (2.534) in mind, Eq. (2.583) can be written as

$$\frac{U}{L}\left(\frac{\partial \mathsf{u}}{\partial \mathsf{x}}+\frac{\partial \mathsf{v}}{\partial \mathsf{y}}\right)+\frac{W}{H}\frac{\partial \mathsf{w}}{\partial \mathsf{z}}=0$$

whence, after division by U/L, the non-dimensional version

$$\frac{\partial \mathsf{u}}{\partial \mathsf{x}}+\frac{\partial \mathsf{v}}{\partial \mathsf{y}}+\frac{LW}{HU}\frac{\partial \mathsf{w}}{\partial \mathsf{z}}=0 \tag{2.591}$$

follows. In order that each term of (2.591) can give, at least a priori, a contribution of order $O(1)$ to (2.591), we assume that

$$\frac{LW}{HU}=O(1) \tag{2.592}$$

so, (2.591) takes its final form $\operatorname{div}\mathbf{u}=0$, that is,

$$\boxed{\frac{\partial \mathsf{u}}{\partial \mathsf{x}}+\frac{\partial \mathsf{v}}{\partial \mathsf{y}}+\frac{\partial \mathsf{w}}{\partial \mathsf{z}}=0} \tag{2.593}$$

Because of (2.592), the vertical velocity is scaled as

$$w=W\mathsf{w} \qquad \text{where} \qquad W=\frac{HU}{L} \tag{2.594}$$

Strictly speaking, (2.594) gives only an upper bound for the scale W of the vertical velocity; in fact, if the flow is planar or almost planar, the first two terms of (2.591) tend to compensate each other while $LW/HU < O(1)$, that is, (2.594) should be changed into $W < HU/L$.

As the geostrophic current is horizontally non-divergent, Eq. (2.593) is consistent with a vertical velocity w such that $\partial\mathsf{w}/\partial\mathsf{z}=0$ at every depth of the ocean. Thus, w is a constant in each fluid column, but no fluid column can indefinitely bring matter towards the sea floor or towards the sea surface; so, the unique physically admissible integral of $\partial\mathsf{w}/\partial\mathsf{z}=0$ at the geostrophic level of approximation is $\mathsf{w}=0$, in accordance with (2.566).

Incompressibility Equation for the Atmosphere

Introducing non-dimensional variables into Eq. (2.584) yields

$$\rho_s\frac{U}{L}\left(\frac{\partial \mathsf{u}}{\partial \mathsf{x}}+\frac{\partial \mathsf{v}}{\partial \mathsf{y}}\right)+\frac{W}{H}\frac{\partial}{\partial \mathsf{z}}(\rho_s\,\mathsf{w})=0 \tag{2.595}$$

As before, division of (2.595) by U/L produces the non-dimensional version of (2.584), that is,

$$\left(\frac{\partial \mathrm{u}}{\partial \mathrm{x}}+\frac{\partial \mathrm{v}}{\partial \mathrm{y}}\right)+\frac{LW}{HU}\frac{1}{\rho_s}\frac{\partial}{\partial \mathrm{z}}(\rho_s \mathrm{w})=0 \tag{2.596}$$

Assuming again (2.592), from (2.596), the final form

$$\boxed{\frac{\partial \mathrm{u}}{\partial \mathrm{x}}+\frac{\partial \mathrm{v}}{\partial \mathrm{y}}+\frac{1}{\rho_s}\frac{\partial}{\partial \mathrm{z}}(\rho_s \mathrm{w})=0} \tag{2.597}$$

is obtained.

Application of (2.597) to the geostrophic current leads again to the same conclusion as in the ocean case, that is, the validity of (2.566) also for the atmosphere.

Zonal Component of the Momentum Equation

Using non-dimensional variables, the terms in Eq. (2.585) may be scaled as follows:

$$\begin{aligned}
&\frac{\mathrm{D}}{\mathrm{D}t}u=\frac{\partial u}{\partial t}+\mathbf{u}\cdot\nabla u \overset{(2.592)}{=} \frac{U}{T}\frac{\partial \mathrm{u}}{\partial \mathrm{t}}+\frac{U^2}{L}\mathbf{u}\cdot\nabla'\mathrm{u}\\
&-fv=-fU\mathrm{v}\\
&\tilde{f}w=\tilde{f}W\mathrm{w}\overset{(2.594)}{=}\tilde{f}\frac{H}{L}U\mathrm{w}\\
&-\frac{1}{\rho}\frac{\partial p}{\partial x}=-\frac{1}{\rho}\frac{P}{L}\frac{\partial \mathrm{p}}{\partial \mathrm{x}}\overset{(2.548)}{=}-f_0U\frac{\rho_s}{\rho}\frac{\partial p}{\partial x}\overset{(2.554)}{=}-f_0U\frac{\partial \mathrm{p}}{\partial \mathrm{x}}\\
&\mathscr{F}u=\left(A_{\mathrm{H}}\nabla_{\mathrm{H}}^2+A_{\mathrm{V}}\frac{\partial^2}{\partial z^2}\right)u=U\left(\frac{A_{\mathrm{H}}}{L^2}{\nabla'_{\mathrm{H}}}^2+\frac{A_{\mathrm{V}}}{H^2}\frac{\partial^2}{\partial z^2}\right)\mathrm{u}
\end{aligned} \tag{2.598}$$

Substituting (2.598) into (2.585) and dividing by f_0U yields the non-dimensional equation

$$\varepsilon_T\frac{\partial \mathrm{u}}{\partial \mathrm{t}}+\varepsilon\mathbf{u}\cdot\nabla'\mathrm{u}-\frac{f}{f_0}\mathrm{v}+\frac{\tilde{f}}{f_0}\delta\mathrm{w}=-\frac{\partial \mathrm{p}}{\partial \mathrm{x}}+\mathscr{F}'\mathrm{u} \tag{2.599}$$

where

$$\mathscr{F}':=\frac{1}{2}\left(\mathrm{E_H}{\nabla'_{\mathrm{H}}}^2+\mathrm{E_V}\frac{\partial^2}{\partial \mathrm{z}^2}\right) \tag{2.600}$$

is the *non-dimensional eddy-viscosity operator*. In (2.599), several non-dimensional parameters appear (some of them already met); they are:

- The *temporal Rossby number* (2.362)

$$\varepsilon_T:=\frac{1}{f_0T} \tag{2.601}$$

- The *advective Rossby number* (2.363), already met in Sects. 2.3.2 and 2.4.1,

$$\varepsilon := \frac{U}{f_0 L} \tag{2.602}$$

- The *aspect ratio* (2.359)

$$\delta := \frac{H}{L} \tag{2.603}$$

- The *horizontal Ekman number*

$$\mathsf{E}_\mathrm{H} := 2\frac{A_\mathrm{H}}{f_0 L^2} \tag{2.604}$$

- The *vertical Ekman number*

$$\mathsf{E}_\mathrm{V} := 2\frac{A_\mathrm{V}}{f_0 H^2} \tag{2.605}$$

Moreover, we shall soon use

- The *non-dimensional beta parameter*

$$\beta := \frac{\beta_0 L^2}{U} \tag{2.606}$$

By using (2.261) and (2.262), Eq. (2.599) becomes

$$\varepsilon_T \frac{\partial \mathsf{u}}{\partial \mathsf{t}} + \varepsilon \mathbf{u} \cdot \nabla' \mathsf{u} - \left(1 + \frac{\beta_0 L}{f_0}\mathsf{y}\right)\mathsf{v} + \left(\frac{\beta_0 R}{f_0} - \frac{L}{R}\mathsf{y}\right)\delta\,\mathsf{w} = -\frac{\partial \mathsf{p}}{\partial \mathsf{x}} + \mathscr{F}'\mathsf{u} \tag{2.607}$$

The Rossby numbers and the aspect ratio appearing in (2.607) are easily evaluated to be lesser than $O(1)$ in all the cases of geophysical interest, and the same is assumed to hold for the Ekman numbers of the same equation. On the other hand, it is known that the $O(1)$ dynamic balance of (2.607) is expected to consist in the geostrophic balance $\mathsf{v} = \partial \mathsf{p}/\partial \mathsf{x}$, so the question arises whether the orders of magnitude of the remaining non-dimensional quantities

$$\frac{\beta_0 L}{f_0}\mathsf{y} \qquad \text{and} \qquad \left(\frac{\beta_0 R}{f_0} - \frac{L}{R}\mathsf{y}\right)\delta$$

are consistent with the dominance of the above-cited geostrophic balance.

Since $f_0 = 2\Omega \sin(\phi_0)$ and $\beta_0 = (2\Omega/R)\cos(\phi_0)$, we can write $\beta_0 L/f_0$ as $(L/R)\cot(\phi_0)$. Thus, the geostrophic balance dominates if

$$\frac{\beta_0 L}{f_0} < O(1) \tag{2.608}$$

that is,

$$\frac{L}{R}|\cot(\phi_0)| < O(1) \tag{2.609}$$

Inequality (2.609) poses a constraint on the possible values of the central latitude ϕ_0, which was anticipated in dealing with the Cartesian approximation of the spherical coordinates. The lower bound $|\phi_0| \geq \pi/6$ is mostly taken to satisfy condition (2.609), when $L \leq 10^6$ m.

Note that, in terms of the non-dimensional beta parameter defined in (2.606), inequality (2.608) becomes

$$\beta\,\varepsilon < O(1) \tag{2.610}$$

About the expression $(\beta_0 R/f_0)\,\delta$, which appears in (2.607), note that $\beta_0 R/f_0$ is very close to unity, and thus $(\beta_0 R/f_0)\,\delta \approx \delta$. Moreover, $(L/R)\,\delta = H/R$; so

$$\left(\frac{\beta_0 R}{f_0} - \frac{L}{R}\mathsf{y}\right)\delta \approx \delta - \frac{H}{R}\mathsf{y} \tag{2.611}$$

The dependence on y of $\delta - (H/R)\mathsf{y}$ is about δ times weaker than that of $(\beta_0 L/f_0)\,\mathsf{y}$, because

$$\frac{H/R}{\beta_0 L/f_0} = \frac{f_0}{\beta_0 R}\delta \approx \delta$$

Hence, the r.h.s. of (2.611) can be approximated by δ, while retaining $1 + (\beta_0 L/f_0)\,\mathsf{y}$ in (2.607) as it stands.

In conclusion, Eq. (2.607) takes the final form

$$\boxed{\varepsilon_T\,\frac{\partial \mathsf{u}}{\partial \mathsf{t}} + \varepsilon\,\mathbf{u}\cdot\nabla'\mathsf{u} - (1 + \beta\,\varepsilon\,\mathsf{y})\,\mathsf{v} + \delta\,\mathsf{w} = -\frac{\partial \mathsf{p}}{\partial \mathsf{x}} + \mathscr{F}'\,\mathsf{u}} \tag{2.612}$$

Meridional Component of the Momentum Equation

Along the same line as above, the following non-dimensional equation for the meridional motion

$$\boxed{\varepsilon_T\,\frac{\partial \mathsf{v}}{\partial \mathsf{t}} + \varepsilon\,\mathbf{u}\cdot\nabla'\mathsf{v} + (1 + \beta\,\varepsilon\mathsf{y})\,\mathsf{u} = -\frac{\partial \mathsf{p}}{\partial \mathsf{y}} + \mathscr{F}'\,\mathsf{v}} \tag{2.613}$$

can be deduced from (2.586).

Vertical Component of the Momentum Equation

We preliminarily expand the quantity $-\rho^{-1}\partial p/\partial z - g$ that appears at the r.h.s. of (2.587). Recalling (2.554) and (2.556), we have, neglecting higher-order terms in r,

$$-\frac{1}{\rho} \overset{(2.554)}{=} -\frac{1}{\rho_s(1+\mathsf{r}\rho')} = -\frac{1}{\rho_s}(1-\mathsf{r}\rho')$$

and

$$\frac{\partial p}{\partial z} \overset{(2.556)}{=} \frac{dp_s}{dz} + \mathsf{r}g\frac{\partial}{\partial \mathsf{z}}(\rho_s \mathsf{p}) \overset{(2.531)}{=} \rho_s\left(-g + \mathsf{r}\frac{g}{\rho_s}\frac{\partial}{\partial \mathsf{z}}(\rho_s \mathsf{p})\right)$$

Hence, we get

$$-\frac{1}{\rho}\frac{\partial p}{\partial z} - g = -\mathsf{r}g\left(\rho' + \frac{1}{\rho_s}\frac{\partial}{\partial \mathsf{z}}(\rho_s \mathsf{p})\right)$$

that is,

$$-\frac{1}{\rho}\frac{\partial p}{\partial z} - g \overset{(2.546)}{=} -\frac{f_0 UL}{H}\left(\rho' + \frac{1}{\rho_s}\frac{\partial}{\partial \mathsf{z}}(\rho_s \mathsf{p})\right) \tag{2.614}$$

In terms of the non-dimensional quantities (2.533) and (2.534), and using (2.614), Eq. (2.587) becomes

$$\frac{UH}{LT}\frac{\partial \mathsf{w}}{\partial \mathsf{t}} + \frac{U^2H}{L^2}\mathbf{u}\cdot\nabla'\mathsf{w} - \tilde{f}U\mathsf{u} =$$
$$= -\frac{f_0 UL}{H}\left(\rho' + \frac{1}{\rho_s}\frac{\partial}{\partial \mathsf{z}}(\rho_s \mathsf{p})\right) + \frac{A_H UH}{L^3}\nabla_H'^2\mathsf{w} + \frac{A_V U}{HL}\frac{\partial^2 \mathsf{w}}{\partial \mathsf{z}^2} \tag{2.615}$$

Then, multiplication of each term of (2.615) by H/f_0UL makes (2.615) non-dimensional, so positions (2.601)–(2.605) can be applied, together with the known approximation $\delta\tilde{f}/f_0 \approx \delta$. Thus, the non-dimensional vertical equation in its final form is

$$\boxed{\delta^2\left(\varepsilon_T\frac{\partial}{\partial \mathsf{t}} + \varepsilon\mathbf{u}\cdot\nabla'\right)\mathsf{w} - \delta\mathsf{u} = -\frac{1}{\rho_s}\frac{\partial}{\partial \mathsf{z}}(\rho_s \mathsf{p}) - \rho' + \delta^2\mathscr{F}'\mathsf{w}} \tag{2.616}$$

The O(1) balance in (2.616) is the same as (2.561), in accordance with the prevailing hydrostatic equilibrium of large-scale geophysical flows.

Non-dimensional Thermodynamic Equations

In the adiabatic case, the procedures to make non-dimensional the thermodynamic equations for the sea and the atmosphere are amenable to a unified treatment. In fact, not only the form of a conservation principle is common to both of them, but also the form of the conserved quantities (i.e., the density of water and the potential temperature of air) is the same.

Explicitly, the starting dimensional equations derive from (2.588) with (2.554), and from (2.589) with (2.560), to obtain

$$\frac{D}{Dt}[\rho_s(1+r\rho')] = 0 \tag{2.617}$$

$$\frac{D}{Dt}[\theta_s(1+r\theta')] = 0 \tag{2.618}$$

respectively, where $r = \varepsilon F$. The related non-dimensional versions must be determined.

Unified Approach for Ocean and Atmosphere

In order to write (2.617) and (2.618) in non-dimensional form, it is convenient to represent both $\rho_s(z)$ and $\theta_s(z)$ by means of a single symbol, say, $b_s(z)$, and, analogously, $\rho'(x, y, z, t)$ and $\theta'(x, y, z, t)$ through $b(x, y, z, t)$, to obtain

$$\frac{D}{Dt}[b_s(1+rb)] = 0 \tag{2.619}$$

in place of (2.617) and (2.618). Hence,

$$r\frac{b_s}{T}\frac{\partial b}{\partial t} + (1+rb)w\frac{db_s}{dz} + rb_s\frac{U}{L}\mathbf{u}\cdot\nabla' b = 0 \tag{2.620}$$

because

$$\frac{Db_s}{Dt} = w\frac{db_s}{dz}$$

$$\frac{Db}{Dt} = \frac{1}{T}\frac{\partial b}{\partial t} + \frac{U}{L}\mathbf{u}\cdot\nabla' b$$

Using the approximation $1 + rb \approx 1$ and setting $w = (UH/L)w$, Eq. (2.620) becomes

$$r\frac{b_s}{T}\frac{\partial b}{\partial t} + \frac{UH}{L}w\frac{db_s}{dz} + rb_s\frac{U}{L}\mathbf{u}\cdot\nabla' b = 0 \tag{2.621}$$

that is to say, recalling that $r = \varepsilon F = U f_0 L/gH$,

$$\frac{H^2}{f_0^2 L^2}\frac{g}{b_s}\frac{db_s}{dz}w + \varepsilon_T\frac{\partial b}{\partial t} + \varepsilon\mathbf{u}\cdot\nabla' b = 0 \tag{2.622}$$

Differences Between Ocean and Atmosphere

The quantity $(g/b_\mathrm{s})\,\mathrm{d}b_\mathrm{s}/\mathrm{d}z$ in the first term of (2.622) is N_s^2, the buoyancy frequency square of the standard fluid, which is negative definite in the case of the ocean and positive definite in the case of the atmosphere. The notations, already introduced in (2.96) and (2.97),

$$\frac{g}{b_\mathrm{s}}\frac{\mathrm{d}b_\mathrm{s}}{\mathrm{d}z} := \frac{g}{\rho_\mathrm{s}}\frac{\mathrm{d}\rho_\mathrm{s}}{\mathrm{d}z} =: -N_\mathrm{s}^2 \qquad \text{(ocean)} \qquad (2.623)$$

$$\frac{g}{b_\mathrm{s}}\frac{\mathrm{d}b_\mathrm{s}}{\mathrm{d}z} := \frac{g}{\theta_\mathrm{s}}\frac{\mathrm{d}\theta_\mathrm{s}}{\mathrm{d}z} =: N_\mathrm{s}^2 \qquad \text{(atmosphere)} \qquad (2.624)$$

are adopted separately in the two cases; moreover, the so-called *stratification parameter* S is introduced through the position

$$\mathsf{S} := \left(\frac{HN_\mathrm{s}}{f_0 L}\right)^2 \qquad (2.625)$$

Therefore, on the whole, in the ocean case, Eqs. (2.622), (2.623) and (2.625) give

$$\boxed{\mathsf{w} = \frac{\varepsilon_\mathrm{T}}{\mathsf{S}}\frac{\partial\rho'}{\partial\mathsf{t}} + \frac{\varepsilon}{\mathsf{S}}\mathbf{u}\cdot\nabla'\rho'} \qquad \text{(ocean)} \qquad (2.626)$$

while, in the atmospheric case, from (2.622), (2.624) and (2.625), the equation

$$\boxed{\mathsf{w} = -\frac{\varepsilon_\mathrm{T}}{\mathsf{S}}\frac{\partial\theta'}{\partial\mathsf{t}} - \frac{\varepsilon}{\mathsf{S}}\mathbf{u}\cdot\nabla'\theta'} \qquad \text{(atmosphere)} \qquad (2.627)$$

follows. If $\varepsilon_\mathrm{T} = \varepsilon$, Eq. (2.626) takes the concise form

$$\boxed{\mathsf{w} = \frac{\varepsilon}{\mathsf{S}}\frac{\mathsf{D}\rho'}{\mathsf{Dt}}} \qquad \text{(ocean)} \qquad (2.628)$$

and analogously (2.627) can be written as

$$\boxed{\mathsf{w} = -\frac{\varepsilon}{\mathsf{S}}\frac{\mathsf{D}\theta'}{\mathsf{Dt}}} \qquad \text{(atmosphere)} \qquad (2.629)$$

where $\mathsf{D}/\mathsf{Dt} := \partial/\partial\mathsf{t} + \mathbf{u}\cdot\nabla'$.

To summarize, thermodynamic equations (2.626) and (2.628) are non-dimensional versions of (2.617), which is valid for the ocean, while thermodynamic equations (2.627) and (2.629) are non-dimensional versions of (2.618), which is valid for the atmosphere.

Thermodynamic Equation (2.589)

Equation (2.589) is equivalent to

$$\frac{1}{\theta}\frac{\mathrm{D}\theta}{\mathrm{D}t}=\frac{\dot{Q}}{c_P T_\mathrm{a}} \tag{2.630}$$

where T_a denotes the absolute temperature. The l.h.s. of the previous equation may be expanded as follows:

$$\begin{aligned}\frac{1}{\theta}\frac{\mathrm{D}\theta}{\mathrm{D}t} &\overset{(2.560)}{=} \frac{\mathsf{r}}{1+\mathsf{r}\theta'}\frac{U}{L}\frac{\mathsf{D}\theta'}{\mathsf{Dt}}+\frac{1}{\theta_s}\frac{\mathrm{d}\theta_\mathrm{s}}{\mathrm{d}z}w \overset{(2.624)}{=} \\ &= \frac{\mathsf{r}}{1+\mathsf{r}\theta'}\frac{U}{L}\frac{\mathsf{D}\theta'}{\mathsf{Dt}}+\frac{N_\mathrm{s}^2}{g}w \overset{\mathsf{r}\ll 1}{\approx} \\ &\approx \mathsf{r}\frac{U}{L}\frac{\mathsf{D}\theta'}{\mathsf{Dt}}+\frac{N_\mathrm{s}^2}{g}w \overset{(2.625)}{=} \\ &= \frac{HN_\mathrm{s}^2U}{Lg}\left(\mathsf{w}+\frac{\varepsilon}{\mathsf{S}}\frac{\mathsf{D}\theta'}{\mathsf{Dt}}\right)\end{aligned} \tag{2.631}$$

Therefore, Eq. (2.630) can be written as

$$\mathsf{w}+\frac{\varepsilon}{\mathsf{S}}\frac{\mathsf{D}\theta'}{\mathsf{Dt}}=\frac{Lg}{HN_\mathrm{s}^2U}\frac{\dot{Q}}{c_p T_\mathrm{a}} \tag{2.632}$$

Finally, by resorting to the identity

$$\frac{Lg}{HN_\mathrm{s}^2U}=\frac{\varepsilon}{\mathsf{S}}\frac{gH}{f_0U^2}$$

Eq. (2.632) takes the final form

$$\boxed{\mathsf{w}+\frac{\varepsilon}{\mathsf{S}}\left(\frac{\mathsf{D}\theta'}{\mathsf{Dt}}-\frac{gH}{f_0U^2}\frac{\dot{Q}}{c_p T_\mathrm{a}}\right)=0} \tag{2.633}$$

The non-dimensional equation (2.633) is consistent, provided that the term

$$\dot{\mathsf{Q}}=\frac{gH}{f_0U^2}\frac{\dot{Q}}{c_p T_\mathrm{a}} \tag{2.634}$$

be (at most) comparable with the $O(1)$ derivative of the potential temperature appearing in the same equation. Thus, the upper bound

$$\dot{Q}\leq O\left(c_p T_\mathrm{a}\frac{f_0U^2}{gH}\right) \tag{2.635}$$

is expected to hold for the heating rate $\dot{Q}$. In the case of the troposphere, $\dot{Q}$ can be evaluated recalling (see Sect. 2.1.2) that $c_p = 1{,}004.5\,\mathrm{J\cdot kg^{-1}\cdot K^{-1}}$ and using $T_a = 288\,\mathrm{K}$ as a representative temperature at the sea level. Moreover, in accordance with the scaling reported in the following Sect. 5.2.1, the quantities $U = O(10\,\mathrm{m/s})$ and $H = O(10^4\,\mathrm{m})$ are also used into the r.h.s. of (2.635) to yield

$$\dot{Q} \leq O(10^2\,\mathrm{J\cdot kg^{-1}\cdot s^{-1}}) \tag{2.636}$$

Indeed, generally, the accession of heat is far smaller than the $O(1)$ non-dimensional time growth rate of potential vorticity; therefore, the diabatic contribution may be neglected or, alternatively, $\dot{Q}$ may enter into (2.633) as a perturbation of an adiabatic unperturbed system. Just the latter point of view will be followed in Sect. 5.2 in dealing with thermally forced atmospheric waves.

On the Role of the Non-dimensional Governing Equations

General Approach to Parameterized Equations

The non-dimensional governing equations, which have been previously inferred, are concerned with the incompressibility, the evolution and the thermodynamics of the large-scale geophysical flows. In the momentum and thermodynamics equations, besides the coordinates and the fields, also a set of non-dimensional parameters appear; they are:

$$\varepsilon_T := \frac{1}{f_0 T} \qquad \varepsilon := \frac{U}{f_0 L} \qquad \beta\,\varepsilon := \frac{\beta_0 L}{f_0} \qquad \delta := \frac{H}{L} \qquad \mathsf{S} := \left(\frac{N_s H}{f_0 L}\right)^2 \tag{2.637}$$

together with $\mathsf{E_H}$ and $\mathsf{E_V}$. The above-listed parameters turn out to be (or are assumed to be, in the case of $\mathsf{E_H}$ and $\mathsf{E_V}$) lesser than unity. As a consequence, the $O(1)$ momentum balance (subscript 0) consists in the diagnostic (i.e. time-independent) equations

$$\begin{aligned} \mathsf{u_0} &= -\frac{\partial \mathsf{p_0}}{\partial \mathsf{y}} \\ \mathsf{v_0} &= \frac{\partial \mathsf{p_0}}{\partial \mathsf{x}} \\ \mathsf{w_0} &= 0 \\ \rho_0' &= -\frac{1}{\rho_s}\frac{\partial}{\partial \mathsf{z}}(\rho_s\,\mathsf{p_0}) \end{aligned} \tag{2.638}$$

which, as expected, coincide with the geostrophic balance and the hydrostatic equilibrium.

What is the role of the remaining terms of the momentum equations (for instance, that of acceleration) and the remaining set of governing equations (for instance, the conservation of potential temperature)? The role is that of providing the prognostic (i.e. time-dependent) equations for the same current, pressure and density fields appearing in (2.638). In this way, it is possible to determine the evolution laws of the geostrophic fields, and, for this reason, the related dynamics is named *quasi-geostrophic*. Quasi-geostrophic equations are the result of the application of these ideas to (more or less complex) models of physical systems, as the following chapters will show with full details. However, in order to explain with symbols how the method works in principle, here we point out a very simple mathematical analogy, quite disconnected with the physics of fluids.

Usually, each order of magnitude of the parameters in (2.637) may be expressed as an integer non-negative power of one of them, which is then called *ordering parameter*. This fact allows us to substitute each parameter with a power of the ordering parameter and, hence, to integrate the governing equations through an asymptotic expansion in that single parameter. This approach is illustrated in the following simple example, characterized by just one parameter. More realistic applications will be given in Sect. 3.1.1 about shallow-water systems and in Chap. 5 about stratified flows in the ocean and atmosphere.

Example: A One-Parameter Toy Model

We postulate the system of equations

$$\begin{cases} (\mathsf{a}+\mathsf{b})+\varepsilon\frac{\partial \mathsf{a}}{\partial \mathsf{t}}+\varepsilon^2\mathsf{f}(\mathsf{b}) & =0 \\ \frac{\partial}{\partial \mathsf{x}}(\mathsf{a}+\mathsf{b})+\varepsilon\mathsf{a}+\varepsilon^2\mathsf{g}(\mathsf{a}) & =0 \end{cases} \tag{2.639}$$

where f and g are bounded functions, $0<\varepsilon\ll 1$ plays the role of the ordering parameter, while $\mathsf{a}=\mathsf{a}(\mathsf{x},\mathsf{t};\varepsilon)$ and $\mathsf{b}=\mathsf{b}(\mathsf{x},\mathsf{t};\varepsilon)$. We suppose that system (2.639) is so complex as to induce us to look for a solution expanded in powers of ε, that is,

$$\mathsf{a}=\sum_{n=0}^{\infty}\varepsilon^n\,\mathsf{a}_n(\mathsf{x},\mathsf{t}) \tag{2.640}$$

$$\mathsf{b}=\sum_{n=0}^{\infty}\varepsilon^n\,\mathsf{b}_n(\mathsf{x},\mathsf{t}) \tag{2.641}$$

whence

$$\mathsf{f}(\mathsf{b})=\sum_{n=0}^{\infty}\varepsilon^n\,\mathsf{f}_n(\mathsf{x},\mathsf{t}) \tag{2.642}$$

$$\mathsf{g}(\mathsf{a})=\sum_{n=0}^{\infty}\varepsilon^n\,\mathsf{g}_n(\mathsf{x},\mathsf{t}) \tag{2.643}$$

Substitution of (2.640)–(2.643) into system (2.639) yields

$$a_0 + b_0 + \varepsilon \left(\frac{\partial a_0}{\partial t} + a_1 + b_1 \right) + \sum_{m=2}^{\infty} \varepsilon^m \left(f_{m-2} + \frac{\partial a_{m-1}}{\partial t} + a_m + b_m \right) = 0$$

and

$$\frac{\partial}{\partial x}(a_0 + b_0) + \varepsilon \left[a_0 + \frac{\partial}{\partial x}(a_1 + b_1) \right] + \sum_{m=2}^{\infty} \varepsilon^m \left[g_{m-2} + a_{m-1} + \frac{\partial}{\partial x}(a_m + b_m) \right] = 0$$

whence, by equating terms of the same order in ε, we obtain to the leading order

$$\begin{cases} a_0 + b_0 & = 0 \\ \frac{\partial}{\partial x}(a_0 + b_0) & = 0 \end{cases} \tag{2.644}$$

and, to the first order,

$$\begin{cases} \frac{\partial a_0}{\partial t} + a_1 + b_1 & = 0 \\ a_0 + \frac{\partial}{\partial x}(a_1 + b_1) & = 0 \end{cases} \tag{2.645}$$

System (2.644) gives the diagnostic equation

$$a_0 + b_0 = 0 \tag{2.646}$$

which states that a_0 and b_0 are opposite, but no information about their evolution is available from (2.644) or (2.646). Unlike (2.644), system (2.645) allows the elimination of $a_1 + b_1$, as a whole, in favour of a_0 and b_0, thus obtaining the following evolution (i.e., prognostic) equation for a_0:

$$\frac{\partial^2 a_0}{\partial t \partial x} = a_0 \tag{2.647}$$

Moreover, from (2.646) and (2.647), we get the evolution equation for b_0:

$$\frac{\partial^2 b_0}{\partial t \partial x} = b_0 \tag{2.648}$$

which has the same form as (2.647). We stress that elimination of $a_1 + b_1$ in favour of the sole a_0 and b_0 is not possible for every set of starting equations, but it is indeed possible for all the equations involved in GFD.

The general integral of (2.647) is given by a superposition of terms of the kind

$$a_0 = \sum_k a_k \exp\left[i \left(-kx + \frac{t}{k} \right) \right] \tag{2.649}$$

and, in the same way,

$$b_0 = \sum_k B_k \exp\left[i\left(-kx + \frac{t}{k}\right)\right] \tag{2.650}$$

The set of coefficients A_k and B_k is determined by initial conditions $a(x, 0; \varepsilon)$ and $B(x, 0; \varepsilon)$, but these details go beyond the scope of the example reported here.

To summarize, Eq. (2.646) is (only mathematically) the analogues of the geostrophic balance expressed by the first two equations of (2.638), while (2.647) and (2.648) are the analogues of the quasi-geostrophic equations which the reader will meet in subsequent chapters.

Exercises

1. Evaluate the vertical profile of the standard potential temperature $\theta_s(z)$, given by (2.95), and plot the result.
   (*Hint*: Recall (2.91) and (2.119)).

2. Show that the centrifugal acceleration (2.284) can be written as the gradient of a scalar.
   (*Hint*: Prove the identity $\boldsymbol{\Omega} \times (\boldsymbol{\Omega} \times \mathbf{r}) = -(1/2)\nabla|\boldsymbol{\Omega} \times \mathbf{r}|^2$).

3. Introduce in the momentum equation of long gravity waves a frictional term that transforms the singularity of $\bar{E}$ appearing in (2.449) with a maximum of that same quantity.

4. We know that Eq. (2.163) and definition (2.164) derive from (2.136) and (2.137) together with position (2.156); analogously, we know that Eq. (2.557) and definition (2.559) derive from (2.547) and (2.548) together with position (2.546).

   Verify that Eqs. (2.163) and (2.557) can be inferred in a unified way by scaling perturbation pressure as $P = \lambda\, \rho_s L g\, \delta$, and density anomaly as $D = \lambda\, \rho_s$, where $\lambda = U/(g\, \tau\, \delta) < O(1)$ with τ a timescale representative of T and f_0^{-1} while $\delta = H/L$.

   Hence, prove that
   $$\theta \approx \theta_s\,(1 + \lambda\, \theta')$$
   with
   $$\theta' = \frac{\rho_s L g\, \delta}{\gamma\, p_s}\, p - \rho'$$

5. Consider a stationary flow in which the amount of local acceleration is of the same order as Coriolis acceleration, and find the consequent relationship between the length scale L and the velocity U of this flow.

   Can L and U refer to a realistic flow? Give an example.

Bibliographical Note

The constitutive equations presented in this chapter are mainly based on contributions by Batchelor (1967), Holton (1979), Pedlosky (1987) and Salby (1996), but we cite also Cushman-Roisin (1994), Gill (1982), Pond and Pickard (1983) and Peixoto and Oort (1992) on the same subject. Internal gravity waves are a widely treated argument, and many details are found in LeBlond and Mysak (1978) and in Pedlosky (2003), while Cushman-Roisin (1994) and Holton (1979) expound them more concisely; so, these books are useful for a first reading. Rotating flows are a central subject of Geophysical Fluid Dynamics, which is found in almost all of the related books. Here we wish to mention Krauss (1973), where rotating flows are explained in a very detailed way. Long gravity waves are a classical argument of physical oceanography since Proudman (1953); see also Gill (1982), Krauss (1973), Pedlosky (1987) and Pedlosky (2003). Foundations of potential vorticity are given by Ertel (1942a,b,c) and Rossby and collaborators (1939), while applications of Ertel's theorem are revisited by Müller (1995). See also Vallis (2006). Jeffrey's theorem is reported by Hide (1978). The governing equations of geophysical flows are treated in all the books of Geophysical Fluid Dynamics and Climate Dynamics; the approach followed here is close to that of Pedlosky (1987) and Vallis (2006).

Part II
Applications

The atmospheric and oceanic sciences are sometimes thought of as not being "beautiful" in the same way as some branches of theoretical physics. Yet surely quasi-geostrophic theory, and the quasi-geostrophic potential vorticity equation, are quite beautiful, both for their austerity of description and richness of behaviour.

Geoffrey K. Vallis (2006)

Chapter 3
Quasi-Geostrophic Single-Layer Models

Abstract The crudest representation of currents and winds is that given by a single, constant-density fluid layer in relative motion with respect to the rotating Earth. In the case of the ocean, the layer is bounded from below by the sea floor and from above by the free surface of the sea. In the case of the atmosphere, the layer is bounded from below by the ground and from above by a hypothetical surface, above which the density of the atmosphere goes abruptly to zero. In both cases, the horizontal pressure gradient arises from the modulation of such surfaces with respect to the geoid, while the Coriolis force tends to arrange the flow in geostrophic balance with the pressure gradient.

The resulting motion is inertial, and potential vorticity is conserved. If Earth's curvature is taken into account, for instance, by considering the beta-plane approximation, a fundamental consequence of potential vorticity conservation for motions crossing circles of latitude is the formation of Rossby waves. The above-described model is also the first step towards the formulation of the homogeneous model of wind-driven ocean circulation in which both the wind-stress forcing and vorticity dissipation are taken into account. In this case, the modulation of the surfaces which sandwich the geostrophic layer is ascribed, according to the classical Ekman theory, to the convergence/divergence of the marine current just below the free surface, caused by the wind stress, and just above the sea floor in the benthic layer, caused by friction. Also, the lateral diffusion of relative vorticity in the interior may be considered as a dissipative mechanism.

The solutions of the homogeneous model explain many fundamental phenomena of large-scale ocean circulation, in spite of the very simple picture which is usually adopted in describing the structure of the planetary wind field over the oceans. It is worth noting that the Ekman model of the benthic layer applies also, as it stands, to the lower atmosphere from the ground up to about one kilometre. The related convergence/divergence of the wind is responsible of the vertical motion of air masses, whose vicissitudes influence the weather.

F. Cavallini and F. Crisciani, *Quasi-Geostrophic Theory of Oceans and Atmosphere*,
Atmospheric and Oceanographic Sciences Library 45, DOI 10.1007/978-94-007-4691-6_3,

3.1 Shallow-Water Model

The quasi-geostrophic, *shallow-water model* deals with the inertial dynamics of a constant-density single-layer fluid, bounded from above by a free surface, from below by a rigid bottom (in case, spatially modulated) and evolving under the influence of Earth's rotation.

In spite of its simplicity, due to a plain physical idealization, the model is able to give some remarkable insights into the behaviour of large-scale geophysical flows, and it is the natural starting point to explore more sophisticated quasi-geostrophic systems.

The use of the names "water" and "bottom" is conventional, in the sense that a shallow-water "atmosphere" above a certain "topography" can be modelled as well.

3.1.1 The Quasi-Geostrophic, Shallow-Water Model

Basic Assumptions of the Model and Its Governing Equations

A constant fluid density and the absence of dissipative mechanisms imply that, like in long gravity waves, the flow is adiabatic and the thermodynamics is identically satisfied. Moreover, no density anomaly exists so, on the whole, the non-dimensional governing equations inferred from (2.593), (2.612), (2.613) and (2.616) take the form

$$\frac{\partial u}{\partial x}+\frac{\partial v}{\partial y}+\frac{\partial w}{\partial z}=0 \tag{3.1}$$

$$\varepsilon_T\frac{\partial u}{\partial t}+\varepsilon(\mathbf{u}\cdot\nabla')u-(1+\beta\varepsilon y)v+\delta w=-\frac{\partial p}{\partial x} \tag{3.2}$$

$$\varepsilon_T\frac{\partial v}{\partial t}+\varepsilon(\mathbf{u}\cdot\nabla')v+(1+\beta\varepsilon y)u=-\frac{\partial p}{\partial y} \tag{3.3}$$

$$\delta^2\varepsilon_T\frac{\partial w}{\partial t}+\delta^2\varepsilon(\mathbf{u}\cdot\nabla')w-\delta u=-\frac{\partial p}{\partial z} \tag{3.4}$$

The energetics associated to (3.1)–(3.4) is obtained multiplying (3.2) by u, (3.3) by v, (3.4) by w and adding the resulting equations, to obtain

$$\left(\frac{\varepsilon_T}{2}\frac{\partial}{\partial t}+\frac{\varepsilon}{2}\mathbf{u}\cdot\nabla'\right)(u^2+v^2+\delta^2w^2)=-u\frac{\partial p}{\partial x}-v\frac{\partial p}{\partial y}-w\frac{\partial p}{\partial z}$$

that is to say, because of (3.1) and the consequent identity $\mathbf{u}\cdot\nabla'p=\mathrm{div}(p\mathbf{u})$,

$$\left(\frac{\varepsilon_T}{2}\frac{\partial}{\partial t}+\frac{\varepsilon}{2}\mathbf{u}\cdot\nabla'\right)(u^2+v^2+\delta^2w^2)=-\mathrm{div}(p\mathbf{u}) \tag{3.5}$$

The dimensional version of (3.5), with $\varepsilon_T = \varepsilon$, expresses the local energy balance

$$\frac{\rho_s}{2}\frac{D}{Dt}(u^2+v^2+w^2) = -\mathrm{div}\,(\tilde{p}\,\mathbf{u}) \tag{3.6}$$

where $\tilde{p} := p - p_s(z)$. A manipulation of the r.h.s. of (3.6), recalling that $-\mathrm{div}\,(p_s\,\mathbf{u}) = g\rho\,\mathrm{D}z/\mathrm{D}t$, leads to the mechanical energy equation established in (1.175), with $A_V = 0$.

We stress that the same Eqs. (3.5) and (3.6) follow even if both the terms $\delta\,\mathsf{w}$ of (3.2) and $-\delta\,\mathsf{u}$ of (3.4) are removed from these equations (the check is trivial). On the contrary, the removal of only one of them would lead to an incorrect energy balance, involving an unphysical extra term due to the presence of the term still present in (3.2) or in (3.4). Therefore, the possible cancellation of $\delta\,\mathsf{w}$ from (3.2) on scaling grounds demands also the cancellation of $-\delta\,\mathsf{u}$ from (3.4) and vice versa. This fact, known as *traditional approximation*, will be useful to simplify the set of governing equations; but to apply this approximation, a unique ordering parameter[1] must be singled out among ε, ε_T, δ and $\beta\,\varepsilon$ by expressing the remaining parameters as a positive power of the ordering one.

For instance, in the ocean case, the orders of magnitude $L = O(10^5\,\mathrm{m})$, $H = O(10^3\,\mathrm{m})$ and $U = O(10^{-1}\,\mathrm{m/s})$, together with the assumption that the *advective timescale* $T_{\mathrm{adv}} := L/U$ is the same as the local timescale T_{loc}, yield

$$\varepsilon = \varepsilon_T = \delta = 10^{-2} \qquad \beta = 1 \qquad \text{(ocean)} \tag{3.7}$$

For the troposphere, one takes $L = O(10^6\,\mathrm{m})$, $H = O(10^4\,\mathrm{m})$ and $U = O(10\,\mathrm{m/s})$; so, retaining again the assumption $\varepsilon = \varepsilon_T$, one obtains

$$\varepsilon = \varepsilon_T = 10^{-1} \qquad \delta = \varepsilon^2 = 10^{-2} \qquad \beta = 1 \qquad \text{(troposphere)} \tag{3.8}$$

In both cases (ocean and troposphere), the Froude number $\mathsf{F} = O(\varepsilon)$. The difference between (3.7) and (3.8) lies only in the relation between δ and ε, but this is not relevant in what follows; so, only the set (3.7) is hereafter considered. Thus, the governing equations become:

$$\frac{\partial \mathsf{u}}{\partial \mathsf{x}} + \frac{\partial \mathsf{v}}{\partial \mathsf{y}} + \frac{\partial \mathsf{w}}{\partial \mathsf{z}} = 0 \tag{3.9}$$

$$\varepsilon\frac{\partial \mathsf{u}}{\partial \mathsf{t}} + \varepsilon\,(\mathbf{u}\cdot\nabla')\,\mathsf{u} - (1+\beta\,\varepsilon\,\mathsf{y})\,\mathsf{v} + \varepsilon\,\mathsf{w} = -\frac{\partial \mathsf{p}}{\partial \mathsf{x}} \tag{3.10}$$

$$\varepsilon\frac{\partial \mathsf{v}}{\partial \mathsf{t}} + \varepsilon\,(\mathbf{u}\cdot\nabla')\,\mathsf{v} + (1+\beta\,\varepsilon\,\mathsf{y})\,\mathsf{u} = -\frac{\partial \mathsf{p}}{\partial \mathsf{y}} \tag{3.11}$$

$$\varepsilon^3\frac{\partial \mathsf{w}}{\partial \mathsf{t}} + \varepsilon^3\,(\mathbf{u}\cdot\nabla')\,\mathsf{w} - \varepsilon\,\mathsf{u} = -\frac{\partial \mathsf{p}}{\partial \mathsf{z}} \tag{3.12}$$

[1] See p. 153 for a definition of *ordering parameter*.

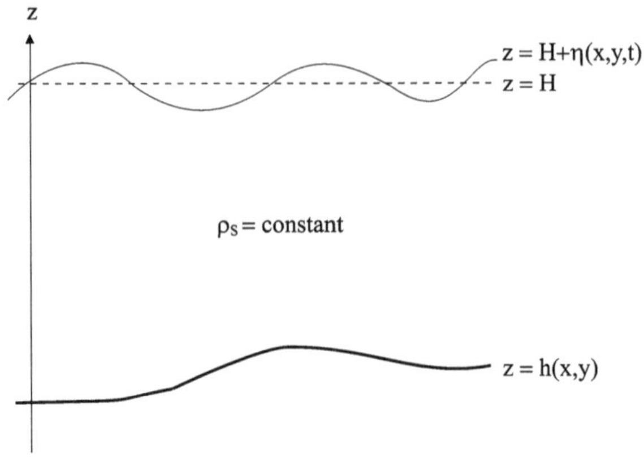

Fig. 3.1 Geometry of the shallow-water model according to (3.13)

Equations (3.9)–(3.12) show that the velocity and pressure fields depend on position, time and the small parameter ε: accordingly, we shall expand these fields in powers of ε once these equations are suitably simplified. We anticipate that, just like in the toy model presented at the end of Sect. 2.4.2, the first two powers of ε are sufficient to derive a prognostic model at the geostrophic level of approximation.

Perturbation Pressure, Free-Surface Elevation and Vertical Velocity

Further simplifications of the momentum equations (3.10) and (3.12) presuppose definite hypotheses about the kinematics of the fluid layer supporting the shallow-water model. To this purpose, we start from a dimensional reference and subsequently consider the non-dimensional counterpart of it. Each fluid column is included in the interval

$$h(x,y) \leq z \leq H + \eta(x,y,t) \tag{3.13}$$

where $z = h(x,y)$ represents the profile of the bottom ($z = 0$ in the case of a flat bottom), H is the typical thickness of the layer, coinciding with the depth of the motion and η (such that $|\eta| \leq H$) is the free-surface elevation at the summit of the column. The spatial modulation of η is responsible of the geostrophic motion, according to (2.436). A sketch of (3.13) is shown in Fig. 3.1. Once z is scaled as

$$z = H\mathsf{z} \tag{3.14}$$

and η by

$$\eta = \frac{f_0 U L}{g}\eta' \tag{3.15}$$

(see (2.386)), the non-dimensional version of interval (3.13) takes the form

$$\frac{h}{H} \leq \mathsf{z} \leq 1 + \varepsilon \mathsf{F} \eta' \tag{3.16}$$

in accordance with (2.388). The vertical velocity at the bottom of the fluid column must be examined with caution. Position

$$\frac{h}{H} = \mathsf{B}\,\mathsf{h} \tag{3.17}$$

where B is representative of the order of magnitude of h/H while h is the non-dimensional bottom modulation, is substituted into boundary condition (1.28) to give $(UH/L)\mathsf{w}(\mathsf{h}) = (UH\mathsf{B}/L)\mathbf{u}\cdot\nabla'\mathsf{h}$, whence

$$\mathsf{w}(\mathsf{h}) = O(\mathsf{B}) \tag{3.18}$$

Since $\mathsf{w}_0 = 0$ (see (2.638)), the expansion of w in powers of ε implies

$$\mathsf{w} \leq O(\varepsilon) \tag{3.19}$$

In the rest of this section, we shall pay special attention to the vertical dynamics, in which

$$\mathsf{w} = O(\varepsilon) \tag{3.20}$$

Thus, (3.18) and (3.20) imply

$$\mathsf{B} = \varepsilon \tag{3.21}$$

whence

$$h = \varepsilon H \mathsf{h} \tag{3.22}$$

Inequality (3.19) is a consequence of the geostrophic dynamics, which does not allow to model flows on marked topography/bathymetry; this is a typical limitation of this kind of approximation. For B even smaller than ε, the flat bottom condition would be again recovered.

Owing to (3.20), the term $\varepsilon\mathsf{w}$ of (3.10) is $O(\varepsilon^2)$ and therefore can be disregarded with respect to the $O(1)$ and $O(\varepsilon)$ terms of the same equation. But, at this point, the traditional approximation demands also the cancellation of $-\varepsilon\mathsf{u}$ from (3.12), so, finally, we have the leading-order equations

$$\frac{\partial \mathsf{u}}{\partial \mathsf{x}} + \frac{\partial \mathsf{v}}{\partial \mathsf{y}} + \frac{\partial \mathsf{w}}{\partial \mathsf{z}} = 0 \tag{3.23}$$

$$\varepsilon\frac{\partial \mathsf{u}}{\partial \mathsf{t}} + \varepsilon(\mathbf{u}\cdot\nabla')\mathsf{u} - (1+\beta\varepsilon\,\mathsf{y})\,\mathsf{v} = -\frac{\partial \mathsf{p}}{\partial \mathsf{x}} \tag{3.24}$$

$$\varepsilon\frac{\partial \mathsf{v}}{\partial \mathsf{t}} + \varepsilon(\mathbf{u}\cdot\nabla')\mathsf{v} + (1+\beta\varepsilon\,\mathsf{y})\,\mathsf{u} = -\frac{\partial \mathsf{p}}{\partial \mathsf{y}} \tag{3.25}$$

$$\frac{\partial \mathsf{p}}{\partial \mathsf{z}} = 0 \tag{3.26}$$

Equation (3.26) shows that, due to the smallness of ε, the vertical acceleration does not affect significantly the vertical derivative of the perturbation pressure; thus, like in long gravity waves, pressure is fully hydrostatic and so the horizontal pressure gradient comes from the free-surface elevation gradient

$$\nabla_{\mathrm{H}}\tilde{p} = \rho\, g\, \nabla_{\mathrm{H}}\eta \tag{3.27}$$

Using (2.555) and (3.15), the non-dimensional version of Eq. (3.27) is written as

$$\nabla'_{\mathrm{H}}\mathsf{p} = \nabla'_{\mathrm{H}}\eta' \tag{3.28}$$

Because of (3.28), the horizontal momentum equations (3.24) and (3.25) take the form

$$\varepsilon\frac{\partial \mathsf{u}}{\partial \mathsf{t}} + \varepsilon(\mathbf{u}\cdot\nabla')\mathsf{u} - (1+\beta\,\varepsilon\,\mathsf{y})\mathsf{v} = -\frac{\partial \eta'}{\partial \mathsf{x}} \tag{3.29}$$

$$\varepsilon\frac{\partial \mathsf{v}}{\partial \mathsf{t}} + \varepsilon(\mathbf{u}\cdot\nabla')\mathsf{v} + (1+\beta\,\varepsilon\,\mathsf{y})\mathsf{u} = -\frac{\partial \eta'}{\partial \mathsf{y}} \tag{3.30}$$

Vertically Integrated Incompressibility Equations

Consider again Eq. (3.26). It implies that the r.h.s.'s of the horizontal momentum equations are independent of z and therefore also the l.h.s.'s of the same equations are so; hence the components u and v of the velocity are depth or height independent as well. Owing to this feature, vertical integration of (3.23) on the interval $\varepsilon\,\mathsf{h} \leq \mathsf{z} \leq 1+\varepsilon\,\mathsf{F}\,\eta'$ yields

$$(1+\varepsilon\,\mathsf{F}\,\eta' - \varepsilon\,\mathsf{h})\,\nabla'_{\mathrm{H}}\cdot\mathbf{u} + \frac{\mathsf{D}}{\mathsf{Dt}}(1+\varepsilon\,\mathsf{F}\,\eta' - \varepsilon\,\mathsf{h}) = 0$$

that is to say

$$\nabla'_{\mathrm{H}}\cdot[(1+\varepsilon\,\mathsf{F}\,\eta' - \varepsilon\,\mathsf{h})\mathbf{u}] + \varepsilon\,\mathsf{F}\frac{\partial \eta'}{\partial \mathsf{t}} = 0 \tag{3.31}$$

Equation (3.31) allows us to express the incompressibility of the flow in terms of η' instead of Eq. (3.23), which involves w. In this way, the set of Eqs. (3.29), (3.30) and (3.31) in the unknowns u, v and η' is integrable.

Since F is $O(\varepsilon)$, the leading-order equation derived from (3.31) is

$$\nabla'_{\mathrm{H}}\cdot[(1-\varepsilon\,\mathsf{h})\mathbf{u}] = 0 \tag{3.32}$$

We stress that (3.32) relies on assumptions $T = T_{\mathrm{adv}}$, $\mathsf{F} = O(\varepsilon)$ and (3.22). Different hypotheses about T, F and h will be used in the following for the linear shallow-water model.

Evolution Equations of the Geostrophic Fields and the Quasi-Geostrophic Vorticity Equation

Equations (3.29), (3.30) and (3.32) depend parametrically on ε; so, we expect that u, v and η' have the same kind of dependence on ε. Therefore, we set:

$$\mathsf{u}(\mathsf{x},\mathsf{y},\mathsf{t};\varepsilon) = \sum_{m\geq 0} \varepsilon^m\, \mathsf{u}_m(\mathsf{x},\mathsf{y},\mathsf{t}) \tag{3.33}$$

$$\mathsf{v}(\mathsf{x},\mathsf{y},\mathsf{t};\varepsilon) = \sum_{m\geq 0} \varepsilon^m\, \mathsf{v}_m(\mathsf{x},\mathsf{y},\mathsf{t}) \tag{3.34}$$

$$\eta'(\mathsf{x},\mathsf{y},\mathsf{t};\varepsilon) = \sum_{m\geq 0} \varepsilon^m\, \eta'_m(\mathsf{x},\mathsf{y},\mathsf{t}) \tag{3.35}$$

Then, expansions (3.33)–(3.35) are substituted into (3.29), (3.30) and (3.32), and, for each power of ε, a set of equations is singled out. As we have anticipated, the first two powers (i.e. $m = 0$ and $m = 1$) are necessary and sufficient to determine a closed system of prognostic equations at the geostrophic level of approximation.

Leading-Order Equations

To the leading order ($m = 0$), Eqs. (3.29), (3.30) and (3.32) give

$$\mathsf{v}_0 = \frac{\partial \eta_0'}{\partial \mathsf{x}} \tag{3.36}$$

$$\mathsf{u}_0 = -\frac{\partial \eta_0'}{\partial \mathsf{y}} \tag{3.37}$$

$$\frac{\partial \mathsf{u}_0}{\partial \mathsf{x}} + \frac{\partial \mathsf{v}_0}{\partial \mathsf{y}} = 0 \tag{3.38}$$

respectively. Equations (3.36) and (3.37) are the components of the non-dimensional geostrophic balance, which in turn imply (3.38).

First-Order Equations

To the first order (i.e. for $m = 1$), Eqs. (3.29), (3.30) and (3.32) yield

$$\frac{\partial \mathsf{u}_0}{\partial \mathsf{t}} + (\mathbf{u}_0 \cdot \nabla')\mathsf{u}_0 - \mathsf{v}_1 - \beta\,\mathsf{y}\,\mathsf{v}_0 = -\frac{\partial \eta_1'}{\partial \mathsf{x}} \tag{3.39}$$

$$\frac{\partial \mathsf{v}_0}{\partial \mathsf{t}} + (\mathbf{u}_0 \cdot \nabla')\mathsf{v}_0 + \mathsf{u}_1 + \beta\,\mathsf{y}\,\mathsf{u}_0 = -\frac{\partial \eta_1'}{\partial \mathsf{y}} \tag{3.40}$$

$$\frac{\partial \mathsf{u}_1}{\partial \mathsf{x}} + \frac{\partial \mathsf{v}_1}{\partial \mathsf{y}} - \mathbf{u}_0 \cdot \nabla' \mathsf{h} = 0 \tag{3.41}$$

respectively, where $\mathbf{u}_0 := \{u_0, v_0\}$ is the non-dimensional geostrophic current in vector notation and use has been made of (3.38) to derive (3.41). It is useful to resort also to the shorthand notation

$$\frac{D_0}{Dt} := \frac{\partial}{\partial t} + u_0 \frac{\partial}{\partial x} + v_0 \frac{\partial}{\partial y} \tag{3.42}$$

in which the advective term of the Lagrangian derivative is expressed in terms of the geostrophic current in place of the total current. Operator (3.42) is named *geostrophic derivative*, and it is linked to the full Lagrangian derivative by the relation

$$\frac{D}{Dt} = \frac{D_0}{Dt} + \varepsilon \mathbf{u}_1 \cdot \nabla' + O(\varepsilon^2) \tag{3.43}$$

By means of (3.42), Eqs. (3.39) and (3.40) take the form

$$\frac{D_0}{Dt} u_0 - v_1 - \beta y v_0 = -\frac{\partial \eta_1'}{\partial x} \tag{3.44}$$

$$\frac{D_0}{Dt} v_0 + u_1 + \beta y u_0 = -\frac{\partial \eta_1'}{\partial y} \tag{3.45}$$

The non-dimensional *geostrophic relative vorticity* is defined as

$$\zeta_0' := \frac{\partial v_0}{\partial x} - \frac{\partial u_0}{\partial y} \tag{3.46}$$

and its evolution equation is obtained by working out the identity

$$\frac{\partial}{\partial x}\left(\frac{D_0}{Dt} v_0 + u_1 + \beta y u_0\right) - \frac{\partial}{\partial y}\left(\frac{D_0}{Dt} u_0 - v_1 - \beta y v_0\right) = 0 \tag{3.47}$$

which follows from (3.44) and (3.45). Hence, a straightforward computation yields, using also (3.38), the vorticity equation

$$\frac{D_0}{Dt} \zeta_0' + \frac{\partial u_1}{\partial x} + \frac{\partial v_1}{\partial y} + \beta v_0 = 0 \tag{3.48}$$

Equation (3.48) involves not only the geostrophic fields ζ_0' and v_0' but also the so-called *ageostrophic velocity* (u_1, v_1). Therefore, Eq. (3.41) is required to eliminate the latter in favour of the former, in order to establish the following evolution equation for the sole geostrophic fields:

$$\frac{D_0}{Dt} \zeta_0' + \mathbf{u}_0 \cdot \nabla' h + \beta v_0 = 0 \tag{3.49}$$

Conservation of Potential Vorticity

A striking aspect of the quasi-geostrophic vorticity equation (3.49) lies in the possibility to restate it as a function of the sole free-surface perturbation η_0' at the geostrophic level of approximation. In fact, using (3.37) and (3.36) into (3.42) and (3.46), one has

$$\frac{D_0}{Dt}\zeta_0' = \left(\frac{\partial}{\partial t} - \frac{\partial \eta_0'}{\partial y}\frac{\partial}{\partial x} + \frac{\partial \eta_0'}{\partial x}\frac{\partial}{\partial y}\right)\nabla'^2\eta_0' \tag{3.50}$$

Moreover,

$$\mathbf{u}_0 \cdot \nabla' h = -\frac{\partial \eta_0'}{\partial y}\frac{\partial h}{\partial x} + \frac{\partial \eta_0'}{\partial x}\frac{\partial h}{\partial y} = \frac{D_0}{Dt}h \tag{3.51}$$

and

$$v_0 = \frac{D_0}{Dt}y \tag{3.52}$$

Thus, substitution of (3.50)–(3.52) into (3.49) yields the equation

$$\frac{D_0}{Dt}(\nabla'^2\eta_0' + h + \beta y) = 0 \tag{3.53}$$

which, expresses the conservation of the quantity

$$\Pi_0' := \nabla'^2\eta_0' + h + \beta y \tag{3.54}$$

following the motion. When deriving Eq. (3.120), we shall ascertain that (3.54) is the potential vorticity at the geostrophic level of approximation.

Two Straightforward Consequences of Conservation Equation (3.53)

1. Equation (3.53) points out the equivalence between the beta effect and the special bathymetric profile $h = \beta y$. In fact, the presence of the beta effect ($\beta \neq 0$) on a flat bottom ($h = 0$) leads to the vorticity equation

$$\frac{D_0}{Dt}(\nabla'^2\eta_0' + \beta y) = 0 \tag{3.55}$$

   which is openly equivalent to the case of a motion on the f-plane ($\beta = 0$) in the presence of the bathymetric profile $h = \beta y$.
2. The explicit version of (3.53) is

$$\frac{\partial}{\partial t}\nabla'^2\eta_0' + \frac{\partial \eta_0'}{\partial x}\frac{\partial \Pi_0'}{\partial y} - \frac{\partial \eta_0'}{\partial y}\frac{\partial \Pi_0'}{\partial x} = 0 \tag{3.56}$$

where Π_0' is given by (3.54). The second and third terms of (3.56) constitute the Jacobian determinant $\mathscr{J}(\eta_0', \Pi_0')$; so, on the whole, each of Eqs. (3.53) and (3.56) is equivalent to

$$\frac{\partial}{\partial \mathsf{t}} \nabla'^2 \eta_0' + \mathscr{J}\left(\eta_0', \Pi_0'\right) = 0 \tag{3.57}$$

In particular, steady motion is governed by

$$\mathscr{J}\left(\eta_0', \Pi_0'\right) = 0 \tag{3.58}$$

The nonlinear part of (3.57) is $\mathscr{J}\left(\eta_0', \nabla'^2 \eta_0'\right)$, which represents the advection of relative vorticity evaluated at the geostrophic level of approximation. In fact, one easily verifies the identity

$$\mathscr{J}\left(\eta_0', \nabla'^2 \eta_0'\right) := \frac{\partial \eta_0'}{\partial \mathsf{x}} \frac{\partial \nabla'^2 \eta_0'}{\partial \mathsf{y}} - \frac{\partial \eta_0'}{\partial \mathsf{y}} \frac{\partial \nabla'^2 \eta_0'}{\partial \mathsf{x}} = \mathbf{u}_0 \cdot \nabla' \zeta_0' \tag{3.59}$$

Linear Quasi-Geostrophic Vorticity Equation

The *local timescale*, when defined as

$$T_{\mathrm{loc}} := \frac{1}{\beta_0 L} \tag{3.60}$$

is of special interest in GFD, since T_{loc} is the period of linear Rossby waves as the reader may verify from the treatement of Rossby waves at the end of this section. The same relevance has the *advective timescale*

$$T_{\mathrm{adv}} := \frac{L}{U} \tag{3.61}$$

Hence, recalling (2.606), we obtain

$$T_{\mathrm{adv}} = \beta \, T_{\mathrm{loc}} \tag{3.62}$$

Note that (3.60) implies

$$\varepsilon_T = \beta \, \varepsilon \tag{3.63}$$

because of (2.601), (2.602) and (2.606).

It is useful to distinguish the non-dimensional time related to (3.60), say $\tilde{\mathsf{t}} := t/T_{\mathrm{loc}}$, from the non-dimensional time $\mathsf{t} := t/T_{\mathrm{adv}}$ that appears in the non-dimensional Lagrangian derivative D/Dt. Thus, the dimensional time coordinate t may be expressed in a twofold way, namely,

$$t = \begin{cases} T_{\mathrm{loc}} \, \tilde{\mathsf{t}} \\ T_{\mathrm{adv}} \, \mathsf{t} \end{cases}$$

and hence, recalling (2.606), we obtain

$$\tilde{\mathsf{t}} = \beta\,\mathsf{t} \tag{3.64}$$

If $\varepsilon = \varepsilon_T$, then $\tilde{\mathsf{t}}$ coincides with t.

Motions with a local timescale $T_{\rm loc}$ much shorter than the advective timescale $T_{\rm adv}$ fulfil the equivalent conditions

$$T_{\rm loc} < O(T_{\rm adv}) \qquad \varepsilon_T > O(\varepsilon) \qquad \beta > O(1) \tag{3.65}$$

which are typical of wave-like flows. In these flows, $T_{\rm loc}$ can be identified with the wave period, while the advective timescale $T_{\rm adv}$ is related to the slow translation of the oscillating field as a whole. A profound consequence of (3.65) is the tendency of the dynamics to linearity. Indeed, as we will ascertain in what follows, dynamics becomes exactly linear at the geostrophic level of approximation.

Example: Atmospheric Flow

Typical values of slowly passing atmospheric flows are:

$$L = O(10^6\,\mathrm{m}) \qquad U = O(1\,\mathrm{m/s}) \qquad T_{\rm loc} = O(10^5\,\mathrm{s}) \qquad T_{\rm adv} = O(10^6\,\mathrm{s})$$

and hence conditions (3.65) are satisfied.

The values above of L and U, together with $H = 10^4\,\mathrm{m}$, allow us to estimate

$$\varepsilon_T = 10^{-1} \qquad \beta = 10 \qquad \varepsilon = \delta = 10^{-2} \tag{3.66}$$

whence

$$\varepsilon = \varepsilon_T^2 \qquad \mathsf{F} = \varepsilon_T \qquad \delta = \varepsilon_T^2 \tag{3.67}$$

Finally, the topography is described by $h = \varepsilon_T H \tilde{\mathsf{h}}$, which is analogous to (3.22) of the nonlinear case.

Derivation of the Linear Dynamics

Because of (3.63) and assuming (3.67), the ordering parameter is ε_T, and, hence, the starting equations of the quasi-geostrophic, shallow-water model (3.1)–(3.4) become

$$\frac{\partial \mathsf{u}}{\partial \mathsf{x}} + \frac{\partial \mathsf{v}}{\partial \mathsf{y}} + \frac{\partial \mathsf{w}}{\partial \mathsf{z}} = 0 \tag{3.68}$$

$$\varepsilon_T \frac{\partial \mathsf{u}}{\partial \tilde{\mathsf{t}}} + \varepsilon_T^2 (\mathbf{u} \cdot \nabla')\mathsf{u} - (1 + \varepsilon_T \mathsf{y})\mathsf{v} + \varepsilon_T^2 \mathsf{w} = -\frac{\partial \mathsf{p}}{\partial \mathsf{x}} \tag{3.69}$$

$$\varepsilon_T \frac{\partial \mathsf{v}}{\partial \tilde{\mathsf{t}}} + \varepsilon_T^2 (\mathbf{u} \cdot \nabla')\mathsf{v} + (1 + \varepsilon_T \mathsf{y})\,\mathsf{u} = -\frac{\partial \mathsf{p}}{\partial \mathsf{y}} \tag{3.70}$$

$$\varepsilon_T^5 \frac{\partial \mathsf{w}}{\partial \tilde{\mathsf{t}}} + \varepsilon_T^6 (\mathbf{u} \cdot \nabla')\mathsf{w} - \varepsilon_T^2 \mathsf{u} = -\frac{\partial \mathsf{p}}{\partial \mathsf{z}} \tag{3.71}$$

By retaining only the terms of order $O(1)$ and $O(\varepsilon_T)$, equations above simplify into

$$\frac{\partial \mathsf{u}}{\partial \mathsf{x}} + \frac{\partial \mathsf{v}}{\partial \mathsf{y}} + \frac{\partial \mathsf{w}}{\partial \mathsf{z}} = 0 \tag{3.72}$$

$$\varepsilon_T \frac{\partial \mathsf{u}}{\partial \tilde{\mathsf{t}}} - (1 + \varepsilon_T \mathsf{y})\,\mathsf{v} = -\frac{\partial \mathsf{p}}{\partial \mathsf{x}} \tag{3.73}$$

$$\varepsilon_T \frac{\partial \mathsf{v}}{\partial \tilde{\mathsf{t}}} + (1 + \varepsilon_T \mathsf{y})\,\mathsf{u} = -\frac{\partial \mathsf{p}}{\partial \mathsf{y}} \tag{3.74}$$

$$\frac{\partial \mathsf{p}}{\partial \mathsf{z}} = 0 \tag{3.75}$$

in which the higher-order nonlinear advective terms present in (3.69)–(3.71) no longer appear. Like in the nonlinear case, (3.75) shows that pressure is fully hydrostatic; so, (3.28) holds again. Thus, Eqs. (3.73)–(3.74) may be written as

$$\varepsilon_T \frac{\partial \mathsf{u}}{\partial \tilde{\mathsf{t}}} - (1 + \varepsilon_T \mathsf{y})\,\mathsf{v} = -\frac{\partial \eta'}{\partial \mathsf{x}} \tag{3.76}$$

$$\varepsilon_T \frac{\partial \mathsf{v}}{\partial \tilde{\mathsf{t}}} + (1 + \varepsilon_T \mathsf{y})\,\mathsf{u} = -\frac{\partial \eta'}{\partial \mathsf{y}} \tag{3.77}$$

Moreover, (3.72) may be integrated vertically to give

$$\nabla'_{\mathrm{H}} \cdot [(1 - \varepsilon_T \tilde{\mathsf{h}})\mathbf{u}] = 0 \tag{3.78}$$

which corresponds to (3.32) of the nonlinear case. Equations (3.76)–(3.78) depend parametrically on ε_T. Therefore, we introduce the following expansions, analogous to (3.33)–(3.35),

$$\mathsf{u}(\mathsf{x}, \mathsf{y}, \tilde{\mathsf{t}}; \varepsilon_T) = \sum_{m \geq 0} \varepsilon_T^m \, {}_m\mathsf{u}(\mathsf{x}, \mathsf{y}, \tilde{\mathsf{t}}) \tag{3.79}$$

$$\mathsf{v}(\mathsf{x}, \mathsf{y}, \tilde{\mathsf{t}}; \varepsilon_T) = \sum_{m \geq 0} \varepsilon_T^m \, {}_m\mathsf{v}(\mathsf{x}, \mathsf{y}, \tilde{\mathsf{t}}) \tag{3.80}$$

$$\eta'(\mathsf{x}, \mathsf{y}, \tilde{\mathsf{t}}; \varepsilon_T) = \sum_{m \geq 0} \varepsilon_T^m \, {}_m\eta'(\mathsf{x}, \mathsf{y}, \tilde{\mathsf{t}}) \tag{3.81}$$

which are substituted into (3.76)–(3.78). According to the same procedure as that applied in the nonlinear case, the vorticity equation

$$\frac{\partial}{\partial \tilde{t}}\,{}_0\zeta' + \frac{\partial\,{}_1u}{\partial x} + \frac{\partial\,{}_1v}{\partial y} + {}_0v = 0 \tag{3.82}$$

is derived, where

$${}_0\zeta' := \frac{\partial\,{}_0v}{\partial x} - \frac{\partial\,{}_0u}{\partial y} \tag{3.83}$$

In addition, from (3.78), we obtain

$$\frac{\partial\,{}_1u}{\partial x} + \frac{\partial\,{}_1v}{\partial y} - {}_0\mathbf{u}\cdot\nabla'\tilde{h} = 0 \tag{3.84}$$

Finally, eliminating the ageostrophic divergence from (3.82) and (3.84) yields the full geostrophic vorticity equation

$$\frac{\partial}{\partial \tilde{t}}\,{}_0\zeta' + {}_0\mathbf{u}\cdot\nabla'\tilde{h} + {}_0v = 0 \tag{3.85}$$

follows.

Using ${}_0u = -\partial\,{}_0\eta'/\partial y$ and ${}_0v = \partial\,{}_0\eta'/\partial x$, which in the linear context are the counterpart of (3.37) and (3.36), Eq. (3.83) yields ${}_0\zeta' = \nabla'^2\,{}_0\eta'$. Therefore, Eq. (3.85) can be entirely written in terms of the sole free-surface elevation at the geostrophic level ${}_0\eta'$ and takes the form

$$\frac{\partial}{\partial \tilde{t}}\nabla'^2\,{}_0\eta' + \frac{\partial\,{}_0\eta'}{\partial x}\frac{\partial}{\partial y}(\tilde{h}+y) - \frac{\partial\,{}_0\eta'}{\partial y}\frac{\partial}{\partial x}(\tilde{h}+y) = 0 \tag{3.86}$$

or, equivalently,

$$\frac{\partial}{\partial \tilde{t}}\nabla'^2\,{}_0\eta' + \mathscr{J}\left({}_0\eta',\, \tilde{h}+y\right) = 0 \tag{3.87}$$

In the case of a flat bottom, Eqs. (3.86) and (3.87) are still valid with $\tilde{h} = 0$.

Dimensional Versions of the Nonlinear and Linear Quasi-Geostrophic Vorticity Equations

The dimensional counterpart of (3.56) is evaluated by resorting to the equations

$$\frac{\partial}{\partial \mathrm{t}} = \frac{L}{U}\frac{\partial}{\partial t} \tag{3.88}$$

$$\mathrm{h} = \frac{f_0 L}{U H} h \tag{3.89}$$

and to the approximation

$$\eta_0' \approx \frac{g}{f_0 U L}\eta \tag{3.90}$$

With the aid of (3.88)–(3.90), the resulting dimensional equation is

$$\frac{\partial}{\partial t}\nabla^2\eta+\frac{\partial\eta}{\partial x}\frac{\partial}{\partial y}\left(\frac{g}{f_0}\nabla^2\eta+\frac{f_0}{H}h+\beta_0 y\right)-\frac{\partial\eta}{\partial y}\frac{\partial}{\partial x}\left(\frac{g}{f_0}\nabla^2\eta+\frac{f_0}{H}h+\beta_0 y\right)=0 \tag{3.91}$$

that is,

$$\frac{\partial}{\partial t}\nabla^2\eta+\mathcal{J}\left(\eta,\,\frac{g}{f_0}\nabla^2\eta+\frac{f_0}{H}h+\beta_0 y\right)=0 \tag{3.92}$$

Unlike the previous case, the dimensional counterpart of (3.86) is evaluated by using the equations

$$\frac{\partial}{\partial\tilde{t}}=\frac{1}{\beta_0 L}\frac{\partial}{\partial t} \tag{3.93}$$

$$\tilde{h}=\frac{f_0}{\beta_0 H L}h \tag{3.94}$$

and the approximation

$${}_0\eta'\approx\frac{g}{f_0 U L}\eta \tag{3.95}$$

Substitution of (3.93)–(3.95) into (3.86) gives the equation

$$\frac{\partial}{\partial t}\nabla^2\eta+\frac{\partial\eta}{\partial x}\frac{\partial}{\partial y}\left(\frac{f_0}{H}h+\beta_0 y\right)-\frac{\partial\eta}{\partial y}\frac{\partial}{\partial x}\left(\frac{f_0}{H}h+\beta_0 y\right)=0 \tag{3.96}$$

which is equivalent to

$$\frac{\partial}{\partial t}\nabla^2\eta+\mathcal{J}\left(\eta,\,\frac{f_0}{H}h+\beta_0 y\right)=0 \tag{3.97}$$

Comparison of (3.92) with (3.97) could induce, incorrectly, to conclude that the latter derives from the former simply by cancellation of the term $\mathcal{J}\left(\eta,\,\nabla^2\eta\right)$ in (3.92). However, setting $\mathcal{J}\left(\eta,\,\nabla^2\eta\right)=0$ has no formal grounds: the above, incorrect, conclusion is originated from using an approximation similar to (3.90) and (3.95) in which the distinction between the fields η_0' and ${}_0\eta'$ is inappropriately ignored.

Formal Developments of the Shallow-Water Vorticity Equation

For convenience, Eqs. (3.56) and (3.86) are reconsidered after substituting position

$$\eta_0'=:\psi \tag{3.98}$$

into the former and

$$ {}_0\eta' =: \Psi \tag{3.99}$$

into the latter. Hence, in terms of the Jacobian determinant, Eqs. (3.56) and (3.86) take the form

$$\frac{\partial}{\partial t}\nabla'^2\psi + \mathscr{J}\left(\psi,\, \nabla'^2\psi + h + \beta y\right) = 0 \tag{3.100}$$

and

$$\frac{\partial}{\partial \tilde{t}}\nabla'^2\Psi + \mathscr{J}\left(\Psi,\, \tilde{h} + y\right) = 0 \tag{3.101}$$

respectively.

Each of the functions $\psi = \psi(x, y, t)$ and $\Psi = \Psi(x, y, \tilde{t})$ is named, in its own context, *stream function* of the related geostrophic flow (see Appendix A, p. 367). Therefore,

$$\mathbf{u}_0' = \hat{\mathbf{k}} \times \nabla\psi \tag{3.102}$$

and

$$ {}_0\mathbf{u}' = \hat{\mathbf{k}} \times \nabla\Psi \tag{3.103}$$

In particular, the geostrophic derivative is written as

$$\frac{D_0}{Dt} = \frac{\partial}{\partial t} + \frac{\partial\psi}{\partial x}\frac{\partial}{\partial y} - \frac{\partial\psi}{\partial y}\frac{\partial}{\partial x} \tag{3.104}$$

so (3.100) is equivalent to the conservation statement

$$\frac{D_0}{Dt}(\nabla'^2\psi + h + \beta y) = 0 \tag{3.105}$$

which is nothing but (3.53) written with ψ in place of η_0'.

Equations (3.100) and (3.101) are openly invariant under the transformation $\psi \mapsto \psi + C$ and $\Psi \mapsto \Psi + C$, respectively, where C is any constant. This holds since the flow is determined only by the gradient of the free-surface modulation, that is, the perturbation pressure in shallow-water systems, which is left unaffected by any additive constant.

Along the rigid boundary of a simply connected planar fluid domain D, whose local normal unit vector is $\hat{\mathbf{n}}$, the no–mass flux condition $\hat{\mathbf{n}} \cdot \mathbf{u} = 0$ holds true, in particular, also for the geostrophic current (3.102) and (3.103). Hence, on the boundary ∂D,

$$\hat{\mathbf{n}} \cdot (\hat{\mathbf{k}} \times \nabla'\psi) = 0 \qquad \text{or} \qquad \hat{\mathbf{n}} \cdot (\hat{\mathbf{k}} \times \nabla'\Psi) = 0$$

that is to say,

$$\hat{\mathbf{t}} \cdot \nabla'\psi = 0 \qquad \text{or} \qquad \hat{\mathbf{t}} \cdot \nabla'\Psi = 0 \tag{3.106}$$

where $\hat{\mathbf{t}} := \hat{\mathbf{k}} \times \hat{\mathbf{n}}$ is the local unit vector tangent to ∂D. Thus, the differentiation of ψ (or Ψ), carried out along ∂D shows that, on the boundary, ψ (or Ψ) takes a constant value. As this constant can be set equal to zero without loss of generality, the no–mass flux condition expressed in terms of the stream function is

$$\psi(\mathsf{x},\mathsf{y},\mathsf{t}) = 0 \quad \forall(\mathsf{x},\mathsf{y}) \in \partial\mathsf{D},\ \forall\mathsf{t}$$

or

$$\Psi(\mathsf{x},\mathsf{y},\tilde{\mathsf{t}}) = 0 \quad \forall(\mathsf{x},\mathsf{y}) \in \partial\mathsf{D},\ \forall\tilde{\mathsf{t}} \tag{3.107}$$

The Status Function and Its Conservation

Equation (3.105) poses the problem of the connection between the conservation principle established by this equation and Ertel's theorem reported in Sect. 2.3.7. To investigate this aspect, it is necessary to ascertain which scalar quantity is conserved, besides $\nabla'^2\psi + \mathsf{h} + \beta\,\mathsf{y}$, in the framework of quasi-geostrophic shallow-water dynamics and denoted, quite in general, by q in Sect. 2.3.7. Vertical integration of the incompressibility equation on the interval $[h, H]$ yields

$$(H-h)\left(\frac{\partial u}{\partial x} + \frac{\partial v}{\partial y}\right) + \frac{\mathrm{D}}{\mathrm{D}t}(H-h) = 0 \tag{3.108}$$

In the same way, vertical integration on the interval $[h, z]$ with $z \le H$ gives

$$(z-h)\left(\frac{\partial u}{\partial x} + \frac{\partial v}{\partial y}\right) + \frac{\mathrm{D}}{\mathrm{D}t}(z-h) = 0 \tag{3.109}$$

Then, elimination of the horizontal divergence from (3.108) and (3.109) leads, after little algebra, to the conservation statement

$$\frac{\mathrm{D}}{\mathrm{D}t}\left(\frac{z-h}{H-h}\right) = 0 \tag{3.110}$$

which establishes the conservation of the so-called *status function*

$$q := \frac{z-h}{H-h} \tag{3.111}$$

following the motion. The conservation principle (3.110) shows that the relative height, from the bottom, of each fluid element (i.e. the actual height expressed in units of the total layer thickness $H-h$) is conserved following the motion. In particular, a fluid element of the upper surface remains indefinitely at the height $z = H$ (the relative height being 1), while a fluid element of the bottom remains indefinitely at the height $z = h$ (the relative height being 0). Whether the free-surface elevation $\eta = E\,\eta'$ enters into (3.110) or not depends on the order of magnitude of

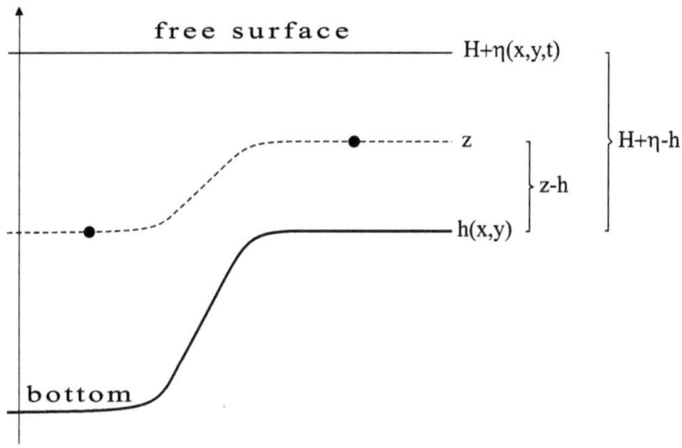

Fig. 3.2 Illustration of (3.111). The fluid element, represented by the *bullet* embedded into the constant-density ambient, moves conserving the relative height $(z-h)/(H-h)$ from the *bottom*. In particular, a fluid element of the free surface remains indefinitely at the height $z=H$ (the relative height is 1), while a fluid element of the *bottom* remains indefinitely at the height $z=h$ (the relative height is 0)

the Froude number F, through the estimate of $E/H = O(\varepsilon\,\mathsf{F})$. In fact, if $\mathsf{F} = O(\varepsilon)$, then $E/H = O(\varepsilon^2)$; so, this quantity may be neglected with respect to unity. On the other hand, if $\mathsf{F} = O(1)$, then $E/H = O(\varepsilon)$ and the free-surface elevation enters explicitly into (3.110), which modifies as (see Fig. 3.2)

$$\frac{\mathrm{D}}{\mathrm{D}t}\left(\frac{z-h}{H+\eta-h}\right)=0 \tag{3.112}$$

Ertel's Theorem in the Framework of the Quasi-Geostrophic Shallow-Water Model

The basic assumptions of the quasi-geostrophic, shallow-water model (i.e. a constant fluid density and the absence of dissipative mechanisms) imply the hypotheses of Ertel's theorem (Sect. 2.3.7); therefore, the potential vorticity of each fluid element is conserved. A straightforward computation based on (3.22) and (3.111) shows that

$$O\left(\frac{\partial q}{\partial x}\right)=O\left(\frac{\partial q}{\partial y}\right)=O\left(\frac{\varepsilon}{L}\right)\ll O\left(\frac{1}{H}\right)=O\left(\frac{\partial q}{\partial z}\right) \tag{3.113}$$

and therefore

$$\nabla q \approx \frac{1}{H-h}\hat{\mathbf{k}} \tag{3.114}$$

Under approximation (3.114), Eq. (2.507) applies; so, the conserved potential vorticity here is

$$\Pi = \frac{\zeta + f_0 + \beta_0 y}{H - h} \tag{3.115}$$

After these preliminaries, we can clarify the connection between (3.105) and the conservation principle given by the equation

$$\frac{\mathrm{D}\Pi}{\mathrm{D}t} = 0 \tag{3.116}$$

with Π given by (3.115).

Potential vorticity (3.115) can be written in terms of ζ', y and h in place of ζ, y and h, respectively, as

$$\Pi = \frac{f_0}{H} \frac{1 + \varepsilon(\zeta' + \beta \mathsf{y})}{1 - \varepsilon \mathsf{h}} \tag{3.117}$$

Moreover, because $\varepsilon < O(1)$ and $\beta = O(1)$, the expansion

$$\begin{aligned} \frac{1 + \varepsilon(\zeta' + \beta \mathsf{y})}{1 - \varepsilon \mathsf{h}} &= [1 + \varepsilon(\zeta' + \beta \mathsf{y})][1 + \varepsilon \mathsf{h} + O(\varepsilon^2)] \\ &= 1 + \varepsilon(\mathsf{h} + \zeta' + \beta \mathsf{y}) + O(\varepsilon^2) \end{aligned} \tag{3.118}$$

holds true; and, up to the first order in ε, from (3.116) and (3.118), the conservation statement

$$\frac{\mathsf{D}}{\mathsf{Dt}}(\mathsf{h} + \zeta' + \beta \mathsf{y}) = 0 \tag{3.119}$$

follows. Finally, recalling (3.43) and setting $\zeta' = \zeta_0' + O(\varepsilon) = \nabla'^2 \psi + O(\varepsilon)$, the leading-order terms of (3.119) produce the equation

$$\frac{\mathsf{D_0}}{\mathsf{Dt}}(\nabla'^2 \psi + \mathsf{h} + \beta \mathsf{y}) = 0$$

which coincides with (3.105). Comparison of (3.54) with (3.115) shows that

$$\Pi = \frac{f_0}{H}[1 + \varepsilon \Pi_0' + O(\varepsilon^2)] \tag{3.120}$$

which allows us to name the non-dimensional quantity Π_0' *potential vorticity at the geostrophic level of approximation*. Hence, (3.100) and (3.101) are called *shallow-water, quasi-geostrophic potential vorticity equations*.

In conclusion, the potential vorticity equations for the quasi-geostrophic shallow-water model are the potential vorticity equations at the geostrophic level of approximation in which the geostrophic and hydrostatic approximations are used.

On the Production of Relative Vorticity

The conservation principle (3.116) leads to an "internal redistribution" of the constituents of the potential vorticity

$$\Pi = \frac{\zeta + f_0 + \beta_0 y}{Z} \tag{3.121}$$

(we have set, in short, $Z := H - h$) in the course of the motion, which bears a noticeable interest from the physical viewpoint. In particular, the presence of the *ambient vorticity* $f_0 + \beta_0 y$ due to Earth's rotation is a necessary condition in order that *relative vorticity* ζ be produced by stretching or squeezing the fluid columns and/or by the latitudinal change of the Coriolis parameter. In fact, conservation of (3.121) from an "initial" state to a "final" state of a certain fluid parcel implies

$$\frac{\zeta_i + f_0 + \beta_0 y_i}{Z_i} = \frac{\zeta_f + f_0 + \beta_0 y_f}{Z_f}$$

whence

$$\zeta_f = \frac{Z_f}{Z_i}\,\zeta_i + \frac{f_0\,(Z_f - Z_i) + \beta_0\,(Z_f\, y_i - Z_i\, y_f)}{Z_i} \tag{3.122}$$

Even if the initial relative vorticity ζ_i is zero, (3.122) shows that relative vorticity ζ_f is produced, that is,

$$\zeta_f = \frac{f_0\,(Z_f - Z_i) + \beta_0\,(Z_f\, y_i - Z_i\, y_f)}{Z_i} \tag{3.123}$$

provided that the fluid thickness changes along a circle of latitude, i.e.,

$$Z_f \neq Z_i \qquad \text{and} \qquad y_f = y_i \tag{3.124}$$

or the fluid thickness is constant in the presence of a displacement with a meridional component, that is,

$$Z_f = Z_i \qquad \text{and} \qquad y_f \neq y_i \tag{3.125}$$

In case (3.124), Eq. (3.123) yields

$$\zeta_f = \frac{\Delta Z}{Z_i}\,(f_0 + \beta_0\, y) \tag{3.126}$$

where $\Delta Z := Z_f - Z_i$ and $y := y_i = y_f$. In case (3.125), Eq. (3.123) gives

$$\zeta_f = -\beta_0\, \Delta y \tag{3.127}$$

where $\Delta y := y_f - y_i$. Note, however, that in a non-rotating system (in which $f_0 = \beta_0 = 0$), Eq. (3.123) implies that, if $\zeta_i = 0$, then $\zeta_f = 0$; thus, in a non-rotating (and non-dissipative) system, relative vorticity cannot be generated but only conserved.

Steady Solutions of the Shallow-Water, Quasi-Geostrophic Model: Flow Over Topography

On the f-plane, steady circulation over topography is governed by the special version of (3.100) given by

$$\mathscr{J}\left(\psi,\, \nabla'^2\psi + \mathsf{h}\right) = 0 \tag{3.128}$$

Circulation is assumed to take place in the square fluid domain

$$\mathsf{D} := [0 \le x \le 1] \times [0 \le y \le 1] \tag{3.129}$$

characterized by the topographic profile

$$\mathsf{h} = \mathsf{h}_0 \sin(\pi \mathsf{x}) \tag{3.130}$$

where h_0 is a $O(1)$ constant.

The putative stream function

$$\psi = \mathsf{A} \sin(\pi \mathsf{x})\,(\mathsf{y}^2 - \mathsf{y}) \tag{3.131}$$

which satisfies the no–mass flux condition at the boundary of (3.129), yields $\nabla'^2\psi = -\pi^2\psi + 2\mathsf{A}\sin(\pi\mathsf{x})$ and, therefore,

$$\nabla'^2\psi + \mathsf{h} = -\pi^2\psi + (\mathsf{h}_0 + 2\mathsf{A})\sin(\pi\mathsf{x}) \tag{3.132}$$

Hence, if $\mathsf{A} = -\mathsf{h}_0/2$, Eq. (3.128) is identically satisfied by ansatz (3.131) in its final form

$$\psi = -\frac{\mathsf{h}_0}{2}\sin(\pi\mathsf{x})\,(\mathsf{y}^2 - \mathsf{y}) \tag{3.133}$$

whose streamlines are plotted in Fig. 3.3 together with the isobaths (3.130) (see Appendix A, p. 370).

Remarks. 1. Note that Eq. (3.128) is equivalent to

$$\hat{\mathbf{k}}\cdot\nabla'\psi \times \nabla'(\nabla'^2\psi) + \hat{\mathbf{k}}\cdot\nabla'\psi \times \nabla'\mathsf{h} = 0 \tag{3.134}$$

whose second term can be written as

$$\hat{\mathbf{k}}\cdot\nabla'\psi \times \nabla'\mathsf{h} = \mathbf{u}_0\cdot\nabla'\mathsf{h} = \pi\,\mathsf{h}_0\,\mathsf{u}_0(\mathsf{x},\mathsf{y})\cos(\pi\mathsf{x}) \tag{3.135}$$

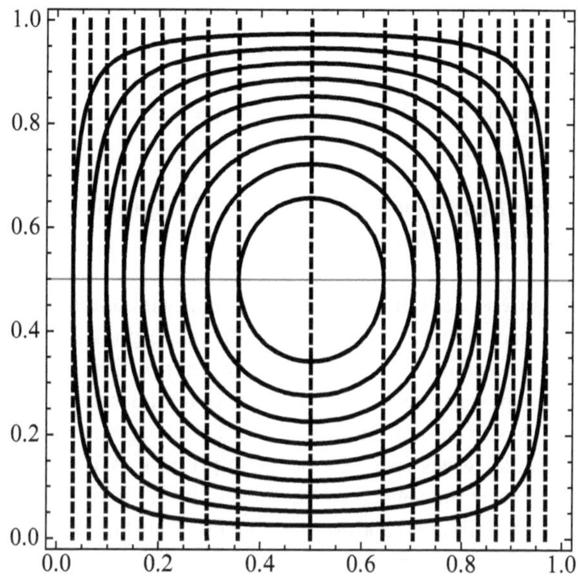

Fig. 3.3 Isobaths (*dashed*) and streamlines (*continuous*) of the shallow-water, quasi-geostrophic model of the flow (3.133) over topography (3.130)

The quantity (3.135) shows that, whenever the flow is meridional (i.e. $u_0 = 0$), the fluid parcels follow the contour lines of h, and the relative vorticity is constant along such branches of the streamlines. On the other hand, wherever the flow has a zonal component (i.e. $u_0 \neq 0$), fluid parcels cross the contour lines of h, and the relative vorticity undergoes an adjustment such that potential vorticity $\Pi_0' := \nabla'^2\psi + h$ is conserved following the motion.

2. Let R be a region of (3.129) whose contour ∂R is a streamline of (3.133), and denote by $\hat{\mathbf{t}}$ the unit vector locally tangent to ∂R. Then, the identity

$$\int_R \nabla'^2\psi \, dx\,dy = \oint_{\partial R} \mathbf{u}_0 \cdot \hat{\mathbf{t}} \, ds \tag{3.136}$$

based on the divergence theorem, holds true for every streamline ∂R. In the case of (3.133), we find

$$\nabla'^2\psi = h_0 \sin(\pi x) \left[\frac{\pi^2}{2}(y^2 - y) - 1\right]$$

and, therefore, $\nabla'^2\psi$ and h_0 have opposite signs. Hence, (3.136) yields

$$h_0 > 0 \quad \Longrightarrow \quad \oint_{\partial R} \mathbf{u}_0 \cdot \hat{\mathbf{t}} \, ds < 0 \quad \Longrightarrow \quad \text{anticyclonic circulation} \tag{3.137}$$

while

$$h_0 < 0 \quad \Longrightarrow \quad \oint_{\partial R} \mathbf{u}_0 \cdot \hat{\mathbf{t}} \, ds > 0 \quad \Longrightarrow \quad \text{cyclonic circulation} \tag{3.138}$$

in the Northern Hemisphere and inversely in the Southern Hemisphere. Thus, the cyclonic or anticyclonic orientation of the circulation in a given hemisphere is controlled by the rising ($h_0 > 0$) or the lowering ($h_0 < 0$) of the topographic profile (3.130), with respect to the flat bottom.

Time-Dependent Solutions of the Shallow-Water, Quasi-Geostrophic Model: The Rossby Waves

In a hypothetic frictionless process in which a material filament of fluid, "initially" aligned along a circle of latitude, undergoes a time-dependent oscillation in the meridional direction around the initial position, a compensation of relative vorticity takes place in such a way that the potential vorticity of each element of the filament is conserved. Iteration in time of the adjustment between planetary and relative vorticity gives rise to a periodic process involving the filament as a whole. This process is the subject of the present investigation.

For a filament endowed with a low translation speed and in the absence of bottom or topographic modulation (i.e. for $\tilde{h} = 0$), Eq. (3.101) rewritten as

$$\frac{\partial}{\partial \tilde{t}} \nabla'^2 \Psi + \frac{\partial \Psi}{\partial x} = 0 \tag{3.139}$$

is expected to govern the time-dependent evolution of the fluid. A putative integral of (3.139) of the kind

$$\Psi(x, y, \tilde{t}) = \cos(kx + ny - \sigma' \tilde{t} + \varphi) \tag{3.140}$$

implies a flow with the meridional velocity

$$v_0 = -k \sin(kx + ny - \sigma' \tilde{t} + \varphi)$$

which has the desired oscillatory character.[2] Linearity of (3.139) leaves arbitrary the $O(1)$ amplitude of (3.140); on the other hand, substitution of (3.140) into (3.139) yields the dispersion relation

$$\sigma' = -\frac{k}{k^2 + n^2} \tag{3.141}$$

which characterizes the wave-like behaviour of the fluid governed by (3.139). Unlike the restoring force of wind waves, internal waves, storm surges, etc., which is due to gravity acceleration, the restoring force of the wave (3.140) arises from the beta effect, that is, from the latitudinal dependence of the Coriolis parameter, represented by $\partial \Psi / \partial x$ in (3.139). Plane waves having this restoring mechanism are called *Rossby waves*, or *planetary waves*. Note that, within the

[2]See the remark at page 181.

f-plane approximation where $\beta_0 = 0$, Rossby waves cannot take place. Physically, this is explained by the fact that the f-plane approximation presupposes meridional displacements small enough to make negligible the influence of the planetary vorticity gradient in the Coriolis acceleration.

Basic features of the Rossby waves derive from (3.141). The zonal phase speed

$$c_x := \frac{\sigma'}{k} = -\frac{1}{k^2 + n^2} \tag{3.142}$$

is always negative, implying a phase propagation to the west, while the meridional phase speed

$$c_y := \frac{\sigma'}{n} = -\frac{k}{n} \frac{1}{k^2 + n^2} \tag{3.143}$$

has an undetermined sign, depending on that of k/n. Thus, lines of constant phase, in particular peaks and troughs, propagate only north-westward, westward or south-westward.

The group velocity

$$(c_{gx}, c_{gy}) = \left(\frac{\partial \sigma'}{\partial k}, \frac{\partial \sigma'}{\partial n} \right)$$

of Rossby wave packets is evaluated from (3.141). Its components are

$$c_{gx} = \frac{k^2 - n^2}{(k^2 + n^2)^2} \tag{3.144}$$

$$c_{gy} = \frac{2kn}{(k^2 + n^2)^2} \tag{3.145}$$

and therefore:

- Wave packets whose components satisfy the inequality $k^2 > n^2$ (“short” in the zonal direction) propagate eastward.
- Wave packets whose components satisfy the inequality $k^2 < n^2$ (“long” in the zonal direction) propagate westward.
- Wave packets whose components satisfy the equation $k^2 = n^2$ are stationary in the zonal direction.

Remark on Ansatz (3.140)

The non-dimensional beta-plane approximation demands the confinement of the flow into a zonal strip, say $0 \le y \le \pi$. Therefore, the stream function Ψ must satisfy the boundary conditions $\Psi(y = 0) = \Psi(y = \pi) = 0$, for instance, by setting

$$\Psi(x, y, \tilde{t}) = \cos(kx + ny - \sigma'\tilde{t} + \varphi) - \cos(kx - ny - \sigma'\tilde{t} + \varphi) \tag{3.146}$$

However, the linearity of (3.139) allows us to consider separately each term of (3.146) and, in particular, (3.140) as representative of a generic Fourier component.

Reflection of Rossby Waves

In the linear regime, the superposition of an incident (Ψ_i) and a reflected (Ψ_r) Rossby wave can be assumed as a putative integral of (3.139) to derive the law of reflection, which is of remarkable interest especially in the oceanographic context, where coastlines actually produce this phenomenon. The reflection of a Rossby wave from a solid boundary, say at $x = 0$, towards the half-space $x > 0$, can be described in terms of the overall stream function

$$\Psi = A_i \cos(k_i x + n_i y - \sigma_i' \tilde{t}) + A_r \cos(k_r x + n_r y - \sigma_r' \tilde{t} + \varphi) \tag{3.147}$$

where the subscript "i" refers to the incident component, while the subscript "r" refers to the reflected one. For a given σ_i' and n_i, the zonal wave number k_i is determined by the dispersion relation

$$\sigma_i' = -\frac{k_i}{k_i^2 + n_i^2} \tag{3.148}$$

under the further condition that wave packets composed with incident components of the kind $A_i \cos(k_i x + n_i y - \sigma_i' \tilde{t})$ propagate westward, that is,

$$k_i^2 < n_i^2 \tag{3.149}$$

Equation (3.148) in the unknown k_i has two roots, that is,

$$k_{i\pm} = -\frac{1}{2\sigma_i'} \pm \sqrt{\left(\frac{1}{2\sigma_i'}\right)^2 - n_i^2} \tag{3.150}$$

but (3.149) selects

$$k_i = k_{i-} \tag{3.151}$$

The reflected wave satisfies the dispersion relation

$$\sigma_r' = -\frac{k_r}{k_r^2 + n_r^2} \tag{3.152}$$

To determine σ_r', k_r and n_r, the boundary condition

$$\Psi = 0 \quad \text{if} \quad x = 0 \qquad \forall y \quad \forall \tilde{t} \tag{3.153}$$

must be applied to (3.147) and, in general, to each component of the wave packet. Condition (3.153) means

$$\mathsf{A_i}\cos(\mathsf{n_i}\,\mathsf{y}-\sigma_i'\tilde{\mathsf{t}})+\mathsf{A_r}\cos(\mathsf{n_r}\,\mathsf{y}-\sigma_r'\tilde{\mathsf{t}}+\varphi)=0 \qquad \forall \mathsf{y} \quad \forall \tilde{\mathsf{t}}$$

and the latter equation is verified for

$$\sigma_i'=\sigma_r' \qquad \mathsf{n_i}=\mathsf{n_r} \qquad \mathsf{A_i}=\mathsf{A_r} \qquad \varphi=\pi \tag{3.154}$$

Substitution of the first two conditions of (3.154) into (3.152) determines the roots

$$\mathsf{k}_{r\pm}=-\frac{1}{2\sigma_{i'}}\pm\sqrt{\left(\frac{1}{2\sigma_i'}\right)^2-\mathsf{n}_i^2} \tag{3.155}$$

Eastward propagation of the reflected wave demands $\mathsf{k}_r^2>\mathsf{n}_r^2$, so the sole root

$$\mathsf{k}_r=\mathsf{k}_{r+} \tag{3.156}$$

is singled out. Hence, the change in the zonal wave number k due to the reflection, evaluated from (3.150), (3.151), and (3.155) and (3.156) is

$$\Delta\mathsf{k}:=\mathsf{k}_{r+}-\mathsf{k}_{i-}=2\sqrt{\left(\frac{1}{2\sigma_i'}\right)^2-\mathsf{n}_i^2} \tag{3.157}$$

while both n and σ' are left unchanged by reflection. Finally, the last two conditions of (3.154) mean that the phase of the incident wave is flipped by 180°, while the amplitude is preserved in the reflection. Thus, (3.147) takes the final form

$$\Psi=\mathsf{A_i}\left\{\cos(\mathsf{k_i}\,\mathsf{x}+\mathsf{n_i}\,\mathsf{y}-\sigma_i'\tilde{\mathsf{t}})-\cos[(\mathsf{k_i}+\Delta\mathsf{k})\mathsf{x}+\mathsf{n_i}\,\mathsf{y}-\sigma_i'\tilde{\mathsf{t}}]\right\} \tag{3.158}$$

where σ_i' and $\Delta\mathsf{k}$ are given by (3.148) and (3.157), respectively (see Fig. 3.4).

If the reflection from $x=0$ takes place towards $x<0$, then $\mathsf{k}_i^2>\mathsf{n}_i^2$ and $\mathsf{k}_r^2<\mathsf{n}_r^2$. Hence, $\mathsf{k_i}=\mathsf{k}_{i+}$ and $\mathsf{k_r}=\mathsf{k}_{r-}$; so, $\mathsf{k}_{r-}-\mathsf{k}_{i+}=-\Delta\mathsf{k}$ and

$$\Psi=\mathsf{A_i}\left\{\cos(\mathsf{k_i}\,\mathsf{x}+\mathsf{n_i}\,\mathsf{y}-\sigma_i'\tilde{\mathsf{t}})-\cos[(\mathsf{k_i}-\Delta\mathsf{k})\mathsf{x}+\mathsf{n_i}\,\mathsf{y}-\sigma_i'\tilde{\mathsf{t}}]\right\} \tag{3.159}$$

To summarize, in an ocean filled up with Rossby waves, small-scale components (with zonal wave number $\mathsf{k}=\mathsf{k_i}$) move to the eastern boundary where they are reflected as components with large east-west scale (with $\mathsf{k}=\mathsf{k_i}-\Delta\mathsf{k}$). On the other hand, the large-scale components (with $\mathsf{k}=\mathsf{k_i}$) move towards the western boundary, where they are reflected as small-scale motions (with $\mathsf{k}=\mathsf{k_i}+\Delta\mathsf{k}$).

In dimensional variables, Eq. (3.157) becomes

$$\Delta k=2\sqrt{\left(\frac{\beta_0}{2\sigma_i}\right)^2-n_i^2}$$

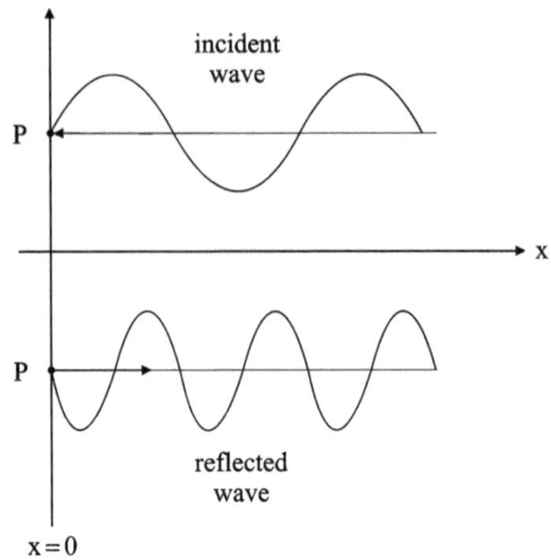

Fig. 3.4 Incident and reflected waves that illustrate Eq. (3.158)

Rossby Waves in a Square Basin

Integration of (3.139), that is,

$$\frac{\partial}{\partial \tilde{t}} \nabla'^2 \Psi + \frac{\partial \Psi}{\partial x} = 0 \tag{3.160}$$

on the square domain

$$D = [0 \le x \le 1] \times [0 \le y \le 1] \tag{3.161}$$

is here carried out under the usual no–mass flux boundary conditions

$$\left.\begin{aligned} \Psi(0, y, \tilde{t}) = \Psi(1, y, \tilde{t}) = 0 \\ \Psi(x, 0, \tilde{t}) = \Psi(x, 1, \tilde{t}) = 0 \end{aligned}\right\} \quad \forall \tilde{t} \tag{3.162}$$

We shall now show the noticeable result that Ψ can be expanded in normal modes, and each mode is the superposition of four Rossby waves: two of which are incident, while the other two arise from reflections at the meridional walls of D.

Consider a candidate solution of the kind

$$\Psi(x, y, \tilde{t}) := \mathrm{Re}[\varphi(x, y, \tilde{t})] \tag{3.163}$$

where

$$\varphi(x, y, \tilde{t}) := \exp[i\,(a x + b \tilde{t})]\, \phi(x, y) \tag{3.164}$$

with a and b real parameters to be conveniently determined. Substitution of (3.164) into (3.160) yields

$$i b \nabla'^2 \phi + i a (1 - a b)\, \phi + (1 - 2 a b) \frac{\partial \phi}{\partial x} = 0 \tag{3.165}$$

By means of the choice $2ab = 1$, the term containing $\partial\phi/\partial x$ is eliminated from (3.165), to obtain

$$\nabla'^2\phi + a^2\phi = 0 \tag{3.166}$$

where, according to (3.162),

$$\begin{aligned} \phi(0, y) &= \phi(1, y) = 0 \qquad \forall y \in [0, 1] \\ \phi(x, 0) &= \phi(x, 1) = 0 \qquad \forall x \in [0, 1] \end{aligned} \tag{3.167}$$

Eigenvalue problem (3.166)–(3.167) yields

$$a = \pi\sqrt{k^2 + n^2} \qquad k, n \in \mathbb{N}^+ \tag{3.168}$$

that is,

$$a = \pi K \qquad \left(K := \sqrt{k^2 + n^2}\right) \tag{3.169}$$

and the related eigenfunctions are

$$\phi_{k,n} = \sin(\pi k x)\sin(\pi n y) \tag{3.170}$$

Moreover, (3.169) yields $b = 1/(2\pi K)$; so, according to (3.164),

$$\varphi_{k,n}(x, y, \tilde{t}) = \exp\left[i\left(\pi K x + \frac{\tilde{t}}{2\pi K}\right)\right]\sin(\pi k x)\sin(\pi n y) \tag{3.171}$$

and, because of (3.163), Ψ is the superposition of the modes

$$\Psi_{k,n}(x, y, \tilde{t}) = \cos\left(\pi K x + \frac{\tilde{t}}{2\pi K}\right)\sin(\pi k x)\sin(\pi n y) \tag{3.172}$$

Thus, each mode $\Psi_{k,n}$ consists of a carrier wave $\cos[\pi K x + \tilde{t}/(2\pi K)]$ modulated by an envelope of sine functions, which assures the no–mass flux boundary condition on ∂D. Equation (3.172) is equivalent to

$$\begin{aligned} \Psi_{k,n}(x, y, \tilde{t}) = &\frac{1}{4}\cos\left(\pi K x + \frac{\tilde{t}}{2\pi K} - \pi k x + \pi n y\right) \\ &+ \frac{1}{4}\cos\left(\pi K x + \frac{\tilde{t}}{2\pi K} + \pi k x - \pi n y\right) \\ &- \frac{1}{4}\cos\left(\pi K x + \frac{\tilde{t}}{2\pi K} - \pi k x - \pi n y\right) \\ &- \frac{1}{4}\cos\left(\pi K x + \frac{\tilde{t}}{2\pi K} + \pi k x + \pi n y\right) \end{aligned} \tag{3.173}$$

so, each term of (3.173) is proportional to

$$\cos\left[\pi(\mathsf{K}\pm\mathsf{k})\mathsf{x}\pm\pi\mathsf{n}\mathsf{y}+\frac{\tilde{\mathsf{t}}}{2\pi\mathsf{K}}\right] \tag{3.174}$$

where signs are not correlated. The dispersion relation satisfied by (3.174) coincides with the propagation condition of the Rossby waves. To prove this, it is useful to put, in short,

$$\mu' := \pi(\mathsf{K}\pm\mathsf{k}) \qquad \nu' := \pm\pi\mathsf{n} \tag{3.175}$$

and verify that, in accordance with (3.141), the frequency σ' is given by

$$-\sigma' = \frac{1}{2\pi\mathsf{K}} = \frac{\mu'}{\mu'^2+\nu'^2} \tag{3.176}$$

In fact, Eq. (3.175) yields

$$\frac{\mu'}{\mu'^2+\nu'^2} = \frac{\pi(\mathsf{K}\pm\mathsf{k})}{\pi^2(\mathsf{K}\pm\mathsf{k})^2+\pi^2\mathsf{n}^2} \overset{(3.169)}{=} \frac{\pi(\mathsf{K}\pm\mathsf{k})}{2\pi^2\mathsf{K}(\mathsf{K}\pm\mathsf{k})} = \frac{1}{2\pi\mathsf{K}}$$

Equation (3.173) can be written also as

$$\begin{aligned}\Psi_{\mathsf{k},\mathsf{n}}(\mathsf{x},\mathsf{y},\tilde{\mathsf{t}}) =& \frac{1}{4}\left\{\cos\left[\pi(\mathsf{K}^-\mathsf{x}+\mathsf{n}\mathsf{y})+\frac{\tilde{\mathsf{t}}}{2\pi\mathsf{K}}\right]-\cos\left[\pi(\mathsf{K}^+\mathsf{x}+\mathsf{n}\mathsf{y})+\frac{\tilde{\mathsf{t}}}{2\pi\mathsf{K}}\right]\right\}\\ &+\frac{1}{4}\left\{\cos\left[\pi(\mathsf{K}^+\mathsf{x}-\mathsf{n}\mathsf{y})+\frac{\tilde{\mathsf{t}}}{2\pi\mathsf{K}}\right]-\cos\left[\pi(\mathsf{K}^-\mathsf{x}-\mathsf{n}\mathsf{y})+\frac{\tilde{\mathsf{t}}}{2\pi\mathsf{K}}\right]\right\}\end{aligned} \tag{3.177}$$

where $\mathsf{K}^\pm := \mathsf{K}\pm\mathsf{k}$. In this way, one recognizes in (3.177) the superposition of the incident wave at $x=0$

$$\frac{1}{4}\cos\left[\pi(\mathsf{K}^-\mathsf{x}+\mathsf{n}\mathsf{y})+\frac{\tilde{\mathsf{t}}}{2\pi\mathsf{K}}\right]$$

with the reflected one

$$-\frac{1}{4}\cos\left[\pi(\mathsf{K}^+\mathsf{x}+\mathsf{n}\mathsf{y})+\frac{\tilde{\mathsf{t}}}{2\pi\mathsf{K}}\right]$$

and the superposition of the incident wave at $x=1$

$$\frac{1}{4}\cos\left[\pi(\mathsf{K}^+\mathsf{x}-\mathsf{n}\mathsf{y})+\frac{\tilde{\mathsf{t}}}{2\pi\mathsf{K}}\right]$$

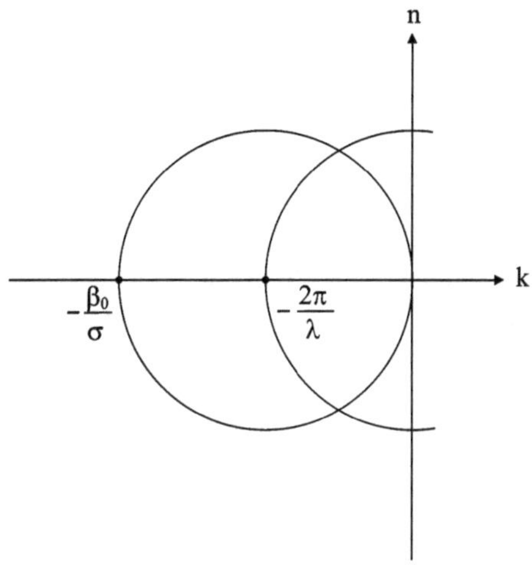

Fig. 3.5 Plot of circles (3.179)–(3.180) on the half plane $k < 0$, to explain inequality (3.181)

with the reflected one

$$-\frac{1}{4}\cos\left[\pi\left(\mathsf{K}^{-}\mathsf{x}-\mathsf{n}\,\mathsf{y}\right)+\frac{\tilde{\mathsf{t}}}{2\pi\mathsf{K}}\right]$$

Because of the reflection at $\mathsf{x}=0$ and at $\mathsf{x}=1$, the wave number changes of the amount $\pm 2\pi\mathsf{k}$, in accordance with (3.157), once the substitutions $\sigma_i' \mapsto -1/(2\pi\mathsf{K})$, $\mathsf{k}_i \mapsto \pi\mathsf{k}$ and $\mathsf{n}_i \mapsto \pi\mathsf{n}$ have been put into it.

A Bound on the Frequency of Rossby Waves

The evolution of a Rossby wave is slow relatively to the Earth's rotation period: this can be pointed out by showing that the dimensional frequency σ (>0) has an upper bound of order $O(\Omega)$. To achieve this, consider the dimensional dispersion relation

$$\sigma = -\beta_0 \frac{k}{k^2+n^2} \tag{3.178}$$

which can be derived from (3.97) with $h=0$ (flat topography), and write (3.178) in the equivalent form

$$\left(k+\frac{\beta_0}{2\sigma}\right)^2+n^2=\left(\frac{\beta_0}{2\sigma}\right)^2 \tag{3.179}$$

where the dimensional wave numbers k and n are linked to the wavelength λ by

$$k^2+n^2=\frac{4\pi^2}{\lambda^2} \tag{3.180}$$

According to Fig. 3.5, wave numbers selected by (3.179)–(3.180) actually exist on the (k,n)-plane, provided that $\beta_0/\sigma \geq 2\pi/\lambda$, that is to say, only if

$$\sigma \leq \frac{\beta_0 \lambda}{2\pi} \tag{3.181}$$

Consider, first, the case of the atmosphere and assume that the Rossby wave oscillates around the central latitude ϕ_0, so that the length of the related circle is $2\pi R\cos(\phi_0)$, where R is Earth's radius. Thus, for some integer n,

$$\lambda = \frac{2\pi}{n} R\cos(\phi_0) \tag{3.182}$$

Recalling that $\beta_0 = 2\Omega\cos(\phi_0)/R$, substitution of the latter equation, together with (3.182), into (3.181) yields $\sigma \leq 2\Omega\cos^2(\phi_0)/n$ and, a fortiori,

$$\sigma < 2\Omega\cos^2(\phi_0) \tag{3.183}$$

For low and midlatitudes, we have $2\cos^2(\phi_0) = O(1)$, and hence (3.183) implies

$$\sigma < O(\Omega) \tag{3.184}$$

Consider now the case of the ocean. The presence of meridional barriers does not allow us to resort to (3.182); however, the inequality

$$\lambda < L \tag{3.185}$$

can be used as an alternative, L being the horizontal length scale of the motion under consideration. Then, from (3.181) and (3.185), the inequality

$$\sigma < \frac{\beta_0 L}{2\pi} \tag{3.186}$$

follows. Moreover, the β-plane approximation demands $\beta_0 L < f_0$; so, (3.186) in turn implies

$$\sigma < \frac{f_0}{2\pi} = \frac{\sin(\phi_0)}{\pi}\Omega \tag{3.187}$$

As $\sin(\phi_0)/\pi < O(1)$, inequality (3.186) again implies (3.184). Indeed, Rossby waves were first proposed by the Swedish meteorologist Carl-Gustaf Arvid Rossby to explain the evolution of mid-latitude weather patterns, whose tendency to persist for several days, consistently with (3.184), is well known.

3.1.2 Developments of the Shallow-Water Equations for Modelling Wind-Driven Ocean Circulation

Wind-driven ocean circulation involves the motion of water bodies extending from the sea surface down to hundreds of metres and having velocities of a few centimetres per second. Far from the coastal regions, the horizontal length scale is that of the forcing, generated mechanically by the planetary wind field; hence, the horizontal scale of the flow is of the order of 1,000 km.

Due to the adiabatic nature of these systems, the simplifying assumption of a constant water density can be introduced to model them starting from a special version of the quasi-geostrophic shallow-water dynamics. In view of this goal, the quasi-geostrophic shallow-water model is suitably reformulated in what follows.

A Generalized Shallow-Water Model

In the model under investigation, the below listed features are taken into account:

1. The orders of magnitude of length, depth and velocity, that is,

$$L = O(10^6\,\text{m}) \qquad H = O(10^3\,\text{m}) \qquad U = O(10^{-2}\,\text{m/s}) \tag{3.188}$$

2. The advective and local timescales, that is,

$$T_{\text{adv}} := \frac{L}{U} = O(10^8\,\text{s}) \qquad T_{\text{loc}} := \frac{1}{\beta_0 L} = O(10^5\,\text{s}) \tag{3.189}$$

3. The turbulent dissipation in the momentum equations
4. A flat bottomed ocean
5. The confinement of the fluid column into a dimensionless interval $\mathrm{z_B} \leq \mathrm{z} \leq \mathrm{z_E}$ such that

$$0 \leq \mathrm{z_B} \leq \mathrm{z} \leq \mathrm{z_E} \leq 1 + \frac{E}{H}\eta' \tag{3.190}$$

where

$$\mathrm{z_E} - \mathrm{z_B} = O(1) \tag{3.191}$$

The surfaces $\mathrm{z} = \mathrm{z_E}$ and $\mathrm{z} = \mathrm{z_B}$ are, in general, spatially modulated and play the role of forcing against turbulent dissipation. In any case, the assumptions $0 \leq \mathrm{z} \leq 1$ and $0 \leq \mathrm{z} \leq 1 + (E/H)\,\eta'$ will be explored as well.

Therefore, in place of (3.1)–(3.4), the starting equations are (2.593), (2.612), (2.613) and (2.616) with $\rho_s = \text{const}$.

The Generalised Quasi-Geostrophic Nonlinear Model

Based on (3.188) and under the assumption of the advective timescale, one finds

$$\varepsilon = \varepsilon_T = 10^{-4} \qquad \delta = O(\varepsilon) \qquad \beta = O(10^3) \tag{3.192}$$

About the Ekman numbers, it is convenient to resort to the position

$$\frac{\mathsf{E_H}}{2} = \frac{\varepsilon}{\mathsf{Re}} \tag{3.193}$$

where

$$\mathsf{Re} := \frac{UL}{A_{\mathrm{H}}}$$

is *Reynolds number.*

Once the advective Rossby number is taken as the ordering parameter and (3.20) is again assumed, by using also the traditional approximation, the leading-order equations derived from the above listed equations are (3.1) together with

$$\varepsilon \frac{\partial \mathsf{u}}{\partial \mathsf{t}} + \varepsilon (\mathbf{u} \cdot \nabla') \mathsf{u} - (1 + \beta \varepsilon \mathsf{y}) \mathsf{v} = -\frac{\partial \mathsf{p}}{\partial \mathsf{x}} + \frac{\varepsilon}{\mathsf{Re}} \left(\frac{\partial^2 \mathsf{u}}{\partial \mathsf{x}^2} + \frac{\partial^2 \mathsf{u}}{\partial \mathsf{y}^2} \right) \tag{3.194}$$

$$\varepsilon \frac{\partial \mathsf{v}}{\partial \mathsf{t}} + \varepsilon (\mathbf{u} \cdot \nabla') \mathsf{v} + (1 + \beta \varepsilon \mathsf{y}) \mathsf{u} = -\frac{\partial \mathsf{p}}{\partial \mathsf{y}} + \frac{\varepsilon}{\mathsf{Re}} \left(\frac{\partial^2 \mathsf{v}}{\partial \mathsf{x}^2} + \frac{\partial^2 \mathsf{v}}{\partial \mathsf{y}^2} \right) \tag{3.195}$$

$$\frac{\partial \mathsf{p}}{\partial \mathsf{z}} = 0 \tag{3.196}$$

Note that, because of (3.196), the horizontal current is independent of z; so, the terms $(\mathsf{E_V}/2)\,\partial^2 \mathsf{u}/\partial \mathsf{z}^2$ and $(\mathsf{E_V}/2)\,\partial^2 \mathsf{v}/\partial \mathsf{z}^2$ included into the last terms of (2.612) and (2.613) do not appear in (3.194) and (3.195). Estimation of the Reynolds number is quite problematic because of the wide range into which the eddy viscosity coefficient A_{H} may, a priori, vary. Hence, in the comparison of ε with Re, it is not possible to fix a unique order of magnitude, although (3.194) and (3.195) seem to request so; therefore, in the momentum equations, the factor $1/\mathsf{Re}$ is retained at each order in ε. This fact leads to single out several dynamic regimes, as some explicit models will show in what follows. Analogously to (3.33), (3.34) and (3.35), the expansions

$$\mathsf{u}(\mathsf{x}, \mathsf{y}, \mathsf{t}; \varepsilon) = \sum_{m \geq 0} \varepsilon^m \, \mathsf{u}_m(\mathsf{x}, \mathsf{y}, \mathsf{t}) \tag{3.197}$$

$$\mathsf{v}(\mathsf{x}, \mathsf{y}, \mathsf{t}; \varepsilon) = \sum_{m \geq 0} \varepsilon^m \, \mathsf{v}_m(\mathsf{x}, \mathsf{y}, \mathsf{t}) \tag{3.198}$$

$$\mathsf{w}(\mathsf{x}, \mathsf{y}, \mathsf{t}; \varepsilon) = \sum_{m \geq 1} \varepsilon^m \, \mathsf{w}_m(\mathsf{x}, \mathsf{y}, \mathsf{t}) \tag{3.199}$$

$$\mathsf{p}(\mathsf{x}, \mathsf{y}, \mathsf{t}; \varepsilon) = \sum_{m \geq 0} \varepsilon^m \, \mathsf{p}_m(\mathsf{x}, \mathsf{y}, \mathsf{t}) \tag{3.200}$$

are substituted into (3.1) and (3.194)–(3.196); then, the equations at the first two levels of approximation are singled out as usual.

At the leading order, we find

$$u_0 = -\frac{\partial p_0}{\partial y} \qquad v_0 = \frac{\partial p_0}{\partial x}$$

whence (3.38) again follows together with

$$\frac{\partial p_0}{\partial z} = 0$$

The first-order equations turn out to be

$$\frac{\partial u_1}{\partial x} + \frac{\partial v_1}{\partial y} + \frac{\partial w_1}{\partial z} = 0 \tag{3.201}$$

$$\frac{D_0}{Dt} u_0 - v_1 - \beta y v_0 = -\frac{\partial p_1}{\partial x} + \frac{1}{Re}\left(\frac{\partial^2 u_0}{\partial x^2} + \frac{\partial^2 u_0}{\partial y^2}\right) \tag{3.202}$$

$$\frac{D_0}{Dt} v_0 + u_1 + \beta y u_0 = -\frac{\partial p_1}{\partial y} + \frac{1}{Re}\left(\frac{\partial^2 v_0}{\partial x^2} + \frac{\partial^2 v_0}{\partial y^2}\right) \tag{3.203}$$

Working out the identity

$$\frac{\partial}{\partial y}\left[\frac{D_0}{Dt} u_0 - v_1 - \beta y v_0 - \frac{1}{Re}\left(\frac{\partial^2 u_0}{\partial x^2} + \frac{\partial^2 u_0}{\partial y^2}\right)\right] =$$
$$= \frac{\partial}{\partial x}\left[\frac{D_0}{Dt} v_0 + u_1 + \beta y u_0 - \frac{1}{Re}\left(\frac{\partial^2 v_0}{\partial x^2} + \frac{\partial^2 v_0}{\partial y^2}\right)\right]$$

derived from (3.202) to (3.203) with the aid of (3.38), the vorticity equation

$$\frac{D_0}{Dt}\zeta_0' + \frac{\partial u_1}{\partial x} + \frac{\partial v_1}{\partial y} + \beta v_0 = \frac{1}{Re}\left(\frac{\partial^2 \zeta_0'}{\partial x^2} + \frac{\partial^2 \zeta_0'}{\partial y^2}\right) \tag{3.204}$$

follows. Equation (3.204) is the generalization of (3.48) in the presence of turbulent dissipation; in fact, in the limit for an infinitely large Reynolds number, the former equation reduces to the latter. Using (3.201), Eq. (3.204) may be written in terms of the stream function $\psi := p_0$ and the vertical velocity w_1 as

$$\frac{\partial}{\partial t}\nabla'^2\psi + \mathscr{J}\left(\psi, \nabla'^2\psi\right) + \beta\frac{\partial \psi}{\partial x} = \frac{\partial w_1}{\partial z} + \frac{1}{Re}\nabla'^4\psi \tag{3.205}$$

where $\nabla'^4\psi := \nabla'^2(\nabla'^2\psi)$. Equation (3.205) can be integrated vertically on the interval $z_B \le z \le z_E$; whence, assuming $z_E - z_B \approx 1$ in accordance with (3.191), one obtains

$$\frac{\partial}{\partial t}\nabla'^2\psi + \mathscr{J}\left(\psi,\, \nabla'^2\psi\right) + \beta\frac{\partial\psi}{\partial x} = w_1(z_E) - w_1(z_B) + \frac{1}{Re}\nabla'^4\psi \tag{3.206}$$

Note that, by introducing the length

$$\delta_M := (A_H/\beta_0)^{1/3} \tag{3.207}$$

the identity $1/Re = \beta\,(\delta_M/L)^3$ results and may be used into the last term of Eq. (3.206). Integration of (3.206) requires the explicit form of the vertical velocities at $z = z_B$ and at $z = z_E$; in general, they will be achieved by investigating what happens into the thin layers $0 \le z \le z_B$ and $z_E \le z \le 1 + \varepsilon F \eta'$, named *Ekman layers*.

The Generalized Quasi-Geostrophic Linear Model

The assumption of the local timescale $T_{loc} := 1/(\beta_0 L)$ together with (3.188) implies

$$\varepsilon_T = O\left(10^{-1}\right) \quad \varepsilon = O(\varepsilon_T^4) \quad \delta = O(\varepsilon_T^3) \quad \beta\,\varepsilon = \varepsilon_T \tag{3.208}$$

Hence, ε_T is the ordering parameter, and the leading-order equations are

$$\varepsilon_T\frac{\partial u}{\partial \tilde{t}} - (1 + \varepsilon_T\, y)v = -\frac{\partial p}{\partial x} + \varepsilon_T\frac{E_H}{2\,\varepsilon_T}\left(\frac{\partial^2 u}{\partial x^2} + \frac{\partial^2 u}{\partial y^2}\right) \tag{3.209}$$

$$\varepsilon_T\frac{\partial v}{\partial \tilde{t}} + (1 + \varepsilon_T\, y)u = -\frac{\partial p}{\partial y} + \varepsilon_T\frac{E_H}{2\,\varepsilon_T}\left(\frac{\partial^2 v}{\partial x^2} + \frac{\partial^2 v}{\partial y^2}\right) \tag{3.210}$$

$$\frac{\partial p}{\partial z} = 0 \tag{3.211}$$

Substitution of expansions

$$u(x, y, \tilde{t}; \varepsilon_T) = \sum_{m\ge 0} \varepsilon_T^m\, {}_m u(x, y, \tilde{t}) \tag{3.212}$$

$$v(x, y, \tilde{t}; \varepsilon_T) = \sum_{m\ge 0} \varepsilon_T^m\, {}_m v(x, y, \tilde{t}) \tag{3.213}$$

$$w(x, y, \tilde{t}; \varepsilon_T) = \sum_{m\ge 1} \varepsilon_T^m\, {}_m v(x, y, \tilde{t}) \tag{3.214}$$

$$p(x, y, \tilde{t}; \varepsilon_T) = \sum_{m\ge 0} \varepsilon_T^m\, {}_m p(x, y, \tilde{t}) \tag{3.215}$$

into (3.1) and (3.209)–(3.211) gives, to the leading order, again the well-known geostrophic balance and the related equations. Moreover, in terms of the length (3.207), the identity $E_H/2\varepsilon_T = (\delta_M/L)^3$ can be introduced into (3.209)–(3.211), so that the first-order equations turn out to be

$$\frac{\partial\, {}_1u}{\partial x} + \frac{\partial\, {}_1v}{\partial y} + \frac{\partial\, {}_1w}{\partial z} = 0 \tag{3.216}$$

$$\frac{\partial\, {}_0u}{\partial \tilde{t}} - {}_1v - y\, {}_0v = -\frac{\partial\, {}_1p}{\partial x} + \left(\frac{\delta_M}{L}\right)^3 \left(\frac{\partial^2}{\partial x^2} + \frac{\partial^2}{\partial y^2}\right) {}_0u \tag{3.217}$$

$$\frac{\partial\, {}_0v}{\partial \tilde{t}} + {}_1u + y\, {}_0u = -\frac{\partial\, {}_1p}{\partial y} + \left(\frac{\delta_M}{L}\right)^3 \left(\frac{\partial^2}{\partial x^2} + \frac{\partial^2}{\partial y^2}\right) {}_0v \tag{3.218}$$

Analogously to the nonlinear case, from (3.217) and (3.218), the vorticity equation

$$\frac{\partial\, {}_0\zeta}{\partial \tilde{t}} + \frac{\partial\, {}_1u}{\partial x} + \frac{\partial\, {}_1v}{\partial y} + {}_0v = \left(\frac{\delta_M}{L}\right)^3 \left(\frac{\partial^2}{\partial x^2} + \frac{\partial^2}{\partial y^2}\right) {}_0\zeta \tag{3.219}$$

can be derived. Using (3.216), Eq. (3.219) is written in terms of the stream function $\Psi := {}_0p$ as

$$\frac{\partial}{\partial \tilde{t}}\nabla'^2\Psi + \frac{\partial \Psi}{\partial x} = \frac{\partial\, {}_1w}{\partial z} + \left(\frac{\delta_M}{L}\right)^3 \nabla'^4\Psi \tag{3.220}$$

Finally, vertical integration of (3.220) on the interval $z_B \le z \le z_E$, where $z_E - z_B \approx 1$, gives

$$\frac{\partial}{\partial \tilde{t}}\nabla'^2\Psi + \frac{\partial \Psi}{\partial x} = {}_1w(z_E) - {}_1w(z_B) + \left(\frac{\delta_M}{L}\right)^3 \nabla'^4\Psi \tag{3.221}$$

Like (3.206), integration of (3.221) presupposes, in general, the explicit form of the vertical velocity at the lower and upper boundaries, which are determined by means of the above quoted Ekman layers.

Special Model Solutions of (3.206) and (3.221)

Special model solutions of (3.206) and (3.221) can be derived in an autonomous way, that is, without the resort to the Ekman layers dynamics, but under definite hypotheses about the vertical velocities appearing in (3.206) and (3.221). This is the case of the so-called *Fofonoff mode* and of a class of Rossby waves, which will be treated below.

A Steady and Inertial Solution of (3.206): The Fofonoff Mode

By definition, in the steady version of (3.206), we have

$$\frac{\partial}{\partial t}\nabla'^2\psi = 0$$

Moreover, the inertial circulation of the fluid included into the layer $0 \le z \le 1$ is realized by setting $w_1(z=0) = w_1(z=1) = 0$ and assuming $1/Re$ vanishingly small. Hence, (3.206) simplifies into

$$\mathscr{J}\left(\psi, \nabla'^2\psi + \beta y\right) = 0 \qquad \forall (x,y) \in D \tag{3.222}$$

with the boundary condition (3.107), that is,

$$\psi(x,y) = 0 \qquad \forall (x,y) \in \partial D \tag{3.223}$$

where

$$D := [0 \le x \le 1] \times [0 \le y \le 1] \tag{3.224}$$

Equation (3.222) is equivalent to (see Appendix A, p. 370)

$$\nabla'^2\psi + \beta y = G(\psi) \tag{3.225}$$

where $G(\psi)$ is some differentiable function of ψ. This indeterminateness is linked to the indefiniteness of the initial condition whose time evolution leads to the solution of the steady problem (3.222) and (3.223) for a certain $G(\psi)$. Moreover, because the fluid domain is spatially finite, the lack of possible asymptotic stream functions, having a known shape, prevents to infer $G(\psi)$; so, the only guideline to single out $G(\psi)$ is mathematical simplicity. Equation 3.225 is equivalent to

$$\beta^{-1}\nabla'^2\psi + y = \beta^{-1}G(\psi) \tag{3.226}$$

where $\beta^{-1} = U/\beta_0 L^2$. With the values reported in (3.188), we find $\beta^{-1} = O(10^{-3})$. At this point, it is useful to introduce the length δ_I, named *inertial boundary layer width* and whose role will be clarified in the following, through the equation

$$\beta^{-1} = \left(\frac{\delta_I}{L}\right)^2 \tag{3.227}$$

whence, again according to the values reported in (3.188),

$$\delta_I \approx 30 \text{ km} \tag{3.228}$$

More generally, we note that for $10^{-2}\,\mathrm{m/s} < U < 5\times10^{-2}\,\mathrm{m/s}$ we obtain $30\,\mathrm{km} < \delta_\mathrm{I} < 70\,\mathrm{km}$, approximately. By using (3.227), Eq. (3.226) becomes $(\delta_\mathrm{I}/L)^2\nabla'^2\psi + \mathrm{y} = (\delta_\mathrm{I}/L)^2 G(\psi)$; and the sole criterion of mathematical simplicity suggests to take $(\delta_\mathrm{I}/L)^2 G(\psi) = \psi$, as Fofonoff did. Hence, the governing equation of the Fofonoff model is

$$\left(\frac{\delta_\mathrm{I}}{L}\right)^2 \nabla'^2\psi + \mathrm{y} = \psi \tag{3.229}$$

the boundary condition being still (3.223). Note that assumption $G(\psi) \propto \psi$ has transformed the nonlinear governing equation (3.222) into the linear equation (3.229). An interior solution $\psi_\mathrm{int} = \psi_\mathrm{int}(\mathrm{y})$ is assumed to satisfy (3.229) for any $\mathrm{y} \in [0,\,1]$, but far enough from the meridional boundaries $\mathrm{x} = 0$ and $\mathrm{x} = 1$.

The interior streamlines, determined by the equation $\psi_\mathrm{int} = \text{constant}$, are of the kind $\mathrm{y} = \text{constant}$. In particular, far from the zonal boundaries $\mathrm{y} = 0$ and $\mathrm{y} = 1$, the smallness of $(\delta_\mathrm{I}/L)^2$ justifies the further approximation $\psi_\mathrm{int} = \mathrm{y}$; so, in such region of the fluid domain, the conservation of potential vorticity $\Pi'_0 = \beta\,\mathrm{y}$ is realized by motions along circles of latitude where $\Delta\mathrm{y} = 0$.

With ψ_int in place of ψ in (3.229), the ordinary differential problem for ψ_int turns out to be

$$\left(\frac{\delta_\mathrm{I}}{L}\right)^2 \frac{\mathrm{d}^2\psi_\mathrm{int}}{\mathrm{dy}^2} - \psi_\mathrm{int} + \mathrm{y} = 0 \qquad 0 \le \mathrm{y} \le 1 \tag{3.230}$$

$$\psi_\mathrm{int}(0) = \psi_\mathrm{int}(1) = 0 \tag{3.231}$$

The solution of problem (3.230) and (3.231) is (Fig. 3.6, top left)

$$\psi_\mathrm{int}(\mathrm{y}) = \mathrm{y} - \frac{\sinh[(L/\delta_\mathrm{I})\,\mathrm{y}]}{\sinh(L/\delta_\mathrm{I})} \tag{3.232}$$

Hence, the zonal current $\mathrm{u}_{0\,\mathrm{int}}(\mathrm{y}) = -\mathrm{d}\psi_\mathrm{int}/\mathrm{dy}$ of the interior is evaluated to be (Fig. 3.6, top right)

$$\mathrm{u}_{0\,\mathrm{int}}(\mathrm{y}) = -1 + \frac{L}{\delta_\mathrm{I}}\frac{\cosh[(L/\delta_\mathrm{I})\,\mathrm{y}]}{\sinh(L/\delta_\mathrm{I})} \tag{3.233}$$

and, in particular, the currents along the zonal boundaries at $\mathrm{y} = 0$ and $\mathrm{y} = 1$ are

$$\mathrm{u}_{0\,\mathrm{int}}(0) = -1 + \frac{L}{\delta_\mathrm{I}}\frac{1}{\sinh(L/\delta_\mathrm{I})} \tag{3.234}$$

$$\mathrm{u}_{0\,\mathrm{int}}(1) = -1 + \frac{L}{\delta_\mathrm{I}}\coth[(L/\delta_\mathrm{I})] \tag{3.235}$$

Condition $L/\delta_\mathrm{I} \gg 1$ implies

$$\frac{L}{\delta_\mathrm{I}}\frac{1}{\sinh(L/\delta_\mathrm{I})} \ll 1 \qquad\qquad \coth[(L/\delta_\mathrm{I})\,\mathrm{y}] \approx 1$$

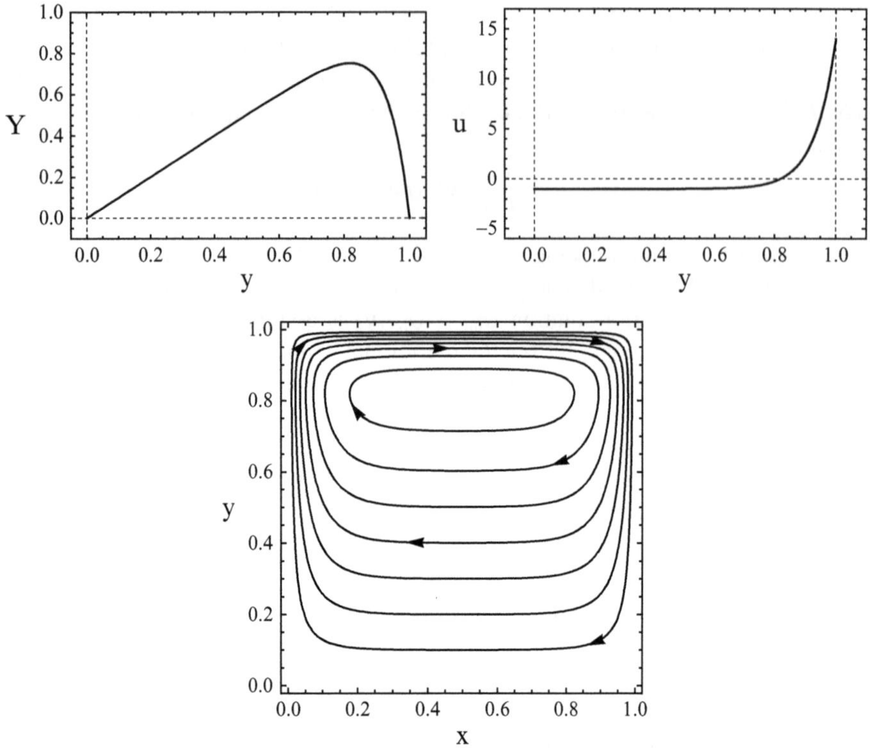

Fig. 3.6 Fofonoff mode. *Top left*: plot of the function $Y(\mathrm{y}) = \psi_{\mathrm{int}}(\mathrm{y})$ given by (3.232), for $L/\delta_{\mathrm{I}} = 15$. *Top right*: plot of $\mathrm{d}\psi_{\mathrm{int}}/\mathrm{dy}$ given by (3.233). *Bottom:* streamlines arising from (3.261)

Therefore, (3.234) yields

$$\mathrm{u}_{0\,\mathrm{int}}(0) \approx -1 \tag{3.236}$$

while (3.235) gives

$$\mathrm{u}_{0\,\mathrm{int}}(1) \approx -1 + \frac{L}{\delta_{\mathrm{I}}} \tag{3.237}$$

Thus, the interior current close to the boundary $\mathrm{y} = 0$ is $O(1)$ and westward, while that close to the boundary $\mathrm{y} = 1$ is $O(L/\delta_{\mathrm{I}}) \gg O(1)$ and eastward.

The relative vorticity of the interior $\zeta'_{0_{\mathrm{int}}} = -\mathrm{d}u_{0\,\mathrm{int}}/\mathrm{dy}$ is evaluated from (3.233):

$$\zeta'_{0\,\mathrm{int}}(\mathrm{y}) = -\left(\frac{L}{\delta_{\mathrm{I}}}\right)^2 \frac{\sinh[(L/\delta_{\mathrm{I}})\mathrm{y}]}{\sinh(L/\delta_{\mathrm{I}})} \tag{3.238}$$

whence

$$\zeta'_{0\,\mathrm{int}}(0) = 0 \tag{3.239}$$

while

$$\zeta'_{0\,\mathrm{int}}(1) = -\left(\frac{L}{\delta_{\mathrm{I}}}\right)^2 \tag{3.240}$$

These results can be explained also on the basis of (3.229), referred to the sole interior region and written as

$$\left(\frac{\delta_{\mathrm{I}}}{L}\right)^2 \zeta'_{0\,\mathrm{int}}(\mathrm{y}) + \mathrm{y} = \psi_{\mathrm{int}}(\mathrm{y}) \tag{3.241}$$

and (3.231). In fact, at $\mathrm{y} = 0$, (3.231) and (3.241) give $(\delta_{\mathrm{I}}/L)^2\,\zeta'_{0\,\mathrm{int}}(0) = 0$, that is, (3.239); on the other hand, at $\mathrm{y} = 1$, (3.231) and (3.241) give $(\delta_{\mathrm{I}}/L)^2\,\zeta'_{0\,\mathrm{int}}(1) + 1 = 0$, that is, (3.240).

To derive the model solution in the full fluid domain, it is convenient to analyse separately the western and the eastern regions of it. Consider first the western one. Inside a thin longitudinal strip, extending from the westernmost interior region up to the western boundary (at $\mathrm{x} = 0$), the streamlines coming from the interior change their zonal orientation and, to satisfy the no–mass flux condition at $\mathrm{x} = 0$, bend and place themselves, in the proximity of the western boundary, almost parallel to the boundary itself. This folding of streamlines can be achieved by means of the superposition of a suitable corrective stream function, say ϕ_{W}, to that of the interior, $\psi_{\mathrm{int}}(\mathrm{y})$; thus,

$$\psi_{\mathrm{W}} = \psi_{\mathrm{int}} + \phi_{\mathrm{W}} \tag{3.242}$$

In order that (3.242) be able to produce the desired folding, it is necessary to redefine a new horizontal length scale, much smaller than L, consistent with the abrupt change of direction of the streamlines in the proximity of the boundary. To this purpose, the new non-dimensional coordinate ξ', called *boundary layer coordinate*, is introduced through the equation

$$L\mathrm{x} = \delta_{\mathrm{I}}\,\xi' \tag{3.243}$$

which holds true because $x = L\mathrm{x}$ and $x = \delta_{\mathrm{I}}\,\xi'$ refer to a unique dimensional longitude x. Thus, (3.242) becomes

$$\psi_{\mathrm{W}} = \psi_{\mathrm{int}}(\mathrm{y}) + \phi_{\mathrm{W}}(\xi', \mathrm{y}) \tag{3.244}$$

Substitution of (3.244) into (3.229) with the aid of the differentiation rule

$$\frac{\partial}{\partial x} = \frac{L}{\delta_{\mathrm{I}}}\frac{\partial}{\partial \xi'} \tag{3.245}$$

determined by (3.243), yields the vorticity equation for the sole ϕ_{W}:

$$\frac{\partial^2 \phi_{\mathrm{W}}}{\partial \xi'^2} + \left(\frac{\delta_{\mathrm{I}}}{L}\right)^2 \frac{\partial^2 \phi_{\mathrm{W}}}{\partial \mathrm{y}^2} - \phi_{\mathrm{W}} = 0 \tag{3.246}$$

In (3.246), the $O(1)$-terms $\partial^2\phi_W/\partial\xi'^2$ and $-\phi_W$ balance each other, while

$$\left(\frac{\delta_I}{L}\right)^2\frac{\partial^2\phi_W}{\partial y^2}=O\left[\left(\frac{\delta_I}{L}\right)^2\right]$$

is a higher-order term, which is negligible in the boundary layer approximation. Therefore, within this approximation, (3.246) simplifies into

$$\frac{\partial^2\phi_W}{\partial\xi'^2}-\phi_W=0 \tag{3.247}$$

The general integral of (3.247) is

$$\phi_W(\xi',y)=C_1(y)\exp(\xi')+C_2(y)\exp(-\xi') \tag{3.248}$$

where $C_1(y)$ and $C_2(y)$ must be determined in order that the following two conditions be verified:

- The no–mass flux condition at $x=0$ applied to ψ_W, that is,

$$\psi_{int}(y)+\phi_W(0,y)=0 \tag{3.249}$$

- The asymptotic merging of ψ_W with $\psi_{int}(y)$ for large ξ', which is realized through the limit relation

$$\lim_{\xi'\to+\infty}\phi_W(\xi',y)=0 \tag{3.250}$$

Equation (3.249) and relation (3.250) allow to single out $C_1(y)=0$ and $C_2(y)=-\psi_{int}(y)$. In the western and in the interior region of the domain, the total solution ψ_W may written entirely in terms of x [using (3.243)] and y:

$$\psi_W(x,y)=\psi_{int}(y)\left[1-\exp\left(-\frac{L}{\delta_I}x\right)\right] \tag{3.251}$$

The meridional current $v_0(0,y)$ is easily evaluated from (3.251). One finds

$$v_0(0,y)=\left(\frac{\partial\psi_W}{\partial x}\right)_{x=0}=\psi_{int}(y)\frac{L}{\delta_I} \tag{3.252}$$

Thus, $v_0(0,y)$ is northward and, according to (3.237), of the same order as $u_{0\,int}(1)$.

In the eastern region of the fluid domain, the situation is quite similar to the previous one. Inside a thin longitudinal strip, extending from the eastern boundary (at $x=1$) to the easternmost interior region, the streamlines coming from the eastern region change their orientation almost parallel to the eastern boundary and bend to take a zonal orientation in the interior. This behaviour is achieved by means of the stream function

$$\psi_E = \psi_{int}(y) + \phi_E(\lambda', y) \tag{3.253}$$

in which the total solution ψ_E is obtained by means of the corrective term ϕ_E superimposed to that of the interior. In (3.253), the boundary layer coordinate λ', positive eastward, is linked to x by

$$L(x-1) = \delta_I \lambda' \tag{3.254}$$

Substitution of (3.253) into (3.229) yields, within the boundary layer approximation, the equation

$$\frac{\partial^2 \phi_E}{\partial \lambda'^2} - \phi_E = 0 \tag{3.255}$$

for the sole ϕ_E. The general integral of (3.255) is

$$\phi_E(\lambda', y) = K_1(y)\exp(\lambda') + K_2(y)\exp(-\lambda') \tag{3.256}$$

where $K_1(y)$ and $K_2(y)$ are determined by the following requirements:

- The no–mass flux condition at $x = 1$ applied to ψ_E, that is,

$$\psi_{int}(y) + \phi_E(0, y) = 0 \tag{3.257}$$

- The asymptotic merging of ψ_E with $\psi_{int}(y)$ for large negative λ', which is realized through the limit relation

$$\lim_{\lambda' \to -\infty} \phi_E(\lambda', y) = 0 \tag{3.258}$$

From (3.256) to (3.258), one finds $K_1(y) = -\psi_{int}(y)$ and $K_2(y) = 0$. In the eastern and in the interior region of the domain, the total solution ψ_E may written entirely in terms of x [using (3.243)] and y [using (3.254)]:

$$\psi_E(x, y) = \psi_{int}(y)\left\{1 - \exp\left[\frac{L}{\delta_I}(x-1)\right]\right\} \tag{3.259}$$

The meridional current $v_0(1, y)$ is easily evaluated from (3.259), that is,

$$v_0(1, y) = -\psi_{int}(y)\frac{L}{\delta_I} \tag{3.260}$$

Equation (3.260) shows that $v_0(1, y)$ is southward and exactly opposite to $v_0(0, y)$.

Finally, because both $\psi_W(x, y) \approx \psi_{int}(y)$ and $\psi_E(x, y) \approx \psi_{int}(y)$ for x approaching the interior, an appropriate representation of the so-called *Fofonoff mode* in the whole basin is given by the composite form (Fig. 3.6, bottom)

$$\psi(x, y) = \psi_{int}(y) \left[1 - \exp\left(-\frac{L}{\delta_I} x\right)\right] \left[1 - \exp\left(\frac{L}{\delta_I}(x - 1)\right)\right] \tag{3.261}$$

which follows from (3.251) and (3.259). Circulation is anticyclonic and intensified in the proximity of the boundary, in the northern part of the domain.

Remarks. 1. The choice $(\delta_I/L)^2 G(\psi) = -\psi$ in place of $(\delta_I/L)^2 G(\psi) = \psi$ leads to the equations

$$\frac{\partial^2 \phi_W}{\partial \xi'^2} + \phi_W = 0 \tag{3.262}$$

and

$$\frac{\partial^2 \phi_E}{\partial \lambda'^2} + \phi_E = 0 \tag{3.263}$$

in place of (3.247) and (3.255), respectively. However, both the general integrals of (3.262) and of (3.263) are oscillating, and, therefore, they cannot merge with ψ_{int} in the interior. For this reason, hypothesis $(\delta_I/L)^2 G(\psi) = -\psi$ must be discarded. Therefore, the formation of a zonal flow in the interior is consistent only with the anti-cyclonic orientation of the circulation field.

2. Consider any streamline and a fluid parcel moving on it. In running westward along the zonal branch of the streamline, say at the latitude $y = y_1$, the potential vorticity of the parcel consists mainly in *planetary vorticity*, that is, $f_0 + \beta y_1$. In running eastward along the upper branch of the same streamline, at $y = y_2 > y_1$, the potential vorticity of the parcel is $\zeta_0' + f_0 + \beta y_2$. Conservation of potential vorticity implies

$$\zeta_0' = \beta (y_1 - y_2) < 0 \tag{3.264}$$

and (3.264) shows that the parcel has acquired negative relative vorticity such that $\zeta_0' \approx \zeta_{0\,int}(1) = -(L/\delta_I)^2$ according to (3.240), while migrating to a higher latitude. Then, in the course of the subsequent migration from $y = y_2$ to $y = y_1$, the same parcel acquires a positive relative vorticity that compensates ζ_0', so potential vorticity takes again the sole initial planetary value $f_0 + \beta y_1$.

3. The Fofonoff mode is east-west invariant; more precisely, it is invariant under the mirror reflection

$$(x, y) \mapsto (1 - x, y) \tag{3.265}$$

which is a one-to-one map. In fact, one can verify immediately that (3.261) transforms into itself under (3.265). However, this property is independent of the boundary layer character of solution (3.261), and it can be inferred directly from the model equations (3.223) and (3.229). First of all, note that the interval $[0 \leq x \leq 1]$ transforms into itself under (3.265); so, the same happens for the fluid domain as a whole. Hence, the boundary condition

$$\psi(1 - x, y) = 0 \qquad \forall (x, y) \in \partial D \tag{3.266}$$

is equivalent to (3.223). Then, define the functions

$$\psi_S(x,y) := \frac{1}{2}[\psi(x,y) + \psi(1-x,y)] \tag{3.267}$$

and

$$\psi_A(x,y) := \frac{1}{2}[\psi(x,y) - \psi(1-x,y)] \tag{3.268}$$

which, under the mirror reflection (3.265), behave as follows:

$$\psi_S(1-x,y) = \psi_S(x,y) \tag{3.269}$$

$$\psi_A(1-x,y) = -\psi_A(x,y) \tag{3.270}$$

In terms of (3.267) and (3.268), the model solution ψ can be written as

$$\psi = \psi_S + \psi_A \tag{3.271}$$

Substitution of (3.271) into (3.229) yields

$$\left(\frac{\delta_I}{L}\right)^2 \nabla'^2(\psi_S + \psi_A) + y = \psi_S + \psi_A \tag{3.272}$$

and application of (3.265) to (3.272) gives

$$\left(\frac{\delta_I}{L}\right)^2 \nabla'^2(\psi_S - \psi_A) + y = \psi_S - \psi_A \tag{3.273}$$

By subtracting (3.273) from (3.272), one obtains, after division by 2,

$$\left(\frac{\delta_I}{L}\right)^2 \nabla'^2 \psi_A = \psi_A \tag{3.274}$$

where

$$\psi_A(x,y) = 0 \qquad \forall (x,y) \in \partial D \tag{3.275}$$

because of (3.266). Problem (3.274)–(3.275) has only the identically null solution; so, (3.271) gives

$$\psi = \psi_S \qquad \forall (x,y) \in \partial D \tag{3.276}$$

and the invariance of ψ under (3.265) is thus proved.

Wave-Like Integrals of (3.221): A Class of Rossby Waves

With reference to (3.221), we explore the case in which $z_B = 0$ is the flat bottom, and z_E is the summit of the whole fluid column, that is,

$$\mathsf{z}_\mathrm{E} = 1 + \frac{E}{H}\eta'$$

The vertical velocity at the bottom is, at the first order,

$$_1\mathsf{w}(\mathsf{z}_\mathrm{B}) = 0 \tag{3.277}$$

while the vertical velocity $_1\mathsf{w}(\mathsf{z}_\mathrm{E})$ is evaluated as follows.

In general,

$$w = \frac{UH}{L}\left[\varepsilon_\mathrm{T}\ {}_1\mathsf{w} + O\left(\varepsilon_\mathrm{T}^2\right)\right] \tag{3.278}$$

and, at the top of the fluid column where (1.31) holds, Eq. (3.278) becomes

$$\frac{\mathrm{D}\eta}{\mathrm{D}t} = \frac{UH}{L}\left[\varepsilon_\mathrm{T}\ {}_1\mathsf{w}(\mathsf{z}_\mathrm{E}) + O\left(\varepsilon_\mathrm{T}^2\right)\right]$$

that is to say, up to the first order,

$$\frac{E}{T}\frac{\partial\ {}_0\eta'}{\partial \tilde{\mathsf{t}}} = \frac{UH}{L}\varepsilon_\mathrm{T}\ {}_1\mathsf{w}(\mathsf{z}_\mathrm{E}) \tag{3.279}$$

where

$$T = \frac{1}{\beta_0 L} \qquad E = \frac{f_0 U L}{g} \qquad \varepsilon_\mathrm{T} = \frac{\beta_0 L}{f_0} \qquad {}_0\eta' = \Psi$$

From (3.279), we obtain

$$_1\mathsf{w}(\mathsf{z}_\mathrm{E}) = \mathsf{F}\frac{\partial\Psi}{\partial\tilde{\mathsf{t}}} \tag{3.280}$$

where $\mathsf{F} := f_0^2 L^2/gH$ is the Froude number.

Then, with the aid of (3.277) and (3.280), the governing equation (3.221) takes the form

$$\frac{\partial}{\partial\tilde{\mathsf{t}}}(\nabla'^2\Psi - \mathsf{F}\Psi) + \frac{\partial\Psi}{\partial\mathsf{x}} = \left(\frac{\delta_\mathrm{M}}{L}\right)^3 \nabla'^4\Psi \tag{3.281}$$

Equation (3.281) is a generalization of (3.139) in which both the oscillation in time of the free-surface elevation, represented by $-\mathsf{F}\partial\Psi/\partial\tilde{\mathsf{t}}$, and the possible lateral diffusion of relative vorticity, represented by the r.h.s. of the same equation, are taken into account. In any case, the basic physics of (3.139) is retained in (3.281), whose wave-like integrals are thus recognized to be Rossby waves.

Putative integrals of (3.281) are conveniently written as

$$\Psi(\mathsf{x},\mathsf{y},\tilde{\mathsf{t}}) = \mathrm{Re}\{\exp[\mathrm{i}(\mathsf{k}\mathsf{x} + \mathsf{n}\mathsf{y} - \sigma'\tilde{\mathsf{t}})]\} \tag{3.282}$$

to take into account possible complex values of σ'. The remark at page 181 also holds in the case of ansatz (3.282). If σ' is real, then (3.282) trivially coincides with (3.140) with $\varphi = 0$. Substitution of $\exp[i(kx+ny-\sigma'\tilde{t})]$ into (3.281) yields the complex dispersion relation

$$\sigma' = -\frac{k}{k^2+n^2+F} - i\left(\frac{\delta_M}{L}\right)^3 \frac{(k^2+n^2)^2}{k^2+n^2+F} \tag{3.283}$$

which is the starting point to elucidate the main features of (3.282). First of all, setting

$$\sigma'_{\Re} = -\frac{k}{k^2+n^2+F} \qquad \sigma'_{\Im} = -\left(\frac{\delta_M}{L}\right)^3 \frac{(k^2+n^2)^2}{k^2+n^2+F} \tag{3.284}$$

Equation (3.282) gives

$$\Psi(x,y,\tilde{t}) = \exp(\sigma'_{\Im}\tilde{t})\cos(kx+ny-\sigma'_{\Re}\tilde{t}) \tag{3.285}$$

and the two cases $\sigma'_{\Im} = 0$ and $\sigma'_{\Im} < 0$ can be investigated separately.

In the first case ($\sigma'_{\Im} = 0$), the effect of turbulence is negligibly small; and the consequent dispersion relation

$$\sigma = \sigma'_{\Re} = -\frac{k}{k^2+n^2+F} \tag{3.286}$$

looks close to (3.141). Hence, the zonal and meridional phase speeds are given by

$$c_x = -\frac{1}{k^2+n^2+F} \qquad c_y = -\frac{k}{n}\frac{1}{k^2+n^2+F} \tag{3.287}$$

respectively, and, because of the free-surface fluctuation, they are slightly smaller than (3.142) and (3.143). Moreover, the zonal and meridional group velocities of wave packets evaluated from (3.286) are

$$c_{gx} = \frac{k^2-(n^2+F)}{(k^2+n^2+F)^2} \tag{3.288}$$

and

$$c_{gy} = 2\frac{kn}{(k^2+n^2+F)^2} \tag{3.289}$$

respectively. Group velocities (3.288) and (3.289) generalize (3.144) and (3.145), respectively. In the context of (3.288), short waves satisfy, by definition, inequality $k^2 > n^2+F$, while long waves satisfy $k^2 < n^2+F$.

The case $\sigma'_{\Im} < 0$ shows that the decay rate $|\sigma'_{\Im}|$, which is the reciprocal of the mean lifetime of the wave $1/|\sigma'_{\Im}|$, is sensitive to the (dimensional) wavelength of (3.285). In fact, by resorting to the non-dimensional version of (3.180), the lifetime can be written as

$$\frac{1}{|\sigma_3'|} = \frac{1}{4\pi^2}\left(\frac{L}{\delta_M}\right)^3\left(\frac{\lambda^2}{L^2} + \frac{F}{4\pi^2}\frac{\lambda^4}{L^4}\right) \tag{3.290}$$

where $L/\lambda > O(1)$ and $F/(4\pi^2) = O(10^{-2})$. Relation (3.290) shows that the lifetime of Rossby waves with relatively short wavelengths, that is for $\lambda \ll L$, is shorter than the lifetime of Rossby waves with relatively long wavelengths, that is, for λ comparable with L. In other words, Rossby waves with short wavelengths decay more rapidly than those with long wavelengths.

For example, consider a packet of large-scale Rossby waves travelling westwards: when it reaches the western boundary, it originates reflected small-scale Rossby waves that will tend to decay completely before reaching the eastern boundary.

A Bound on the Frequency of Non-dissipative Rossby Waves with a Free Surface

The evaluation of the upper bound of the dimensional frequency $\sigma\,(> 0)$ in the presence of a free-surface oscillation (i.e. for $F > 0$) starts in analogy with the case of a rigid lid, formerly derived. Thus, consider the dimensional version of

$$\frac{\partial}{\partial \mathsf{\tilde{t}}}(\nabla'^2\Psi - \mathsf{F}\Psi) + \frac{\partial \Psi}{\partial \mathsf{x}} = 0$$

that is,

$$\frac{\partial}{\partial t}\left(\nabla^2\eta - \frac{f_0^2}{gH}\eta\right) + \beta_0\frac{\partial \eta}{\partial x} = 0$$

and the related dispersion relation

$$\sigma = -\beta_0\frac{k}{k^2 + n^2 + L_{\mathrm{ext}}^{-2}} \tag{3.291}$$

where the external radius of deformation $L_{\mathrm{ext}} := \sqrt{gH}/|f_0|$ has been already introduced in (2.390). Equation (3.291) can be written as the Cartesian equation of a circle on the (k, n)-plane

$$\left(k + \frac{\beta_0}{2\sigma}\right)^2 + n^2 = r^2 \tag{3.292}$$

where the radius is

$$r := \sqrt{\left(\frac{\beta_0}{2\sigma}\right)^2 - \frac{1}{L_{\mathrm{ext}}^2}} \tag{3.293}$$

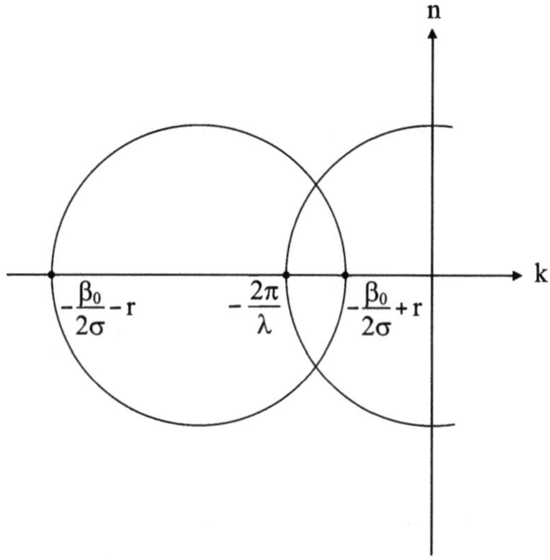

Fig. 3.7 Plot of circles (3.292)–(3.296) on the half plane $k < 0$, to explain inequality (3.297)

The obvious condition

$$\left(\frac{\beta_0}{2\sigma}\right)^2 - \frac{1}{L_{\text{ext}}^2} > 0 \tag{3.294}$$

demands

$$\sigma < \frac{1}{2}\beta_0 L_{\text{ext}} \tag{3.295}$$

Once (3.295) is satisfied, the admissible couples (k, n) belong both to the circle (3.292) and to the circle (3.180), that is,

$$k^2 + n^2 = \frac{4\pi^2}{\lambda^2} \tag{3.296}$$

According to Fig. 3.7, circles (3.292) and (3.296) intersect each other if and only if

$$\left|\frac{\beta_0}{2\sigma} - \frac{2\pi}{\lambda}\right| \leq r \tag{3.297}$$

Inequality (3.297) implies

$$\sigma \leq \frac{\beta_0 \lambda}{2\pi + (\lambda/L_{\text{ext}})^2/(2\pi)} \tag{3.298}$$

so, on the whole, (3.295) and (3.298) yield

$$\sigma \leq \frac{\beta_0}{2} \min\left[L_{\text{ext}}, \frac{\lambda}{\pi + (\lambda/L_{\text{ext}})^2/(4\pi)}\right] \tag{3.299}$$

As

$$L_{\text{ext}} \geq \frac{\lambda}{\pi + (\lambda/L_{\text{ext}})^2/(4\pi)} \qquad \forall \lambda > 0 \tag{3.300}$$

since (3.300) is equivalent to $(2\pi - \Lambda)^2 \geq 0$ with $\Lambda := \lambda/L_{\text{ext}}$, inequality (3.299) gives

$$\sigma \leq \frac{\beta_0 \lambda}{2\pi + (\lambda/L_{\text{ext}})^2/(2\pi)} \tag{3.301}$$

which, in turn, implies

$$\sigma < \frac{\beta_0 \lambda}{2\pi} \tag{3.302}$$

Thus, the upper bound (3.181) is again recovered.

3.2 Homogeneous Model

The homogeneous model of wind-driven ocean circulation has a long history, which starts from the early observations of the trajectories of the icebergs, in polar seas, by F. Nansen (at the end of nineteenth century), but is still under investigation although the core of the model is well established for some time.

The input is the planetary wind-stress field over the ocean or, better, an adequately simplified representation of it. The output is the transport that takes place into the mid-depth layer and is forced by the Ekman pumping, due to wind stress, above this layer. Steady circulation is then maintained by bottom friction and/or lateral diffusion of relative vorticity. The first mechanism is realized by means of a secondary circulation into a thin benthic layer; the second one is mainly based on a standard parameterization of turbulence.

Thus, the theory of surface and benthic Ekman layers is the preliminary step in the development of the homogeneous model.

3.2.1 The Ekman Layers

The wind-driven ocean circulation is the result of the almost stationary response of the water body of the ocean to the wind forcing and to the turbulent friction in the presence of Earth's rotation. The action of the wind at the free surface of the ocean is transmitted through the wind-stress $\boldsymbol{\tau}$, whose usual parameterization is

$$\boldsymbol{\tau} = C_{\text{D}}\, \rho_{\text{air}}\, |\mathbf{U}_{10}|\, \mathbf{U}_{10} \tag{3.303}$$

In (3.303), $C_{\text{D}} = O(10^{-3})$ is the so-called *drag coefficient*, $\rho_{\text{air}} = O(1\,\text{kg/m}^3)$ is air density at the sea level, and $\mathbf{U}_{10}(x, y) = O(10\,\text{m/s})$ is the wind velocity vector, conventionally detected 10 m above the sea surface.

For the purposes of GFD, (3.303) is preferably written as

$$\boldsymbol{\tau} = \tau_0 \boldsymbol{\tau}'(\mathsf{x}, \mathsf{y}) \tag{3.304}$$

where

$$\tau_0 = O(0.1\,\mathrm{kg} \cdot \mathrm{m}^{-1} \cdot \mathrm{s}^{-2}) \tag{3.305}$$

is its characteristic intensity, and

$$\boldsymbol{\tau}'(\mathsf{x}, \mathsf{y}) = \tau_\mathsf{x}(\mathsf{x}, \mathsf{y})\hat{\mathbf{i}} + \tau_\mathsf{y}(\mathsf{x}, \mathsf{y})\hat{\mathbf{j}} \tag{3.306}$$

is the non-dimensional counterpart of (3.303). We now investigate the effect of wind-stress below the sea surface, in the thin layer included in the interval (see (3.190))

$$\mathsf{z}_\mathrm{E} \le \mathsf{z} \le 1 + \frac{E}{H}\eta' \tag{3.307}$$

Apart from wind waves, we find an important sign in (1.182), which states that at the air-sea interface (i.e., at $z = H + \eta$) the wind stress is

$$\boldsymbol{\tau} = \rho_\mathrm{s} A_\mathrm{V} \left(\frac{\partial \mathbf{u}}{\partial z}\right)_{z=H+\eta} \tag{3.308}$$

Equation (3.308) shows that the vertical profile of the horizontal current decreases with depth; so, the related dynamics essentially differs from the shallow-water model, even if the assumption of a constant density ρ_s is retained. On the other hand, on continuity grounds, the current and pressure fields are expected to match the geostrophic interior at great depths.

Due to the relatively small thickness of the upper Ekman layer (i.e. the layer included in (3.307)), the motion depth H_E is of necessity far smaller than H. This fact demands a rescaling of the non-dimensional vertical coordinate: let ξ' be the new coordinate, while, as we know, z is the old one. Customarily, ξ' points downward and is zero at the level of the free surface, at $\mathsf{z} = 1$. Thus,

$$\xi' \propto 1 - \mathsf{z} \tag{3.309}$$

On the other hand, as any dimensional displacement Δz is independent of the depth of the motion, then

$$\Delta z = H \Delta \mathsf{z} = -H_\mathrm{E} \Delta \xi' \tag{3.310}$$

Hence, from (3.309) and (3.310), one concludes

$$\xi' = \frac{H}{H_\mathrm{E}}(1 - \mathsf{z}) \tag{3.311}$$

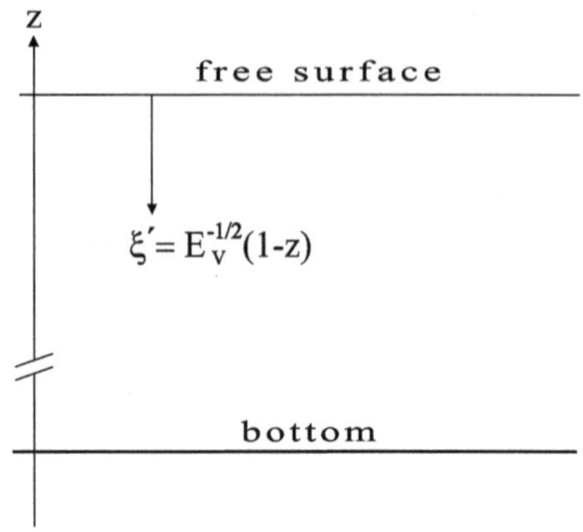

Fig. 3.8 Illustration of stretched coordinate of the *upper* Ekman layer

and, from (3.311), the differentiation rule

$$\frac{\partial}{\partial \mathsf{z}} = -\frac{H}{H_{\mathrm{E}}}\frac{\partial}{\partial \xi'} \tag{3.312}$$

follows. Besides wind waves, the wind stress generates turbulence, extending from the free surface to a depth $O(H_{\mathrm{E}})$, which is strong enough to make $O(1)$ the non-dimensional terms

$$\frac{\mathsf{E_V}}{2}\frac{\partial^2 \mathsf{u}}{\partial \mathsf{z}^2} \qquad \frac{\mathsf{E_V}}{2}\frac{\partial^2 \mathsf{v}}{\partial \mathsf{z}^2} \tag{3.313}$$

We stress that quantities (3.313) just parameterize the effect of turbulence on the momentum equations (2.612), (2.613) and (2.616). Now, substitution of (3.312) into (3.313) gives

$$\frac{\mathsf{E_V}}{2}\frac{\partial^2 \mathsf{u}}{\partial \mathsf{z}^2} = \frac{\mathsf{E_V}}{2}\left(\frac{H}{H_{\mathrm{E}}}\right)^2\frac{\partial^2 \mathsf{u}}{\partial \xi'^2} \qquad \frac{\mathsf{E_V}}{2}\frac{\partial^2 \mathsf{v}}{\partial \mathsf{z}^2} = \frac{\mathsf{E_V}}{2}\left(\frac{H}{H_{\mathrm{E}}}\right)^2\frac{\partial^2 \mathsf{v}}{\partial \xi'^2} \tag{3.314}$$

so the smallness of the vertical Ekman number $\mathsf{E_V}$ is compensated in (3.314) by the factor $(H/H_{\mathrm{E}})^2$, provided that

$$H_{\mathrm{E}} = H\sqrt{\mathsf{E_V}} \tag{3.315}$$

Hence, from (3.311), $\xi' = (1-\mathsf{z})/\sqrt{\mathsf{E_V}}$ as Fig. 3.8 shows; and, from (3.312), the transformation law

$$\frac{\partial}{\partial \mathsf{z}} = -\frac{1}{\sqrt{\mathsf{E_V}}}\frac{\partial}{\partial \xi'} \tag{3.316}$$

implies

$$\frac{\mathsf{E_V}}{2}\frac{\partial^2 \mathsf{u}}{\partial \mathsf{z}^2} = \frac{1}{2}\frac{\partial^2 \mathsf{u}}{\partial \xi'^2} \qquad \frac{\mathsf{E_V}}{2}\frac{\partial^2 \mathsf{v}}{\partial \mathsf{z}^2} = \frac{1}{2}\frac{\partial^2 \mathsf{v}}{\partial \xi'^2} \tag{3.317}$$

Note that $|\Delta\xi'| = |\Delta \mathsf{z}|/\sqrt{\mathsf{E_V}} \gg |\Delta \mathsf{z}|$, and for this reason, ξ' is named *stretched coordinate*.

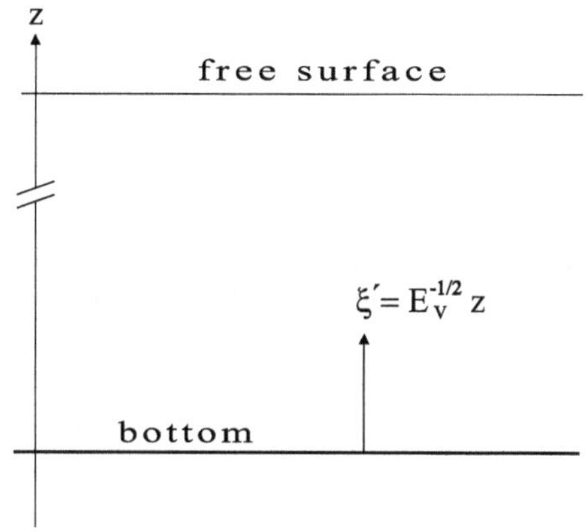

Fig. 3.9 Illustration of stretched coordinate of the benthic Ekman layer

Equation (3.315) shows that the vertical Ekman number can be quantified if the depth of the motion in the upper Ekman layer can be somehow estimated. In principle, current metre data should be able to answer the question, but the practical realization is made difficult by the noise due to the wind waves activity. For instance, the orders of magnitude $H_E = O(50\,\text{m})$ and $H = O(10^3\,\text{m})$ yield the estimate $\mathsf{E_V} = O(10^{-3})$, which is only indicative.

In the benthic Ekman layer, the current decreases to zero because of the turbulent friction produced by the fluid in motion in the proximity of the bottom, here taken flat for simplicity. This damping of the current is determined by the terms (3.313), which become $O(1)$ because of the smallness of the motion depth H_E with respect to H. Here we denote by H_E the depth of the motion in the benthic layer, just as in the case of the upper Ekman layer.

The non-dimensional coordinate of the layer (Fig. 3.9)

$$\xi' = \mathsf{z}/\sqrt{\mathsf{E_V}}$$

where $\mathsf{E_V}$ is still given by (3.315), generates the terms (3.317), which are present in the $O(1)$ momentum equation. On continuity grounds, the current and pressure fields of the benthic layer are expected to match the geostrophic interior for large ξ', while the no-slip boundary condition is applied at the bottom (i.e. for $\xi' = 0$) as effect of the damping of the flow due to friction.

Dynamics of the Upper Ekman Layer

On the basis of the previous discussion, the leading-order terms of the momentum equations in the upper Ekman layer are

$$\mathsf{v} = \frac{\partial \mathsf{p}}{\partial \mathsf{x}} - \frac{1}{2}\frac{\partial^2 \mathsf{u}}{\partial \xi'^2} \tag{3.318}$$

$$u = -\frac{\partial p}{\partial y} + \frac{1}{2}\frac{\partial^2 v}{\partial \xi'^2} \tag{3.319}$$

$$\frac{\partial p}{\partial \xi'} = 0 \tag{3.320}$$

where (3.316) has been taken into account. In (3.318)–(3.320) $u = u(x, y, \xi')$, $v = v(x, y, \xi')$ and $p = p(x, y, \xi')$. In particular, (3.320) is nothing but the equation $\partial p/\partial z = 0$, written in terms of ξ in place of z, which is the leading-order balance of (2.616) for a constant-density fluid.

Both a definite boundary condition at $\xi' = 0$ and an asymptotic condition for $\xi' \to +\infty$ are necessary to single out a unique solution from (3.318) to (3.320). The boundary condition at $\xi' = 0$ comes from the non-dimensional version of (3.308), which may be written as

$$\tau_0 \boldsymbol{\tau}' = -\frac{1}{2}\rho_s f_0 U H \sqrt{E_V}\left(\frac{\partial \mathbf{u}}{\partial \xi'}\right)_{\xi'=0} \tag{3.321}$$

Setting, in short,

$$\alpha := \frac{2\,\tau_0}{\rho_s f_0 U H \sqrt{E_V}} \tag{3.322}$$

the components of (3.321) are

$$\left(\frac{\partial u}{\partial \xi'}\right)_{\xi'=0} = -\,\alpha\,\tau_x \tag{3.323}$$

and

$$\left(\frac{\partial v}{\partial \xi'}\right)_{\xi'=0} = -\,\alpha\,\tau_y \tag{3.324}$$

In accordance with the validity of the $O(1)$-equations (2.638) in the geostrophic fluid body below the upper Ekman layer, the matching conditions of this layer with the interior (i.e. for $\xi' \to +\infty$) are

$$\lim_{\xi'\to+\infty} u(\xi') = u_0 \tag{3.325}$$

$$\lim_{\xi'\to+\infty} v(\xi') = v_0 \tag{3.326}$$

$$\lim_{\xi'\to+\infty} p(\xi') = p_0 \tag{3.327}$$

where u_0, v_0 and p_0 are the non-dimensional velocity and pressure fields of the geostrophic interior.

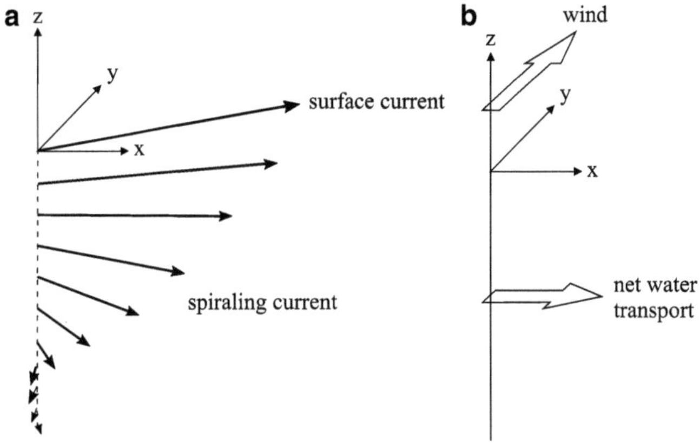

Fig. 3.10 Panel (**a**): detailed space structure of the current (3.331) which decreases exponentially with the depth and, at the same time, rotates clockwise. Panel (**b**): the net water transport in the upper Ekman layer is orthogonal to the wind stress and lies on the right with respect to the latter in the northern hemisphere

To solve problem (3.318)–(3.320) with (3.323), (3.324) and (3.325)–(3.327), we start from (3.320) and (3.327), which imply

$$\mathsf{p} = \mathsf{p}_0 \qquad\qquad \forall \xi' \in [0, +\infty[\tag{3.328}$$

Thus, (3.318) and (3.319) can be written as

$$\mathsf{v} = \mathsf{v}_0(x, y) - \frac{1}{2}\frac{\partial^2 \mathsf{u}}{\partial \xi'^2} \tag{3.329}$$

$$\mathsf{u} = \mathsf{u}_0(x, y) + \frac{1}{2}\frac{\partial^2 \mathsf{v}}{\partial \xi'^2} \tag{3.330}$$

respectively. Note that the coupled differential equations (3.329) and (3.330) are actually ordinary since they do not involve differentiation with respect to x or y. At this point, problem (3.329) and (3.330) with (3.323), (3.324), (3.325) and (3.326) in the unknowns u and v can be solved analytically (see the Appendix at the end of this Section). The result, written in non-dimensional vector form, is

$$\mathbf{u} = \mathbf{u}_0 + \frac{\alpha}{2}\exp(-\xi')\,[(\cos\xi' - \sin\xi')\,\boldsymbol{\tau}' - (\cos\xi' + \sin\xi')\,\hat{\mathbf{k}} \times \boldsymbol{\tau}'] \tag{3.331}$$

where $\mathbf{u}_0 = (\mathsf{u}_0, \mathsf{v}_0)$, while $\boldsymbol{\tau}'$ is the non-dimensional wind stress, and α is given by (3.322). Equation (3.331) generates the celebrated *Ekman spiral*, which is probably the graph most frequently reported in the books of physical oceanography (see Fig. 3.10, panel a).

The transport $\mathbf{M}$ induced by the ageostrophic part $\mathbf{u}-\mathbf{u}_0$ of (3.331) is

$$\mathbf{M}=\int_0^1(\mathbf{u}-\mathbf{u}_0)\,\mathrm{d}z$$

that is to say, in terms of the stretched coordinate $\xi'=(1-z)/\sqrt{E_V}$,

$$\mathbf{M}=\frac{1}{\sqrt{E_V}}\int_0^{+\infty}(\mathbf{u}-\mathbf{u}_0)\,\mathrm{d}\xi' \tag{3.332}$$

In (3.332), the integration limit $\xi'=+\infty$ is representative of $1/\sqrt{E_V}$ since the exponential damping of $\mathbf{u}-\mathbf{u}_0$ with respect to ξ' makes practically equivalent the use of $+\infty$ in place of $1/\sqrt{E_V}$; of course, such limit simplifies the computation of (3.332). Integration of (3.332) with the aid of (3.331) gives

$$\mathbf{M}=\frac{\alpha\sqrt{E_V}}{2}\boldsymbol{\tau}'\times\hat{\mathbf{k}} \tag{3.333}$$

while the dimensional counterpart of (3.333) is

$$\mathbf{M}=\frac{1}{\rho_s f_0}\boldsymbol{\tau}\times\hat{\mathbf{k}} \tag{3.334}$$

Equations (3.333) and (3.334) show, respectively, that:

- The ageostrophic transport is orthogonal to the wind stress (Fig. 3.10, panel b).
- The dimensional ageostrophic transport is independent of the intensity of turbulence.

The Vertical Velocity at the Transition Depth Between the Upper Ekman Layer and the Geostrophic Interior

Consider the incompressibility equation $\mathrm{div}\,\mathbf{u}+\partial w/\partial z=0$ and (3.316), that is,

$$\frac{\partial w}{\partial \xi'}=\sqrt{E_V}\,\mathrm{div}\,\mathbf{u} \tag{3.335}$$

From (3.335) and the request that the vertical velocity at the free surface is negligibly small, one finds

$$w(\xi')=\sqrt{E_V}\int_0^{\xi'}\mathrm{div}\,\mathbf{u}\,\mathrm{d}\xi' \tag{3.336}$$

Integration of (3.336) is executed by means of (3.331) and with the use of the identity $\mathrm{div}\,(\hat{\mathbf{k}}\times\boldsymbol{\tau}')=-\hat{\mathbf{k}}\cdot\mathrm{rot}\,\boldsymbol{\tau}'$ to obtain

$$\mathsf{w}(\xi') = \frac{\alpha\sqrt{\mathsf{E_V}}}{2}\left\{\mathrm{e}^{-\xi'}(\sin\xi')\,\mathsf{div}\,\boldsymbol{\tau}' + \left[1-\mathrm{e}^{-\xi'}\cos\xi'\right]\hat{\mathbf{k}}\cdot\mathsf{rot}\,\boldsymbol{\tau}'\right\} \tag{3.337}$$

The vertical velocity $\mathsf{w_E}$ at the transition depth between the upper Ekman layer and the geostrophic interior is given by

$$\mathsf{w_E} = \lim_{\xi'\to\infty}\mathsf{w}(\xi')$$

that is,

$$\mathsf{w_E} = \frac{\alpha\sqrt{\mathsf{E_V}}}{2}\hat{\mathbf{k}}\cdot\mathsf{rot}\,\boldsymbol{\tau}' \tag{3.338}$$

Substitution of (3.322) for α into (3.338) yields

$$\mathsf{w_E} = \frac{\tau_0}{\rho_s f_0 U H}\hat{\mathbf{k}}\cdot\mathsf{rot}\,\boldsymbol{\tau}' \tag{3.339}$$

The equation

$$\mathsf{div}\,\mathbf{M} = \mathsf{w_E} \tag{3.340}$$

which follows from (3.333) and (3.338), shows how the convergence/divergence of the transport $\mathbf{M}$, in the proximity of the free surface, determines the vertical velocity $\mathsf{w_E}$ by means of which the wind forcing is transmitted to the ocean interior. This fundamental mechanism will be treated in more detail in the framework of the so-called "homogeneous model" of wind-driven circulation in Sect. 3.2.2. The dimensional version of (3.338) is

$$w_E = \frac{1}{\rho_s f_0}\hat{\mathbf{k}}\cdot\mathrm{rot}\,\boldsymbol{\tau} \tag{3.341}$$

as one easily verifies multiplying both sides of (3.338) by $U H/L$ and using (3.322).

From (3.334) and (3.341), one obtains

$$\mathrm{div}\,\mathbf{M} = w_E \tag{3.342}$$

which is the dimensional counterpart of (3.340).

The Lower Ekman Layer

In the lower Ekman layer, the horizontal flow undergoes a frictional retardation, which brings it to zero at the bottom depth. With reference to the stretched coordinate

$$\xi' = \frac{\mathsf{z}}{\sqrt{\mathsf{E_V}}} \tag{3.343}$$

at the bottom ($z = 0$, i.e. $\xi' = 0$), the flow

$$\mathbf{u} = u(x, y, \xi')\hat{\mathbf{i}} + v(x, y, \xi')\hat{\mathbf{j}} \tag{3.344}$$

satisfies the no-slip boundary condition, that is,

$$u(x, y, 0) = 0 \qquad v(x, y, 0) = 0 \tag{3.345}$$

while, for $\xi' \to +\infty$, (3.344) matches the geostrophic flow above it. Because of (3.343), the leading-order momentum equations still are (3.329) and (3.330). To summarize, the circulation problem consists in (3.329) and (3.330) with (3.325), (3.326) and (3.345). Its solution is (see Sect. 3.2.1)

$$\mathbf{u} = (1 - e^{-\xi'}\cos\xi')\mathbf{u}_0 + (e^{-\xi'}\sin\xi')\hat{\mathbf{k}} \times \mathbf{u}_0 \tag{3.346}$$

The ageostrophic transport in the benthic Ekman layer has the same form as (3.332), where $\mathbf{u} - \mathbf{u}_0$ is now given by (3.346). A straightforward computation yields

$$\mathbf{M} = \frac{\sqrt{E_V}}{2}(\hat{\mathbf{k}} \times \mathbf{u}_0 - \mathbf{u}_0) \tag{3.347}$$

while the dimensional version of (3.347) is

$$\mathbf{M} = \frac{H_E}{2}(\hat{\mathbf{k}} \times \mathbf{u}_0 - \mathbf{u}_0) \tag{3.348}$$

where $\mathbf{u}_0$ is the dimensional geostrophic current.

The Vertical Velocity at the Transition Depth Between the Benthic Ekman Layer and the Geostrophic Interior

Because of (3.343), the incompressibility equation here is

$$\frac{\partial w}{\partial \xi'} = -\sqrt{E_V}\,\mathrm{div}\,\mathbf{u} \tag{3.349}$$

and the vertical integration of (3.349), with the no–mass flux boundary condition at the bottom, gives

$$w(\xi') = -\sqrt{E_V}\int_0^{\xi'} \mathrm{div}\,\mathbf{u}\,d\xi' \tag{3.350}$$

With the aid of (3.346) and the identity $\mathrm{div}(\hat{\mathbf{k}} \times \mathbf{u}_0) = -\hat{\mathbf{k}} \cdot \mathrm{rot}\,\mathbf{u}_0$, (3.350) yields

$$w(\xi') = \frac{\sqrt{E_V}}{2}\left[1 - e^{-\xi'}(\sin\xi' - \cos\xi')\right]\hat{\mathbf{k}} \cdot \mathrm{rot}\,\mathbf{u}_0 \tag{3.351}$$

The vertical current at the transition depth between the benthic Ekman layer and the geostrophic interior is denoted by w_B, and is given by

$$w_B = \lim_{\xi' \to \infty} w(\xi')$$

that is,

$$w_B = \frac{\sqrt{E_V}}{2} \hat{\mathbf{k}} \cdot \mathrm{rot}\,\mathbf{u}_0 \tag{3.352}$$

The equation

$$\mathrm{div}\,\mathbf{M} = -w_B \tag{3.353}$$

which follows from (3.347) and (3.352), shows how the convergence/divergence of the transport $\mathbf{M}$, in the proximity of the bottom, determines the vertical velocity w_B by means of which the relative vorticity is dissipated by bottom friction. The dimensional counterpart of (3.352) is

$$w_B = \frac{H_E}{2} \zeta \tag{3.354}$$

where $\zeta = \hat{\mathbf{k}} \cdot \mathrm{rot}\,\mathbf{u}_0$.

Appendix: Evaluation of the Flow in the Ekman Layers

In both the Ekman layers, the momentum equations are

$$v = v_0(x, y) - \frac{1}{2} \frac{\partial^2 u}{\partial \xi'^2} \tag{3.355}$$

$$u = u_0(x, y) + \frac{1}{2} \frac{\partial^2 v}{\partial \xi'^2} \tag{3.356}$$

They can be decoupled to give

$$\frac{\partial^4 u}{\partial \xi'^4} + 4u = 4u_0 \tag{3.357}$$

$$\frac{\partial^4 v}{\partial \xi'^4} + 4v = 4v_0 \tag{3.358}$$

The general integrals of (3.357) and (3.358) are

$$u = u_0 + \sum_{k=1}^{4} C_k \exp(\alpha_k \xi') \tag{3.359}$$

$$\mathsf{v} = \mathsf{v}_0 + \sum_{k=1}^{4} \mathsf{D}_k \exp(\alpha_k \xi') \tag{3.360}$$

respectively, where

$$\alpha_1 = -1 - \mathrm{i} \qquad \alpha_2 = -1 + \mathrm{i} \qquad \alpha_3 = 1 + \mathrm{i} \qquad \alpha_4 = 1 - \mathrm{i}$$

while $\mathsf{C}_k = \mathsf{C}_k(\mathsf{x}, \mathsf{y})$ and $\mathsf{D}_k = \mathsf{D}_k(\mathsf{x}, \mathsf{y})$. The matching conditions (3.325) and (3.326) demand $\mathsf{C}_3 = \mathsf{C}_4 = \mathsf{D}_3 = \mathsf{D}_4 = 0$, and therefore (3.359) and (3.360) simplify into

$$\mathsf{u} = \mathsf{u}_0 + \mathrm{e}^{-\xi'}(\mathsf{K}_1 \sin\xi' + \mathsf{K}_2 \cos\xi') \tag{3.361}$$

$$\mathsf{v} = \mathsf{v}_0 + \mathrm{e}^{-\xi'}(\mathsf{K}_3 \sin\xi' + \mathsf{K}_4 \cos\xi') \tag{3.362}$$

respectively, where $\mathsf{K}_i = \mathsf{K}_i(\mathsf{x}, \mathsf{y})$. Substitution of (3.361) and (3.362) into (3.355) and (3.356) gives K_3 and K_4 in terms of K_1 and K_2, which implies

$$\mathsf{u} = \mathsf{u}_0 + \mathrm{e}^{-\xi'}(\mathsf{K}_1 \sin\xi' + K_2 \cos\xi') \tag{3.363}$$

$$\mathsf{v} = \mathsf{v}_0 + \mathrm{e}^{-\xi'}(-\mathsf{K}_2 \sin\xi' + K_1 \cos\xi') \tag{3.364}$$

The coefficients K_1 and K_2 are selected by the remaining boundary condition, as follows.

We define the vector function

$$\mathbf{V} = \mathsf{K}_1 \hat{\mathbf{i}} - \mathsf{K}_2 \hat{\mathbf{j}} \tag{3.365}$$

which implies

$$\hat{\mathbf{k}} \times \mathbf{V} = \mathsf{K}_2 \hat{\mathbf{i}} + \mathsf{K}_1 \hat{\mathbf{j}} \tag{3.366}$$

Using (3.365) and (3.366), Eqs. (3.363) and (3.364) are summarized in vector notation as

$$\mathbf{u} = \mathbf{u}_0 + \mathrm{e}^{-\xi'}(\sin\xi' \, \mathbf{V} + \cos\xi' \, \hat{\mathbf{k}} \times \mathbf{V}) \tag{3.367}$$

Consider, first, the upper Ekman layer. Equation (3.367) implies

$$\left(\frac{\partial \mathbf{u}}{\partial \xi'}\right)_{\xi'=0} = \mathbf{V} - \hat{\mathbf{k}} \times \mathbf{V} \tag{3.368}$$

while the boundary condition (3.323) and (3.324) at the free surface ($\xi = 0$) is, in vector form,

$$\left(\frac{\partial \mathbf{u}}{\partial \xi'}\right)_{\xi'=0} = -\alpha \boldsymbol{\tau}' \tag{3.369}$$

By equating (3.368) with (3.369)

$$\mathbf{V}-\hat{\mathbf{k}}\times\mathbf{V}=-\alpha\boldsymbol{\tau}' \tag{3.370}$$

which implies, once the cross product $\hat{\mathbf{k}}\times$ is applied to (3.370),

$$\mathbf{V}+\hat{\mathbf{k}}\times\mathbf{V}=-\alpha\hat{\mathbf{k}}\times\boldsymbol{\tau}' \tag{3.371}$$

Then, from (3.370) and (3.371), both $\mathbf{V}$ and $\hat{\mathbf{k}}\times\mathbf{V}$ are separately singled out:

$$\mathbf{V}=-\frac{\alpha}{2}(\hat{\mathbf{k}}\times\boldsymbol{\tau}'+\boldsymbol{\tau}') \qquad \hat{\mathbf{k}}\times\mathbf{V}=-\frac{\alpha}{2}(\hat{\mathbf{k}}\times\boldsymbol{\tau}'-\boldsymbol{\tau}') \tag{3.372}$$

Finally, after a trivial rearrangement, substitution of (3.372) into (3.367) gives (3.331).

Consider, then, the lower Ekman layer. The no-slip boundary condition at the bottom ($\xi'=0$) in terms of (3.367) is given by

$$\hat{\mathbf{k}}\times\mathbf{V}+\mathbf{u}_0=0 \tag{3.373}$$

whence

$$\mathbf{V}=\hat{\mathbf{k}}\times\mathbf{u}_0 \tag{3.374}$$

Substitution of (3.373) and (3.374) into (3.367) produces (3.346).

3.2.2 *The Homogeneous Model of the Wind-Driven Oceanic Circulation*

The outstanding role of the homogeneous model of the wind-driven oceanic circulation lies in its ability to explain, to a remarkable extent, the phenomenology inherent the ocean gyres at the basin scale in terms of a single fluid layer forced by the wind stress and subject to vorticity dissipation.

Moreover, the homogeneous model is fit for subsequent developments towards the layered models, which are fundamental to investigate the baroclinic aspects of the ocean in motion.

The Vorticity Equations of the Homogeneous Models

The vorticity equations of the homogeneous models are obtained from the steady version of (3.206) in the nonlinear case or (3.221) in the linear one, once the vertical velocities at the top and the bottom of the fluid layer are determined by (3.339) and (3.352), respectively. In this way, the sole unknown of the model equation is the stream function ψ, by means of which all the dynamical variables can be evaluated.

Consider, first, the steady version of the nonlinear equation (3.206):

$$\mathscr{J}\left(\psi,\, \nabla'^2\psi\right) + \beta\,\frac{\partial\psi}{\partial \mathsf{x}} = \mathsf{w}_1(\mathsf{z}=\mathsf{z}_\mathrm{E}) - \mathsf{w}_1(\mathsf{z}=\mathsf{z}_\mathrm{B}) + \frac{1}{\mathsf{Re}}\nabla'^4\psi \tag{3.375}$$

The relation between $\mathsf{w}_1(\mathsf{z}=\mathsf{z}_\mathrm{E})$ and w_E given by (3.339) follows from the truncation up to the leading order in ε of the expansion $\mathsf{w}_\mathrm{E} = \varepsilon\,\mathsf{w}_1(\mathsf{z}=\mathsf{z}_\mathrm{E}) + O(\varepsilon^2)$, whence

$$\mathsf{w}_1(\mathsf{z}=\mathsf{z}_\mathrm{E}) = \frac{\tau_0}{\varepsilon\,\rho_\mathrm{s}\,f_0\,U\,H}\,\hat{\mathbf{k}}\cdot\mathrm{rot}\,\boldsymbol{\tau}' \tag{3.376}$$

The analogous procedure applied to (3.352), with $\nabla'^2\psi$ in place of $\hat{\mathbf{k}}\cdot\mathrm{rot}\,\mathbf{u}_0$, yields

$$\mathsf{w}_1(\mathsf{z}=\mathsf{z}_\mathrm{B}) = \frac{\sqrt{\mathsf{E}_\mathrm{V}}}{2\,\varepsilon}\,\nabla'^2\psi \tag{3.377}$$

Then, substitution of (3.376) and (3.377) into (3.375) divided by β gives

$$\frac{1}{\beta}\,\mathscr{J}\left(\psi,\, \nabla'^2\psi\right) + \frac{\partial\psi}{\partial \mathsf{x}} = \frac{\tau_0}{\beta\,\varepsilon\,\rho_\mathrm{s}\,f_0\,U\,H}\,\hat{\mathbf{k}}\cdot\mathrm{rot}\,\boldsymbol{\tau}' - \frac{\sqrt{\mathsf{E}_\mathrm{V}}}{2\,\beta\,\varepsilon}\,\nabla'^2\psi + \frac{1}{\beta\,\mathsf{Re}}\,\nabla'^4\psi \tag{3.378}$$

In terms of the lengths

$$\delta_\mathrm{I} := \frac{L}{\sqrt{\beta}} \qquad \delta_\mathrm{S} := \frac{f_0\,\sqrt{\mathsf{E}_\mathrm{V}}}{2\,\beta_0} \qquad \delta_\mathrm{M} := \left(\frac{A_\mathrm{H}}{\beta_0}\right)^{1/3} \tag{3.379}$$

Eq. (3.378) takes the form

$$\left(\frac{\delta_\mathrm{I}}{L}\right)^2 \mathscr{J}\left(\psi,\, \nabla'^2\psi\right) + \frac{\partial\psi}{\partial \mathsf{x}} = \frac{\tau_0}{\beta\,\varepsilon\,\rho_\mathrm{s}\,f_0\,U\,H}\,\hat{\mathbf{k}}\cdot\mathrm{rot}\,\boldsymbol{\tau}' - \frac{\delta_\mathrm{S}}{L}\,\nabla'^2\psi + \left(\frac{\delta_\mathrm{M}}{L}\right)^3 \nabla'^4\psi \tag{3.380}$$

since

$$\frac{1}{\beta\,\mathsf{Re}} = \left(\frac{\delta_\mathrm{M}}{L}\right)^3$$

Moreover, estimation of

$$\frac{\tau_0}{\beta\,\varepsilon\,\rho_\mathrm{s}\,f_0\,U\,H} = \frac{\tau_0}{\beta_0\,\rho_\mathrm{s}\,L\,U\,H}$$

by means of (3.188) with $\tau_0 = O(10^{-1}\mathrm{kg}\cdot\mathrm{m}^{-1}\cdot\mathrm{s}^{-2})$ shows that

$$\frac{\tau_0}{\beta_0\,\rho_\mathrm{s}\,L\,U\,H} = O(1) \tag{3.381}$$

Therefore, (3.380) can be simplified into its final form

$$\left(\frac{\delta_I}{L}\right)^2 \mathscr{J}\left(\psi, \nabla'^2\psi\right) + \frac{\partial\psi}{\partial x} = \hat{\mathbf{k}}\cdot \mathrm{rot}\,\boldsymbol{\tau}' - \frac{\delta_S}{L}\nabla'^2\psi + \left(\frac{\delta_M}{L}\right)^3 \nabla'^4\psi \tag{3.382}$$

Then, consider the steady version of the linear equation (3.221)

$$\frac{\partial\Psi}{\partial x} = {}_1w(z=z_E) - {}_1w(z=z_B) + \left(\frac{\delta_M}{L}\right)^3 \nabla'^4\Psi \tag{3.383}$$

The relation between ${}_1w(z=z_E)$ and w_E is given by the truncation up to the leading order in ε_T of the expansion $w_E = \varepsilon_T\ {}_1w(z=z_E) + O(\varepsilon_T^2)$ whence, by resorting again to (3.381), we obtain

$${}_1w(z=z_E) = \frac{w_E}{\varepsilon_T} = \hat{\mathbf{k}}\cdot \mathrm{rot}\,\boldsymbol{\tau}' \tag{3.384}$$

and, recalling definitions (3.379),

$${}_1w(z=z_B) = \frac{\sqrt{E_V}}{2\,\varepsilon_T}\nabla'^2\Psi = \frac{\delta_S}{L}\nabla'^2\Psi \tag{3.385}$$

On the whole, from (3.383) to (3.385), the linear vorticity equation

$$\frac{\partial\Psi}{\partial x} = \hat{\mathbf{k}}\cdot \mathrm{rot}\,\boldsymbol{\tau}' - \frac{\delta_S}{L}\nabla'^2\Psi + \left(\frac{\delta_M}{L}\right)^3 \nabla'^4\Psi \tag{3.386}$$

finally follows. Equations (3.382) and (3.386) govern the single-layer, wind-driven ocean circulation in different regimes.

To single out model solutions from (3.382) and (3.386), we must:

- Establish the wind-stress field $\boldsymbol{\tau}'$ as an input, which determines the forcing

$$\hat{\mathbf{k}}\cdot \mathrm{rot}\,\boldsymbol{\tau}' \tag{3.387}$$

- Fix linear homogeneous additional boundary conditions demanded to single out a unique solution when the fourth-order vorticity sinks

$$\left(\frac{\delta_M}{L}\right)^3 \nabla'^4\psi \quad \text{and} \quad \left(\frac{\delta_M}{L}\right)^3 \nabla'^4\Psi$$

  are present.

Energy arguments based on the time-dependent version of (3.382) and (3.386) suggest, as we will ascertain in the following, to select additional conditions consistent with the inequalities

$$\oint_{\partial D} (\hat{\mathbf{n}} \cdot \nabla' \psi) \nabla'^2 \psi \, ds \leq 0 \qquad \text{or} \qquad \oint_{\partial D} (\hat{\mathbf{n}} \cdot \nabla' \Psi) \nabla'^2 \Psi \, ds \leq 0 \tag{3.388}$$

In (3.388), $\hat{\mathbf{n}}$ is the unit vector locally normal to ∂D and pointing outside D, while ds is the differential arclength along ∂D, oriented anticlockwise. In the case of the standard square fluid domain

$$D = [0 \leq x \leq 1] \times [0 \leq y \leq 1] \tag{3.389}$$

a straightforward computation gives (3.388) in the form

$$\int_0^1 \left[\frac{\partial^2 \psi}{\partial y^2} \frac{\partial \psi}{\partial y} \right]_{y=0}^{y=1} dx + \int_0^1 \left[\frac{\partial^2 \psi}{\partial x^2} \frac{\partial \psi}{\partial x} \right]_{x=0}^{x=1} dy \leq 0 \tag{3.390}$$

or

$$\int_0^1 \left[\frac{\partial^2 \Psi}{\partial y^2} \frac{\partial \Psi}{\partial y} \right]_{y=0}^{y=1} dx + \int_0^1 \left[\frac{\partial^2 \Psi}{\partial x^2} \frac{\partial \Psi}{\partial x} \right]_{x=0}^{x=1} dy \leq 0 \tag{3.391}$$

respectively.

The Sverdrup Balance

Owing to the smallness of the coefficients $(\delta_I/L)^2$, δ_S/L and $(\delta_M/L)^3$ (which are far smaller than $O(1)$ whenever the terms $\mathscr{J}(\psi, \nabla'^2\psi)$, $\nabla'^2\psi$ and $\nabla'^4\psi$ are $O(1)$), the $O(1)$ dynamic equilibrium, derived from both (3.382) and (3.386), relies on the so-called *Sverdrup balance*

$$\frac{\partial \psi_{\text{int}}}{\partial x} = \hat{\mathbf{k}} \cdot \operatorname{rot} \boldsymbol{\tau}'(x, y) \tag{3.392}$$

On the other hand, the Sverdrup balance ceases to hold in the regions of the basin where at least one of the terms $(\delta_I/L)^2 \mathscr{J}(\psi, \nabla'^2\psi)$, $(\delta_S/L)\nabla'^2\psi$ or $(\delta_M/L)^3\nabla'^4\psi$ is $O(1)$, which happens if one or more of the terms $\mathscr{J}(\psi, \nabla'^2\psi)$, $\nabla'^2\psi$ or $\nabla'^4\psi$ take values large enough. The Sverdrup flow $\psi_{\text{int}}(x, y)$ is determined by integrating (3.392) under the no–mass flux boundary condition, which – *a priori* – may be imposed either at $x = 0$ or at $x = 1$ relatively to the domain (3.389). The alternative

$$\text{either} \qquad \psi_{\text{int}}(0, y) = 0 \qquad \text{or} \qquad \psi_{\text{int}}(1, y) = 0 \tag{3.393}$$

will be decided on energy arguments; in any case, we anticipate that the correct condition is the second one of (3.393), and hence

$$\psi_{\text{int}}(x, y) = -\int_x^1 \hat{\mathbf{k}} \cdot \operatorname{rot} \boldsymbol{\tau}'(s, y) \, ds \tag{3.394}$$

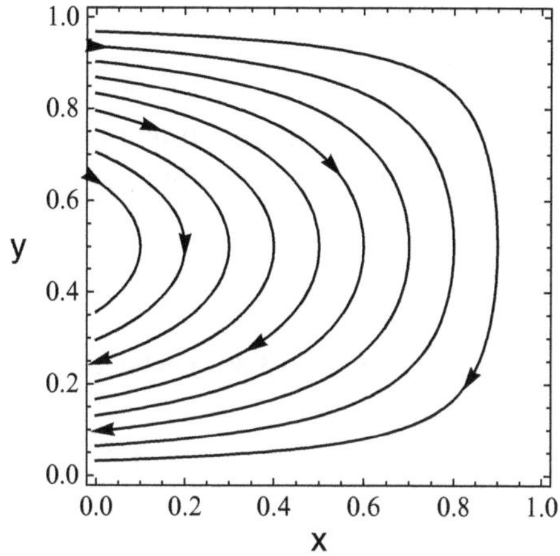

Fig. 3.11 Streamlines of (3.398). These Streamlines mimic the structure of a subtropical gyre and imply a southward motion of the fluid columns in the interior and eastern part of the basin

An immediate consequence of (3.392) is that, if at certain latitude the wind-stress curl vanishes (i.e., $\hat{\mathbf{k}} \cdot \mathrm{rot}\,\boldsymbol{\tau}' = 0$), then the meridional current $v_{int} = \partial \psi_{int}/\partial x$ vanishes at the same latitude, thus yielding a locally zonal flow. This is the reason why the actual wind-driven oceanic circulation arranges itself in gyres, each being included between a couple of circles of latitude which constitute two consecutive zeros of $\hat{\mathbf{k}} \cdot \mathrm{rot}\,\boldsymbol{\tau}'$. Due to this fact, if a wind-driven, single-gyre flow is modelled in the usual fluid domain (3.389) then the wind forcing has to satisfy the condition

$$\hat{\mathbf{k}} \cdot \mathrm{rot}\,\boldsymbol{\tau}' = 0 \qquad \text{at } y = 0 \text{ and at } y = 1 \tag{3.395}$$

For instance, to account for the wind-driven subtropical gyre of the Atlantic Ocean (where the westerlies and the trades flow in opposite directions), the wind-stress field

$$\boldsymbol{\tau}' = -\frac{\cos(\pi y)}{\pi}\hat{\mathbf{i}} \tag{3.396}$$

can be introduced with reference to (3.389), whence

$$\hat{\mathbf{k}} \cdot \mathrm{rot}\,\boldsymbol{\tau}' = -\sin(\pi y) \tag{3.397}$$

Substitution of (3.397) into (3.394) gives

$$\psi_{int}(x, y) = (1 - x)\sin(\pi y) \tag{3.398}$$

A set of streamlines of (3.398) are reported in Fig. 3.11. In linear homogeneous models, the streamlines of the Sverdrup balance, for instance, those given in Fig. 3.11, represent a part of the actual patterns of the fluid parcels, but the total solution must be supplemented by a further term, say $\phi(x, y)$, in order that the streamlines of

$$\Psi(x, y) = \psi_{int}(x, y) + \phi(x, y) \tag{3.399}$$

be closed lines. In (3.399), $\phi(\mathsf{x}, \mathsf{y})$ is appreciably different from zero only in a thin region in the proximity of the western boundary. In nonlinear homogeneous models, a superposition of the kind (3.399) does not hold, and the pattern of streamlines of the Sverdrup balance may strongly differ from that of the total solution.

The dimensional version of (3.392) is obtained by setting

$$v_{\rm int} := U \frac{\partial \psi_{\rm int}}{\partial \mathsf{x}}$$

and using (3.381) in (3.392), whence the dimensional Sverdrup balance

$$\beta_0 H v_{\rm int} = \frac{1}{\rho_s} \hat{\mathbf{k}} \cdot \mathrm{rot}\, \boldsymbol{\tau} \tag{3.400}$$

follows after little algebra. Note that the quantity $H v_{\rm int}$ is a transport, so $\psi_{\rm int}$ should be named, more literally, *transport stream function* rather than only *stream function*. The physics on which balance (3.400) is based can be explained not only by resorting to scaling arguments, as we did at the beginning of this paragraph, but also by means of potential vorticity conservation together with the dynamics of the upper Ekman layer. Consider in fact a material volume of fluid included, at a certain "initial" time, in a cylinder between two planes, the first being the ocean floor (say, for simplicity, at $z = 0$), and the second one located at the transition region between the upper Ekman layer and the geostrophic interior (say, at $z = z_E$). Usually, this kind of material volume is named *fluid column*. The "initial" height of the cylinder is $z_E - 0 = H$. The transport (3.334) induces the vertical velocity (3.341) at the summit of the fluid column according to (3.342). Assume that relative vorticity ζ is negligible with respect to planetary vorticity $f(y)$, as actually happens in the ocean interior where $\zeta / f(y) = O(\varepsilon)$. Then, conservation of the potential vorticity of the fluid column [i.e. $D\Pi_{\rm int}/Dt = 0$ with $\Pi_{\rm int} := (f_0 + \beta_0 y)/H$] implies

$$\beta_0 H v_{\rm int} - \left(1 + \frac{\beta_0 y}{f_0}\right) f_0 w_E = 0 \tag{3.401}$$

In (3.401), $v_{\rm int} = Dy/Dt$ while $w_E = DH/Dt$ is given by (3.341); so, from the approximation $1 + \beta_0 y / f_0 \approx 1$, Eq. (3.401) takes immediately the form (3.400). Thus, the fluid column migrates meridionally with velocity

$$v_{\rm int} = \frac{1}{\beta_0 H \rho_s} \hat{\mathbf{k}} \cdot \mathrm{rot}\, \boldsymbol{\tau} \tag{3.402}$$

In the course of time, the fluid column tends to flatten against the sea bottom (by definition no mass flux is possible across the surface of a material volume) and translates southward as a whole; in the meantime, a new fluid column, identical to the former at the "initial" time, takes its place. And things are repeated in endless way.

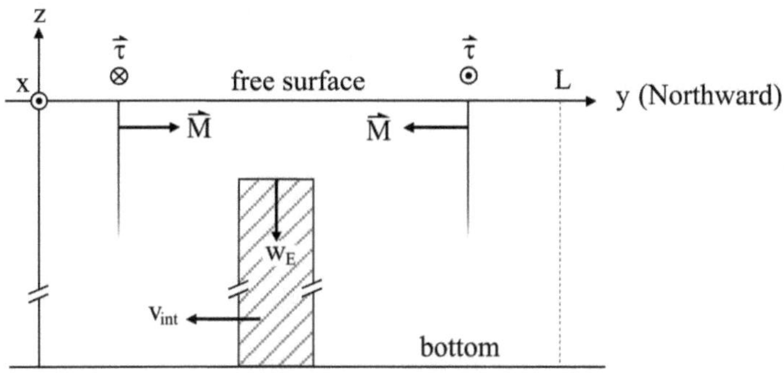

Fig. 3.12 Sketch of the model governed by (3.403), (3.404), (3.405) and (3.406). Since the wind stress curl is negative in $0 \le y \le L$, in the same stripe the transport **M** is convergent and generates the downward vertical velocity w_E. Then, because of conservation of potential vorticity, every material fluid column migrates southward with velocity v_{int}

For instance, if the wind stress is the dimensional version of (3.396), that is,

$$\boldsymbol{\tau}(y) = -\frac{\tau_0}{\pi} \cos\left(\pi \frac{y}{L}\right) \hat{\mathbf{i}} \tag{3.403}$$

then the subsurface transport (3.334) has the form

$$\mathbf{M} = \frac{\tau_0}{\pi \rho_s f_0} \cos\left(\pi \frac{y}{L}\right) \hat{\mathbf{j}} \tag{3.404}$$

and points northward for $0 \le y < L/2$, southward for $L/2 < y \le L$. Thus, the convergence of **M** yields a downward vertical velocity (3.341)

$$w_E = -\frac{\tau_0}{\rho_s f_0 L} \sin\left(\pi \frac{y}{L}\right) \tag{3.405}$$

extending on the whole latitude $0 < y < L$ of the basin.

In the special case (3.405), the meridional velocity (3.402) is southward and given by

$$v_{int} = -\frac{\tau_0}{\beta_0 \rho_s H L} \sin\left(\pi \frac{y}{L}\right) \tag{3.406}$$

The meaning of the physical quantities involved in this instance, namely, $\boldsymbol{\tau}$, **M**, w_E, and v_{int}, is clarified in Fig. 3.12.

The Sverdrup Balance and the Equatorial Countercurrent

We momentarily release hypothesis (3.396) in favour of the more general stress

$$\boldsymbol{\tau}'(\mathsf{y}) = \mathsf{S}(\mathsf{y})\hat{\mathbf{i}} \tag{3.407}$$

In the following discussion, we assume the presence of a current field in the proximity of the western boundary – which is able to close the streamlines of the interior. Owing to (3.407), stream function (3.394) gives

$$\psi_{int}(x, y) = (1 - x)\frac{dS}{dy} \tag{3.408}$$

and the Sverdrup flow

$$\mathbf{u}_{int} = \left(-\frac{\partial \psi_{int}}{\partial y}, \frac{\partial \psi_{int}}{\partial x}\right)$$

can be written as

$$\mathbf{u}_{int} = (x - 1)\frac{d^2S}{dy^2}\hat{\mathbf{i}} - \frac{dS}{dy}\hat{\mathbf{j}} \tag{3.409}$$

where $x - 1 \leq 0$. If, in a certain latitudinal interval, we have

$$S(y)\frac{d^2S}{dy^2} > 0 \tag{3.410}$$

then the wind stress and the Sverdrup flow are almost antiparallel (apart from a small meridional component of (3.409) due to dS/dy); that is to say, the wind blows contrary to the underlying current. While this configuration is not realized by (3.396) and (3.398), relationship (3.410) is able to explain the existence of the observed countercurrent in the Pacific at a latitude between 5 °N and 10 °N, flowing against the trade winds. Indeed, just this phenomenon led Sverdrup, in 1947, to derive his celebrated Eq. (3.400), which governs such counterintuitive wind driven model.

Numerical Example

We now consider a certain set of observed values of $\tau(\phi)$ relative to the Pacific Ocean, where the latitude ϕ varies between a few degrees above the equator and about 50 °N. Further, we normalize these values so that we deal with a function $\tau = \tau'(y)$, where $-1/\pi \leq \tau' \leq 1/\pi$ and $0 \leq y \leq 1$ are linked by (3.407). Finally, we compute the best fit to these data by using the function

$$S(y) = \begin{cases} f(a_1, b_1, c_1, y_1; y) & \text{if } 0 \leq y \leq Y \\ f(a_2, b_2, c_2, y_2; y) & \text{if } Y < y \leq 1 \end{cases} \tag{3.411}$$

where f is the function defined by

$$f(a, b, c, y_0; y) := a + b(y - y_0) * \exp[-c * (y - y_0)^2]$$

as shown in Fig. 3.13(a).

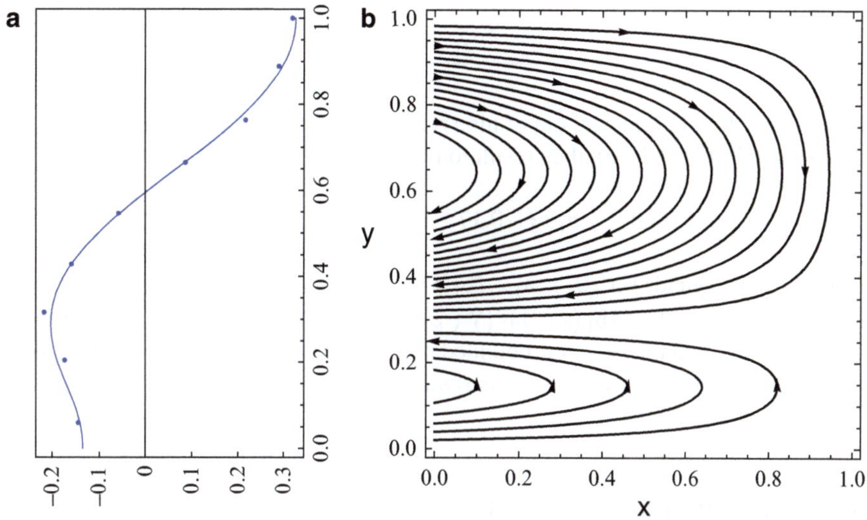

Fig. 3.13 Equatorial countercurrent. Panel (**a**): wind stress versus latitude. Panel (**b**): streamlines of Sverdrup flow due to the wind stress shown in panel (*a*). Note that, following Munk (1950), the orientation of the τ'-axis in (*a*) is not the standard one for a Cartesian graph of $\tau'(y)$

Assumptions

$$\left[\frac{d}{dy} f(a_1, b_1, c_1, y_1; y)\right]_{y=0} = 0$$

$$\left[\frac{d}{dy} f(a_1, b_1, c_1, y_1; y)\right]_{y=Y} = 0$$

yield

$$c_1 = \frac{2}{Y^2} \qquad y_1 = \frac{Y}{2}$$

Likewise, assumptions

$$\left[\frac{d}{dy} f(a_2, b_2, c_2, y_2; y)\right]_{y=1} = 0$$

$$\left[\frac{d}{dy} f(a_2, b_2, c_2, y_2; y)\right]_{y=Y} = 0$$

yield

$$c_2 = \frac{2}{(1-Y)^2} \qquad y_2 = \frac{1+Y}{2}$$

Finally, the continuity condition

$$f(a_1, b_1, c_1, y_1; Y) = f(a_2, b_2, c_2, y_2; Y)$$

implies

$$\mathsf{a}_2 = \frac{\mathsf{b}_2 + 2\,\mathsf{a}_1\sqrt{\mathsf{e}} + \mathsf{b}_1\,\mathsf{Y} - \mathsf{b}_2\,\mathsf{Y}}{2\sqrt{\mathsf{e}}}$$

Hence, only four parameters out of nine remain to be estimated numerically; for the data shown in Fig. 3.13(a), they result to be

$$\mathsf{Y} \approx 0.29 \qquad \mathsf{a}_1 \approx -0.17 \qquad \mathsf{b}_1 \approx -0.38 \qquad \mathsf{b}_2 \approx 1.2$$

Hence, (3.407) yields $\hat{\mathbf{i}}\cdot\boldsymbol{\tau}'(0) = \mathsf{a}_1 - (\mathsf{b}_1\,\mathsf{Y})/(2\sqrt{\mathsf{e}}) \approx -0.14 < 0$, while, according to (3.409), $\hat{\mathbf{i}}\cdot\mathbf{u}_{\text{int}}(0) = 4\,\mathsf{b}_1\,(\mathsf{x}-1)/(\mathsf{Y}\sqrt{\mathsf{e}}) > 0$. This means that the wind stress is westward in a suitable interval bounded from below by $\mathsf{y} = 0$, whereas the zonal component of the interior flow is eastward. The flow field relative to (3.411) is depicted in Fig. 3.13(b).

Energetics of the Homogeneous Model

In the present subsection, we shall take into account only the time-dependent counterpart of (3.386) derived by using (3.221), that is,

$$\frac{\partial}{\partial \tilde{t}}\nabla'^2\Psi + \frac{\partial \Psi}{\partial \mathsf{x}} = \hat{\mathbf{k}}\cdot\text{rot}\,\boldsymbol{\tau}' - \frac{\delta_{\text{S}}}{L}\nabla'^2\Psi + \left(\frac{\delta_{\text{M}}}{L}\right)^3\nabla'^4\Psi \tag{3.412}$$

because the Jacobian term in (3.382) does not contribute to the energy equation integrated over D (see Appendix A, p. 372). Multiplication of (3.412) by Ψ and the subsequent integration over D with the aid of the no–mass flux condition along ∂D yields, after little algebra,

$$\frac{1}{2}\frac{\text{d}}{\text{d}\tilde{t}}\int_{\text{D}}|\nabla'\Psi|^2\,\text{d}x\,\text{d}y + \int_{\text{D}}\hat{\mathbf{k}}\cdot(\Psi\,\text{rot}\,\boldsymbol{\tau}')\,\text{d}x\,\text{d}y =$$

$$= -\frac{\delta_{\text{S}}}{L}\int_{\text{D}}|\nabla'\Psi|^2\,\text{d}x\,\text{d}y + \left(\frac{\delta_{\text{M}}}{L}\right)^3\int_{\text{D}}(\nabla'\Psi)\cdot\nabla'(\nabla'^2\Psi)\,\text{d}x\,\text{d}y \tag{3.413}$$

In particular, using Stokes' theorem and standard differential identities, one may express the second term in (3.413) as

$$\begin{aligned}\int_{\text{D}}\hat{\mathbf{k}}\cdot(\Psi\,\text{rot}\,\boldsymbol{\tau}')\,\text{d}x\,\text{d}y &= \int_{\text{D}}\left[\hat{\mathbf{k}}\cdot\text{rot}\,(\Psi\boldsymbol{\tau}') - \hat{\mathbf{k}}\cdot\nabla'\Psi\times\boldsymbol{\tau}'\right]\text{d}x\,\text{d}y \\ &= \int_{\partial\text{D}}\Psi\boldsymbol{\tau}'\cdot\hat{\mathbf{t}}\,\text{d}s - \int_{\text{D}}\boldsymbol{\tau}'\cdot\hat{\mathbf{k}}\times\nabla'\Psi\,\text{d}x\,\text{d}y \\ &= -\int_{\text{D}}\boldsymbol{\tau}'\cdot\mathbf{u}_0\,\text{d}x\,\text{d}y \end{aligned} \tag{3.414}$$

where $\hat{\mathbf{t}}$ is the versor tangent to ∂D. Moreover, using the divergence theorem, the last integral in (3.413) may be expressed as

$$\int_{\mathrm{D}} (\nabla'\Psi)\cdot\nabla'(\nabla'^2\Psi)\,\mathrm{d}x\mathrm{d}y = \int_{\mathrm{D}} \mathrm{div}\left[(\nabla'^2\Psi)\nabla'\Psi\right]\mathrm{d}x\mathrm{d}y - \int_{\mathrm{D}} (\nabla'^2\Psi)^2\,\mathrm{d}x\mathrm{d}y$$
$$= \int_{\partial\mathrm{D}} (\nabla'^2\Psi)\hat{\mathbf{n}}\cdot\nabla'\Psi\mathrm{d}s - \int_{\mathrm{D}} (\nabla'^2\Psi)^2\,\mathrm{y}\mathrm{d}x\mathrm{d}y \tag{3.415}$$

With the aid of (3.414) and (3.415), Eq. (3.413) can be written as

$$\frac{1}{2}\frac{\mathrm{d}}{\mathrm{d}\tilde{t}}\int_{\mathrm{D}} |\nabla'\Psi|^2\,\mathrm{d}x\mathrm{d}y = \int_{\mathrm{D}} \boldsymbol{\tau}'\cdot\mathbf{u}_0\,\mathrm{d}x\mathrm{d}y - \frac{\delta_{\mathrm{S}}}{L}\int_{\mathrm{D}} |\nabla'\Psi|^2\,\mathrm{d}x\mathrm{d}y$$
$$-\left(\frac{\delta_{\mathrm{M}}}{L}\right)^3\int_{\mathrm{D}} (\nabla'^2\Psi)^2\,\mathrm{d}x\mathrm{d}y + \left(\frac{\delta_{\mathrm{M}}}{L}\right)^3\int_{\partial\mathrm{D}} (\nabla'^2\Psi)\hat{\mathbf{n}}\cdot\nabla'\Psi\mathrm{d}s \tag{3.416}$$

Condition (3.388) applies to the last term of (3.416): for instance, if the wind stress is suddenly turned off, then (3.388) implies

$$\frac{1}{2}\frac{\mathrm{d}}{\mathrm{d}\tilde{t}}\int_{\mathrm{D}} |\nabla'\Psi|^2\,\mathrm{d}x\mathrm{d}y < -\frac{\delta_{\mathrm{S}}}{L}\int_{\mathrm{D}} |\nabla'\Psi|^2\,\mathrm{d}x\mathrm{d}y - \left(\frac{\delta_{\mathrm{M}}}{L}\right)^3\int_{D} (\nabla'^2\Psi)^2\,\mathrm{d}x\mathrm{d}y \tag{3.417}$$

Because of the inequality (see Appendix A, p. 374)

$$-\int_{\mathrm{D}} (\nabla'^2\Psi)^2\,\mathrm{d}x\mathrm{d}y < -2\pi^2\int_{\mathrm{D}} |\nabla'\Psi|^2\,\mathrm{d}x\mathrm{d}y$$

(where D is given by (3.389)), quite reasonably inequality (3.417) implies that the kinetic energy of the flow decays faster than exponentially in time, while this conclusion cannot be established if the inequality opposite to (3.388) is assumed.

On the Boundary Condition of the Sverdrup Flow

On the basis of above results, we easily ascertain that the steady version of (3.413) implies

$$\int_{\mathrm{D}} \hat{\mathbf{k}}\cdot(\Psi\,\mathrm{rot}\,\boldsymbol{\tau}')\,\mathrm{d}x\mathrm{d}y < 0 \tag{3.418}$$

that is to say, recalling (3.392),

$$\int_{\mathrm{D}} \Psi\frac{\partial\psi_{\mathrm{int}}}{\partial\mathrm{x}}\,\mathrm{d}x\mathrm{d}y < 0 \tag{3.419}$$

Owing to (3.399), inequality (3.419) is equivalent to

$$\int_{\mathsf{D}} (\psi_{\mathrm{int}} + \phi) \frac{\partial \psi_{\mathrm{int}}}{\partial \mathsf{x}} \, \mathsf{dx}\,\mathsf{dy} < 0$$

that is to say,

$$\int_0^1 \left[\psi_{\mathrm{int}}^2(1,\mathsf{y}) - \psi_{\mathrm{int}}^2(0,\mathsf{y})\right] \mathsf{dy} + 2\int_{\mathsf{D}} \phi \frac{\partial \psi_{\mathrm{int}}}{\partial \mathsf{x}} \, \mathsf{dx}\,\mathsf{dy} < 0 \tag{3.420}$$

The task of the component ϕ of the total stream function (3.399) is to bring to zero Ψ at the meridional boundary, where ψ_{int} is unable to satisfy the no–mass flux condition: thus, ϕ is appreciably different from zero only in a thin region of the kind $\overline{\mathsf{D}} = [\mathsf{x}_0 \le \mathsf{x} \le \mathsf{x}_0 + \delta_{\mathrm{F}}/L] \times [0 \le \mathsf{y} \le 1]$, where either $\mathsf{x}_0 = 0$ or $\mathsf{x}_0 = 1 - \delta_{\mathrm{F}}/L$ and $\delta_{\mathrm{F}}/L < O(1)$, δ_{F} being a dimensional length determined by the kind of dissipation, as we will see when dealing with some definite models. In both cases,

$$\frac{\partial \psi_{\mathrm{int}}}{\partial \mathsf{x}} \phi = \begin{cases} O(1) & \text{inside } \overline{\mathsf{D}} \subset \mathsf{D} \\ \text{almost zero} & \text{outside } \overline{\mathsf{D}} \end{cases}$$

Hence,

$$2\int_{\mathsf{D}} \phi \frac{\partial \psi_{\mathrm{int}}}{\partial \mathsf{x}} \, \mathsf{dx}\,\mathsf{dy} \approx 2\int_{\overline{\mathsf{D}}} \phi \frac{\partial \psi_{\mathrm{int}}}{\partial \mathsf{x}} \, \mathsf{dx}\,\mathsf{dy} = O\left(\frac{\delta_{\mathrm{F}}}{L}\right) \tag{3.421}$$

Therefore, the leading term at the l.h.s. of (3.420) is $\int_0^1 [\psi_{\mathrm{int}}^2(1,\mathsf{y}) - \psi_{\mathrm{int}}^2(0,\mathsf{y})]\,\mathsf{dy}$, and the latter is negative only if

$$\psi_{\mathrm{int}}(1,\mathsf{y}) = 0 \tag{3.422}$$

Equation (3.422) selects the correct boundary condition to be applied to the Sverdrup balance (3.392) to obtain (3.394).

Linear Models of the Steady Wind-Driven Circulation

The governing equations of the linear models are special versions of (3.386) with a definite forcing (3.387). The stream function Ψ is requested to satisfy the no–mass flux condition along the boundary $\partial\mathsf{D}$ of the fluid domain D. In the case in which $\nabla'^4\Psi \neq 0$, further additional boundary conditions, consistent with (3.388), must be imposed to Ψ. To be concrete, forcing (3.397) will be applied; and the free-slip additional condition

$$\nabla'^2\Psi = 0 \qquad \text{on } \partial\mathsf{D} \tag{3.423}$$

that is,

$$\left(\frac{\partial^2\Psi}{\partial x^2}\right)_{x=0} = \left(\frac{\partial^2\Psi}{\partial x^2}\right)_{x=1} = 0$$

$$\left(\frac{\partial^2\Psi}{\partial y^2}\right)_{y=0} = \left(\frac{\partial^2\Psi}{\partial y^2}\right)_{y=1} = 0 \tag{3.424}$$

will be assumed. Note that boundary conditions (3.424) imply, via (3.391),

$$\oint_{\partial D} (\hat{\mathbf{n}} \cdot \nabla'\Psi)\, \nabla'^2\Psi = 0$$

Thus, the second of inequalities (3.388) is verified. With (3.399) in mind, and D given by (3.389), the main problem consists in the determination of ϕ and can be solved analytically with the aid of boundary layer techniques applied to the meridional boundary of equation $x = 0$. This approach is quite similar to that already used in dealing with the Fofonoff mode.

3.2.3 Classical Solutions of the Homogeneous Model

Stommel's Model

Stommel's model (1948) was conceived with the specific purpose to explain the dynamic cause of westward intensification, which is a common feature of all the oceanic gyres of the world ocean. The governing equation is (3.386) without the term $(\delta_M/L)^3\, \nabla'^4\Psi$, that is,

$$\frac{\partial\Psi}{\partial x} = -\sin(\pi y) - \frac{\delta_S}{L}\, \nabla'^2\Psi \tag{3.425}$$

According to (3.399) the total stream function is

$$\Psi = \psi_{\text{int}}(x, y) + \phi(\xi', y) \tag{3.426}$$

and ψ_{int} is given by (3.398). The appropriate boundary layer coordinate ξ' appearing in (3.426) satisfies the equation

$$Lx = \delta_S\, \xi' \tag{3.427}$$

whence, in particular,

$$\frac{\partial}{\partial x} = \frac{L}{\delta_S}\frac{\partial}{\partial \xi'} \qquad \text{and} \qquad \nabla'^2 = \left(\frac{L}{\delta_S}\right)^2 \frac{\partial^2}{\partial \xi'^2} + \frac{\partial^2}{\partial y^2} \tag{3.428}$$

Substitution of (3.426) into (3.425), with the use of (3.428) and a trivial rearrangement, yields the boundary layer equation for $\phi = \phi(\xi', y)$

$$\frac{\partial \phi}{\partial \xi'} = -\left(\frac{\delta_S}{L}\right)^2 \left(\nabla'^2 \Psi + \frac{\partial^2 \phi}{\partial y^2}\right) - \frac{\partial^2 \phi}{\partial \xi'^2} \tag{3.429}$$

Since $\delta_S < O(1)$, in the boundary layer approximation the square ratio $(\delta_S/L)^2 \ll 1$ is neglected in (3.429) with respect to unity, and thus the $O(1)$ equation

$$\frac{\partial \phi}{\partial \xi'} = -\frac{\partial^2 \phi}{\partial \xi'^2} \tag{3.430}$$

is taken into account. Boundary condition $\Psi(0, y) = 0$ written in terms of (3.426) gives

$$\phi(\xi' = 0, y) = -\sin(\pi y) \tag{3.431}$$

On the other hand, $\phi(x = 1, y) = 0$ should be written as

$$\phi\left(\xi' = \frac{L}{\delta_S}, y\right) = 0 \tag{3.432}$$

but, since $L/\delta_S \gg 1$, (3.432) is substituted by

$$\lim_{\xi' \to \infty} \phi(\xi', y) = 0 \tag{3.433}$$

The general integral of (3.430) is

$$\phi(\xi', y) = C_1(y) + C_2(y)\exp(-\xi') \tag{3.434}$$

and the functions $C_1(y)$ and $C_2(y)$ are singled out by resorting to (3.431) and (3.433). The latter implies $C_1(y) = 0$, whence the former yields $C_2(y) = -\sin(\pi y)$. Thus,

$$\phi = -\sin(\pi y)\exp(-\xi') \tag{3.435}$$

and hence the model solution written in terms of the original variables x and y is

$$\Psi(x, y) = \left[1 - x - \exp\left(-\frac{L}{\delta_S}x\right)\right]\sin(\pi y) \tag{3.436}$$

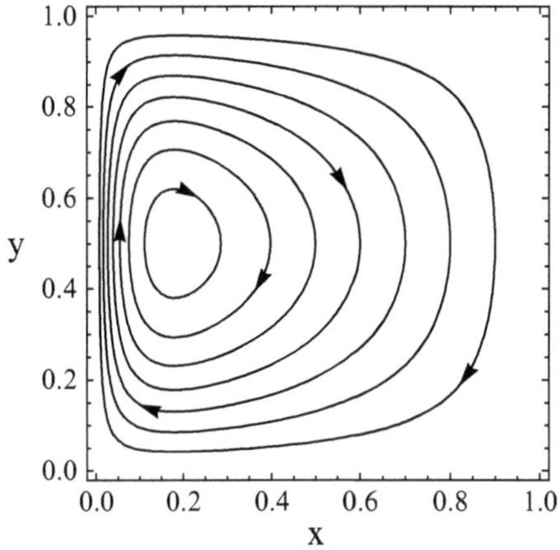

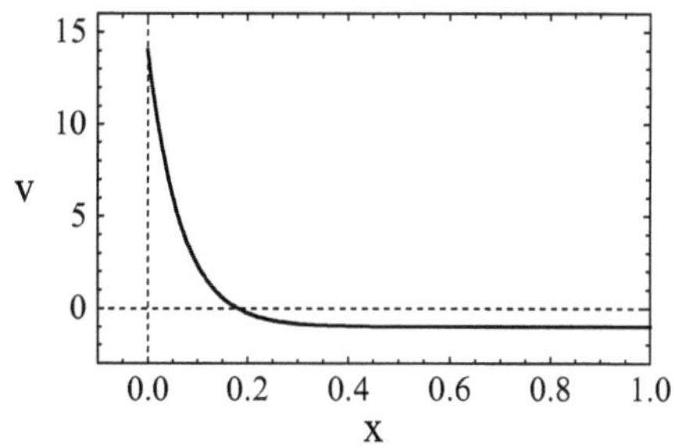

Fig. 3.14 Stommel's model. *Left*: streamlines. *Right*: mid-basin horizontal profile of meridional velocity

Using (3.436), it is possible to evaluate the intensity of the meridional current $v_0 = \partial\Psi/\partial x$ along the eastern and the western boundaries, and to check whether the latter is intensified or not. Indeed, one easily obtains from (3.436)

$$v_0(0,y) = \left(\frac{L}{\delta_S} - 1\right)\sin(\pi y) = O\left(\frac{L}{\delta_S}\right) \qquad \text{(northward)} \qquad (3.437)$$

while, neglecting $\exp(-L/\delta_S)$ with respect to unity in evaluating $(\partial\Psi/\partial x)_{x=1}$,

$$v_0(1,y) = -\sin(\pi y) = O(1) \qquad \text{(southward)} \qquad (3.438)$$

Thus, $v_0(0,y)/v_0(1,y) = O(L/\delta_S) \gg O(1)$, and the westward intensification is quite evident. See Fig. 3.14.

We now want to show, in a most direct way, that the responsible of westward intensification is the beta effect. To this aim, we reconsider the vorticity equation without $\partial\Psi/\partial x$, and solve the related Dirichlet boundary value problem, that is,

$$r\nabla'^2\Psi + \sin(\pi y) = 0 \qquad \forall(x,y) \in D \qquad (3.439)$$

$$\Psi(x,y) = 0 \qquad \forall(x,y) \in \partial D \qquad (3.440)$$

In (3.439), the constant r is introduced in place of δ_S/L to point out that (3.439) implies $r\nabla'^2\Psi = O(1)$, but this estimate cannot be achieved if $r = O(\delta_S/L)$.

By means of the position

$$\Psi = X(x)\sin(\pi y) \qquad (3.441)$$

the partial differential problem (3.439) and (3.440) is easily transformed into the ordinary differential problem

$$\mathsf{X}''(\mathsf{x}) - \pi^2 \mathsf{X}(\mathsf{x}) + \frac{1}{\mathsf{r}} = 0 \tag{3.442}$$

$$\mathsf{X}(0) = \mathsf{X}(1) = 0 \tag{3.443}$$

whose solution is

$$\mathsf{X}(\mathsf{x}) = \frac{1}{\pi^2 \mathsf{r}} \left\{ 1 - \frac{\cosh[\pi(\mathsf{x} - 1/2)]}{\cosh(\pi/2)} \right\} \tag{3.444}$$

From (3.444), the meridional currents at $\mathsf{x} = 0$ and $\mathsf{x} = 1$ are evaluated to be

$$\mathsf{v}_0(0, \mathsf{y}) = \frac{1}{\pi \mathsf{r}} \tanh\left(\frac{\pi}{2}\right) \sin(\pi \mathsf{y}) \qquad \text{(western boundary)} \tag{3.445}$$

$$\mathsf{v}_0(1, \mathsf{y}) = -\mathsf{v}_0(0, \mathsf{y}) \qquad \text{(eastern boundary)} \tag{3.446}$$

and correspond to northward and southward flow, respectively. Relation (3.446) shows that the westward intensification is clearly absent in model (3.439)–(3.440).

We return to (3.436) and evaluate relative vorticity $\nabla'^2\Psi$. A straightforward computation gives

$$\nabla'^2\Psi = \left[\pi^2(\mathsf{x} - 1) + \left(\pi^2 - \frac{L^2}{\delta_S^2}\right) \exp\left(-\frac{L}{\delta_S}\mathsf{x}\right)\right] \sin(\pi \mathsf{y}) \tag{3.447}$$

Note that $\mathsf{x} - 1 < 0$ and $\pi^2 - L^2/\delta_S^2 < 0$ imply $\nabla'^2\Psi < 0$ on the whole fluid domain. As a consequence, the vertical velocity w_B at the top of the benthic Ekman layer is also negative (see (3.354)); so, the related transport is divergent, in accordance with (3.353).

From (3.447) we have, in particular,

$$(\nabla'^2\Psi)_{\mathsf{x}=0} = -\left(\frac{L}{\delta_S}\right)^2 \sin(\pi \mathsf{y}) \gg O(1) \tag{3.448}$$

$$(\nabla'^2\Psi)_{\mathsf{x}=1} = \left(\pi^2 - \frac{L^2}{\delta_S^2}\right) \exp\left(-\frac{L}{\delta_S}\right) \sin(\pi \mathsf{y}) \ll O(1) \tag{3.449}$$

Equations (3.448) and (3.449) agree with the fact that ambient rotation allows the production of relative vorticity: in fact, in the course of the motion along its own streamline, the relative vorticity of a fluid parcel is very low far from the western boundary, but becomes $O(L^2/\delta_S^2)$ in the western boundary layer, where each streamline necessarily passes.

We can also check, a posteriori, the validity of inequality (3.421) for $\delta_F = \delta_S$. In fact, recalling (3.434) in the form (3.435) and using (3.427), we have

$$\frac{\partial \psi_{\text{int}}}{\partial \mathsf{x}}\,\phi\,\mathrm{d}\mathsf{x} = \frac{\delta_S}{L}\,\sin^2(\pi \mathsf{y})\,\exp(-\xi')\,\mathrm{d}\xi'$$

and therefore

$$2\int_D \frac{\partial \psi_{\text{int}}}{\partial \mathsf{x}}\,\phi\,\mathrm{d}\mathsf{x}\,\mathrm{d}\mathsf{y} = 2\,\frac{\delta_S}{L}\int_0^1 \sin^2(\pi \mathsf{y})\,\mathrm{d}\mathsf{y}\int_0^{+\infty}\exp(-\xi')\,\mathrm{d}\xi' = \frac{\delta_S}{L}$$

Remark About Westward Intensification

Consider again the equation

$$\frac{\partial \Psi}{\partial \mathsf{x}} = \hat{\mathbf{k}}\cdot \mathrm{rot}\,\boldsymbol{\tau}' - \frac{\delta_s}{L}\,\nabla'^2\Psi \tag{3.450}$$

Let A be a region of D encircled by a streamline of Ψ; so, Ψ takes a constant value along the boundary ∂A of A. Then, using Stokes' and divergence theorems, integration of (3.450) over A yields the equation

$$\oint_{\partial A}\left(\boldsymbol{\tau}' - \frac{\delta_S}{L}\,\mathbf{u}_0\right)\cdot\hat{\mathbf{t}}\,\mathrm{d}s = 0 \tag{3.451}$$

In (3.451), $\hat{\mathbf{t}}$ is the unit vector locally tangent to ∂A, while ds is the differential arc length along ∂A, and $\mathbf{u}_0 = \hat{\mathbf{k}} \times \nabla'\Psi$. If $\mathbf{u}_0$ were $O(1)$ along the whole circuit ∂A, then integral (3.451) would yield

$$\oint_{\partial A}\boldsymbol{\tau}'\cdot\hat{\mathbf{t}}\,\mathrm{d}s = 0 \tag{3.452}$$

owing to the smallness of δ_S/L compared to $\boldsymbol{\tau}' = O(1)$. But the constraint (3.452) on the wind-stress field is not physically justified. To violate (3.452), the geostrophic current $\mathbf{u}_0$ must take, somewhere along ∂A, values high enough to make $O(1)$ the product $(\delta_S/L)\mathbf{u}_0$, which becomes comparable with $\boldsymbol{\tau}'$; so, (3.451) is satisfied without resorting to (3.452). This, whatever the streamline ∂A may be. Thus, the intensification takes place on each streamline or, equivalently, each streamline must pass across a region where the current is intensified. This fact is evident in Fig. 3.14 (left panel), where each streamline exhibits a branch that looks "constricted" close to the western boundary.

Munk's Model

The governing equation of Munk's model (1950) is (3.386) without the dissipative term $-(\delta_S/L)\nabla'^2\Psi$, that is,

$$\frac{\partial \Psi}{\partial x} = -\sin(\pi y) + \left(\frac{\delta_M}{L}\right)^3 \nabla'^4 \Psi \tag{3.453}$$

The total stream function Ψ has again the form (3.426) with ψ_{int} given by (3.398). Thus, the appropriate boundary layer coordinate ξ' satisfies the equation

$$Lx = \delta_M \xi' \tag{3.454}$$

whence the differentiation rule

$$\frac{\partial}{\partial x} = \frac{L}{\delta_M}\frac{\partial}{\partial \xi'} \tag{3.455}$$

follows. Equation (3.453), written by means of $\phi = \phi(\xi', y)$, takes the form

$$\frac{\partial \phi}{\partial \xi'} = \frac{\partial^4 \phi}{\partial \xi'^4} + 2\left(\frac{\delta_M}{L}\right)^2 \frac{\partial^4 \phi}{\partial \xi'^2 \partial y^2} + \left(\frac{\delta_M}{L}\right)^4 \left(\nabla^4 \Psi + \frac{\partial^4 \phi}{\partial y^4}\right) \tag{3.456}$$

Within the boundary layer approximation, in which $(\delta_M/L)^2 \ll 1$ is neglected with respect to unity, (3.456) simplifies to

$$\frac{\partial \phi}{\partial \xi'} = \frac{\partial^4 \phi}{\partial \xi'^4} \tag{3.457}$$

The boundary conditions are again (3.431) and (3.433), which are here reported for convenience:

$$\phi(0, y) = -\sin(\pi y) \tag{3.458}$$

$$\lim_{\xi' \to \infty} \phi(\xi', y) = 0 \tag{3.459}$$

but an *additional condition* is necessary to single out a unique solution. Several additional conditions are candidate, as the so-called *free-slip condition* defined by

$$\left(\frac{\partial^2 \phi}{\partial \xi'^2}\right)_{\xi'=0} = 0 \tag{3.460}$$

The general integral of (3.457) is

$$\phi(\xi', y) = \sum_{k=1}^{4} C_k(y) \exp(\lambda_k \xi') \tag{3.461}$$

where

$$\lambda_1 = 0 \qquad \lambda_2 = 1 \qquad \lambda_3 = \frac{-1+i\sqrt{3}}{2} \qquad \lambda_4 = \frac{-1-i\sqrt{3}}{2}$$

and conditions (3.458)–(3.460) demand

$$C_1(y) = C_2(y) = 0 \qquad C_3(y) = \lambda_4 \sin(\pi y) \qquad C_4(y) = \lambda_3 \sin(\pi y) \tag{3.462}$$

Substitution of (3.462) into (3.461) yields

$$\phi(\xi', y) = \sin(\pi y)\, \varphi(\xi') \exp\left(-\frac{1}{2}\xi'\right) \tag{3.463}$$

where

$$\varphi(\xi') := \frac{1}{\sqrt{3}} \sin\left(\frac{\sqrt{3}}{2}\xi'\right) - \cos\left(\frac{\sqrt{3}}{2}\xi'\right)$$

and the total solution Ψ follows from (3.426), (3.461), (3.462) and (3.454):

$$\Psi(x, y) = \left[1 - x + \varphi\left(\frac{L}{\delta_M} x\right) \exp\left(-\frac{1}{2}\frac{L}{\delta_M} x\right)\right] \sin(\pi y) \tag{3.464}$$

from which a noticeable similarity of the latter with (3.436) can be observed (see Fig. 3.15). A set of streamlines of (3.464) are reported in panel (a) of Fig. 3.15, while the meridional velocity at the mid-basin latitude is shown in panel (b). Analogously to Stommel's model, all the features of Munk's model can be inferred from (3.464). In both cases, the qualitative features of the circulation are basically the same, with the exception of some details relative to the western boundary layer.

The boundary layer approximation is very useful not only to derive analytical model solutions in the linear regimen, but also to point out the different dynamic balances that take place in the central and eastern region of the fluid domain with respect to the western boundary layer. In fact, in the former region the dominance of the Sverdrup balance shows that the frictionless flow is governed only by the wind forcing; on the contrary, in the western boundary layer the motion, mainly meridional, is entirely controlled by dissipation according to (3.430) for the model of Stommel and to (3.457) for Munk's model. In this layer, the interior solution enters only as a boundary condition, as (3.431) shows for both the models.

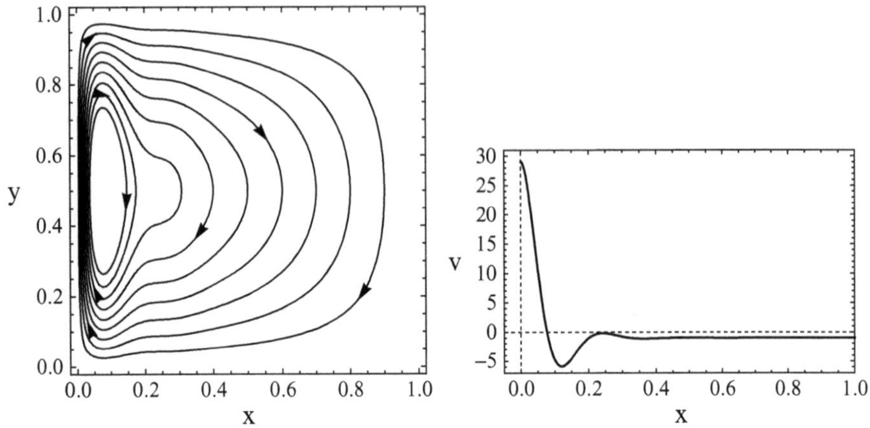

Fig. 3.15 Solution of Munk's model (3.457)–(3.459) under the assumption (3.460). *Left*: streamlines. *Right*: velocity profile at $y = 1/2$

The total solution (3.464) must satisfy (3.423) along the whole boundary; thus, also at $x = 1$, $y = 0$, and $y = 1$. Now, (3.464) implies

$$\nabla'^2\Psi = -\pi^2(1-x)\sin(\pi y) + \nabla'^2\phi \tag{3.465}$$

but, far from the western boundary, $\nabla'^2\Psi \approx -\pi^2(1-x)\sin(\pi y)$. Hence, it is trivial to check that $\nabla'^2\Psi$ at $x = 1$, $y = 0$, and $y = 1$; so (3.423) is verified throughout ∂D. The fact that the part of the solution coming from the Sverdrup balance, that is, $(1-x)\sin(\pi y)$, is able to satisfy the additional boundary conditions at $x = 1$, $y = 0$, and $y = 1$ is somehow fortuitous; if this circumstance does not happen, suitable boundary layer correction terms must be introduced also at the other boundaries and added to $\psi_{\text{int}}(x, y)$. Once the wind-stress curl is fixed, an alternative may be to look for additional conditions consistent with (3.390) and with the interior stream function. For instance, if

$$\hat{\mathbf{k}}\cdot \operatorname{rot}\boldsymbol{\tau}' = -\sin(\pi x)\sin(\pi y) \tag{3.466}$$

then (3.394) yields

$$\psi_{\text{int}}(x, y) = \frac{1+\cos(\pi x)}{\pi}\sin(\pi y) \tag{3.467}$$

whence

$$\left(\frac{\partial \psi_{\text{int}}}{\partial x}\right)_{x=1} = 0 \qquad \left(\frac{\partial^2 \psi_{\text{int}}}{\partial y^2}\right)_{y=0} = \left(\frac{\partial^2 \psi_{\text{int}}}{\partial y^2}\right)_{y=1} = 0 \tag{3.468}$$

Therefore, the total stream function $\Psi = \psi_{\text{int}} + \phi$ satisfies the additional conditions

$$\left(\frac{\partial \Psi}{\partial x}\right)_{x=1} = 0 \qquad \left(\frac{\partial^2 \Psi}{\partial y^2}\right)_{y=0} = \left(\frac{\partial^2 \Psi}{\partial y^2}\right)_{y=1} = 0$$

Thus, here the no-slip condition replaces the free-slip one (recall (3.424)) at the eastern boundary.

The validity of (3.421), with $\delta_F = \delta_M$, can be again verified. In fact,

$$2\int_D \frac{\partial \psi_{\text{int}}}{\partial x}\phi\, dx\, dy = 2\frac{\delta_M}{L}\mathfrak{I}\int_0^1 \sin^2(\pi y)\, dy \tag{3.469}$$

where

$$\mathfrak{I} := \int_0^\infty \exp(-\xi'/2)\left[\frac{1}{\sqrt{3}}\sin\left(\frac{\sqrt{3}}{2}\xi'\right) - \cos\left(\frac{\sqrt{3}}{2}\xi'\right)\right]$$

but a straightforward computation shows that $\mathfrak{I} = 0$, whence

$$2\int_D \frac{\partial \psi_{\text{int}}}{\partial x}\phi\, dx\, dy = 0 \tag{3.470}$$

Two Remarks About the Boundary Layer Method in Solving Stommel's and Munk's Models

- With reference to Stommel's model, the balance of the $O(1)$-term at the l.h.s of (3.430) with another $O(1)$ term at the r.h.s. of the same equation can be achieved only by means of (3.427), in the sense that a length scale different from δ_S does not allow to derive (3.430).

  The same happens for Munk's model, in which the derivation of (3.457) strictly relies on (3.454) where the length scale of the boundary layer is δ_M.

  The check is left to the reader.
- If the boundary layer method is applied also to the eastern boundary, then, owing to (3.422), the resulting correction term turns out to be identically zero in both models, as the reader can check.

About Nonlinear Models of the Steady Wind-Driven Circulation

Nonlinear models are governed by the vorticity equation (3.382) and are not amenable to analytical treatment. The numerical approach, whose details go beyond the scopes of these notes, allow to run the model once all its input have been fixed numerically. Thus, in the framework of (3.382), the usual strategy is to select the fluid domain, typically the square $D = [0 \le x \le 1] \times [0 \le y \le 1]$, a certain forcing (for instance, (3.397)), and to explore, in practice on the basis of a suitable set of streamlines, different output, each being related to a given choice of $(\delta_I/L)^2$, δ_S/L

and $(\delta_M/L)^3$, where usually $L = 10^6$ m is understood. Obviously, also all the linear regimes can be simulated, setting $(\delta_I/L)^2 = 0$ and keeping different from zero δ_S/L or $(\delta_M/L)^3$. A set of streamlines produced by the vorticity equation

$$\left(\frac{\delta_I}{L}\right)^2 \mathscr{J}(\psi, \nabla'^2\psi) + \frac{\partial \psi}{\partial x} = -\sin(\pi y) + \left(\frac{\delta_M}{L}\right)^3 \nabla'^4\psi \tag{3.471}$$

in different dynamic regimes are reported below (courtesy of Gualtiero Badin PhD, Liverpool University). The linear case is obtained from (3.471) setting, formally, $\delta_I = 0$ and choosing suitable values of δ_M. In nonlinear regimes, each of them is characterized by the relative amplitude of δ_I with respect to δ_M. We shall now briefly consider the following cases: $\delta_I = 0$, $\delta_I = \delta_M$, $\delta_I \gg \delta_M$.

- If $\delta_I = 0$ and $\delta_M = 7 \times 10^4$ m, whence $(\delta_M/L)^3 = 3.43 \times 10^{-4}$, a set of streamlines of the model solution is reported in Fig. 3.16. The patterns are symmetrical with respect to the mid-latitude mirror reflection $y \leftrightarrow 1 - y$, the westward intensification takes place in the proximity of the sole western boundary while, in the eastern region of the basin, the streamlines of (3.398) adequately represent those of the full equation (3.471).
- If the value $\delta_I = 7 \times 10^4$ m is retained but $\delta_I = \delta_M$, then $(\delta_M/L)^3 = 3.43 \times 10^{-4}$ and $(\delta_I/L)^2 = 5 \times 10^{-3}$. In this case, the model is affected by a moderate amount of nonlinearity, and the streamlines behave like in Fig. 3.17. Thus, the symmetry under the mid-latitude mirror reflection is broken, and the centre of the gyre migrates northward. As a consequence, the intensification of the flow involves mainly the north-western corner of the basin, while the streamlines of (3.398) are not able to represent those of the full equation (3.471) even in the eastern region.
- A highly nonlinear configuration is obtained setting $\delta_M = 7 \times 10^3$ m and δ_I one order of magnitude greater than δ_M, that is, $\delta_I = 7 \times 10^4$ m. Hence, $(\delta_M/L)^3 = 3.43 \times 10^{-7}$ and $(\delta_I/L)^2 = 5 \times 10^{-3}$. In this case, the dissipative term of (3.471) is almost negligible with respect to the nonlinear one, and an almost inertial behaviour is expected. In fact, the related streamlines, depicted in Fig. 3.18, tend to produce a configuration roughly symmetric under the mid-longitude mirror reflection, which is reminiscent of the Fofonoff mode. The similitude is reinforced by the prevalence of the intensification in the proximity of the northern boundary.

3.2.4 Basic Dynamics of the Atmospheric Ekman Layer

So far, we concentrated our attention to the oceanic case, but now we shortly go back to the theory of the benthic Ekman layer to illustrate an application of the model in the atmospheric framework.

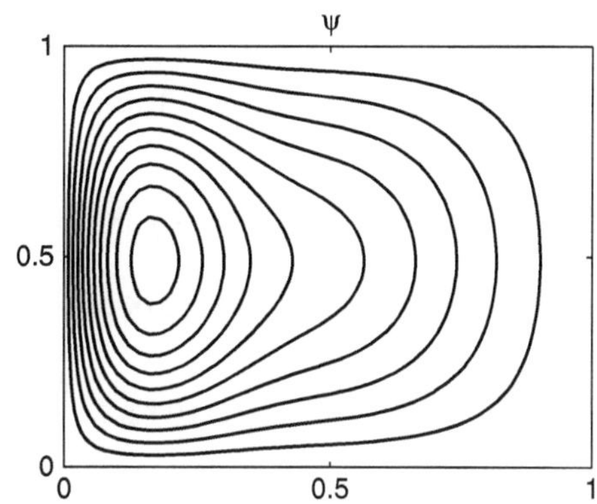

Fig. 3.16 Streamlines of the numerical solution of the homogeneous model (3.471) with $\delta_I = 0$, $\delta_M = 7 \times 10^4$ m and $L = 10^6$ m

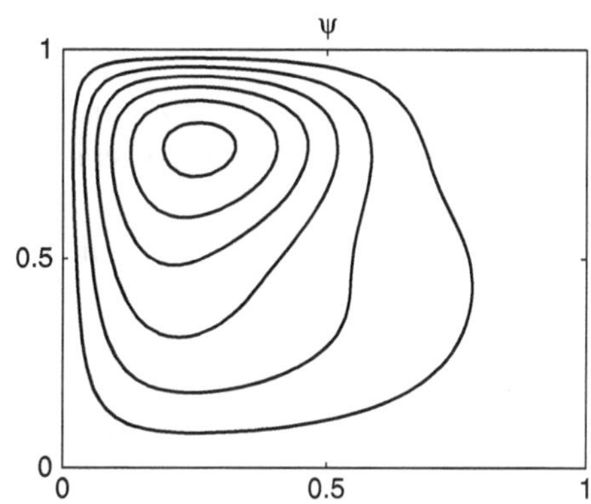

Fig. 3.17 Streamlines of the numerical solution of the homogeneous model (3.471) with $\delta_I = \delta_M = 7 \times 10^4$ m and $L = 10^6$ m

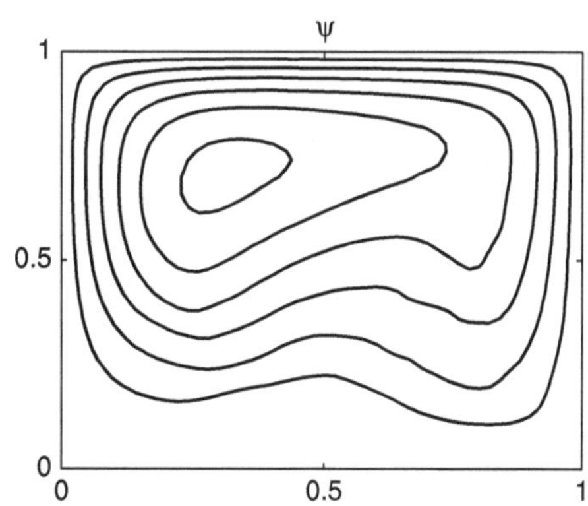

Fig. 3.18 Streamlines of the numerical solution of the homogeneous model (3.471) with $\delta_I = 7 \times 10^4$ m, $\delta_M = 7 \times 10^3$ m and $L = 10^6$ m

The atmospheric Ekman layer extends from the ground to about 1 km above it, where the geostrophic circulation takes place. Quite analogously to the benthic layer of the ocean, the geostrophic wind $\mathbf{u}_0$ of the free atmosphere undergoes frictional retardation, which is communicated through the secondary circulation that develops in the atmospheric Ekman layer. The vertical velocity w_B at the top of this layer is given by (3.354). This equation, multiplied by ρ, can be written more conveniently as

$$\rho\, w_B = \frac{H_E}{2}\,[\hat{\mathbf{k}}\cdot \mathrm{rot}\,(\rho\,\mathbf{u}_0) - \hat{\mathbf{k}}\cdot \nabla\rho \times \mathbf{u}_0)] \tag{3.472}$$

Quite in general, $\nabla\rho \approx (\partial\rho/\partial z)\hat{\mathbf{k}}$ or ρ is even constant, so (3.472) simplifies into

$$\rho\, w_B = \frac{H_E}{2}\,\hat{\mathbf{k}}\cdot \mathrm{rot}\,(\rho\,\mathbf{u}_0) \tag{3.473}$$

Based on the estimates $H_E = O(1\,\mathrm{km})$ and $\hat{\mathbf{k}}\cdot \mathrm{rot}\,\mathbf{u}_0 = O(10^{-5}\,\mathrm{s}^{-1})$, valid at the atmospheric synoptic scale, Eq. (3.473) yields $w_B = O(5\,\mathrm{mm/s})$. Hence, the related vertical acceleration $\partial w/\partial t = O(w_B\,U/L) = O(5\times 10^{-8}\mathrm{m/s}^2)$ is negligibly small with respect to gravity.

Consider a certain region D of the (flat) ground, which will be specified in the following, and integrate (3.473) over it. With the aid of Stokes' theorem, one obtains

$$\int_D \rho\, w_B\, \mathrm{d}x\mathrm{d}y = \frac{H_E}{2}\oint_{\partial D} \rho\,\hat{\mathbf{t}}\cdot \mathbf{u}_0\,\mathrm{d}s \tag{3.474}$$

Since, according to (2.536),

$$\rho\,\mathbf{u}_0 = \frac{1}{f_0}\hat{\mathbf{k}}\times\nabla p \tag{3.475}$$

substitution of (3.475) into (3.474) gives, after little algebra,

$$\int_D \rho\, w_B\, \mathrm{d}x\mathrm{d}y = \frac{H_E}{2 f_0}\oint_{\partial D} \frac{\partial p}{\partial n}\,\mathrm{d}s \tag{3.476}$$

where $\partial/\partial n := \hat{\mathbf{n}}\cdot\nabla$, with $\hat{\mathbf{n}}$ the unit vector locally normal to ∂D and pointing outward D. Relationship (3.476) links the vertical mass flux $\int_D \rho\, w_B\,\mathrm{d}x\mathrm{d}y$ with the pressure gradient flux $\oint_{\partial D}(\partial p/\partial n)\,\mathrm{d}s$ across the boundary ∂D of the region D.

Consider first the northern hemisphere ($f_0 > 0$). If ∂D encircles a high-pressure area, then we have $\partial p/\partial n < 0$ on ∂D, and, because of (3.476), the vertical mass flux at the top of the Ekman layer is negative. Thus, inside the fluid column $0 < z < H_E$ with cross section D, the flux is downward and decreases from $\int_D \rho\, w_B\,\mathrm{d}x\mathrm{d}y$ in the proximity of $z = H_E$ to zero at the ground. On the contrary, if ∂D encircles a low-pressure area, then $\partial p/\partial n > 0$ on ∂D and so the vertical mass flux is upward and increasing with height. In the case of high pressure, the suspended matter is transported to the ground so that the sky looks clear and the typical good weather condition is realized. In the case of low pressure, wet air raises towards colder layers where it condenses. Hence, the typical rainy weather follows. These scenarios are depicted in Fig. 3.19, which clearly shows the relative position of the geostrophic

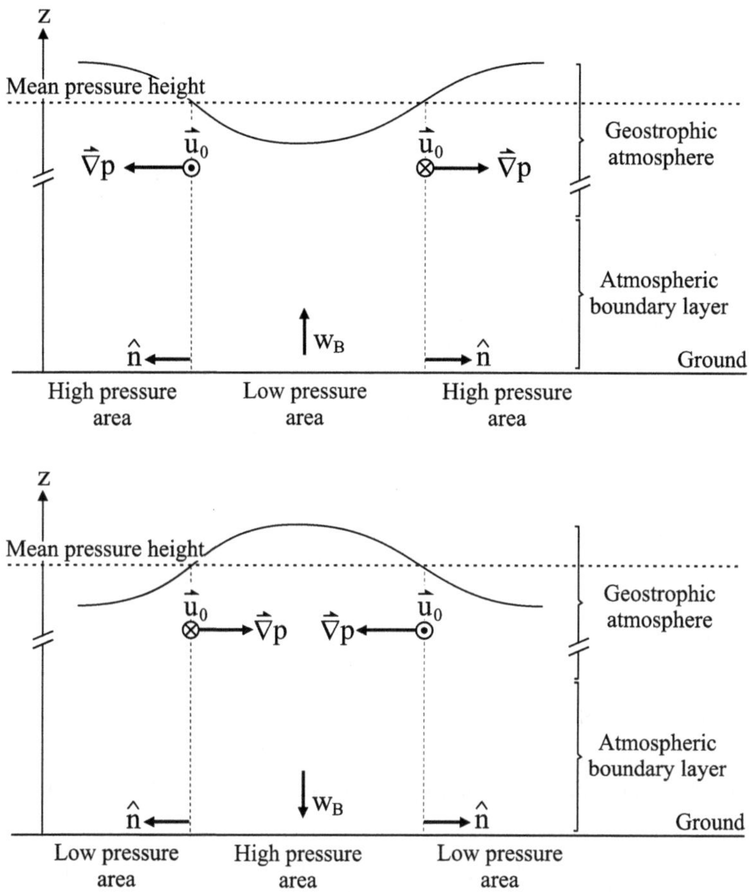

Fig. 3.19 Pictorial explanation of Eq. (3.476). *Upper panel* refers to the case in which ∂D encircles a low-pressure area, while *lower panel* refers to a high-pressure area. In the first case, $\partial p/\partial n = |\nabla p| > 0$, so w_B is *upward*; in the second case, $\partial p/\partial n = -|\nabla p| < 0$, so w_B is *downward*. In the boundary layer, the horizontal wind is given by (3.346): note the cross-isobaric component due to friction, proportional to $\hat{\mathbf{k}} \times \mathbf{u}_0$

wind $\mathbf{u}_0$, the pressure gradient ∇p and the vertical velocity of the flow w_B in the atmospheric boundary layer.

About the southern hemisphere, it is known that a mirror change $\phi \to -\phi$ of latitude with respect to the equatorial plane implies

$$f_0 \to -f_0 \tag{3.477}$$

and, hence, the geostrophic balance in turn implies

$$\mathbf{u}_0 \cdot \hat{\mathbf{t}} \to -\mathbf{u}_0 \cdot \hat{\mathbf{t}} \tag{3.478}$$

On the other hand, application of (3.477) and (3.478) to the equation $\partial p/\partial n = \rho f_0 \mathbf{u}_0 \cdot \hat{\mathbf{t}}$ proves the invariance

$$\frac{\partial p}{\partial n} \to \frac{\partial p}{\partial n} \tag{3.479}$$

Therefore, under $\phi \to -\phi$ and because of (3.478), the l.h.s. of Eq. (3.474) transforms according to

$$\int_D \rho\, w_B \, dx dy \to -\int_D \rho\, w_B \, dx dy \tag{3.480}$$

Moreover, because of (3.477) and 3.479, we also obtain

$$\frac{1}{f_0} \oint_{\partial D} \frac{\partial p}{\partial n} ds \to -\frac{1}{f_0} \oint_{\partial D} \frac{\partial p}{\partial n} ds \tag{3.481}$$

Finally, the transformation laws (3.480) and 3.481 assure that Eq. (3.476) is left unchanged under $\phi \to -\phi$.

Exercises

1. Draw the Ekman spiral for the benthic layer starting from (3.346).
2. A classical approach in solving Munk's model is based on the method of the boundary layer approximation. In general, the resulting model solution is

$$\psi = \psi_{\mathrm{int}}(x, y) + \exp\left(-\frac{Lx}{2\,\delta_M}\right)\left[C(y)\sin\left(\frac{\sqrt{3}Lx}{2\,\delta_M}\right) - \psi_{\mathrm{int}}(0, y)\cos\left(\frac{\sqrt{3}Lx}{2\,\delta_M}\right)\right]$$

where the function $C(y)$ is selected by means of the additional boundary conditions at the western wall (say, at $x = 0$).

Verify that, in the case of partial-slip boundary conditions, that is,

$$A(y)\left(\frac{\partial \psi}{\partial x}\right)_{x=0} + B(y)\left(\frac{\partial^2 \psi}{\partial x^2}\right)_{x=0} = 0$$

the function $C(y)$ turns out to be

$$C(y) = \frac{1}{\sqrt{3}} \frac{LB(y) + \delta_M A(y)}{LB(y) - \delta_M A(y)} \psi_{\mathrm{int}}(0, y)$$

Trace back solution (3.464) as a special case of the general solution reported above.

3. Show that, although the dynamics of the Ekman layers cannot be derived with friction $(-k\,u,\,-k\,v)$ in place of

$$\left(\frac{E_V}{2}\frac{\partial^2 u}{\partial z^2},\frac{E_V}{2}\frac{\partial^2 v}{\partial z^2}\right)$$

Stommel's model (3.425) retains in any case its form.

4. Evaluate the integrated energy equation associated to Eq. (3.281) along the same lines as (3.412). Does a new term appear?

5. Verify that

$$\psi = -\mathsf{U}\,\mathsf{y} + \mathsf{A}\sin(\mathsf{k}\mathsf{x} + \mathsf{n}\mathsf{y} - \sigma'\mathsf{t})$$

is an exact integral of Eq. (3.100) for all U and A, provided that the dispersion relation

$$\sigma' = \mathsf{U}\,\mathsf{k} - \frac{\beta\,\mathsf{k}}{\mathsf{k}^2 + \mathsf{n}^2}$$

is satisfied. Show that $\mathsf{U} > c_{\mathsf{x}}$, where c_{x} is the zonal phase speed of the sinusoidal wave. Can ψ represent an eastward propagating disturbance, like most of mid-latitude atmospheric weather systems?

Bibliographical Note

Shallow-water dynamics and related models, such as Rossby waves and circulation on topography, are extensively treated by Cushman-Roisin (1994), Dijkstra (2005), Hendershott (1987), Kamenkovich et al. (1986), Pedlosky (1987), Salmon (1998) and Vallis (2006). In this framework, we also cite the classical paper of Fofonoff (1954). The development of shallow-water dynamics in view of the wind-driven circulation is considered by Pedlosky (1987) and Pedlosky (1996).

In a seminal paper, Charney (1955) investigated the dynamic connection between the upper Ekman layer and the establishment of permanent currents in the ocean interior. Ekman layers are now a standard topic of physical oceanography and are expounded in most of the books of this field. In particular, the effect of bottom topography on the benthic Ekman layer is considered by Cushman-Roisin (1994) and Pedlosky (1987) while our treatment is a bit more general from the formal point of view in order to hold also in the two-layer model. The Ekman layer on the ground is a typical argument of dynamic meteorology: the reader may consult, for instance, Holton (1979) and Stull (1988). The basic equation known as Sverdrup balance Sverdrup (1947) and the homogeneous model of wind-driven circulation are cornerstones of physical oceanography since Stommel (1948) and Munk (1950). A detailed review with the related bibliography is found in Pedlosky (1996).

In their phenomenological study, Hellerman and Rosenstein (1983) refer to the observed variation of the wind stress with latitude at the planetary scale; but, for single-gyre models, this is usually approximated by a sinusoidal wind-stress curl. A still open problem is related to the choice of suitable additional boundary conditions to close the circulation model as pointed out, for example, by Carnevale et al. (2001), Crisciani and Cavallini (2007), and Crisciani et al. (2007). In alternative to the approach of Pedlosky (1965), the boundary condition satisfied by the Sverdrup transport at the eastern coast used in the present book has been derived following Crisciani and Purini (1997). Among the many notable numeric investigations concerning the homogeneous model, we point out the classical papers by Bryan (1963) and Veronis (1966), together with the more recent ones by Boning (1986) and Crisciani and Özgökmen (2001).

Chapter 4
Quasi-Geostrophic Two-Layer Model

Abstract The two-layer model is the first step towards the picture of the dynamics of a continuously stratified ocean and atmosphere, which are closer to common intuition. Models with a finite, and greater than two, number of layers are just generalizations of the simpler two-layer one – and their formulation is left to the interested reader.

In each layer, the dynamics is basically that of the homogeneous model illustrated in Chap. 3, with the exception of the interface between them, whose motion makes the layers mutually interacting. Thus, the quasi-geostrophic governing equations are coupled and may also include a reciprocal frictional retardation produced at the interface.

In deriving the energetics of this kind of system, a new quantity appears, the available potential energy. This is given by the difference between the potential energy of the flow, which presupposes a modulated interface, and the potential energy of the two-layered fluid in the absence of motion, which implies a flat interface. Thus, available potential energy can be transformed into kinetic energy and vice versa according to the thermal wind relation, while the potential energy in the absence of motion cannot be transformed into kinetic energy. For instance, the production of eddies and vortices in the ocean and the atmosphere is based on the transformation of the available potential energy of a certain "basic state" into the kinetic energy of time-dependent "disturbances", which are shaped just like eddies and vortices.

4.1 Basic QG Equations for the Two-Layer Model

The starting point of this section is the derivation of the pressure field in each layer, by means of which the profile of the interface is expressed. In turn, the motion of the interface yields the vertical velocity of the fluid at the same height (for the atmosphere) or depth (for the ocean). Furthermore, the possible influence of a frictional interaction between the layers is also taken into account. This approach

F. Cavallini and F. Crisciani, *Quasi-Geostrophic Theory of Oceans and Atmosphere*, Atmospheric and Oceanographic Sciences Library 45, DOI 10.1007/978-94-007-4691-6_4,

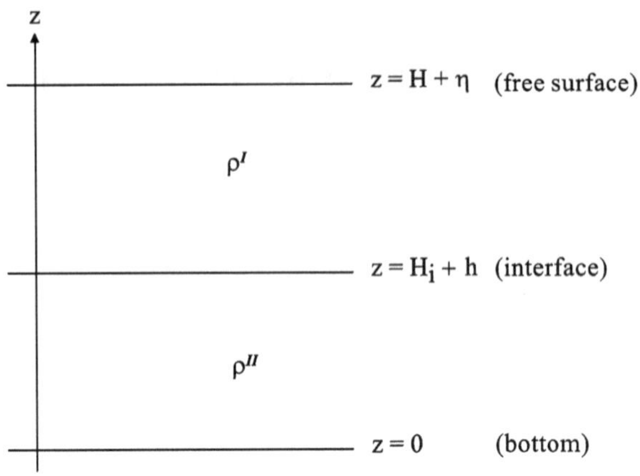

Fig. 4.1 Basic scheme of the two-layer model

allows the vertical integration of the vorticity equation, separately in each layer, to obtain a set of coupled quasi-geostrophic equations. As an application of the results above, the dynamics of Rossby waves in a two-layer, unbounded system is investigated.

4.1.1 The Quasi-Geostrophic Two-Layer Model

The simplest model that exhibits baroclinic features is the two-layer model. Consider a fluid included between the ground (or the sea floor), located at $z = 0$ (a flat bottom is assumed for simplicity) and the upper surface

$$z = H + \eta(x, y, t)$$

where the constant H represents the mean thickness of the full layer, while η is its fluctuating part. Dimensional quantities are understood. Unlike the single-layer model, an impermeable and deformable interface placed at

$$z = H_{\rm i} + h(x, y, t) \tag{4.1}$$

separates the lighter fluid ($\rho = \rho^I$) of the upper layer ($H_{\rm i} + h < z \leq H + \eta$) from the heavier fluid ($\rho = \rho^{II} > \rho^I$) of the lower layer ($0 \leq z < H_{\rm i} + h$). The geometry of the two-layer model is sketched in Fig. 4.1 together with the related symbols. If the fluid is at rest, (4.1) is simply $z_{\rm i} = H_{\rm i}$; otherwise, the deformation h of the interface influences the potential vorticity of each layer and induces a mutual interaction between them. Note that the related upper and lower non-dimensional layers are defined by $(H_{\rm i} + h)/H < \mathrm{z} \leq 1 + \eta/H$ and $0 \leq \mathrm{z} < (H_{\rm i} + h)/H$, respectively. The non-dimensional vorticity equations are obtained from (3.205) in the non-linear context, or from (3.220) in the linear one, by means of a vertical integration extended to each layer. At the top of the upper layer and at the bottom of the lower layer, the Ekman vertical velocities (3.339) and (3.352) apply, respectively, but the assumption of an unforced rigid lid and/or a frictionless bottom can be taken into account as well. The situation is less obvious at the interface, whose non-dimensional height (depth) is

$$\mathsf{z}_{\mathrm{i}} := \frac{H_{\mathrm{i}} + h}{H} \tag{4.2}$$

where the connection between h and the pressure perturbations $\tilde{p}^I$ and $\tilde{p}^{II}$ must be clarified, while the possible formation of a turbulent layer in the proximity of the interface cannot be ignored a priori.

Interface and Pressure

The basic problem is to derive the relation between $h(x,y,t)$ appearing in (4.1) and the pressure perturbations $\tilde{p}^I$ and $\tilde{p}^{II}$, which are prevalently in geostrophic balance with the related horizontal velocities.

Consider, first, a generic height (or depth) z in the upper layer (where the density is ρ^I), that is, assume

$$H_{\mathrm{i}} + h < z \leq H + \eta \tag{4.3}$$

At such height z (see Fig. 4.2, panel (a)), the pressure p is (apart, in the marine case, from the atmospheric pressure which is taken here as constant)

$$p = \rho^I g\,(H + \eta - z) \tag{4.4}$$

that is to say,

$$p = \rho^I g\,(H - z) + \rho^I g\,\eta \tag{4.5}$$

In (4.5), the term

$$p^I_{\mathrm{s}}(z) := \rho^I g\,(H - z) \tag{4.6}$$

is the standard hydrostatic contribution to total pressure, while

$$\tilde{p}^I(x, y, t) := \rho^I g\eta \tag{4.7}$$

is the pressure perturbation due to the modulation of the upper surface. The latter is in geostrophic balance with the current of the upper layer.

Consider, then (see Fig. 4.2, panel (b)), a generic height (or depth) z where the density is ρ^{II}, that is, assume

$$0 \leq z < H_{\mathrm{i}} + h \tag{4.8}$$

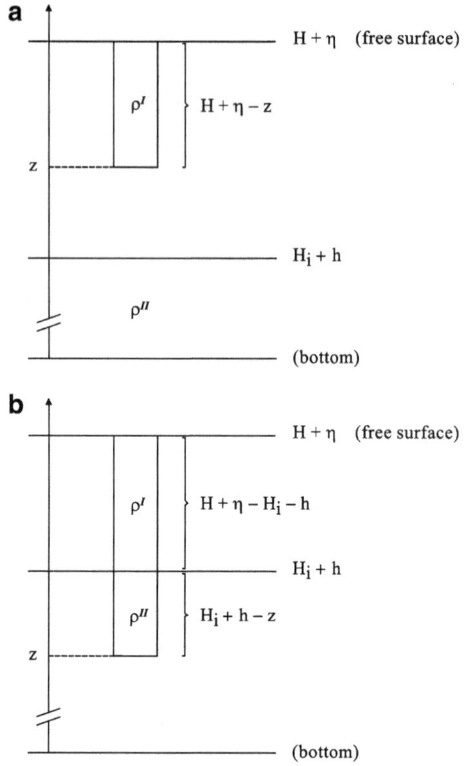

Fig. 4.2 Panel (**a**): The pressure at the depth z of the *upper layer* is given by the fluid column of density ρ^I and thickness $H+\eta-z$. Panel (**b**): The pressure at the depth z of the *lower layer* is given by the contribution of the fluid column of density ρ^I and thickness $H+\eta-H_\mathrm{i}-h$ (fully included in the upper layer) plus the contribution of the fluid column of density ρ^{II} and thickness $H_\mathrm{i}+h-z$

At such height z, the pressure p is given by the contribution $\rho^I g\,(H+\eta-H_\mathrm{i}-h)$ of the full upper layer, plus the partial contribution $\rho^{II} g\,(H_\mathrm{i}+h-z)$ of the lower layer between z and $H_\mathrm{i}+h$. On the whole,

$$\begin{aligned} p &= \rho^I g\,(H+\eta-H_\mathrm{i}-h)+\rho^{II} g\,(H_\mathrm{i}+h-z) \\ &= p_\mathrm{s}^{II}(z)+\tilde{p}^{II}(x,y,t) \end{aligned} \tag{4.9}$$

where

$$p_\mathrm{s}^{II}(z) := \rho^I g\,(H-H_\mathrm{i})+\rho^{II} g\,(H_\mathrm{i}-z) \tag{4.10}$$

is the standard hydrostatic component, and

$$\tilde{p}^{II}(x,y,t) := \rho^I g\,(\eta-h)+\rho^{II} g h \tag{4.11}$$

is the related pressure perturbation, which is in geostrophic balance with the current of the lower layer. Note that (4.6) and (4.10) imply $p_\mathrm{s}^I(H_\mathrm{i}) = p_\mathrm{s}^{II}(H_\mathrm{i}) = \rho^I g\,(H-H_\mathrm{i})$, consistently with the fact that hydrostatic pressure is continuous across the interface. Setting $\Delta\rho := \rho^{II}-\rho^I$ and using (4.7), the pressure perturbation (4.11) can be written as

$$\tilde{p}^{II}(x,y,t) = g\,\Delta\rho\; h(x,y,t)+\tilde{p}^I(x,y,t)$$

whence the relation

$$h = \frac{\tilde{p}^{II} - \tilde{p}^{I}}{g\Delta\rho} \tag{4.12}$$

follows. Therefore,

$$\frac{\mathrm{D}h}{\mathrm{D}t} = \frac{1}{g\Delta\rho}\frac{\mathrm{D}}{\mathrm{D}t}(\tilde{p}^{II} - \tilde{p}^{I}) \tag{4.13}$$

Vertical Velocity of the Interface

For future applications, it is useful to evaluate the non-dimensional vertical velocity $\mathsf{w}(\mathsf{z}_\mathrm{i})$ of the interface, starting from (4.2) and (4.13). About the l.h.s. of (4.13), we note that $w(H_\mathrm{i} + h) = Dh/Dt$ by definition, and hence,

$$\frac{\mathrm{D}h}{\mathrm{D}t} = \frac{U}{L}\frac{\mathsf{D}}{\mathsf{Dt}}(H\,\mathsf{z}_\mathrm{i}) = \frac{UH}{L}\mathsf{w}(\mathsf{z}_\mathrm{i}) \tag{4.14}$$

In the non-linear context, at the geostrophic level of approximation, in each layer, $\tilde{p} = f_0\,\rho_\mathrm{s}\,U\,L\,\psi$ according to (2.555). Hence, the r.h.s, of (4.13) becomes

$$\frac{\mathrm{D}}{\mathrm{D}t}\frac{1}{g\Delta\rho}(\tilde{p}^{II} - \tilde{p}^{I}) = \frac{U}{L}\frac{f_0\,U\,L}{g'}\frac{\mathsf{D}_0}{\mathsf{Dt}}(\psi^{II} - \psi^{I}) \tag{4.15}$$

where $g' := g\Delta\rho/\rho_\mathrm{s} = O(10^{-2}\,\mathrm{m}\cdot\mathrm{s}^{-2})$ is *reduced gravity* and

$$\frac{\mathsf{D}_0}{\mathsf{Dt}}(\psi^{II} - \psi^{I}) = \frac{\partial}{\partial\mathsf{t}}(\psi^{II} - \psi^{I}) + \mathscr{J}(\psi^{I},\,\psi^{II}) \tag{4.16}$$

with D_0/Dt the operator appearing in (3.104). By equating (4.14) with (4.15), we get

$$\mathsf{w}(\mathsf{z}_\mathrm{i}) = \frac{f_0\,U\,L}{g'H}\frac{\mathsf{D}_0}{\mathsf{Dt}}(\psi^{II} - \psi^{I}) \tag{4.17}$$

In terms of the rotational Froude number

$$\mathsf{F} = \frac{f_0^2\,L^2}{g'H} \tag{4.18}$$

Equation (4.17) yields the non-dimensional vertical velocity at the interface in the form

$$\mathsf{w}(\mathsf{z}_\mathrm{i}) = \mathsf{F}\,\varepsilon\,\frac{\mathsf{D}_0}{\mathsf{Dt}}(\psi^{II} - \psi^{I}) \tag{4.19}$$

In the linear context, starting from

$$\frac{\mathrm{D}}{\mathrm{D}t} = \frac{U}{L}\left(\frac{L}{U\,T_\mathrm{loc}}\frac{\partial}{\partial\mathsf{t}} + \mathbf{u}\cdot\nabla'\right) \tag{4.20}$$

and assuming the local time scale $T_{\text{loc}} = (\beta_0 L)^{-1}$, (4.20) becomes

$$\frac{\mathrm{D}}{\mathrm{D}t} = \frac{U}{L}\left(\beta\frac{\partial}{\partial\tilde{t}} + \mathbf{u}\cdot\nabla'\right) \tag{4.21}$$

where $\beta := \beta_0 L^2/U > O(1)$. Hence, because $\mathbf{u}\cdot\nabla' = O(1)$, (4.21) can be simplified into

$$\frac{\mathrm{D}}{\mathrm{D}t} = \frac{U}{L}\beta\frac{\partial}{\partial\tilde{t}} \tag{4.22}$$

Using (4.22) and recalling that $T_{\text{loc}} = (\beta_0 L)^{-1}$ implies $\beta\,\varepsilon = \varepsilon_{\mathrm{T}}$, the r.h.s. of (4.13) becomes

$$\frac{\mathrm{D}}{\mathrm{D}t}\frac{1}{g\Delta\rho}(\tilde{p}^{II} - \tilde{p}^{I}) = \frac{U}{L}\beta\frac{f_0\,\rho_{\mathrm{s}}\,U\,L}{g\Delta\rho}\frac{\partial}{\partial\tilde{t}}(\Psi^{II} - \Psi^{I}) = \frac{U}{L}H\,\mathsf{F}\,\varepsilon_{\mathrm{T}}\frac{\partial}{\partial\tilde{t}}(\Psi^{II} - \Psi^{I}) \tag{4.23}$$

Finally, from (4.14) and (4.23), the non-dimensional vertical velocity at the interface

$$\mathsf{w}(\mathsf{z}_{\mathrm{i}}) = \mathsf{F}\,\varepsilon_{\mathrm{T}}\frac{\partial}{\partial\tilde{t}}(\Psi^{II} - \Psi^{I}) \tag{4.24}$$

follows.

4.1.2 Frictional Interaction Between the Layers

The difference between the geostrophic velocities $\mathbf{u}^{I}$ and $\mathbf{u}^{II}$ of the two-layer model can be ascribed to a dissipative mechanism quite similar to the Ekman layers considered in Sect. 3.2.1, which is assumed to take place above and below the interface, in the proximity of it. In this picture, a special role is played by the velocity w^{I} at the transition height (or depth) between the geostrophic interior of the upper layer and the underlying Ekman sublayer, and by the velocity w^{II} at the transition height (or depth) between the geostrophic interior of the lower layer and the overhanging Ekman sublayer. To evaluate w^{I} and w^{II}, it is necessary to determine preliminarily both the horizontal current $\mathbf{u}_{\mathrm{f}}^{I}$ (where subscript $_{\mathrm{f}}$ stands for "friction") that matches $\mathbf{u}^{I}$ with the current at the upper side of the interface and the horizontal current $\mathbf{u}_{\mathrm{f}}^{II}$ that matches $\mathbf{u}^{II}$ with the current at the lower side of the same interface. However, unlike the classical Ekman theory, neither the current at the upper side nor that at the lower side of the interface is known, but in spite of this, the assumption that friction disappears in the case in which $\mathbf{u}^{I} = \mathbf{u}^{II}$ is sufficient to parameterize the frictional interaction between the layers.

In the usual non-dimensional framework, according to (4.2),

$$\mathsf{z} = \mathsf{z}_{\mathrm{i}} = \frac{H_{\mathrm{i}} + h}{H} \tag{4.25}$$

is the coordinate of the interface. Starting from (4.25), we introduce the *upward stretched coordinate*

$$\xi' := \frac{\mathrm{z} - \mathrm{z_i}}{\sqrt{E_\mathrm{V}}} \tag{4.26}$$

and the *downward stretched coordinate*

$$\lambda' := \frac{\mathrm{z_i} - \mathrm{z}}{\sqrt{E_\mathrm{V}}} \tag{4.27}$$

where E_V is the vertical Ekman number.

In the proximity of the interface, above it, the horizontal current $\mathbf{u}_\mathrm{f}^I$ depends on ξ', while below it, the horizontal current $\mathbf{u}_\mathrm{f}^{II}$ depends on λ'. In the following two subsections, we shall derive the vertical velocities w^I and w^{II} due to the horizontal convergence/divergence of $\mathbf{u}_\mathrm{f}^I$ and $\mathbf{u}_\mathrm{f}^{II}$ (see Fig. 4.3).

Frictional Upper Layer

With reference to the upper layer, and in full analogy with Eq. (3.367) of Sect. 3.2.1, the current $\mathbf{u}_\mathrm{f}^I(\xi')$ is

$$\mathbf{u}_\mathrm{f}^I(\xi') = \mathbf{u}^I + \mathrm{e}^{-\xi'}\left[\sin\xi'\,\mathbf{V}^I + \cos\xi'\,\hat{\mathbf{k}}\times\mathbf{V}^I\right] \tag{4.28}$$

and, in particular,

$$\mathbf{u}_\mathrm{f}^I(0) = \mathbf{u}^I + \hat{\mathbf{k}}\times\mathbf{V}^I \tag{4.29}$$

From (4.29), we obtain $\mathbf{V}^I = -\hat{\mathbf{k}}\times\left[\mathbf{u}_\mathrm{f}^I(0) - \mathbf{u}^I\right]$, whence (4.28) yields

$$\mathbf{u}_\mathrm{f}^I(\xi') = \mathbf{u}^I + \mathrm{e}^{-\xi'}\left\{-\sin\xi'\,\hat{\mathbf{k}}\times\left[\,\mathbf{u}_\mathrm{f}^I(0) - \mathbf{u}^I\right] + \cos\xi'\,\left[\mathbf{u}_\mathrm{f}^I(0) - \mathbf{u}^I\right]\right\} \tag{4.30}$$

Equation (4.30) implies

$$\mathrm{div}\,\mathbf{u}_\mathrm{f}^I = \mathrm{e}^{-\xi'}\,\sin\xi'\,\hat{\mathbf{k}}\cdot\mathrm{rot}\left[\mathbf{u}_\mathrm{f}^I(0) - \mathbf{u}^I\right] \tag{4.31}$$

and, using the continuity equation

$$\frac{\partial \mathrm{w}_\mathrm{f}^I}{\partial \xi'} + \sqrt{E_\mathrm{V}}\,\mathrm{div}\,\mathbf{u}_\mathrm{f}^I = 0$$

the vertical velocity in the Ekman sublayer is

$$\mathrm{w}_\mathrm{f}^I(\xi') = \mathrm{w}_\mathrm{f}^I(0) - \sqrt{E_\mathrm{V}}\,\mathscr{W}(\xi')\,\hat{\mathbf{k}}\cdot\mathrm{rot}\left[\mathbf{u}_\mathrm{f}^I(0) - \mathbf{u}^I\right] \tag{4.32}$$

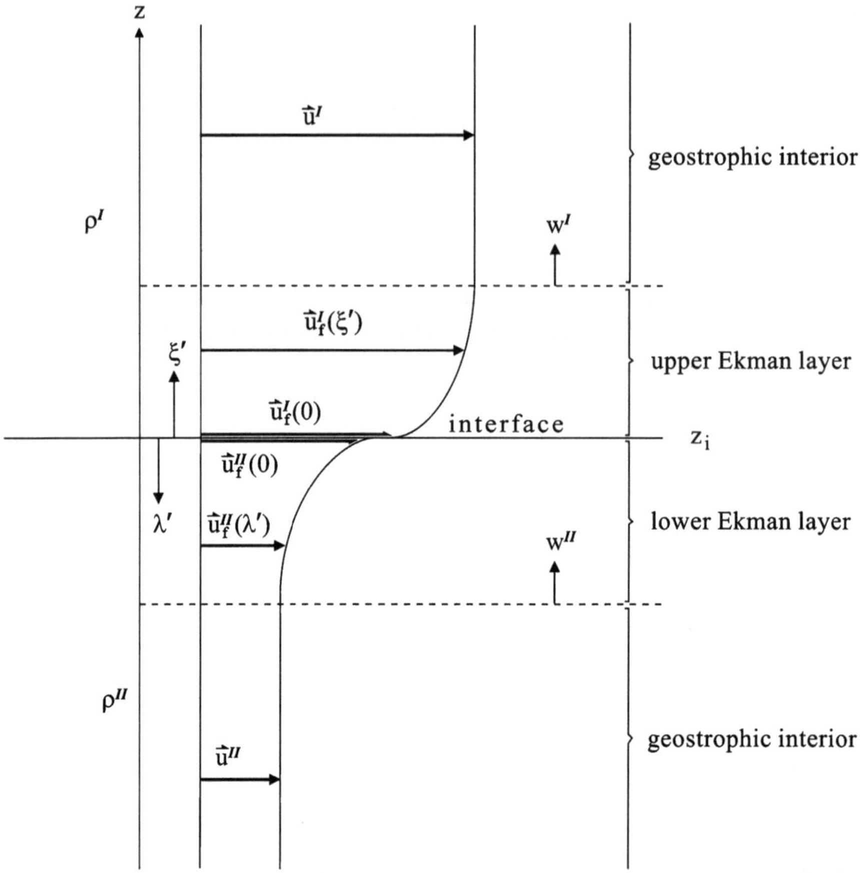

Fig. 4.3 Structure of the non-dimensional flow in the proximity of the interface, at $z = z_i$. Just above and below z_i, two frictional Ekman layers are present, in which the horizontal currents are $\mathbf{u}^I_f(\xi')$ and $\mathbf{u}^{II}_f(\lambda')$, respectively. Far from the interface, these currents match the geostrophic currents $\mathbf{u}^I$ and $\mathbf{u}^{II}$, respectively. The weak vertical velocity w^I at the transition depth between $\mathbf{u}^I_f(\xi')$ and $\mathbf{u}^I$ links friction with the potential vorticity of the *upper layer*; the same does w^{II} in the *lower layer*

where

$$\mathscr{W}(\xi') := \frac{1}{2}\left[1 - e^{-\xi'}(\cos\xi' + \sin\xi')\right]$$

In (4.32), we identify $w^I_f(0)$ with the vertical velocity of the density interface, that is, with (4.19) or (4.24) in the non-linear (linear) case, respectively; in symbols,

$$w^I_f(0) = \begin{cases} F\,\varepsilon\,\frac{D_0}{Dt}(\psi^{II} - \psi^I) & \text{(non-linear)} \\ \\ F\,\varepsilon_T\,\frac{\partial}{\partial t}(\Psi^{II} - \Psi^I) & \text{(linear)} \end{cases} \tag{4.33}$$

Finally, the vertical velocity w^I at the transition height (or depth) $\mathsf{z} = \mathsf{z}^I$ between the geostrophic interior and the underlying Ekman sublayer is given by

$$\mathsf{w}^I := \lim_{\xi' \to \infty} w_{\mathrm{f}}^I(\xi')$$

that is to say, recalling (4.32),

$$\mathsf{w}^I = \mathsf{w}_{\mathrm{f}}^I(0) - \frac{1}{2}\sqrt{E_{\mathrm{V}}}\,\hat{\mathbf{k}} \cdot \mathrm{rot}\,[\mathbf{u}_{\mathrm{f}}^I(0) - \mathbf{u}^I] \tag{4.34}$$

Frictional Lower Layer

Consider then the lower layer, where (4.27) holds. Here, the current $\mathbf{u}_{\mathrm{f}}^{II}(\lambda')$ is

$$\mathbf{u}_{\mathrm{f}}^{II}(\lambda') = \mathbf{u}^{II} + \mathrm{e}^{-\lambda}\,[\sin\lambda'\,\mathbf{V}^{II} + \cos\lambda'\,\hat{\mathbf{k}} \times \mathbf{V}^{II}] \tag{4.35}$$

where

$$\mathbf{V}^{II} := -\hat{\mathbf{k}} \times [\mathbf{u}_{\mathrm{f}}^{II}(0) - \mathbf{u}^{II}]$$

so (4.35) is equivalent to

$$\mathbf{u}_{\mathrm{f}}^{II}(\lambda') = \mathbf{u}^{II} - \mathrm{e}^{-\lambda'}\,\{\sin\lambda'\,\hat{\mathbf{k}} \times [\mathbf{u}_{\mathrm{f}}^{II}(0) - \mathbf{u}^{II}] + \cos\lambda'\,[\mathbf{u}_{\mathrm{f}}^{II}(0) - \mathbf{u}^{II}]\} \tag{4.36}$$

From the continuity equation

$$\frac{\partial \mathsf{w}_{\mathrm{f}}^{II}}{\partial \lambda'} - \sqrt{E_{\mathrm{V}}}\,\mathrm{div}\,\mathbf{u}_{\mathrm{f}}^{II} = 0$$

with $\mathrm{div}\,\mathbf{u}_{\mathrm{f}}^{II}$ evaluated from (4.36), the vertical velocity in the Ekman sublayer is

$$\mathsf{w}_{\mathrm{f}}^{II}(\lambda') = \mathsf{w}_{\mathrm{f}}^I(0) + \sqrt{E_{\mathrm{V}}}\,\mathscr{W}(\lambda')\,\hat{\mathbf{k}} \cdot \mathrm{rot}\,[\mathbf{u}_{\mathrm{f}}^{II}(0) - \mathbf{u}^{II}] \tag{4.37}$$

Note that the vertical velocity of the interface is $\mathsf{w}_{\mathrm{f}}^I(0) = \mathsf{w}_{\mathrm{f}}^{II}(0)$. Finally, the vertical velocity w^{II} at the transition height (or depth) $\mathsf{z} = \mathsf{z}^{II}$ between the geostrophic interior and the overhanging Ekman sublayer is given by

$$\mathsf{w}^{II} := \lim_{\lambda' \to \infty} \mathsf{w}_{\mathrm{f}}^{II}(\lambda')$$

that is to say,

$$\mathsf{w}^{II} = \mathsf{w}_{\mathrm{f}}^I(0) + \frac{1}{2}\sqrt{E_{\mathrm{V}}}\,\hat{\mathbf{k}} \cdot \mathrm{rot}\,[\mathbf{u}_{\mathrm{f}}^{II}(0) - \mathbf{u}^{II}] \tag{4.38}$$

Vertical Velocity

The simplest way to achieve the disappearance of the above dissipative mechanism, in the case $\mathbf{u}^I = \mathbf{u}^{II}$, is by putting

$$\mathrm{rot}\,\mathbf{u}_\mathrm{f}^I(0) = \mathrm{rot}\,\mathbf{u}^{II} \tag{4.39}$$

and

$$\mathrm{rot}\,\mathbf{u}_\mathrm{f}^{II}(0) = \mathrm{rot}\,\mathbf{u}^{I} \tag{4.40}$$

Note that $\mathbf{u}_\mathrm{f}^I(0)$ differs from $\mathbf{u}^{II}$ in the gradient of a scalar, and the same holds for $\mathbf{u}_\mathrm{f}^{II}(0)$ and $\mathbf{u}^I$. Under assumptions (4.39) and (4.40), the velocities (4.34) and (4.38) become

$$\mathrm{w}^I = \mathrm{w}_\mathrm{f}^I(0) + \frac{1}{2}\sqrt{E_\mathrm{V}}\,\hat{\mathbf{k}}\cdot\mathrm{rot}\,(\mathbf{u}^I - \mathbf{u}^{II}) \tag{4.41}$$

and

$$\mathrm{w}^{II} = \mathrm{w}_\mathrm{f}^I(0) + \frac{1}{2}\sqrt{E_\mathrm{V}}\,\hat{\mathbf{k}}\cdot\mathrm{rot}\,(\mathbf{u}^I - \mathbf{u}^{II}) \tag{4.42}$$

respectively. By using (4.33) and expressing $\mathbf{u}^I$, $\mathbf{u}^{II}$ in terms of the geostrophic stream functions, Eqs. (4.41) and (4.42) take the form

$$\mathrm{w}^I = \mathrm{w}^{II} = \begin{cases} \mathsf{F}\,\varepsilon\,\frac{\mathrm{D}_0}{\mathrm{D}t}(\psi^{II} - \psi^I) + \frac{1}{2}\sqrt{E_\mathrm{V}}\,\nabla'^2(\psi^I - \psi^{II}) & \text{(non} - \text{linear)} \\ \\ \mathsf{F}\,\varepsilon_\mathrm{T}\,\frac{\partial}{\partial \tilde{t}}(\Psi^{II} - \Psi^I) + \frac{1}{2}\sqrt{E_\mathrm{V}}\,\nabla'^2(\Psi^I - \Psi^{II}) & \text{(linear)} \end{cases} \tag{4.43}$$

We remark that the time derivatives at the r.h.s. of (4.43) are related to the inertial behaviour of the interface, while the terms involving the Laplacian account for the friction in a neighbourhood of the interface.

4.1.3 The Vorticity Equations of the Two-Layer Model

Preliminaries

At this point, we have all the ingredients to integrate vertically (3.205) and (3.220), but a preliminary summary of the system under investigation is in order. Only non-dimensional quantities are used.

The thickness of the upper geostrophic layer is

$$1 + \frac{\eta}{H} - \frac{H_\mathrm{i} + h}{H} \approx \frac{H - H_\mathrm{i}}{H}$$

and that of the lower geostrophic layer is

$$\frac{H_i + h}{H} \approx \frac{H_i}{H}$$

Thus, vertical integration of quantities independent of z in the geostrophic layers means their multiplication by $(H - H_i)/H$ in the case of the upper layer, and by H_i/H in the case of the lower layer. Moreover, we denote by z_E the transition depth between the upper Ekman layer and the upper geostrophic layer, while we know that z^I is the transition depth between the upper geostrophic layer and the underlying Ekman sublayer (in the proximity of the interface). Thus, the upper geostrophic layer is included between z_E and z^I, with $z_E - z^I \approx (H - H_i)/H$. We also know that z^{II} is the transition depth between the lower geostrophic layer and the overhanging Ekman sublayer (in the proximity of the interface), while we denote by z_B the transition depth between the lower geostrophic layer and the bottom Ekman layer. Hence, $z^{II} - z_B \approx H_i/H$.

Now, in order to integrate vertically Eqs. (3.205) and (3.220), the vertical velocities

$$w_1(z_E), \quad w_1(z^I), \quad w_1(z^{II}), \quad w_1(z_B)$$

and

$${}_1w(z_E), \quad {}_1w(z^I), \quad {}_1w(z^{II}), \quad {}_1w(z_B)$$

must be preliminarily evaluated, with the expansions

$$w = \varepsilon w_1 + O(\varepsilon^2) \qquad \text{(non-linearcase)} \tag{4.44}$$

$$w = \varepsilon_T \, {}_1w + O(\varepsilon_T^2) \qquad \text{(linearcase)} \tag{4.45}$$

kept in mind.

Consider first $w_1(z_E)$. According to (4.44) and up to the first order in ε, we have $w_1(z_E) = w_E/\varepsilon$, where w_E is given by (3.339); hence,

$$w_1(z_E) = \frac{\tau_0}{\varepsilon \rho^I f_0 U H} \hat{\mathbf{k}} \cdot \mathrm{rot}\, \boldsymbol{\tau}' \tag{4.46}$$

Analogously, up to the first order in ε, we have $w_1(z^I) = w^I/\varepsilon$, where w^I is given by the first line of (4.43). Thus,

$$w_1(z^I) = F \frac{D_0}{Dt}(\psi^{II} - \psi^I) + \frac{\sqrt{E_v}}{2\varepsilon} \nabla'^2 (\psi^I - \psi^{II}) \tag{4.47}$$

In the same way, using (4.16), one obtains

$$w_1(z^{II}) = F \frac{D_0}{Dt}(\psi^{II} - \psi^I) + \frac{\sqrt{E_v}}{2\varepsilon} \nabla'^2 (\psi^I - \psi^{II}) \tag{4.48}$$

Evaluation of $\mathsf{w}_1(\mathsf{z}_B) \approx \mathsf{w}_B/\varepsilon$, where w_B is given by (3.352), yields, up to the first order in ε,

$$\mathsf{w}_1(\mathsf{z}_B) = \frac{\sqrt{E_{vB}}}{2\varepsilon}\nabla'^2\psi^{II} \tag{4.49}$$

In (4.49), the possibility that the bottom Ekman number E_{vB} be different from the interface Ekman number E_v has been taken into account.

Consider now ${}_1\mathsf{w}(\mathsf{z}_E)$. Because of (4.45), up to the first order in ε_T and recalling again (3.339), we have

$${}_1\mathsf{w}(\mathsf{z}_E) = \frac{\tau_0}{\varepsilon_T\,\rho^I f_0 U H}\hat{\mathbf{k}}\cdot \mathrm{rot}\,\boldsymbol{\tau}' \tag{4.50}$$

The vertical velocity ${}_1\mathsf{w}(\mathsf{z}^I) \approx \mathsf{w}^I/\varepsilon_T$ is evaluated from the second line of (4.43), thus yielding

$${}_1\mathsf{w}(\mathsf{z}^I) = \mathsf{F}\frac{\partial}{\partial\tilde{\mathsf{t}}}(\Psi^{II} - \Psi^I) + \frac{\sqrt{E_v}}{2\,\varepsilon_T}\nabla'^2(\Psi^I - \Psi^{II}) \tag{4.51}$$

In the same way, we obtain

$${}_1\mathsf{w}(\mathsf{z}^{II}) = \mathsf{F}\frac{\partial}{\partial\tilde{\mathsf{t}}}(\Psi^{II} - \Psi^I) + \frac{\sqrt{E_v}}{2\,\varepsilon_T}\nabla'^2(\Psi^I - \Psi^{II}) \tag{4.52}$$

and finally, analogously to (4.49),

$${}_1\mathsf{w}(\mathsf{z}_B) = \frac{\sqrt{E_{vB}}}{2\,\varepsilon_T}\nabla'^2\Psi^{II} \tag{4.53}$$

Non-linear Vorticity Equations

Upper Layer

Consider (3.205) divided by β and with ψ^I in place of ψ. Integration from z^I to z_E with the aid of (4.46) and (4.47) gives, after division by $(H - H_i)/H$,

$$\begin{aligned}
&\frac{1}{\beta}\left[\frac{\partial}{\partial \mathsf{t}}\nabla'^2\psi^I + \mathscr{J}\left(\psi^I,\,\nabla'^2\psi^I\right)\right] + \frac{\partial\psi^I}{\partial \mathsf{x}} \\
&= \frac{1}{\beta}\frac{H}{H - H_i}\left[\frac{\tau_0}{\varepsilon\rho^I f_0 U H}\hat{\mathbf{k}}\cdot\mathrm{rot}\,\boldsymbol{\tau}' + \left(\mathsf{F}\frac{\mathsf{D}_0}{\mathsf{Dt}} - \frac{\sqrt{E_v}}{2\varepsilon}\nabla'^2\right)(\psi^I - \psi^{II})\right] \\
&+ \frac{1}{\beta\,\mathrm{Re}}\nabla'^4\psi^I
\end{aligned} \tag{4.54}$$

We now pose

$$\mathsf{F}^I := \frac{f_0^2 L^2}{g'(H-H_\mathrm{i})} \qquad \delta_\mathrm{S}^I := \frac{H}{H-H_\mathrm{i}} \frac{f_0\sqrt{E_\mathrm{v}}}{2\beta_0} \qquad \frac{1}{\beta\,\mathsf{Re}} := \left(\frac{\delta_\mathrm{M}}{L}\right)^3 \tag{4.55}$$

and point out the estimate

$$\frac{\tau_0}{\rho^I \beta_0 L U (H-H_\mathrm{i})} = O(1)$$

which is requested in order that the Sverdrup balance holds in the basin interior. Then, with the aid of (4.16), Eq. (4.54) can be rearranged in its final form, that is,

$$\left(\frac{\delta_\mathrm{I}}{L}\right)^2 \frac{\partial}{\partial \mathsf{t}}[\nabla'^2\psi^I - \mathsf{F}^I(\psi^I-\psi^{II})] + \left(\frac{\delta_\mathrm{I}}{L}\right)^2 \mathscr{J}(\psi^I,\, \nabla'^2\psi^I + \mathsf{F}^I\psi^{II}) + \frac{\partial \psi^I}{\partial \mathsf{x}}$$
$$= \hat{\mathbf{k}}\cdot \mathrm{rot}\,\boldsymbol{\tau}' - \frac{\delta_\mathrm{S}^I}{L}\nabla'^2(\psi^I-\psi^{II}) + \left(\frac{\delta_\mathrm{M}}{L}\right)^3 \nabla'^4\psi^I \tag{4.56}$$

where we have used (3.379) in the form $1/\beta = (\delta_\mathrm{I}/L)^2$.

Lower Layer

In the lower layer, Eq. (4.54) is restated with ψ^{II} in place of ψ^I at the l.h.s., with H/H_i in place of $H/(H-H_\mathrm{i})$, and by using (4.48) and (4.49) in place of (4.46) and (4.47), respectively. Thus, we obtain

$$\frac{1}{\beta}\left[\frac{\partial}{\partial \mathsf{t}}\nabla'^2\psi^{II} + \mathscr{J}(\psi^{II},\, \nabla'^2\psi^{II})\right] + \frac{\partial \psi^{II}}{\partial \mathsf{x}}$$
$$= \frac{1}{\beta}\frac{H}{H_\mathrm{i}}\left[\mathsf{F}\frac{\mathsf{D}_0}{\mathsf{Dt}}(\psi^{II}-\psi^I) + \frac{\sqrt{E_\mathrm{v}}}{2\varepsilon}\nabla'^2(\psi^I-\psi^{II}) - \frac{\sqrt{E_\mathrm{vB}}}{2\varepsilon}\nabla'^2\psi^{II}\right] + \frac{1}{\beta\,\mathsf{Re}}\nabla'^4\psi^{II} \tag{4.57}$$

With the aid of (4.16) and positions

$$\mathsf{F}^{II} := \frac{f_0^2 L^2}{g' H_\mathrm{i}} \qquad \delta_\mathrm{S}^{II} := \frac{H}{H_\mathrm{i}} \frac{f_0\sqrt{E_\mathrm{v}}}{2\beta_0} \qquad \delta_\mathrm{SB} := \frac{H}{H_\mathrm{i}} \frac{f_0\sqrt{E_\mathrm{vB}}}{2\beta_0} \tag{4.58}$$

after some rearrangements and using again (3.379), Eq. (4.57) takes the final form

$$\left(\frac{\delta_\mathrm{I}}{L}\right)^2 \frac{\partial}{\partial \mathsf{t}}[\nabla'^2\psi^{II} - \mathsf{F}^{II}(\psi^{II}-\psi^I)] + \left(\frac{\delta_\mathrm{I}}{L}\right)^2 \mathscr{J}(\psi^{II},\, \nabla'^2\psi^{II} + \mathsf{F}^{II}\psi^I) + \frac{\partial \psi^{II}}{\partial \mathsf{x}}$$
$$= -\frac{\delta_\mathrm{S}^{II}}{L}\nabla'^2(\psi^{II}-\psi^I) - \frac{\delta_\mathrm{SB}}{L}\nabla'^2\psi^{II} + \left(\frac{\delta_\mathrm{M}}{L}\right)^3 \nabla'^4\psi^{II} \tag{4.59}$$

A Special Case: Single-Layer Motion on a Quiescent Abyss

Equations (4.56) and (4.59) can be applied in several contexts: for instance, Rossby waves (Crisciani and Purini 2011) or to investigate conditions under which the forced motion of the upper layer propagates downward into the lower layer (Rhines and Young 1982).

Here, we wish to point out the limit case of a single layer in motion on the quiescent abyss: we shall see that its mathematical formulation is very close to that for the homogeneous ocean. This model presupposes that the lower layer is "infinitely" deep and at rest. Thus, because H_{i} diverges to infinity, positions (4.58) imply that both F^{II} and δ_{S}^{II} go to zero; at the same time, the absence of motion in the lower layer implies $\psi^{II}=0$ everywhere. Hence, (4.59) is identically satisfied, while (4.56) simplifies into

$$\left(\frac{\delta_{\mathrm{I}}}{L}\right)^2\frac{\partial}{\partial \mathrm{t}}(\nabla'^2\psi^I-\mathsf{F}^I\psi^I)+\left(\frac{\delta_{\mathrm{I}}}{L}\right)^2\mathscr{J}(\psi^I,\nabla'^2\psi^I)+\frac{\partial\psi^I}{\partial\mathrm{x}}$$
$$=\hat{\mathbf{k}}\cdot\mathrm{rot}\,\boldsymbol{\tau}'-\frac{\delta_{\mathrm{S}}^I}{L}\nabla'^2\psi^I+\left(\frac{\delta_{\mathrm{M}}}{L}\right)^3\nabla'^4\psi^I \tag{4.60}$$

The only difference between (4.60) and the vorticity equation of the homogeneous models (apart from ψ^I in place of ψ) lies in the presence, in (4.60), of the further term $-(\delta_{\mathrm{I}}/L)^2\mathsf{F}^I\partial\psi^I/\partial\mathrm{t}$, which describes the time evolution of the interface separating the fluid in motion from the quiescent abyss. From the point of view of the vorticity of the flow, the time evolution of the interface has exactly the same role as the free surface oscillation of a single-layer system (recall the derivation of (3.281)); in fact, both the mechanisms lead to the same kind of deformation of fluid columns. Note also that the dissipative term $-(\delta_S^I/L)\nabla'^2(\psi^I-\psi^{II})$ of (4.56), which refers to the frictional interaction between the currents in the two layers, transforms in (4.60) into the bottom-like friction term $-(\delta_S^I/L)\nabla'^2\psi^I$.

Linear Vorticity Equations

Upper Layer

Consider (3.220) with Ψ^I in place of Ψ. Integration from z^I to z_{E} with the aid of (4.50) and (4.51) gives

$$\frac{\partial}{\partial\tilde{\mathrm{t}}}\nabla'^2\Psi^I+\frac{\partial\Psi^I}{\partial\mathrm{x}}=\frac{H}{H-H_{\mathrm{i}}}\left[\frac{\tau_0}{\varepsilon_{\mathrm{T}}\rho^I f_0 U H}\hat{\mathbf{k}}\cdot\mathrm{rot}\,\boldsymbol{\tau}'\right.$$
$$\left.-\left(\mathsf{F}\frac{\partial}{\partial\tilde{\mathrm{t}}}-\frac{\sqrt{E_{\mathrm{v}}}}{2\varepsilon_{\mathrm{T}}}\nabla'^2\right)(\Psi^{II}-\Psi^I)\right]+\left(\frac{\delta_{\mathrm{M}}}{L}\right)^3\nabla'^4\Psi^I \tag{4.61}$$

With the aid of positions (4.55), Eq. (4.61) can be restated in its final form, that is,

$$\frac{\partial}{\partial \tilde{\mathsf{t}}}\left[\nabla'^2\Psi^I - \mathsf{F}^I(\Psi^I - \Psi^{II})\right] + \frac{\partial \Psi^I}{\partial \mathsf{x}} = \hat{\mathbf{k}}\cdot \mathrm{rot}\,\boldsymbol{\tau}' - \frac{\delta_{\mathrm{S}}^I}{L}\nabla'^2(\Psi^I - \Psi^{II}) + \left(\frac{\delta_{\mathrm{M}}}{L}\right)^3 \nabla'^4\Psi^I \tag{4.62}$$

Lower Layer

In the lower layer, Eq. (4.61) is rewritten with Ψ^{II} in place of Ψ^I at the l.h.s., with H/H_{i} in place of $H/(H-H_{\mathrm{i}})$, and by using (4.52), (4.53) in place of (4.50), and (4.51), respectively. Hence, we have

$$\frac{\partial}{\partial \tilde{\mathsf{t}}}\nabla'^2\Psi^{II} + \frac{\partial \Psi^{II}}{\partial \mathsf{x}} = \frac{H}{H_{\mathrm{i}}}\left[\mathsf{F}\frac{\partial}{\partial \tilde{\mathsf{t}}}(\Psi^{II} - \Psi^I) + \frac{\sqrt{E_{\mathrm{v}}}}{2\,\varepsilon_{\mathrm{T}}}\nabla'^2(\Psi^I - \Psi^{II}) - \frac{\sqrt{E_{\mathrm{vB}}}}{2\,\varepsilon_{\mathrm{T}}}\nabla'^2\Psi^{II}\right] + \left(\frac{\delta_{\mathrm{M}}}{L}\right)^3 \nabla'^4\Psi^{II} \tag{4.63}$$

By resorting to (4.58), Eq. (4.63) can be written finally as

$$\frac{\partial}{\partial \tilde{\mathsf{t}}}\left[\nabla'^2\Psi^{II} - \mathsf{F}^{II}(\Psi^{II} - \Psi^I)\right] + \frac{\partial \Psi^{II}}{\partial \mathsf{x}} = -\frac{\delta_{\mathrm{S}}^{II}}{L}\nabla'^2(\Psi^{II} - \Psi^I) - \frac{\delta_{\mathrm{SB}}}{L}\nabla'^2\Psi^{II} + \left(\frac{\delta_{\mathrm{M}}}{L}\right)^3 \nabla'^4\Psi^{II} \tag{4.64}$$

Remarks. A simplified version of (4.62) and (4.64) can be obtained under the following assumptions:

- The motion is unforced, so $\hat{\mathbf{k}}\cdot \mathrm{rot}\,\boldsymbol{\tau}' = 0$ in (4.62).
- Both the layers have the same thickness, so $\mathsf{F}^I = \mathsf{F}^{II} = 2\,\mathsf{F}$ and $\delta_{\mathrm{S}}^I = \delta_{\mathrm{S}}^{II} =: \mathsf{r}L$.
- Bottom friction and lateral diffusion of relative vorticity are absent in both the layers.

Hence, Eqs. (4.62) and (4.64) take the form

$$\frac{\partial}{\partial \tilde{\mathsf{t}}}\left[\nabla'^2\Psi^I - 2\,\mathsf{F}(\Psi^I - \Psi^{II})\right] + \frac{\partial \Psi^I}{\partial \mathsf{x}} = -\mathsf{r}\nabla'^2(\Psi^I - \Psi^{II}) \tag{4.65}$$

$$\frac{\partial}{\partial \tilde{\mathsf{t}}}\left[\nabla'^2\Psi^{II} - 2\,\mathsf{F}(\Psi^{II} - \Psi^I)\right] + \frac{\partial \Psi^{II}}{\partial \mathsf{x}} = -\mathsf{r}\nabla'^2(\Psi^{II} - \Psi^I) \tag{4.66}$$

Equations (4.65) and (4.66) clearly illustrate, without too many mathematical intricacies, the role of the time evolution of the interface (term proportional to F in the l.h.s.) and of friction in a neighbourhood of the interface (r.h.s.).

4.1.4 Wave-Like Solution of the Model

Because of the linearity of (4.65) and (4.66), if

$$\Phi^I = \mathsf{A}^I \exp[\mathrm{i}(\mathsf{kx}+\mathsf{ny}-\sigma'\tilde{\mathsf{t}})] \tag{4.67}$$

and

$$\Phi^{II} = \mathsf{A}^{II} \exp[\mathrm{i}(\mathsf{kx}+\mathsf{ny}-\sigma'\tilde{\mathsf{t}})] \tag{4.68}$$

satisfy (4.65) and (4.66), then also

$$\Psi^I := \mathrm{Re}\,\Phi^I \tag{4.69}$$

and

$$\Psi^{II} := \mathrm{Re}\,\Phi^{II} \tag{4.70}$$

satisfy the same equations. In the present context, (4.69) and (4.70) are the putative stream functions of the model. The remark at Page 181 also holds in the case of trial functions (4.67) and (4.68).

In (4.67) and (4.68), the amplitudes A^I and A^{II} are constant parameters, while the frequency σ' is determined by substituting (4.69) and (4.70) in (4.65) and (4.66), to obtain a system of algebraic equations, written here in matrix form as

$$\begin{pmatrix} \mathsf{a} & \mathsf{b} \\ \mathsf{b} & \mathsf{a} \end{pmatrix} \begin{pmatrix} \mathsf{A}^I \\ \mathsf{A}^{II} \end{pmatrix} = \begin{pmatrix} 0 \\ 0 \end{pmatrix} \tag{4.71}$$

where

$$\begin{aligned} \mathsf{a} &= -\mathsf{r}(\mathsf{k}^2+\mathsf{n}^2)+\mathrm{i}[\mathsf{k}+\sigma'(2\mathsf{F}+\mathsf{k}^2+\mathsf{n}^2)] \\ \mathsf{b} &= \mathsf{r}(\mathsf{k}^2+\mathsf{n}^2)-2\mathrm{i}\mathsf{F}\sigma' \end{aligned}$$

Nontrivial solutions of (4.71) demand that the determinant of the matrix in (4.71) be zero, that is,

$$\mathsf{a} = \mathsf{b} \tag{4.72}$$

or

$$\mathsf{a} = -\mathsf{b} \tag{4.73}$$

From (4.72) and (4.73), we find that a wave is admissible if it satisfies the complex dispersion relation

$$\sigma' = \frac{-\mathsf{k}-2\mathrm{i}\mathsf{r}(\mathsf{k}^2+\mathsf{n}^2)}{4\mathsf{F}+\mathsf{k}^2+\mathsf{n}^2} =: \sigma'_+ \tag{4.74}$$

or the real dispersion relation

$$\sigma' = -\frac{\mathsf{k}}{\mathsf{k}^2+\mathsf{n}^2} =: \sigma'_- \tag{4.75}$$

respectively. The real part of σ' shows that Rossby waves are concerned. Solution (4.74) implies, via (4.71) and (4.72),

$$\mathsf{A}^I = -\mathsf{A}^{II} \tag{4.76}$$

while (4.75) implies, via (4.71) and (4.73),

$$\mathsf{A}^I = \mathsf{A}^{II} \tag{4.77}$$

Therefore, with reference to (4.76), stream functions (4.69) and (4.70) are given by

$$\begin{aligned} \Psi^I &= \mathrm{Re}\{\mathsf{A}^I \exp[\mathrm{i}(\mathsf{kx}+\mathsf{ny}-\sigma'_+\tilde{\mathsf{t}})]\} \\ &\overset{(4.74)}{=} \mathsf{A}^I \exp\left[-\frac{2\mathsf{r}(\mathsf{k}^2+\mathsf{n}^2)}{4\mathsf{F}+\mathsf{k}^2+\mathsf{n}^2}\tilde{\mathsf{t}}\right]\cos\left(\mathsf{kx}+\mathsf{ny}-\frac{\mathsf{k}}{4\mathsf{F}+\mathsf{k}^2+\mathsf{n}^2}\tilde{\mathsf{t}}\right) \end{aligned} \tag{4.78}$$

and

$$\Psi^{II} = -\Psi^I \tag{4.79}$$

Using (3.180) and (4.18), the non-dimensional lifetime appearing in (4.78) is

$$\frac{4\mathsf{F}+\mathsf{k}^2+\mathsf{n}^2}{2\mathsf{r}(\mathsf{k}^2+\mathsf{n}^2)} = \frac{1}{2\mathsf{r}}\left(1+\frac{1}{\pi^2}\frac{\lambda^2}{L_{\mathrm{int}}^2}\right)$$

where λ is the dimensional wavelength and $L_{\mathrm{int}} := \sqrt{g'H}/f_0$ is the internal deformation radius. Therefore, for wavelengths much smaller than the internal deformation radius, the lifetime is almost constant and close to $1/(2\mathsf{r})$. In the opposite case, the decay is slower.

With reference to (4.77), stream functions (4.69) and (4.70) coincide and are given by

$$\begin{aligned} \Psi^I = \Psi^{II} &= \mathrm{Re}\{\mathsf{A}^I \exp[\mathrm{i}(\mathsf{kx}+\mathsf{ny}-\sigma'_-\tilde{\mathsf{t}})]\} \\ &\overset{(4.75)}{=} \mathsf{A}^I \cos\left(\mathsf{kx}+\mathsf{ny}-\frac{\mathsf{k}}{\mathsf{k}^2+\mathsf{n}^2}\tilde{\mathsf{t}}\right) \end{aligned} \tag{4.80}$$

In the latter case, the so-called *barotropic* one $\nabla'^2(\Psi^{II}-\Psi^I)$ is trivially zero, so the system does not decay in time. In the *baroclinic* case, represented by (4.78) and (4.79), also the interface is involved in the motion.

4.1.5 Behaviour of the Interface

The time evolution of the interface (4.2) can be easily derived from (4.12) with the perturbation pressure $\tilde{p}$ expressed in terms of Ψ, to obtain

$$\mathsf{z}_{\mathrm{i}} = \frac{H_{\mathrm{i}}}{H} + \frac{f_0 U L}{g' H} (\Psi^{II} - \Psi^{I}) \tag{4.81}$$

In the case of the model governed by (4.65) and (4.66), hypothesis $H = 2H_{\mathrm{i}}$ leads, in the baroclinic configuration, to relation (4.79), that is, $\Psi^{II} = -\Psi^{I}$. Hence, owing to the identity $f_0 U L / g' H = \varepsilon \mathsf{F}$, Eq. (4.81) takes the form

$$\mathsf{z}_{\mathrm{i}} = \frac{1}{2} + 2\varepsilon \mathsf{F} \Psi^{II} \tag{4.82}$$

or, equivalently,

$$\mathsf{z}_{\mathrm{i}} = \frac{1}{2} - 2\varepsilon \mathsf{F} \Psi^{I} \tag{4.83}$$

In the course of the motion, and in the presence of dissipation, we have both $\Psi^{I} \to 0$ and $\Psi^{II} \to 0$, so (4.82) and (4.83) show that $\mathsf{z}_{\mathrm{i}} \to 1/2$ or, in dimensional variables, $z_{\mathrm{i}} \to H/2$. However, due to the homogeneity of the governing vorticity equations, the amplitude of the oscillating interface cannot be determined.

The structure of the interface given by (4.82) or (4.83) influences the mass transport of a two-layer model. Consider, in fact, the dimensional mass transport $\mathbf{M}$ extended to the whole fluid layer, from $z = 0$ up to $z = H$, that is to say,

$$\mathbf{M} = \rho^{I} \int_{H_{\mathrm{i}}+h}^{H} \mathbf{u}^{I}\, \mathrm{d}z + \rho^{II} \int_{0}^{H_{\mathrm{i}}+h} \mathbf{u}^{II}\, \mathrm{d}z \tag{4.84}$$

In terms of the non-dimensional variable $\mathsf{z} := z/H$ and using (4.2), Eq. (4.84) takes the form

$$\mathbf{M} = \rho^{I} U H (1 - \mathsf{z}_{\mathrm{i}}) \mathsf{u}^{I} + \rho^{II} U H \mathsf{z}_{\mathrm{i}} \mathsf{u}^{II} \tag{4.85}$$

that is to say, by using (4.82), (4.83) and expressing the non-dimensional currents in terms of the stream functions,

$$\mathbf{M} = \rho^{I} U H \left(\frac{1}{2} + 2\varepsilon \mathsf{F} \Psi^{I}\right) \hat{\mathbf{k}} \times \nabla' \Psi^{I} + (\rho^{I} + \Delta\rho)\, U H \left(\frac{1}{2} + 2\varepsilon \mathsf{F} \Psi^{II}\right) \hat{\mathbf{k}} \times \nabla' \Psi^{II} \tag{4.86}$$

As the dimensional quantity $\rho^{I} U H$ can be taken as the representative order of magnitude of $\mathbf{M}$, the related non-dimensional mass transport M turns out to be

$$\mathsf{M} = \left(\frac{1}{2} + 2\varepsilon \mathsf{F} \Psi^{I}\right) \hat{\mathbf{k}} \times \nabla' \Psi^{I} + \left(1 + \frac{\Delta\rho}{\rho^{I}}\right) \left(\frac{1}{2} + 2\varepsilon \mathsf{F} \Psi^{II}\right) \hat{\mathbf{k}} \times \nabla' \Psi^{II} \tag{4.87}$$

Due to the smallness of $\varepsilon \mathsf{F}$ and $\Delta\rho/\rho^I$ with respect to unity, the $O(1)$ contribution to the geostrophic mass transport $\mathbf{M}$ inferred from (4.87) is

$$\mathbf{M} = \frac{1}{2}\hat{\mathbf{k}} \times \nabla' \left(\Psi^I + \Psi^{II}\right) \tag{4.88}$$

but, because of (4.79), the r.h.s. of (4.88) is zero. In other words, the $O(1)$ geostrophic mass transport of coupled Rossby waves, extended to the whole depth of the motion, is zero. One could ascertain that this conclusion is independent of the relative thickness of the fluid layers.

4.2 Energetics of the Two-Layer Model

In this section, the concept of available potential energy for a two-layer system is preliminarily introduced for a fluid at rest. This is obtained by setting a fictitious non-flat interface between the upper and lower fluids and comparing the so derived potential energy with that evaluated in the case of a flat interface. Then, in the context of a uniformly rotating ambient, the sloping interface is ascribed to a couple of quasi-geostrophic flows, and in this way, the link between available potential energy and geostrophic stream functions is inferred. Subsequently, by evaluating the integrated energetics of the two-layer model, the same form of available potential energy is again recovered from the potential-vorticity equations.

4.2.1 Energetics Associated to the Quasi-Geostrophic Two-Layer Model

In the framework of the two-layer model, the modulation $h(x, y, t)$ of the interface $H_{\mathrm{i}} + h$ separating the fluids in the upper and lower layers makes potential energy $\mathscr{U}$ dependent on h, say $\mathscr{U} = \mathscr{U}(h)$. Potential energy is minimum if the interface is flat, that is, if $\mathscr{U} = \mathscr{U}(0)$. The so-called *available potential energy* (APE) is just the difference

$$\mathrm{APE} := \mathscr{U}(h) - \mathscr{U}(0) \tag{4.89}$$

If APE > 0, two-layer systems allow the transformation of the potential energy $\mathscr{U}(h) - \mathscr{U}(0)$ into kinetic energy, and conversely, kinetic energy can be partially transformed into available potential energy.

For a two-layer uniformly rotating flow in geostrophic balance, the thermal wind equation (2.576) applies: the gradient $\nabla\rho$ of the density anomaly is orthogonal to the interface, and unless the latter is flat, the r.h.s. of (2.576) is different from zero. Thus, the same equation states that the geostrophic current of the upper layer is different from that of the lower, so the relative difference between the current in each

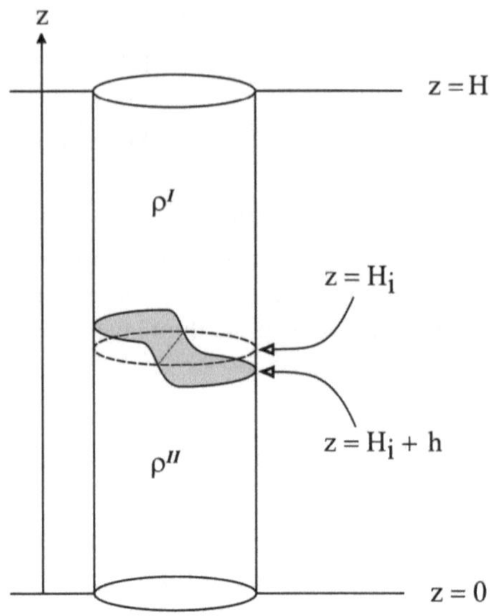

Fig. 4.4 Available potential energy is produced by the deformation of the interface between fluid columns $h \neq 0$ of different density $\rho^I < \rho^{II}$ with respect to the flat configuration $h = 0$

layer is able to support the sloping interface. In the context of the quasi-geostrophic dynamics, (4.89) can be expressed by means of the perturbation pressures $\tilde{p}^I$ and $\tilde{p}^{II}$ or by means of the stream functions Ψ^I and Ψ^{II}, already introduced in Sect. 4.1.1. The energy equation of the quasi-geostrophic two-layer model shows that the available potential energy naturally arises in deriving the equation itself, more precisely, by expanding the terms related to the oscillation of the interface. This section expounds these concepts.

4.2.2 Available Potential Energy of a Two-Layer Fluid

Consider a fluid column of constant cross section, say D, extending from $z = 0$ up to $z = H$ and with the density distribution

$$\rho = \begin{cases} \rho^I & \text{if} \quad H_\mathrm{i} + h < z \leq H \\ \rho^{II} & \text{if} \quad 0 \leq z < H_\mathrm{i} + h \end{cases} \tag{4.90}$$

where $\rho^{II} > \rho^I$ ensures static stability (see Fig. 4.4).

Assume that total mass is conserved in the volume $D \times H$. Since the interface is impermeable, it influences only the distribution of the density inside the column but does not change total mass. Therefore,

$$\int_D \left(\int_0^{H_\mathrm{i}+h} \rho^{II}\,\mathrm{d}z + \int_{H_\mathrm{i}+h}^{H} \rho^I\,\mathrm{d}z \right) \mathrm{d}x\,\mathrm{d}y = \int_D \left(\int_0^{H_\mathrm{i}} \rho^{II}\,\mathrm{d}z + \int_{H_\mathrm{i}}^{H} \rho^I\,\mathrm{d}z \right) \mathrm{d}x\,\mathrm{d}y \tag{4.91}$$

whence

$$\int_D h\,\mathrm{d}x\,\mathrm{d}y = 0 \tag{4.92}$$

Equation (4.92) means that, in the mean, h is zero on D, and this fact will be useful to evaluate (4.89).

According to (4.89),

$$\begin{aligned}\mathrm{APE} &= \int_D \left(\int_0^{H_\mathrm{i}+h} g\rho^{II} z\,\mathrm{d}z + \int_{H_\mathrm{i}+h}^{H} g\rho^{I} z\,\mathrm{d}z\right)\mathrm{d}x\,\mathrm{d}y \\ &\quad - \int_D \left(\int_0^{H_\mathrm{i}} g\rho^{II} z\,\mathrm{d}z + \int_{H_\mathrm{i}}^{H} g\rho^{I} z\,\mathrm{d}z\right)\mathrm{d}x\,\mathrm{d}y\end{aligned} \tag{4.93}$$

and a straightforward computation of (4.93) yields

$$\mathrm{APE} = \int_D g\left(H_\mathrm{i}\,h + \frac{1}{2}h^2\right)(\rho^{II} - \rho^{I})\,\mathrm{d}x\,\mathrm{d}y \tag{4.94}$$

Because of (4.92), Eq. (4.94) simplifies into

$$\mathrm{APE} = \frac{g\Delta\rho}{2}\int_D h^2\,\mathrm{d}x\,\mathrm{d}y \tag{4.95}$$

and the latter shows that $\mathrm{APE} \geq 0$ and $\mathrm{APE} = 0 \Leftrightarrow h = 0$. Substitution of (4.12) into (4.95) yields

$$\mathrm{APE} = \frac{1}{2g\Delta\rho}\int_D \left(\tilde{p}^{II} - \tilde{p}^{I}\right)^2\mathrm{d}x\,\mathrm{d}y \tag{4.96}$$

Equation (4.96) is fit for deriving the non-dimensional APE, say APE, starting from

$$\tilde{p}^n \approx f_0\,\rho_\mathrm{s}\,U\,L\,\Psi^n \tag{4.97}$$

for $n = I, II$. Substitution of (4.97) in (4.95) gives

$$\mathrm{APE} = \frac{f_0^2\,\rho_\mathrm{s}^2\,U^2 L^4}{2g\Delta\rho}\int_\mathsf{D}(\Psi^{II} - \Psi^{I})^2\,\mathrm{d}x\,\mathrm{d}y \tag{4.98}$$

Equation (4.98) is equivalent to

$$\mathrm{APE} = \rho_\mathrm{s} H L^2 U^2\,\frac{f_0^2 L^2}{2g' H}\int_\mathsf{D}(\Psi^{II} - \Psi^{I})^2\,\mathrm{d}x\,\mathrm{d}y \tag{4.99}$$

where $g' := (\Delta\rho/\rho_\mathrm{s})g$ is the reduced gravity. The factor $\rho_\mathrm{s} H L^2 U^2$ of (4.99) constitutes the dimensional component of APE, while

$$\mathsf{APE} = \frac{f_0^2 L^2}{2g' H}\int_\mathsf{D}(\Psi^{II} - \Psi^{I})^2\,\mathrm{d}x\,\mathrm{d}y \tag{4.100}$$

expresses the non-dimensional available potential energy. In terms of the Froude number (4.18), Eq. (4.100) takes the concise form

$$\mathsf{APE} = \frac{\mathsf{F}}{2}\int_{\mathsf{D}} (\Psi^{II} - \Psi^{I})^2 \,\mathsf{d}x\,\mathsf{d}y \tag{4.101}$$

which naturally arises in evaluating the integrated energetics of any two-layer system, as we will ascertain in the next subsection.

The orders of magnitude of the relative amounts of kinetic, available and potential energy of the two-layered flow can be estimated as follows. The integrated kinetic energy of the upper layer is

$$\mathrm{KE}^I = O\left(\frac{1}{2}\rho^I U^2 L^2 (H - H_{\mathrm{i}})\right) \tag{4.102}$$

while the integrated kinetic energy of the lower layer is

$$\mathrm{KE}^{II} = O\left(\frac{1}{2}\rho^{II} U^2 L^2 H_{\mathrm{i}}\right) \tag{4.103}$$

Moreover, (4.98) gives immediately

$$\mathrm{APE} = O\left(\frac{f_0^2 \rho_{\mathrm{s}}^2 U^2 L^4}{2g\Delta\rho}\right) \tag{4.104}$$

while the potential energy $\mathscr{U}(0)$ appearing in (4.89) is given by

$$\begin{aligned} \mathscr{U}(0) &= \int_D \left(\int_0^{H_{\mathrm{i}}} g\rho^{II} z\,\mathrm{d}z + \int_{H_{\mathrm{i}}}^{H} g\rho^I z\,\mathrm{d}z\right)\mathrm{d}x\,\mathrm{d}y \\ &= O\left(\frac{1}{2} g\rho_{\mathrm{s}} H^2 L^2\right) \end{aligned} \tag{4.105}$$

According to (4.55), we have

$$\mathsf{F}^I := \left(\frac{L}{L_{\mathrm{int}}^I}\right)^2$$

where

$$L_{\mathrm{int}}^I := \frac{\sqrt{g'(H - H_{\mathrm{i}})}}{f_0} \tag{4.106}$$

is the internal deformation radius of the upper layer. Hence, (4.102) and (4.104) yield

$$\frac{\mathrm{KE}^I}{\mathrm{APE}} = O\left(\frac{1}{\mathsf{F}^I}\right) \tag{4.107}$$

Analogously, we introduce the internal deformation radius of the lower layer, namely,

$$L_{\text{int}}^{II} := \frac{\sqrt{g' H_{\text{i}}}}{f_0} \tag{4.108}$$

and hence, (4.103) and (4.104) give

$$\frac{\text{KE}^{II}}{\text{APE}} = O\left(\frac{1}{\mathsf{F}^{II}}\right) \tag{4.109}$$

where, recalling (4.58),

$$\mathsf{F}^{II} := \left(\frac{L}{L_{\text{int}}^{II}}\right)^2$$

Moreover, from (4.104) and (4.105), one obtains

$$\frac{\text{APE}}{\mathscr{U}(0)} = O\left(\mathsf{F}\frac{U^2}{gH}\right) \tag{4.110}$$

where, from (4.18),

$$\mathsf{F} := \frac{f_0^2 L^2}{g' H}$$

Equations (4.107) and (4.109) show that, in both layers, the available potential energy predominates over the kinetic energy at horizontal scales larger than the internal deformation radius. Thus, estimation of (4.106) and (4.108) for the troposphere and the ocean may be interesting to ascertain when this happens.

Mid-latitude troposphere can be approximated by two layers of equal thickness:

$$H - H_{\text{i}} = H_{\text{i}} \approx 5 \times 10^3\,\text{m}$$

with a reduced gravity

$$g' \approx 2\,\text{m} \cdot \text{s}^{-2}$$

Hence, for $f_0 \approx 10^{-4}\,\text{s}^{-1}$, Eqs. (4.106) and (4.108) give

$$L_{\text{int}}^{I} = L_{\text{int}}^{II} \approx 10^6\,\text{m} \tag{4.111}$$

so the amounts of kinetic and available potential energy are comparable on the synoptic scale where $L = O(10^6\,\text{m})$.

In the ocean case, the values

$$H - H_{\text{i}} = 500\,\text{m} \qquad\qquad H_{\text{i}} = 3{,}000\,\text{m}$$

may be somehow realistic in some areas of the subtropical North Atlantic, where

$$g' \approx 2 \times 10^{-2}\,\text{m} \cdot \text{s}^{-2}$$

Hence, for $f_0 \approx 10^{-4}\,\mathrm{s}^{-1}$, Eqs. (4.106) and (4.108) give

$$L^{I}_{\mathrm{int}} \approx 3.2 \times 10^4\,\mathrm{m} \tag{4.112}$$

and

$$L^{II}_{\mathrm{int}} \approx 7.7 \times 10^4\,\mathrm{m} \tag{4.113}$$

respectively. For instance, at the basin scale where $L = O(10^6\,\mathrm{m})$, owing to (4.112) and (4.113), the amount of kinetic energy is $O(10^{-3})$ times smaller than the amount of available potential energy.

Consider now ratio (4.110); one may check that available potential energy is a small fraction of total potential energy, both for the atmospheric and the ocean case. In the former case, for $L = O(10^6\,\mathrm{m})$, $U = O(10\,\mathrm{m \cdot s^{-1}})$ and using again previous values, one obtains

$$\frac{\mathrm{APE}}{\mathscr{U}(0)} = O(5 \times 10^{-4}) \qquad \text{(atmosphere)}$$

In the latter case, for $L = O(10^6\,\mathrm{m})$ and $U = O(10^{-2}\,\mathrm{m \cdot s^{-1}})$, the estimate

$$\frac{\mathrm{APE}}{\mathscr{U}(0)} = O(4 \times 10^{-7}) \qquad \text{(ocean)}$$

follows. Note that, in the evaluation of (4.110), in any case, the ratio U^2/gH is a damping factor, since the horizontal advective velocity U is much smaller than the phase speed $\sqrt{gH}$ of long waves (see Eqs. (2.407) and (2.421)).

4.2.3 *Integrated Energetics of the Two-Layer Model (4.65)–(4.66)*

To see how available potential energy is an intrinsic component of the energy balance of two-layer models and to avoid too many mathematical intricacies, we refer here to the simplified model governed by Eqs. (4.65)–(4.66).

The starting point to derive the energetics of a non-dimensional QG two-layer model is multiplication of $(H - H_{\mathrm{i}})\,\Psi^I/H$ by the vorticity equation of the upper layer, multiplication of $H_{\mathrm{i}}\,\Psi^{II}/H$ by the vorticity equation of the lower layer and the subsequent integration of their sum over the horizontal fluid domain D. Accordingly, since $H_{\mathrm{i}} = H/2$ in the special case of Eqs. (4.65)–(4.66), the first is multiplied by $\Psi^I/2$ and the second by $\Psi^{II}/2$ to obtain

$$\frac{1}{2}\,\mathsf{div}\left(\Psi^I \frac{\partial}{\partial \tilde{\mathsf{t}}} \nabla' \Psi^I\right) - \frac{1}{4}\frac{\partial}{\partial \tilde{\mathsf{t}}}\left|\nabla' \Psi^I\right|^2 - \mathsf{F}\,\Psi^I \frac{\partial}{\partial \tilde{\mathsf{t}}}\left(\Psi^I - \Psi^{II}\right) + \frac{1}{4}\frac{\partial}{\partial \mathsf{x}}(\Psi^I)^2$$
$$= \frac{\mathsf{r}}{2}\,\Psi^I\,\nabla'^2\left(\Psi^{II} - \Psi^I\right) \tag{4.114}$$

and

$$\frac{1}{2}\mathsf{div}\left(\Psi^{II}\frac{\partial}{\partial\tilde{\mathsf{t}}}\nabla'\Psi^{II}\right)-\frac{1}{4}\frac{\partial}{\partial\tilde{\mathsf{t}}}\left|\nabla'\Psi^{II}\right|^2-\mathsf{F}\,\Psi^{II}\frac{\partial}{\partial\tilde{\mathsf{t}}}\left(\Psi^{II}-\Psi^{I}\right)+\frac{1}{4}\frac{\partial}{\partial\mathsf{x}}(\Psi^{II})^2$$
$$=\frac{\mathsf{r}}{2}\Psi^{II}\nabla'^2\left(\Psi^{I}-\Psi^{II}\right) \tag{4.115}$$

Note that interchanging Ψ^I with Ψ^{II} in (4.114) yields (4.115) and conversely. The sum of (4.114) with (4.115) takes, after some rearrangement, the form

$$\frac{1}{2}\mathsf{div}\left(\Psi^{I}\frac{\partial}{\partial\tilde{\mathsf{t}}}\nabla'\Psi^{I}+\Psi^{II}\frac{\partial}{\partial\tilde{\mathsf{t}}}\nabla'\Psi^{II}\right)-\frac{1}{4}\frac{\partial}{\partial\tilde{\mathsf{t}}}\left(\left|\nabla'\Psi^{I}\right|^2+\left|\nabla'\Psi^{II}\right|^2\right)$$
$$+\mathsf{F}\left(\Psi^{II}-\Psi^{I}\right)\frac{\partial}{\partial\tilde{\mathsf{t}}}\left(\Psi^{I}-\Psi^{II}\right)+\frac{1}{4}\frac{\partial}{\partial\mathsf{x}}\left[(\Psi^{I})^2+(\Psi^{II})^2\right]$$
$$=\frac{\mathsf{r}}{2}\left(\Psi^{I}-\Psi^{II}\right)\nabla'^2\left(\Psi^{II}-\Psi^{I}\right) \tag{4.116}$$

Integration of (4.116) on the horizontal fluid domain D, with the aid of the no-mass flux boundary conditions

$$\Psi^I=\Psi^{II}=0 \qquad \forall\mathsf{x},\mathsf{y}\in\partial\mathsf{D} \qquad \forall\mathsf{t}$$

together with the divergence and Green's theorems, yields

$$\frac{1}{4}\frac{\mathrm{d}}{\mathrm{d}\tilde{\mathsf{t}}}\int_{\mathsf{D}}\left(\left|\nabla'\Psi^{I}\right|^2+\left|\nabla'\Psi^{II}\right|^2\right)\mathrm{d}\mathsf{x}\,\mathrm{d}\mathsf{y}+\frac{\mathsf{F}}{2}\frac{\mathrm{d}}{\mathrm{d}\tilde{\mathsf{t}}}\int_{\mathsf{D}}\left(\Psi^{I}-\Psi^{II}\right)^2\mathrm{d}\mathsf{x}\,\mathrm{d}\mathsf{y}$$
$$=-\frac{\mathsf{r}}{2}\int_{\mathsf{D}}\left|\nabla'\left(\Psi^{I}-\Psi^{II}\right)\right|^2\mathrm{d}\mathsf{x}\,\mathrm{d}\mathsf{y} \tag{4.117}$$

Because, in the case under investigation, $\mathsf{r}=\delta_{\mathsf{S}}^{I}/L=\delta_{\mathsf{S}}^{II}/L$ while both $(H-H_{\mathrm{i}})/H$ and H_{i}/H are equal to 1/2, Eq. (4.117) can be written in a more meaningful way as

$$\frac{\mathrm{d}}{\mathrm{d}\tilde{\mathsf{t}}}\int_{\mathsf{D}}\left[\frac{H-H_{\mathrm{i}}}{2H}\left|\nabla'\Psi^{I}\right|^2+\frac{H_{\mathrm{i}}}{2H}\left|\nabla'\Psi^{II}\right|^2+\frac{\mathsf{F}}{2}\left(\Psi^{I}-\Psi^{II}\right)^2\right]\mathrm{d}\mathsf{x}\,\mathrm{d}\mathsf{y}$$
$$=-\frac{\delta_{\mathsf{S}}}{2L}\int_{\mathsf{D}}\left|\nabla'\left(\Psi^{I}-\Psi^{II}\right)\right|^2\mathrm{d}\mathsf{x}\,\mathrm{d}\mathsf{y} \tag{4.118}$$

where $\delta_{\mathsf{S}}:=\delta_{\mathsf{S}}^{I}=\delta_{\mathsf{S}}^{II}$. Notation above shows that the first two integrals actually represent quantities resulting from a 3-D integration, the vertical integrations being substituted by multiplication of each integrand by the related non-dimensional layer thickness.

The kinetic energy

$$\mathsf{KE}=\int_{\mathsf{D}}\left[\frac{H-H_{\mathrm{i}}}{2H}\left|\nabla'\Psi^{I}\right|^2+\frac{H_{\mathrm{i}}}{2H}\left|\nabla'\Psi^{II}\right|^2\right]\mathrm{d}\mathsf{x}\,\mathrm{d}\mathsf{y} \tag{4.119}$$

and the available potential energy (recall definition (4.101))

$$\mathsf{APE} = \frac{\mathsf{F}}{2}\int_{\mathsf{D}} \left(\Psi^{I} - \Psi^{II}\right)^{2} \mathsf{d}x\mathsf{d}y \tag{4.120}$$

constitute the integrated mechanical energy of the coupled Rossby waves, whose time rate of change is determined, according to (4.118), by the energy sink S due to the friction at the interface between the layers, namely,

$$\mathsf{S} = -\frac{\delta_{\mathsf{S}}}{2L}\int_{\mathsf{D}} \left|\nabla'\left(\Psi^{I} - \Psi^{II}\right)\right|^{2} \mathsf{d}x\mathsf{d}y \tag{4.121}$$

which is negative definite. Note that the integral appearing in (4.121) is equal to

$$\int_{\mathsf{D}} \left[\left(\mathsf{u}^{I} - \mathsf{u}^{II}\right)^{2} + \left(\mathsf{v}^{I} - \mathsf{v}^{II}\right)^{2}\right] \mathsf{d}x\mathsf{d}y = \|\mathbf{u}^{I} - \mathbf{u}^{II}\|^{2}$$

The norm $\|\mathbf{u}^{I} - \mathbf{u}^{II}\|$ represents a kind of distance between the geostrophic currents of the upper and lower layers. Thus, a large distance corresponds to an intense dissipation. In the opposite limit case, that is, for $\mathbf{u}^{I} = \mathbf{u}^{II}$, dissipation cancels out.

In short, Eq. (4.118) means

$$\frac{\mathrm{d}}{\mathrm{d}\tilde{t}}(\mathsf{KE} + \mathsf{APE}) = \mathsf{S} \tag{4.122}$$

This result points out that:

- Available potential energy naturally arises in the presence of a stratified flow, ultimately because density and velocity stratification are linked by the thermal wind relation.
- According to (4.121), inequality $\mathsf{S} < 0$ proves the dissipative effect of the vertical velocities w^{I} and w^{II} already met in (4.43), which determine the vorticity equations of the two-layer models.

If we use the general equation (4.56) and (4.59) in place of system (4.65)–(4.66) to derive the energetics, then we find an energy balance that differs from (4.122) for the presence of the energy source due to wind forcing and of the energy sinks due to bottom friction and lateral diffusion of relative vorticity, just like in Eq. (3.416).

Exercises

1. With reference to (4.62) and (4.64), consider the simplified coupled equations

$$\frac{\partial}{\partial \tilde{t}}\left[\nabla'^{2}\Psi^{I} - \mathsf{F}^{I}\left(\Psi^{I} - \Psi^{II}\right)\right] + \frac{\partial \Psi^{I}}{\partial \mathsf{x}} = 0$$

$$\frac{\partial}{\partial \tilde{t}}\left[\nabla'^2\Psi^{II} - \mathsf{F}^{II}\left(\Psi^{II} - \Psi^{I}\right)\right] + \frac{\partial \Psi^{II}}{\partial \mathsf{x}} = 0$$

Discuss the physical meaning of their integrals for $\mathsf{F}^I \gg \mathsf{F}^{II}$ and for $\mathsf{F}^I \ll \mathsf{F}^{II}$.

2. Generalize the model above, without special hypotheses on the Froude numbers, to a three-layer system within the rigid-lid and flat-bottom approximations.

3. A cubic container $L \times L \times L$, with four walls parallel to gravity acceleration, is divided into two equal volumes by a vertical, impermeable and removable interface. One of the volumes contains a liquid of density ρ_1 and the other a liquid of density ρ_2. These liquids are immiscible. Keeping in mind the possibility to remove the interface, compute the available potential energy of this system.
(*Solution*: $APE = (gL^4/8)(\rho_2 - \rho_1)$.)

4. Consider the steady versions of (4.56) and (4.59), and assume that the leading terms of these equations give

$$\left(\frac{\delta_\mathrm{I}}{L}\right)^2 \mathscr{J}\left(\psi^I,\, \mathsf{F}^I\psi^{II}\right) + \frac{\partial \psi^I}{\partial \mathsf{x}} = \hat{\mathbf{k}} \cdot \mathrm{rot}\,\boldsymbol{\tau}'$$

$$\left(\frac{\delta_\mathrm{I}}{L}\right)^2 \mathscr{J}\left(\psi^{II},\, \mathsf{F}^{II}\psi^{I}\right) + \frac{\partial \psi^{II}}{\partial \mathsf{x}} = 0$$

Starting from the equations above and using the known estimate

$$\frac{\tau_0}{LU\,(H - H_\mathrm{i})} = O(\rho^I \beta_0)$$

show that the dimensional Sverdrup balance

$$\beta_0 H\left(\frac{H - H_\mathrm{i}}{H}\, v^I_{\mathrm{int}} + \frac{H_\mathrm{i}}{H}\, v^{II}_{\mathrm{int}}\right) = \frac{1}{\rho^I}\,\hat{\mathbf{k}} \cdot \mathrm{rot}\,\boldsymbol{\tau}$$

follows, where v^I_{int} is the meridional velocity of the interior in the upper layer, while v^{II}_{int} is the meridional velocity of the interior in the lower layer and ρ_I is the density of the fluid in the upper layer. Note the analogy with (3.400) through the correspondence

$$\frac{H - H_\mathrm{i}}{H}\, v^I_{\mathrm{int}} + \frac{H_\mathrm{i}}{H}\, v^{II}_{\mathrm{int}} \mapsto v_{\mathrm{int}} \qquad\qquad \rho^I \mapsto \rho_\mathrm{s}$$

5. In the quasi-geostrophic framework, the interaction between the layers is described in terms of the Jacobian determinant $\mathscr{J}\left(\psi^I,\, \psi^{II}\right)$. Verify the identity

$$\mathscr{J}\left(\psi^I,\, \psi^{II}\right) = \hat{\mathbf{k}} \cdot \left(\mathbf{u}^I \times \mathbf{u}^{II}\right)$$

where $\mathbf{u}^I$ and $\mathbf{u}^{II}$ are the geostrophic currents of the upper and lower layer, respectively. Explain the interaction in terms of the geostrophic currents.

Bibliographical Note

Multilayer and especially two-layer models are considered in most of the books of Geophysical Fluid Dynamics, for instance, in Apel (1987), Cushman-Roisin (1994), Gill (1982), Young (1987), LeBlond and Mysak (1978), Pedlosky (1987, 1996), Salmon (1998), Vallis (2006) and Csanady Csanady (1982). We mention also the historic paper of Lorenz (1955) on available potential energy and the seminal paper of Rhines and Young (1982) on the vertical propagation of wind-driven circulation.

Chapter 5
Quasi-Geostrophic Models of Continuously Stratified Flows

Abstract The transition from layered flows to continuously stratified flows requires the use of thermodynamic equations basically because, in the latter models, neither horizontal current is depth or height independent nor density fluid is a constant. Thus, vertical integration of the incompressibility equation is no longer fit for explicitly yielding the vertical velocity field, and the latter must be inferred on the basis of further assumptions about large-scale geophysical flows.

These assumptions concern the resort to the thermodynamic equations already expounded in Sec. 2.4.2. In particular, in the core of the oceanic water body, sea water density is conserved up to a high degree of approximation, so that (2.588) applies there; on the other hand, near the ground, the atmosphere undergoes thermal forcing because of the long-wave radiation emitted by the Earth and the greenhouse effect, so that (2.589) applies in the troposphere.

In Chap. 5, the adiabatic quasi-geostrophic circulation on two typical ocean scales will be developed together with the analysis of steady, quasi-geostrophic atmospheric waves under the effect of a spatially modulated thermal forcing. Moreover, the effect of bathymetry on ocean currents and of topography on adiabatic winds are also taken into account in some typical cases.

5.1 QG Continuously Stratified Flows in the Ocean

Two scales are separately taken into account: the oceanic mesoscale and the basin scale. In the first case, the quasi-geostrophic vorticity dynamics so derived is applied to two physically relevant systems. The first one deals with a steady, inertially rotating flow in a bounded domain, whose model solution is solved by a variational method; the second system consists in free baroclinic Rossby waves, governed by a linear vorticity dynamics. Finally, the concept of available potential energy, formerly introduced in the two-layer model, is extended to a continuously stratified flow, and subsequently, the quasi-geostrophic vorticity equation is used to establish

F. Cavallini and F. Crisciani, *Quasi-Geostrophic Theory of Oceans and Atmosphere*, Atmospheric and Oceanographic Sciences Library 45, DOI 10.1007/978-94-007-4691-6_5,

the integrated energy balance at the oceanic mesoscale. In particular, the available potential energy is again recovered in this balance.

The vorticity dynamics at the basin scale takes into account both wind-driven flows and inertial oscillations. Wind-driven flows are governed by the baroclinic Sverdrup balance, which is a generalization of the homonymous balance already found in the homogeneous model. Then, it is shown that the closure of the transport streamlines can be obtained by introducing a dissipative mechanism analogous to that considered in Munk's model, thus obtaining a kind of Munk's model for the transport. The effect of a cross-stream bathymetric modulation on basin-scale currents is also analysed in the case of a ridge and of an escarpment. Then, the behaviour of the resulting flows is explained in terms of potential-vorticity conservation. Inertial oscillations are realized by a class of non-dispersive and non-linear Rossby waves, which constitute a limit case of baroclinic Rossby waves for flows with negligibly small relative vorticity.

5.1.1 *The Quasi-Geostrophic Oceanic Mesoscale*

In continuously stratified flows, the density anomaly of the fluid varies with continuity in space and time. Thus, besides the pressure and current fields, also the distribution of density anomaly must be determined in dealing with these flows. In a density-conserving ocean, the density anomaly is governed by the non-dimensional equation (2.628), that is,

$$w = \frac{\varepsilon}{S}\frac{D\rho'}{Dt} \tag{5.1}$$

Under the approximation of a frictionless flow, the full set of governing equations is constituted by (5.1) and, according to (2.593), (2.612), (2.613) and (2.616), by the continuity and momentum equations

$$\frac{\partial u}{\partial x} + \frac{\partial v}{\partial y} + \frac{\partial w}{\partial z} = 0 \tag{5.2}$$

$$\varepsilon_T \frac{\partial u}{\partial t} + \varepsilon \mathbf{u} \cdot \nabla' u - (1 + \beta \varepsilon y) v + \delta w = -\frac{\partial p}{\partial x} \tag{5.3}$$

$$\varepsilon_T \frac{\partial v}{\partial t} + \varepsilon \mathbf{u} \cdot \nabla' v + (1 + \beta \varepsilon y) u = -\frac{\partial p}{\partial y} \tag{5.4}$$

$$\delta^2 \varepsilon_T \frac{\partial w}{\partial t} + \delta^2 \varepsilon \mathbf{u} \cdot \nabla' w - \delta u = -\frac{1}{\rho_s}\frac{\partial}{\partial z}(\rho_s p) - \rho' \tag{5.5}$$

The estimate of the stratification parameter S, appearing in (5.1) and given by (2.625), is based on the buoyancy frequency square $N_S^2(z)$ of the rest-state ocean; so, also an estimate of the latter quantity is necessary for scaling (5.1).

Inertial Mesoscale Dynamics

The inertial mesoscale dynamics is characterized by orders of magnitude such that $\varepsilon < O(1)$ and

$$\beta = O(1) \qquad \delta = O(\varepsilon)$$
$$\mathsf{S} = O(1) \qquad \mathsf{F} = O(\varepsilon) \tag{5.6}$$

Relations (5.6) are verified, for instance, by

$$L = 100\,\mathrm{km} \qquad H = 2{,}000\,\mathrm{m} \qquad U = 0.1\,\mathrm{m/s} \qquad N_{\mathrm{S}} = 5\times 10^{-3}\,\mathrm{s}^{-1} \tag{5.7}$$

where the buoyancy frequency refers to the depth of the pycnocline. Hence,

$$\varepsilon = O(10^{-2}) \tag{5.8}$$

thus, each relation of (5.6) is verified. Assuming the advective timescale, one has that the first of (5.6) implies

$$\varepsilon_{\mathrm{T}} = \varepsilon \tag{5.9}$$

and therefore, the Rossby number ε can be taken as the ordering parameter of the governing equations (5.2)–(5.5) and (5.1), which now take the form

$$\frac{\partial \mathsf{u}}{\partial \mathsf{x}} + \frac{\partial \mathsf{v}}{\partial \mathsf{y}} + \frac{\partial \mathsf{w}}{\partial \mathsf{z}} = 0 \tag{5.10}$$

$$\varepsilon\left(\frac{\partial}{\partial \mathsf{t}} + \mathbf{u}\cdot\nabla'\right)\mathsf{u} - (1+\beta\varepsilon\mathsf{y})\mathsf{v} + \varepsilon\mathsf{w} = -\frac{\partial \mathsf{p}}{\partial \mathsf{x}} \tag{5.11}$$

$$\varepsilon\left(\frac{\partial}{\partial \mathsf{t}} + \mathbf{u}\cdot\nabla'\right)\mathsf{v} + (1+\beta\varepsilon\mathsf{y})\mathsf{u} = -\frac{\partial \mathsf{p}}{\partial \mathsf{y}} \tag{5.12}$$

$$\varepsilon^3\left(\frac{\partial}{\partial \mathsf{t}} + \mathbf{u}\cdot\nabla'\right)\mathsf{w} - \varepsilon\mathsf{u} = -\frac{1}{\rho_{\mathrm{s}}}\frac{\partial}{\partial \mathsf{z}}(\rho_{\mathrm{s}}\mathsf{p}) - \rho' \tag{5.13}$$

$$\mathsf{w} = \frac{\varepsilon}{\mathsf{S}}\left(\frac{\partial}{\partial \mathsf{t}} + \mathbf{u}\cdot\nabla'\right)\rho' \tag{5.14}$$

respectively. The smallness of ε allows us to introduce the expansions

$$\mathbf{u}(\mathbf{x},\mathsf{t};\varepsilon) = \mathbf{u}_0(\mathbf{x},\mathsf{t}) + \varepsilon\,\mathbf{u}_1(\mathbf{x},\mathsf{t}) + O(\varepsilon^2) \tag{5.15}$$

$$\mathsf{p}(\mathbf{x},\mathsf{t};\varepsilon) = \mathsf{p}_0(\mathbf{x},\mathsf{t}) + \varepsilon\,\mathsf{p}_1(\mathbf{x},\mathsf{t}) + O(\varepsilon^2) \tag{5.16}$$

$$\rho'(\mathbf{x},\mathsf{t};\varepsilon) = \rho_0'(\mathbf{x},\mathsf{t}) + \varepsilon\,\rho_1'(\mathbf{x},\mathsf{t}) + O(\varepsilon^2) \tag{5.17}$$

quite analogous to shallow-water expansions (3.33)–(3.35) with $\mathbf{u} = (u, v, w)$ and $\mathbf{x} = (x, y, z)$. Once (5.15)–(5.17) are substituted into (5.10)–(5.14), the resulting equations to the leading and the first order in ε turn out to be the following:

Zeroth order:

$$\frac{\partial u_0}{\partial x} + \frac{\partial v_0}{\partial y} + \frac{\partial w_0}{\partial z} = 0 \tag{5.18}$$

$$v_0 = \frac{\partial p_0}{\partial x} \tag{5.19}$$

$$u_0 = -\frac{\partial p_0}{\partial y} \tag{5.20}$$

$$\rho_0' = -\frac{\partial p_0}{\partial z} \tag{5.21}$$

$$w_0 = 0 \tag{5.22}$$

The derivation of Eq. (5.21) requires some explanation. The zeroth order of (5.13) is

$$\frac{1}{\rho_s} H \frac{\partial \rho_s}{\partial z} p_0 + \frac{\partial p_0}{\partial z} + \rho_0' = 0 \tag{5.23}$$

where, in the first term,

$$\frac{1}{\rho_s} H \frac{\partial \rho_s}{\partial z} = -\frac{N_s^2 H}{g}$$

is $O(5 \times 10^{-3})$ because of (5.7). Thus, the leading-order terms of (5.23) yield (5.21).

Equations (5.19) and (5.20) represent nothing but the geostrophic balance and can be written in vector form as $\mathbf{u}_0 = \hat{\mathbf{k}} \times \nabla' p_0$. The latter equation implies

$$\frac{\partial \mathbf{u}_0}{\partial z} = \hat{\mathbf{k}} \times \nabla' \frac{\partial p_0}{\partial z}$$

whence the useful orthogonality relation

$$\frac{\partial \mathbf{u}_0}{\partial z} \cdot \nabla' \frac{\partial p_0}{\partial z} = 0 \tag{5.24}$$

is derived. From (5.18) and (5.22), one obtains

$$\frac{\partial u_0}{\partial x} + \frac{\partial v_0}{\partial y} = 0 \tag{5.25}$$

in accordance with the fact that the geostrophic current is horizontally non-divergent.

First order:

$$\frac{\partial u_1}{\partial x} + \frac{\partial v_1}{\partial y} + \frac{\partial w_1}{\partial z} = 0 \tag{5.26}$$

$$\frac{D_0}{Dt}u_0 - v_1 - \beta y v_0 = -\frac{\partial p_1}{\partial x} \tag{5.27}$$

$$\frac{D_0}{Dt}v_0 + u_1 + \beta y u_0 = -\frac{\partial p_1}{\partial y} \tag{5.28}$$

$$w_1 = \frac{1}{S}\frac{D_0}{Dt}\rho_0' \tag{5.29}$$

By eliminating ρ_0' from (5.21) and (5.29), one has

$$w_1 = -\frac{1}{S}\frac{D_0}{Dt}\frac{\partial p_0}{\partial z} \tag{5.30}$$

The first-order equation coming from the vertical momentum does not give any further information and can be hereafter neglected. From (5.27) and (5.28), the equation

$$\frac{\partial}{\partial x}\left(\frac{D_0}{Dt}v_0 + u_1 + \beta y u_0\right) - \frac{\partial}{\partial y}\left(\frac{D_0}{Dt}u_0 - v_1 - \beta y v_0\right) = 0$$

is derived. Hence, we can establish the evolution equation

$$\frac{D_0}{Dt}\zeta_0' + \frac{\partial u_1}{\partial x} + \frac{\partial v_1}{\partial y} + \beta v_0 = 0 \tag{5.31}$$

where

$$\zeta_0' = \frac{\partial v_0}{\partial x} - \frac{\partial u_0}{\partial y}$$

is the relative vorticity at the geostrophic level of approximation. Equation (5.31) is nothing but (3.48) obtained in the framework of the shallow-water theory. Here, however, the distinctive point lies in the elimination of the first-order terms

$$\frac{\partial u_1}{\partial x} + \frac{\partial v_1}{\partial y} = -\frac{\partial w_1}{\partial z} \tag{5.32}$$

by resorting to (5.30) rather than to the vertical integration of (5.31), which would be inconclusive because of the baroclinic nature of the system under investigation. Equations (5.30) and (5.32) imply

$$\frac{\partial u_1}{\partial x} + \frac{\partial v_1}{\partial y} = \frac{\partial}{\partial z}\left(\frac{1}{S}\frac{D_0}{Dt}\frac{\partial p_0}{\partial z}\right) \tag{5.33}$$

and substitution of (5.33) into (5.31) yields

$$\frac{D_0}{Dt}\zeta_0' + \frac{\partial}{\partial z}\left(\frac{1}{S}\frac{D_0}{Dt}\frac{\partial p_0}{\partial z}\right) + \beta v_0 = 0 \tag{5.34}$$

Since u_0 and v_0 are in geostrophic balance with p_0, as (5.19) and (5.20) show, Eq. (5.34) has the equivalent form

$$\frac{D_0}{Dt}\nabla_H'^2 p_0 + \frac{\partial}{\partial z}\left(\frac{1}{S}\frac{D_0}{Dt}\frac{\partial p_0}{\partial z}\right) + \beta\frac{\partial p_0}{\partial x} = 0 \tag{5.35}$$

where p_0 (i.e. the perturbation pressure at the geostrophic level of approximation) is the sole unknown and $\nabla_H'^2$ is the horizontal Laplacian operator. The second term of (5.35) can be rewritten as

$$\begin{aligned}
&\frac{\partial}{\partial z}\left(\frac{1}{S}\frac{D_0}{Dt}\frac{\partial p_0}{\partial z}\right)\\
&= \frac{\partial}{\partial t}\frac{\partial}{\partial z}\left(\frac{1}{S}\frac{\partial p_0}{\partial z}\right) + \frac{\partial}{\partial z}\left[\mathbf{u}_0\cdot\nabla'\left(\frac{1}{S}\frac{\partial p_0}{\partial z}\right)\right]\\
&= \frac{\partial}{\partial t}\frac{\partial}{\partial z}\left(\frac{1}{S}\frac{\partial p_0}{\partial z}\right) + \frac{\partial \mathbf{u}_0}{\partial z}\cdot\nabla'\left(\frac{1}{S}\frac{\partial p_0}{\partial z}\right) + \mathbf{u}_0\cdot\nabla'\frac{\partial}{\partial z}\left(\frac{1}{S}\frac{\partial p_0}{\partial z}\right)\\
&= \frac{D_0}{Dt}\frac{\partial}{\partial z}\left(\frac{1}{S}\frac{\partial p_0}{\partial z}\right) + \frac{1}{S}\frac{\partial \mathbf{u}_0}{\partial z}\cdot\nabla'\frac{\partial p_0}{\partial z}
\end{aligned} \tag{5.36}$$

and (5.24) allows us to drop the last term of (5.36), thus obtaining the identity

$$\frac{\partial}{\partial z}\left(\frac{1}{S}\frac{D_0}{Dt}\frac{\partial p_0}{\partial z}\right) = \frac{D_0}{Dt}\frac{\partial}{\partial z}\left(\frac{1}{S}\frac{\partial p_0}{\partial z}\right) \tag{5.37}$$

With the aid of (5.37), Eq. (5.35) becomes

$$\frac{D_0}{Dt}\left[\nabla_H'^2 p_0 + \frac{\partial}{\partial z}\left(\frac{1}{S}\frac{\partial p_0}{\partial z}\right)\right] + \beta\frac{\partial p_0}{\partial x} = 0 \tag{5.38}$$

or, in alternative,

$$\frac{D_0}{Dt}\left[\nabla_H'^2 p_0 + \frac{\partial}{\partial z}\left(\frac{1}{S}\frac{\partial p_0}{\partial z}\right) + \beta y\right] = 0 \tag{5.39}$$

Equation (5.39) expresses the conservation, at the geostrophic level of approximation, of the *non-dimensional potential vorticity*

$$\Pi_0' := \nabla_H'^2 p_0 + \beta y + \frac{\partial}{\partial z}\left(\frac{1}{S}\frac{\partial p_0}{\partial z}\right) \tag{5.40}$$

of each material element of fluid in motion. The quantity

$$\frac{\partial}{\partial z}\left(\frac{1}{S}\frac{\partial p_0}{\partial z}\right)$$

is named *thermal vorticity*, because of its link with the thermodynamic equation (5.1).

In terms of the stream function $\psi = \mathrm{p}_0$ and of the Jacobian determinant, an equivalent formulation of (5.38) is

$$\frac{\partial}{\partial \mathrm{t}}\left(\nabla_{\mathrm{H}}^{\prime 2}\psi + \frac{\partial}{\partial \mathrm{z}}\left(\frac{1}{\mathrm{S}}\frac{\partial \psi}{\partial \mathrm{z}}\right)\right) + \mathscr{J}\left(\psi,\, \nabla_{\mathrm{H}}^{\prime 2}\psi + \frac{\partial}{\partial \mathrm{z}}\left(\frac{1}{\mathrm{S}}\frac{\partial \psi}{\partial \mathrm{z}}\right)\right) + \beta\frac{\partial \psi}{\partial \mathrm{x}} = 0 \tag{5.41}$$

while (5.39) yields the alternative form

$$\frac{\partial}{\partial \mathrm{t}}\left(\nabla_{\mathrm{H}}^{\prime 2}\psi + \frac{\partial}{\partial \mathrm{z}}\left(\frac{1}{\mathrm{S}}\frac{\partial \psi}{\partial \mathrm{z}}\right)\right) + \mathscr{J}\left(\psi,\, \nabla_{\mathrm{H}}^{\prime 2}\psi + \frac{\partial}{\partial \mathrm{z}}\left(\frac{1}{\mathrm{S}}\frac{\partial \psi}{\partial \mathrm{z}}\right) + \beta \mathrm{y}\right) = 0 \tag{5.42}$$

To summarize, Eq. (5.41) or (5.42) governs the inertial vorticity dynamics at the oceanic mesoscale under the hypothesis (5.9).

Linear Dynamics

For very small values of the horizontal velocity, say,

$$U = 5 \times 10^{-3}\ \mathrm{m/s} \tag{5.43}$$

and assuming

$$L = 100\ \mathrm{km} \qquad H = 2{,}000\ \mathrm{m} \qquad N_{\mathrm{S}} = 5 \times 10^{-3}\ \mathrm{s}^{-1} \tag{5.44}$$

as before, the estimates

$$\varepsilon_{\mathrm{T}} = O(10^{-2}) \qquad \varepsilon = O(\varepsilon_{\mathrm{T}}^2) \qquad \delta = O(\varepsilon_{\mathrm{T}}) \qquad \mathrm{S} = O(1) \tag{5.45}$$

follow. Then, the ordering parameter of (5.2)–(5.5) is ε_{T}. Note that $\beta = O(20)$, and therefore, the local timescale $T_{\mathrm{loc}} = (\beta_0 L)^{-1}$ can be taken into account in place of the advective one, $T_{\mathrm{adv}} = L/U$. Hence, the local time derivative is $\partial/\partial\tilde{\mathrm{t}}$, and the identity $\beta\,\varepsilon = \varepsilon_{\mathrm{T}}$ holds. Thus, the full set of governing equations up to the first order in ε_T is

$$\frac{\partial \mathrm{u}}{\partial \mathrm{x}} + \frac{\partial \mathrm{v}}{\partial \mathrm{y}} + \frac{\partial \mathrm{w}}{\partial \mathrm{z}} = 0 \tag{5.46}$$

$$\varepsilon_{\mathrm{T}}\frac{\partial \mathrm{u}}{\partial \tilde{\mathrm{t}}} - (1 + \varepsilon_{\mathrm{T}}\mathrm{y})\,\mathrm{v} = -\frac{\partial \mathrm{p}}{\partial \mathrm{x}} \tag{5.47}$$

$$\varepsilon_{\mathrm{T}}\frac{\partial \mathrm{v}}{\partial \tilde{\mathrm{t}}} + (1 + \varepsilon_{\mathrm{T}}\mathrm{y})\,\mathrm{u} = -\frac{\partial \mathrm{p}}{\partial \mathrm{y}} \tag{5.48}$$

$$\frac{1}{\rho_{\mathrm{s}}}\frac{\partial}{\partial \mathrm{z}}(\rho_{\mathrm{s}}\,\mathrm{p}) + \rho' = 0 \tag{5.49}$$

$$\mathrm{w} = \frac{\varepsilon_T}{S}\frac{\partial \rho'}{\partial \tilde{t}} \tag{5.50}$$

In place of (5.15)–(5.17), the expansions

$$\mathbf{u}(\mathbf{x}, \tilde{t}, \varepsilon_T) = {}_0\mathbf{u}(\mathbf{x}, \tilde{t}) + \varepsilon_T\, {}_1\mathbf{u}(\mathbf{x}, \tilde{t}) + O(\varepsilon_T^2) \tag{5.51}$$

$$\mathrm{p}(\mathbf{x}, \tilde{t}, \varepsilon_T) = {}_0\mathrm{p}(\mathbf{x}, \tilde{t}) + \varepsilon_T\, {}_1\mathrm{p}(\mathbf{x}, \tilde{t}) + O(\varepsilon_T^2) \tag{5.52}$$

$$\rho'(\mathbf{x}, \tilde{t}, \varepsilon_T) = {}_0\rho'(\mathbf{x}, \tilde{t}) + \varepsilon_T\, {}_1\rho'(\mathbf{x}, \tilde{t}) + O(\varepsilon_T^2) \tag{5.53}$$

are substituted into (5.46)–(5.50) to obtain, according to the procedure well known at this point, the vorticity equation

$$\frac{\partial}{\partial \tilde{t}}\,{}_0\zeta' + \frac{\partial}{\partial x}\,{}_1\mathrm{u} + \frac{\partial}{\partial y}\,{}_1\mathrm{v} + {}_0\mathrm{v} = 0 \tag{5.54}$$

in place of (5.31), where

$${}_0\zeta' = \frac{\partial}{\partial x}\,{}_0\mathrm{v} - \frac{\partial}{\partial y}\,{}_0\mathrm{u}$$

Likewise, we obtain

$${}_1\mathrm{w} = \frac{1}{S}\frac{\partial}{\partial \tilde{t}}\,{}_0\rho' \tag{5.55}$$

in place of (5.29), where ${}_0\rho' = -\partial\, {}_0\mathrm{p}/\partial z$ because of (5.49). With the aid of the continuity equation

$$\frac{\partial}{\partial x}\,{}_1\mathrm{u} + \frac{\partial}{\partial y}\,{}_1\mathrm{v} = -\frac{\partial}{\partial z}\,{}_1\mathrm{w}$$

the first-order fields are eliminated from (5.54) in favour of the sole perturbation pressure ${}_0\mathrm{p}$ at the geostrophic level of approximation, to obtain the *linear vorticity equation*

$$\frac{\partial}{\partial \tilde{t}}\left[\nabla_H'^2\, {}_0\mathrm{p} + \frac{\partial}{\partial z}\left(\frac{1}{S}\frac{\partial}{\partial z}\,{}_0\mathrm{p}\right)\right] + \frac{\partial}{\partial x}\,{}_0\mathrm{p} = 0 \tag{5.56}$$

Once the standard notation $\Psi = {}_0\mathrm{p}$ is introduced into (5.56), the latter equation takes the final form

$$\frac{\partial}{\partial \tilde{t}}\left[\nabla_H'^2\Psi + \frac{\partial}{\partial z}\left(\frac{1}{S}\frac{\partial \Psi}{\partial z}\right)\right] + \frac{\partial \Psi}{\partial x} = 0 \tag{5.57}$$

Ertel's Theorem in the Framework of a Quasi-Geostrophic Density-Conserving Ocean

Equation (5.39) shows that, at the geostrophic level of approximation, the potential vorticity (5.40) of a material volume of seawater is conserved following the motion, when the flow is governed by the inertial mesoscale dynamics. This statement presupposes a density-conserving flow, that is,

$$\frac{D\rho}{Dt} = 0 \tag{5.58}$$

where dimensional quantities are understood. Thus, the hypotheses of Ertel's theorem are satisfied, and (5.39) is expected to be a special version of the latter theorem. Here we clarify and develop this idea.

We recall that, under hypothesis (5.58), Ertel's theorem takes the form

$$\frac{D}{Dt}\left(\frac{1}{\rho}(\boldsymbol{\omega} + 2\boldsymbol{\Omega}) \cdot \nabla\rho\right) = 0 \tag{5.59}$$

and the problem arises to find the connection between (5.59) and (5.39). We anticipate that the method consists in (i) establishing the non-dimensional version of (5.59) and (ii) applying a suitable scale analysis to the latter. Since

$$\nabla\rho \approx \frac{\partial\rho}{\partial z}\hat{\mathbf{k}} \tag{5.60}$$

we consider Ertel's theorem in the approximate version (2.507), that is,

$$\frac{D}{Dt}\left(\frac{\zeta + f_0 + \beta_0 y}{\rho}\frac{\partial\rho}{\partial z}\right) = 0 \tag{5.61}$$

where, within the beta-plane approximation, $\zeta + f_0 + \beta_0 y = \hat{\mathbf{k}} \cdot (\boldsymbol{\omega} + 2\boldsymbol{\Omega})$. Consider now, separately, the "factors" D/Dt, $\zeta + f_0 + \beta_0 y$ and $\rho^{-1}\partial\rho/\partial z$ appearing in (5.61). We have

$$\frac{D}{Dt} = \frac{U}{L}\left(\frac{\mathrm{D}_0}{\mathrm{Dt}} + \varepsilon \mathbf{u}_1 \cdot \nabla' + O(\varepsilon^2)\right) \tag{5.62}$$

$$\zeta + f_0 + \beta_0 y = \left(1 + \varepsilon\zeta' + \frac{\beta_0 L}{f_0}\mathrm{y}\right) f_0 = \left[1 + \varepsilon\left(\nabla_{\mathrm{H}}'^2 \mathrm{p}_0 + \beta\mathrm{y}\right)\right] f_0 \tag{5.63}$$

$$\frac{1}{\rho}\frac{\partial\rho}{\partial z} = -\frac{\mathsf{F}}{H}\left(\mathsf{S} + \varepsilon\frac{\partial^2 \mathrm{p}_0}{\partial \mathrm{z}^2}\right) \tag{5.64}$$

Equation (5.64) deserves an explanation. Recalling (2.554) with (2.551), we evaluate

$$\frac{1}{\rho}\frac{\partial\rho}{\partial z} = \frac{\partial}{\partial z}\ln[\rho_s(1 + \varepsilon\mathsf{F}\rho')] \approx \frac{1}{\rho_s}\frac{d\rho_s}{dz} + \frac{\varepsilon\mathsf{F}}{H}\frac{\partial\rho_0'}{\partial \mathrm{z}} = \frac{1}{\rho_s}\frac{d\rho_s}{dz} - \frac{\varepsilon\mathsf{F}}{H}\frac{\partial^2\mathrm{p}_0}{\partial\mathrm{z}^2} \tag{5.65}$$

where use has been made of (5.21) in the last step. Moreover, positions $\mathsf{F} = f_0^2 L^2/gH$ and $\mathsf{S} = N_s^2 H^2/f_0^2 L^2$ with $N_s^2 = -(g/\rho_s)\,d\rho_s/dz$ allow us to conclude that

$$\frac{1}{\rho_s}\frac{d\rho_s}{dz} = -\frac{\mathsf{SF}}{H} \tag{5.66}$$

Then, substitution of (5.66) into (5.65) yields, after a trivial rearrangement, Eq. (5.64).

By resorting to (5.62)–(5.64), Eq. (5.61) implies

$$\left(\frac{\mathsf{D}_0}{\mathsf{Dt}}+\varepsilon\mathbf{u}_1\cdot\nabla'\right)\left\{\left[1+\varepsilon(\nabla_{\mathrm{H}}'^2\mathsf{p}_0+\beta\mathsf{y})\right]\left(\mathsf{S}+\varepsilon\frac{\partial^2\mathsf{p}_0}{\partial\mathsf{z}^2}\right)\right\}=0 \tag{5.67}$$

where $\mathsf{S}=O(1)$, $\beta=O(1)$ and $\varepsilon<O(1)$. Disregarding $O(\varepsilon^2)$-terms, Eq. (5.67) can be conveniently written as

$$\left(\frac{\mathsf{D}_0}{\mathsf{Dt}}+\varepsilon\mathbf{u}_1\cdot\nabla'\right)\left[\mathsf{S}+\varepsilon\frac{\partial^2\mathsf{p}_0}{\partial\mathsf{z}^2}+\varepsilon\mathsf{S}(\nabla_{\mathrm{H}}'^2\mathsf{p}_0+\beta\mathsf{y})\right]=0 \tag{5.68}$$

whence, owing to the obvious identity

$$\frac{\mathsf{D}_0}{\mathsf{Dt}}\mathsf{S}(\mathsf{z})=0$$

we get

$$\frac{\mathsf{D}_0}{\mathsf{Dt}}\left(\frac{\partial^2\mathsf{p}_0}{\partial\mathsf{z}^2}+\mathsf{S}\nabla_{\mathrm{H}}'^2\mathsf{p}_0+\mathsf{S}\beta\mathsf{y}\right)+\mathbf{u}_1\cdot\nabla'\mathsf{S}=0 \tag{5.69}$$

Division of (5.69) by $\mathsf{S}(\mathsf{z})$ gives, after a rearrangement,

$$\frac{\mathsf{D}_0}{\mathsf{Dt}}(\nabla_{\mathrm{H}}'^2\mathsf{p}_0+\beta\mathsf{y})+\frac{1}{\mathsf{S}}\frac{\mathsf{D}_0}{\mathsf{Dt}}\frac{\partial^2\mathsf{p}_0}{\partial\mathsf{z}^2}+\frac{1}{\mathsf{S}}\mathsf{w}_1\frac{\mathsf{dS}}{\mathsf{dz}}=0 \tag{5.70}$$

The vertical velocity w_1 appearing in the last term of (5.70) can be expressed by means of (5.30) to obtain

$$\frac{\mathsf{D}_0}{\mathsf{Dt}}(\nabla_{\mathrm{H}}'^2\mathsf{p}_0+\beta\mathsf{y})+\frac{1}{\mathsf{S}}\frac{\mathsf{D}_0}{\mathsf{Dt}}\frac{\partial^2\mathsf{p}_0}{\partial\mathsf{z}^2}-\frac{1}{\mathsf{S}^2}\frac{\mathsf{dS}}{\mathsf{dz}}\frac{\mathsf{D}_0}{\mathsf{Dt}}\frac{\partial\mathsf{p}_0}{\partial\mathsf{z}}=0 \tag{5.71}$$

that is to say,

$$\frac{\mathsf{D}_0}{\mathsf{Dt}}(\nabla_{\mathrm{H}}'^2\mathsf{p}_0+\beta\mathsf{y})+\frac{\partial}{\partial\mathsf{z}}\left(\frac{1}{\mathsf{S}}\frac{\mathsf{D}_0}{\mathsf{Dt}}\frac{\partial\mathsf{p}_0}{\partial\mathsf{z}}\right)=0 \tag{5.72}$$

Identity (5.37) allows us to write (5.72) as

$$\frac{\mathsf{D}_0}{\mathsf{Dt}}(\nabla_{\mathrm{H}}'^2\mathsf{p}_0+\beta\mathsf{y})+\frac{\mathsf{D}_0}{\mathsf{Dt}}\frac{\partial}{\partial\mathsf{z}}\left(\frac{1}{\mathsf{S}}\frac{\partial\mathsf{p}_0}{\partial\mathsf{z}}\right)=0 \tag{5.73}$$

whence (5.39) finally follows. The conclusion is that vorticity equation (5.39) is the statement of conservation of potential vorticity $\Pi=\rho^{-1}\boldsymbol{\omega}_{\mathrm{a}}\cdot\nabla\rho$ at the geostrophic and hydrostatic level of approximation.

A Steady Baroclinic Flow on the f-Plane

Consider, only to visualize the problem, a widely extended and flat-bottomed lake in which, due to the thermal structure of the water body, a density anomaly is superimposed to the standard water density. Since the lake is sensitive to ambient (Earth's) rotation, in the steady regime a baroclinic current takes place according to the thermal wind equation (2.576).

For mathematical simplicity, let $\mathsf{D} := [0 \leq \mathsf{x} \leq 1] \times [0 \leq \mathsf{y} \leq 1]$ be the non-dimensional horizontal region occupied by the lake, and let $\mathsf{V} := \mathsf{D} \times [-1 \leq \mathsf{z} \leq 0]$ be the whole fluid domain. Under the simplifying hypothesis that the stratification parameter S is approximately an $O(1)$-constant and within the f-plane approximation, the resulting non-dimensional quasi-geostrophic current set-up in the lake is governed by Eq. (5.42) with $\partial/\partial \mathsf{t} = 0$ and $\beta = 0$, that is to say,

$$\mathscr{J}\left(\psi, \Pi_0'\right) = 0 \tag{5.74}$$

where

$$\Pi_0' = \nabla_{\mathrm{H}}'^2 \psi + \frac{1}{\mathsf{S}} \frac{\partial^2 \psi}{\partial \mathsf{z}^2}$$

is the potential vorticity of the flow. The no–mass flux condition

$$\psi = 0 \qquad \text{if} \quad (\mathsf{x}, \mathsf{y}) \in \partial \mathsf{D} \text{ and } -1 < \mathsf{z} \leq 0 \tag{5.75}$$

applies along the coastline. Moreover, the request of zero vertical velocity w_1 at $\mathsf{z} = 0$ and at $\mathsf{z} = -1$ demands, according to the steady version of (5.30),

$$\mathscr{J}\left(\psi, \frac{\partial \psi}{\partial \mathsf{z}}\right) = 0 \quad \text{at } \mathsf{z} = 0 \text{ and at } \mathsf{z} = -1 \tag{5.76}$$

To establish linear and homogeneous boundary conditions for ψ consistent with (5.76), we postulate

$$\frac{\partial \psi}{\partial \mathsf{z}} = 0 \qquad \text{if} \quad (\mathsf{x}, \mathsf{y}) \in \mathsf{D} \text{ and } \mathsf{z} = -1 \tag{5.77}$$

$$\psi = \alpha' \frac{\partial \psi}{\partial \mathsf{z}} \qquad \text{if} \quad (\mathsf{x}, \mathsf{y}) \in \mathsf{D} \text{ and } \mathsf{z} = 0 \tag{5.78}$$

where α' is a given constant. Equation (5.77) means that the non-dimensional density anomaly $\rho' = -\partial \psi / \partial \mathsf{z}$ dies away at the bottom. On the other hand, at the upper surface, neither the stream function nor the density anomaly is expected to be zero; so, (5.78) is a reasonable relation consistent with (5.76).

The crucial point is that Eqs. (5.74), (5.75), (5.77) and (5.78) are not sufficient to single out a unique model solution; in fact, (5.74) is equivalent to a relationship of the kind

$$F(\psi, \Pi_0', z) = 0$$

where $F(.,..,.)$ is any differentiable function of its arguments (see Appendix A, p. 370). However, at least at the present stage of the investigation, no reason for the preference of a definite $F(.,..,.)$ has physical grounds. To overcome this point, we postulate two additional constraints on ψ:

1. The equation

$$\int_{\mathsf{V}} \psi^2 \, \mathsf{dV} = \mathsf{M} \tag{5.79}$$

   where M is a positive constant
2. The relaxation of the system to a minimum of the total mechanical energy $\mathsf{E}(\psi)$ given by

$$\mathsf{E}(\psi) = \frac{1}{2} \int_{\mathsf{V}} \left[|\nabla_{\mathrm{H}}'^2 \psi|^2 + \frac{1}{\mathsf{S}} \left(\frac{\partial \psi}{\partial \mathsf{z}} \right)^2 \right] \mathsf{dV} \tag{5.80}$$

The derivation of (5.80) will be given in the following of this section (p. 294). We stress that (5.79) implies $\mathsf{E}(\psi) > 0$ (see Appendix 1, p. 289); so, we avoid the system to be in the rest state, the latter having, trivially, the absolute minimum of total mechanical energy. In view of future computations, we put in short

$$\mathfrak{I}_0'(\psi) := \int_{\mathsf{V}} \psi^2 \, \mathsf{dV} \tag{5.81}$$

The Variational Problem Based on points (1) and (2) listed above, the variational problem consists in finding a function $\psi(\mathsf{x}, \mathsf{y}, \mathsf{z})$ such that:

1. Function ψ solves the partial differential equation (5.74).
2. Function ψ satisfies the boundary conditions (5.75), (5.77) and (5.78).
3. Function ψ minimizes the functional

$$\mathfrak{I}'(\psi, \lambda) := \mathsf{E}(\psi) + \lambda \left[\mathfrak{I}_0'(\psi) - \mathsf{M} \right] \tag{5.82}$$

   where λ is a Lagrange multiplier.

Hence, a solution of the variational problem must satisfy the system of equations

$$\begin{cases} \dot{\mathsf{E}} + \lambda \dot{\mathfrak{I}}_0' = 0 \\ \mathfrak{I}_0'(\psi) = \mathsf{M} \end{cases} \tag{5.83}$$

where the dots denote differentiation (in the sense of Fréchet or, more generally, Gâteaux) with respect to ψ Zeidler (1985). From (5.83), we find (see Appendix 2, p. 289, for computational details)

$$\begin{cases} \nabla_{\mathrm{H}}'^2 \psi + \dfrac{1}{\mathsf{S}} \dfrac{\partial^2 \psi}{\partial z^2} - 2\lambda\, \psi = 0 \\ \displaystyle\int_{\mathsf{V}} \psi^2 \, \mathsf{dV} = \mathsf{M} \end{cases} \tag{5.84}$$

The first equation of (5.84) shows that if the system under investigation reaches a state of minimum total mechanical energy, then

$$F(\psi, \Pi_0', z) = \Pi_0' - 2\lambda\, \psi \tag{5.85}$$

where the Lagrange multiplier λ is determined by solving problems (5.75), (5.77), (5.78) and (5.84).

Single-Vortex Solutions The inertial motion of the lake over a flat bottom leads us to conceive a circulation similar to that of an almost unviscous liquid in a cup; in other words, a single vortex filling the lake is expected as a likely model solution. Hence, to satisfy (5.75), a putative solution can be sought of the form

$$\psi = \phi(\mathsf{z})\, \sin(\pi \mathsf{x})\, \sin(\pi \mathsf{y}) \tag{5.86}$$

Substitution of (5.86) into (5.77)–(5.79) and (5.84) yields the following ordinary problem for the vertical profile $\phi(\mathsf{z})$ of (5.86) and the Lagrange multiplier λ:

$$\frac{\mathsf{d}^2\phi}{\mathsf{dz}^2} - 2\mathsf{S}\,(\pi^2 + \lambda)\, \phi(\mathsf{z}) = 0 \tag{5.87}$$

$$\phi(0) = \alpha' \left[\frac{\mathsf{d}\phi}{\mathsf{dz}}\right]_{\mathsf{z}=0} \tag{5.88}$$

$$\left[\frac{\mathsf{d}\phi}{\mathsf{dz}}\right]_{\mathsf{z}=-1} = 0 \tag{5.89}$$

$$\int_{-1}^{0} \phi^2(\mathsf{z})\, \mathsf{dz} = 4\,\mathsf{M} \tag{5.90}$$

The alternatives

$$2\mathsf{S}\,(\pi^2 + \lambda) > 0 \tag{5.91}$$

$$2\mathsf{S}\,(\pi^2 + \lambda) < 0 \tag{5.92}$$

will be considered separately, while the case $2\mathsf{S}\,(\pi^2 + \lambda) = 0$ implies the general integral $\phi(\mathsf{z}) = \mathsf{A} + \mathsf{B}\,\mathsf{z}$, which is not consistent with boundary conditions (5.88) and (5.89).

First alternative: $2\mathsf{S}\,(\pi^2 + \lambda) > 0$. For convenience, we put $2\mathsf{S}\,(\pi^2 + \lambda) =: \mathsf{K}^2$, the general integral can be written as

$$\phi(\mathsf{z}) = \mathsf{A}\, \sinh(\mathsf{K}\,\mathsf{z}) + \mathsf{B}\, \cosh(\mathsf{K}\,\mathsf{z}) \tag{5.93}$$

Boundary conditions (5.88) and (5.89) applied to (5.93) constitute the system

$$\begin{cases} \mathsf{A}\cosh \mathsf{K} = \mathsf{B}\sinh \mathsf{K} \\ \mathsf{B} = \alpha' \mathsf{K}\mathsf{A} \end{cases} \tag{5.94}$$

whence the equation in the unknown K

$$\tanh \mathsf{K} = \frac{1}{\alpha' \mathsf{K}} \tag{5.95}$$

follows. Provided that $\alpha' > 0$, Eq. (5.95) has two symmetrical solutions, say,

$$\mathsf{K} = \mathsf{K}_0^{\pm}(\alpha') \tag{5.96}$$

where $\mathsf{K}_0^{-}(\alpha') = -\mathsf{K}_0^{+}(\alpha')$. Finally, a straightforward computation based on (5.90), (5.94), (5.95) and (5.96) yields

$$\lambda = \frac{\mathsf{K}_0^2}{2\mathsf{S}} - \pi^2 \tag{5.97}$$

$$\phi(\mathsf{z}) = \sqrt{\frac{8\mathsf{M}}{1 + [\sinh(2\mathsf{K}_0)]/(2\mathsf{K}_0)}} \cosh[\mathsf{K}_0(\mathsf{z}+1)] \tag{5.98}$$

In (5.97) and (5.98), parameter K_0 represents any one of $\mathsf{K}_0^{+}(\alpha')$ and $\mathsf{K}_0^{-}(\alpha')$, since both λ and ϕ are trivially invariant under the substitution $\mathsf{K}_0^{+}(\alpha') \leftrightarrow \mathsf{K}_0^{-}(\alpha')$. Thus, problem (5.87)–(5.90) has a unique solution under hypothesis (5.91). From the total solution

$$\psi = \sqrt{\frac{8\mathsf{M}}{1 + [\sinh(2\mathsf{K}_0)]/(2\mathsf{K}_0)}} \cosh[\mathsf{K}_0(\mathsf{z}+1)] \sin(\pi \mathsf{x}) \sin(\pi \mathsf{y}) \tag{5.99}$$

one easily evaluates the density anomaly $\rho' = -\partial\psi/\partial\mathsf{z}$ and the related current field $\mathbf{u}_0 = \hat{\mathbf{k}} \times \nabla'_{\mathrm{H}}\psi$. Figure 5.1 illustrates the spatial distribution of ρ' and of the y-component of the velocity $\mathbf{u}_0$, derived from solution (5.99), through the related isolines on the vertical plane $\mathsf{y} = 1/2$. The higher the slope of density anomaly, the higher the absolute value of the shear of the flow; this is in accordance with the non-dimensional version of the thermal wind equation

$$\frac{\partial \mathsf{v}_0}{\partial \mathsf{z}} = -\frac{\partial \rho'}{\partial \mathsf{x}}$$

The fluid rotates clockwise. The horizontal circulation pattern determined by (5.99) is, at any depth, like that depicted in Fig. 1.7 at p. 25.

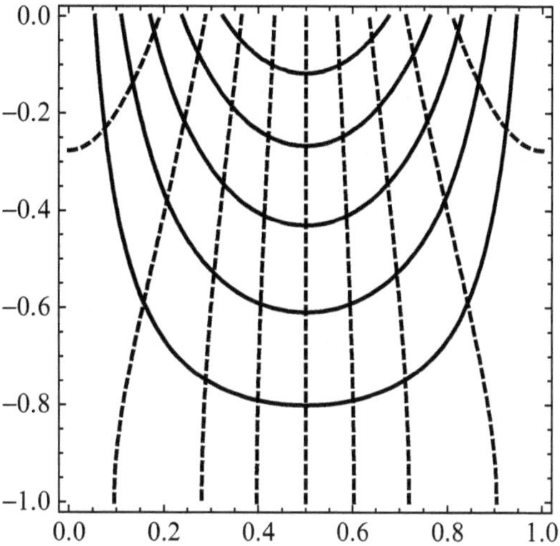

Fig. 5.1 Contour plot of density anomaly (*continuous*) and y-velocity (*dashed*) using (5.99) with $y = 1/2$ (*vertical plane*), while the radicand and K_0 are taken equal to 1. The contour values of density anomaly increase downwards, while the y-velocity increases in absolute value towards the coasts and takes opposite values at points symmetric with respect to $x = 1/2$. The y-velocity is positive for $0 \le x < 1/2$ and negative for $1/2 < x \le 1$

Second alternative: $2S(\pi^2 + \lambda) < 0$. For short, we now put $2S(\pi^2 + \lambda) =: -K^2$; so, the general integral can be written as

$$\phi(z) = A\sin(Kz) + B\cos(Kz) \tag{5.100}$$

Boundary conditions (5.88) and (5.89) applied to (5.100) imply the equation

$$\tan(K) = -\frac{1}{\alpha' K} \tag{5.101}$$

in the unknown K. Equation (5.101) has infinite solutions, say $K_0^{(n)} = K_0^{(n)}(\alpha')$, whatever α' may be. Thus, problem (5.87)–(5.90) has infinite solutions under hypothesis (5.92); they can be summarized as

$$\phi_n(z) = \sqrt{\frac{8M}{1 - [\sin(2K_0^{(n)})]/(2K_0^{(n)})}}\cos[K_0^{(n)}(z+1)] \tag{5.102}$$

and therefore, (5.86) yields

$$\psi_n = \sqrt{\frac{8M}{1 - [\sin(2K_0^{(n)})]/(2K_0^{(n)})}}\cos[K_0^{(n)}(z+1)]\sin(\pi x)\sin(\pi y) \tag{5.103}$$

Moreover, each solution (5.103) is coupled to the Lagrange multiplier

$$\lambda_n = -\pi^2 - \frac{[K_0^{(n)}]^2}{2S} \tag{5.104}$$

Preferred Solution A reason (not the sole, actually) for the preference of solution (5.99) with respect to (5.103) follows from the request of static stability of every geophysical fluid, at every depth, at least in the steady regime. It is known that, in a frame of reference where (as usual) the z-axis is positive "upward," static stability demands

$$\frac{\partial \rho}{\partial z} < 0 \tag{5.105}$$

where $\rho = \rho_s + \tilde{\rho}$. Even if every actual standard density profile $\rho_s(z)$ verifies condition $d\rho_s/dz < 0$, this inequality alone is not sufficient to assure, a priori, the overall static stability condition (5.105). Actually, only if

$$\frac{\partial \tilde{\rho}}{\partial z} < 0 \tag{5.106}$$

a sufficient condition for the validity of (5.105) follows. Now, the non-dimensional version of (5.106), written in terms of ψ, is

$$\frac{\partial^2 \psi}{\partial z^2} > 0 \tag{5.107}$$

so, solutions (5.99) and (5.103) can be tested against (5.107).

In the case of (5.99), we have

$$\frac{\partial^2 \psi}{\partial z^2} = K_0^2 \sqrt{\frac{8\,M}{1 - [\sinh(2\,K_0)]/(2\,K_0)}} \cosh[K_0\,(z+1)]\,\sin(\pi x)\,\sin(\pi y) \tag{5.108}$$

and (5.107) is certainly satisfied $\forall\,(x, y, z) \in V$.

In the case of (5.103), we have

$$\frac{\partial^2 \psi_n}{\partial z^2} = -(K_0^{(n)})^2 \sqrt{\frac{8\,M}{1 - [\sin(2\,K_0^{(n)})]/(2\,K_0^{(n)})}} \cos[K_0^{(n)}\,(z+1)]\,\sin(\pi x)\,\sin(\pi y) \tag{5.109}$$

and (5.107) is not satisfied at every depth. Indeed, for any $K_0^{(n)}$, there exists a suitably small $z_0 > 0$ such that at the depth $z = -1 + z_0$ we have $\cos[K_0^{(n)}\,(z+1)] > 0$, and hence, (5.107) is violated.

This is one of the reasons for considering model solution (5.99) to be physically grounded better than (5.103).

Remark About Constraint (5.79) We know – recall (2.532) – that the dimensional pressure field p is the superposition of the standard pressure $p_s(z)$, arising from the hydrostatic equilibrium, plus a perturbation term $\tilde{p}(x, y, z)$, which is responsible for the motion of the fluid. Thus, one might quantify the strength of the pressure perturbation $\Delta\tilde{p}$, due to the density anomaly $\tilde{\rho}$, in terms of the quantity

$$\Delta\tilde{p} = \sqrt{\frac{1}{V}\int_V (p - p_s)^2 \, dV} = \sqrt{\frac{1}{V}\int_V \tilde{p}^2 \, dV} \tag{5.110}$$

On the other hand, at the geostrophic level of approximation, we have $\tilde{p} = \rho_s f_0 U L \psi$; so, (5.110) yields

$$\Delta\tilde{p} = \rho_s f_0 U L \sqrt{\int_{\mathsf{V}} \psi^2 \, d\mathsf{V}} = \rho_s f_0 U L \sqrt{\mathsf{M}} \tag{5.111}$$

Up to a proportionality constant, relationship (5.111) shows that the constant M appearing in (5.79) is proportional to the square of the pressure perturbation that induces the geostrophic circulation in the water body of our hypothetical lake.

Appendix 1: Proof of the Inequality $\mathsf{E}(\psi) > 0$

Consider

$$\int_{\mathsf{V}} \psi^2 \, d\mathsf{V} = \int_{-1}^{0} d\mathsf{z} \int_{\mathsf{D}} \psi^2 \, dx\,dy$$

Because of (5.75), inequality

$$2\pi^2 \int_{\mathsf{D}} \psi^2 \, dx\,dy \le \int_{\mathsf{D}} |\nabla'_{\mathrm{H}}\psi|^2 \, dx\,dy \tag{5.112}$$

holds true (see Appendix A, p. 373). In turn, (5.79) and (5.112) imply

$$\frac{1}{2}\int_{\mathsf{D}} |\nabla'_{\mathrm{H}}\psi|^2 \, dx\,dy \ge \pi^2 \int_{\mathsf{D}} \psi^2 \, dx\,dy = \pi^2 \mathsf{M} > 0 \tag{5.113}$$

Hence, recalling (5.80), we get

$$\mathsf{E}(\psi) \ge \frac{1}{2}\int_{\mathsf{D}} |\nabla'_{\mathrm{H}}\psi|^2 \, dx\,dy > 0$$

Appendix 2: Derivation of the First Equation of System (5.84)

Consider, with reference to (5.82),

$$\mathfrak{I}'(\psi + h, \lambda) = \mathsf{E}(\psi + h) + \lambda\,[\mathfrak{I}'_0(\psi + h) - \mathsf{M}] =$$

$$= \frac{1}{2}\int_{\mathsf{V}} \left[|\nabla'^2_{\mathrm{H}}(\psi + h)|^2 + \frac{1}{\mathsf{S}}\left(\frac{\partial\psi}{\partial \mathsf{z}} + \frac{\partial h}{\partial \mathsf{z}}\right)^2 \right] d\mathsf{V} + \lambda \left(\int_{\mathsf{V}} (\psi + h)^2 \, d\mathsf{V} - \mathsf{M} \right) \tag{5.114}$$

where $h = 0$ on the boundary of the integration volume V. Evaluation of (5.114), up to the first order in h, yields

$$\mathfrak{I}'(\psi+h,\lambda) = \mathfrak{I}'(\psi,\lambda) + \int_V \left(\nabla'_H\psi \cdot \nabla'_H h + \frac{1}{\mathsf{S}} \frac{\partial\psi}{\partial \mathsf{z}} \frac{\partial h}{\partial \mathsf{z}} + 2\lambda\,\psi h \right) \mathrm{dV} \quad (5.115)$$

Moreover, the divergence theorem implies

$$\int_V \nabla'_H\psi \cdot \nabla'_H h\,\mathrm{dV} = \int_{\partial V} h\mathbf{n} \cdot \nabla'_H\psi\,\mathrm{dS} - \int_V h\nabla'^2_H\psi\,\mathrm{dV}$$

where ∂V is the boundary of V; so,

$$\int_V \nabla'_H\psi \cdot \nabla'_H h\,\mathrm{dV} = -\int_V h\nabla'^2_H\psi\,\mathrm{dV} \quad (5.116)$$

because $h = 0$ on ∂V. Analogously,

$$\int_V \frac{\partial\psi}{\partial \mathsf{z}} \frac{\partial h}{\partial \mathsf{z}}\,\mathrm{dV} = \int_D \left[h\frac{\partial\psi}{\partial \mathsf{z}} \right]_{\mathsf{z}=-1}^{\mathsf{z}=0} \mathrm{dx\,dy} - \int_V \frac{\partial^2\psi}{\partial \mathsf{z}^2} h\,\mathrm{dV}$$

whence

$$\int_V \frac{\partial\psi}{\partial \mathsf{z}} \frac{\partial h}{\partial \mathsf{z}}\,\mathrm{dV} = -\int_V \frac{\partial^2\psi}{\partial \mathsf{z}^2} h\,\mathrm{dV} \quad (5.117)$$

Substitution of (5.116) and (5.117) into (5.115) gives

$$\mathfrak{I}'(\psi+h,\lambda) = \mathfrak{I}'(\psi,\lambda) + \dot{\mathfrak{I}}'(\psi,\lambda)(h) \quad (5.118)$$

where

$$\dot{\mathfrak{I}}'(\psi,\lambda)(h) = -\int_V \left(\nabla'^2_H\psi + \frac{1}{\mathsf{S}} \frac{\partial^2\psi}{\partial \mathsf{z}^2} - 2\lambda\,\psi \right) h\,\mathrm{dV} \quad (5.119)$$

and, therefore, the minimum of $\mathfrak{I}'(\psi,\lambda)$ is obtained when ψ solves the equation

$$\nabla'^2_H\psi + \frac{1}{\mathsf{S}} \frac{\partial^2\psi}{\partial \mathsf{z}^2} - 2\lambda\,\psi = 0 \quad (5.120)$$

which is the first equation of system (5.84).

Baroclinic Rossby Waves

The vorticity equation (5.57), that is,

$$\frac{\partial}{\partial \tilde{\mathsf{t}}} \left[\nabla'^2_H\Psi + \frac{\partial}{\partial \mathsf{z}} \left(\frac{1}{\mathsf{S}} \frac{\partial\Psi}{\partial \mathsf{z}} \right) \right] + \frac{\partial\Psi}{\partial \mathsf{x}} = 0$$

is fit for investigating, for instance, the wave-like behaviour of a flow in a horizontally unbounded fluid layer included, for simplicity, between two rigid boundaries, at $z = 0$ and at $z = 1$. The no–mass flux condition across the planes at $z = 0$ and at $z = 1$ demands

$$ {}_1w = 0 \qquad \text{at } z = 0 \text{ and at } z = 1 \tag{5.121} $$

However, Eq. (5.121) must be restated in terms of Ψ in order to deal with a closed model. Equation (5.55), written as a function of Ψ, takes the form

$$ {}_1w = -\frac{1}{S}\frac{\partial}{\partial \tilde{t}}\frac{\partial \Psi}{\partial z} \tag{5.122} $$

and so (5.121) is satisfied if

$$ \frac{\partial \Psi}{\partial z} = 0 \qquad \text{at } z = 0 \text{ and at } z = 1 \tag{5.123} $$

Therefore, the model under investigation is governed by (5.57) and (5.123), under the further assumption of a wave-like lateral motion. Substitution of the putative stream function

$$ \Psi(x, y, z, \tilde{t}) = \Phi(z)\cos(kx + ny - \sigma'\tilde{t}) \tag{5.124} $$

into (5.57) gives

$$ \frac{d}{dz}\left(\frac{1}{S}\frac{d\Phi}{dz}\right) = \left(\frac{k}{\sigma'} + k^2 + n^2\right)\Phi(z) \tag{5.125} $$

while (5.123) implies

$$ \frac{d\Phi}{dz} = 0 \qquad \text{at } z = 0 \text{ and at } z = 1 \tag{5.126} $$

The remark at p. 181 also holds in the case of ansatz (5.124).

Integration of (5.125) from $z = 0$ to $z = 1$ with the aid of (5.126) yields

$$ \int_0^1 \Phi(z)\,dz = 0 \tag{5.127} $$

Thus, all the baroclinic wave-like solutions of the kind (5.124), that is, wave-like solutions with $\Phi(z)$ not constant, have zero vertically averaged horizontal velocity; in fact,

$$ \int_0^1 \mathbf{u}_0\,dz = \int_0^1 \hat{\mathbf{k}} \times \nabla'_H \Psi\,dz = \left[\int_0^1 \Phi(z)\,dz\right]\hat{\mathbf{k}} \times \nabla'_H \cos(kx + ny - \sigma'\tilde{t}) = 0 $$

because of (5.127).

To simplify the model further on, the stability parameter S is hereafter taken as an $O(1)$ constant, so (5.57) becomes

$$\frac{\partial}{\partial \tilde{t}}\left(\nabla_H'^2\Psi+\frac{1}{S}\frac{\partial^2\Psi}{\partial z^2}\right)+\frac{\partial\Psi}{\partial x}=0 \tag{5.128}$$

and consequently, (5.125) becomes

$$\frac{d^2\Phi}{dz^2}-S\left(\frac{k}{\sigma'}+k^2+n^2\right)\Phi(z)=0 \tag{5.129}$$

One easily checks that

$$\Phi(z)=\cos(a\,z) \tag{5.130}$$

satisfies (5.129) provided that

$$a^2+S\left(\frac{k}{\sigma'}+k^2+n^2\right)=0 \tag{5.131}$$

On the other hand, (5.130) satisfies also (5.126) if and only if

$$a=a_m:=m\pi \tag{5.132}$$

Hence, (5.130) and (5.132) select the admissible modes

$$\Phi(z)=\cos(m\pi z) \tag{5.133}$$

while the dispersion relation of each mode, derived from (5.131) and (5.132), is

$$\sigma_m'=-\frac{k}{k^2+n^2+(m\pi)^2/S}\qquad (m=0,1,2,\ldots) \tag{5.134}$$

Note that barotropic Rossby waves are obtained from (5.124) and (5.133) for $m=0$. In this case, condition (5.134) coincides with the barotropic dispersion relation (3.141), that is, $\sigma'=-k/(k^2+n^2)$. For modes with $m\geq 1$, relation (5.134) can be written as

$$\sigma_m'=-\frac{k}{k^2+n^2+(m\pi)^2L^2/L_{\text{int}}^2} \tag{5.135}$$

where $L_{\text{int}}:=N_s H/f_0$ is the *internal deformation radius* for continuously stratified flows. Equation (5.135) can be compared with the dispersion relation for barotropic Rossby waves in the presence of a fluctuating free surface (3.284), that is,

$$\sigma'=-\frac{k}{k^2+n^2+L^2/L_{\text{ext}}^2} \tag{5.136}$$

where $L_{\text{ext}} := \sqrt{gH}/f_0$. Because of the formal analogy between the quantities $(m\pi/L_{\text{int}})^2$ of (5.135) and L_{ext}^{-2} of (5.136), all the results concerning Rossby waves in a homogeneous layer, and already analysed in the framework of the shallow-water model, can be applied to each vertical baroclinic mode.

Note on the Internal Deformation Radius

The internal deformation radius appearing in (5.135) is defined as

$$L_{\text{int}} := \frac{N_s H}{f_0} \tag{5.137}$$

while, in Sect. 4.1.4 at page 261, the same quantity has been defined as

$$L_{\text{int}} := \frac{\sqrt{g'H}}{f_0} \tag{5.138}$$

Actually, position (5.138) is an approximation of (5.137). In fact, the buoyancy frequency square is

$$N_s^2 = -\frac{g}{\rho_s}\frac{\mathrm{d}\rho_s}{\mathrm{d}z} \approx -\frac{g}{\rho_s}\frac{\Delta\rho_s}{H_\rho} = \frac{g'}{H_\rho} \tag{5.139}$$

but in the two-layer model, we have $H_\rho = H$ and hence $N_s^2 \approx \frac{g'}{H}$. Then, by using this approximation in (5.137), Eq. (5.138) follows.

Energetics of a Quasi-Geostrophic, Continuously Stratified Ocean

Available Potential Energy of a Continuously Stratified Ocean Equation

$$\mathsf{APE} = \frac{f_0^2 L^2}{2g'H}\int_{\mathsf{D}} (\psi^I - \psi^{II})^2\,\mathrm{d}x\mathrm{d}y \tag{5.140}$$

already met in the two-layer quasi-geostrophic model and reported in (4.100), can be extended to continuously stratified flows. To achieve this, consider first the differential contribution to APE, say dAPE, that comes from a thin fluid layer included between z and z + dz. The change of total density from z to z + dz is prevailingly due to the standard density ρ_s, which decreases of the amount $-\mathrm{d}\rho_s$ if z increases of dz. Therefore, if variations along the vertical axis only are taken into account, the substitution of $\rho_0/\Delta\rho$ with $-\rho_s/\mathrm{d}\rho_s$ into (5.140) gives

$$\mathsf{dAPE} = \frac{f_0^2 L^2 \rho_0}{2gH\Delta\rho} \int_{\mathsf{D}} (\psi_2 - \psi_1)^2 \, \mathsf{dx\,dy}$$

$$= -\frac{\rho_s}{\mathsf{d}\rho_s} \frac{f_0^2 L^2}{2gH} \int_{\mathsf{D}} [\psi(\mathsf{x},\mathsf{y},\mathsf{z},\mathsf{t}) - \psi(\mathsf{x},\mathsf{y},\mathsf{z}+\mathsf{dz},\mathsf{t})]^2 \, \mathsf{dx\,dy} \quad (5.141)$$

By using Taylor's truncation

$$\psi(\mathsf{x},\mathsf{y},\mathsf{z}+\mathsf{dz},\mathsf{t}) = \psi(\mathsf{x},\mathsf{y},\mathsf{z},\mathsf{t}) + \frac{\partial \psi}{\partial \mathsf{z}} \mathsf{dz} \quad (5.142)$$

Eq. (5.141) can be written as

$$\mathsf{dAPE} = \left[-\rho_s \frac{f_0^2 L^2}{2gH} \left(\frac{\mathsf{d}\rho_s}{\mathsf{dz}} \right)^{-1} \int_{\mathsf{D}} \left(\frac{\partial \psi}{\partial \mathsf{z}} \right)^2 \mathsf{dx\,dy} \right] \mathsf{dz} \quad (5.143)$$

In the ocean, we have

$$-\frac{g}{\rho_s} \frac{\mathsf{d}\rho_s}{\mathsf{dz}} = H N_s^2 \quad (5.144)$$

and substitution of (5.144) into (5.143) yields

$$\mathsf{dAPE} = \left[\frac{f_0^2 L^2}{2H^2 N_s^2} \int_{\mathsf{D}} \left(\frac{\partial \psi}{\partial \mathsf{z}} \right)^2 \mathsf{dx\,dy} \right] \mathsf{dz} \quad (5.145)$$

Finally, the non-dimensional available potential energy of the flow included into the volume V extending from $\mathsf{z} = 0$ to $\mathsf{z} = 1$ is

$$\mathsf{APE} = \frac{1}{2} \int_{\mathsf{V}} \frac{1}{\mathsf{S}} \left(\frac{\partial \psi}{\partial \mathsf{z}} \right)^2 \mathsf{dV} \quad (5.146)$$

where the stability parameter

$$\mathsf{S} = \left(\frac{H N_s}{f_0 L} \right)^2$$

has been used together with the notation $\int_{\mathsf{V}} := \int_0^1 \mathsf{dz} \int_{\mathsf{D}} \mathsf{dx\,dy}$.

Integrated Energy Equation of the Ocean Like in the single-layer model, the energy equation is obtained multiplying the vorticity equation by $-\psi$ and integrating each product over the three-dimensional fluid domain. Here we consider the case of (5.42) written in the form

$$\frac{\partial}{\partial \mathsf{t}} \left[\nabla_{\mathrm{H}}'^2 \psi + \frac{\partial}{\partial \mathsf{z}} \left(\frac{1}{\mathsf{S}} \frac{\partial \psi}{\partial \mathsf{z}} \right) \right] + \mathscr{J}(\psi, \Pi_0') = 0 \quad (5.147)$$

where

$$\Pi_0' = \nabla_H'^2 \psi + \frac{\partial}{\partial z}\left(\frac{1}{S}\frac{\partial \psi}{\partial z}\right) + \beta y$$

The lateral boundary condition for (5.147) is the usual requirement of no–mass flux across the boundary ∂D of the fluid domain D, that is,

$$\psi = 0 \qquad \forall (x, y) \in \partial D \tag{5.148}$$

while the vertical boundary conditions come from the statement that the vertical velocity is zero at the bottom (i.e. at $z = 0$) and at the free surface (i.e. at $z = 1$) under the rigid-lid approximation. Thus, up to the first order in ε, the vertical boundary conditions are

$$w_1 = 0 \qquad \text{at } z = 0 \text{ and at } z = 1 \tag{5.149}$$

Multiplication of (5.147) by $-\psi$ and the subsequent integration of each product on D, with the aid of (5.148) and the divergence theorem, yield

$$\frac{1}{2}\int_D \frac{\partial}{\partial t}|\nabla_H' \psi|^2 dx dy - \int_D \psi \frac{\partial}{\partial t}\frac{\partial}{\partial z}\left(\frac{1}{S}\frac{\partial \psi}{\partial z}\right) dx dy + \frac{1}{2}\int_D \mathscr{J}\left(\psi^2, \Pi_0'\right) dx dy = 0 \tag{5.150}$$

Due to (5.148) and to the identity (see Appendix A, p. 369)

$$\mathscr{J}(p, q) = \frac{\partial}{\partial x}\left(p\frac{\partial q}{\partial y}\right) - \frac{\partial}{\partial y}\left(p\frac{\partial q}{\partial x}\right)$$

Green's identities allow us to conclude that

$$\int_D \mathscr{J}\left(\psi^2, \Pi_0'\right) dx dy = 0$$

So, only the first two terms of (5.150) enter into the energy equation. Consider, in particular, the second term of (5.150), which looks like the most intricate. Owing to (5.30), the first-order vertical velocity w_1 is given by

$$w_1 = -\frac{1}{S}\left[\frac{\partial}{\partial t}\frac{\partial \psi}{\partial z} + \mathscr{J}\left(\psi, \frac{\partial \psi}{\partial z}\right)\right] \tag{5.151}$$

and therefore, in the identity

$$\psi\left[\frac{\partial}{\partial t}\frac{\partial}{\partial z}\left(\frac{1}{S}\frac{\partial \psi}{\partial z}\right)\right] = \frac{\partial}{\partial z}\left(\psi \underbrace{\frac{1}{S}\frac{\partial}{\partial t}\frac{\partial \psi}{\partial z}}\right) - \frac{1}{S}\frac{\partial \psi}{\partial z}\frac{\partial}{\partial t}\frac{\partial \psi}{\partial z} \tag{5.152}$$

the quantity highlighted at the r.h.s, of (5.152) can be written with the aid of (5.151) as

$$\frac{1}{S}\frac{\partial}{\partial t}\frac{\partial \psi}{\partial z} = -\frac{1}{S}\mathscr{J}\left(\psi, \frac{\partial \psi}{\partial z}\right) - w_1 \tag{5.153}$$

Moreover, the last term of (5.152) can be written as

$$\frac{1}{S}\frac{\partial \psi}{\partial z}\frac{\partial}{\partial t}\frac{\partial \psi}{\partial z} = \frac{1}{2S}\frac{\partial}{\partial t}\left(\frac{\partial \psi}{\partial z}\right)^2 \tag{5.154}$$

and then the substitution of (5.153) and (5.154) into (5.152) yields

$$\psi\left[\frac{\partial}{\partial t}\frac{\partial}{\partial z}\left(\frac{1}{S}\frac{\partial \psi}{\partial z}\right)\right] = -\frac{\partial}{\partial z}\left\{\psi\left[w_1 + \frac{1}{S}\mathscr{J}\left(\psi, \frac{\partial \psi}{\partial z}\right)\right]\right\} - \frac{1}{2S}\frac{\partial}{\partial t}\left(\frac{\partial \psi}{\partial z}\right)^2 \tag{5.155}$$

Because of (5.155), Eq. (5.150) takes the form

$$\frac{1}{2}\int_D \frac{\partial}{\partial t}|\nabla'_H \psi|^2 \,dx\,dy + \frac{\partial}{\partial z}\int_D \left[\psi w_1 + \frac{1}{2S}\mathscr{J}\left(\psi^2, \frac{\partial \psi}{\partial z}\right)\right] dx\,dy$$
$$+ \int_D \frac{1}{2S}\frac{\partial}{\partial t}\left(\frac{\partial \psi}{\partial z}\right)^2 dx\,dy = 0 \tag{5.156}$$

and hence, seeing that (see Appendix A, p. 372)

$$\int_D \mathscr{J}\left(\psi^2, \frac{\partial \psi}{\partial z}\right) dx\,dy = 0$$

we get

$$\frac{1}{2}\int_D \frac{\partial}{\partial t}|\nabla'_H \psi|^2 \,dx\,dy + \frac{\partial}{\partial z}\int_D \psi w_1 \,dx\,dy + \int_D \frac{1}{2S}\frac{\partial}{\partial t}\left(\frac{\partial \psi}{\partial z}\right)^2 dx\,dy = 0 \tag{5.157}$$

Vertical integration of (5.157) with the aid of (5.149) gives the equation of mechanical energy conservation in its final non-dimensional form:

$$\frac{dE}{dt} = 0 \tag{5.158}$$

where

$$E(\psi) = \frac{1}{2}\int_V \left[|\nabla'_H \psi|^2 + \frac{1}{S}\left(\frac{\partial \psi}{\partial z}\right)^2\right] dV \tag{5.159}$$

The r.h.s. of (5.159) is the sum of the integrated kinetic energy plus the available potential energy, the latter coinciding with (5.146).

5.1.2 *Quasi-Geostrophic Dynamics at the Oceanic Basin Scale*

The basin scale is characterized by an horizontal length L of the same order as the atmospheric synoptic scale, while the horizontal velocity U is smaller than that of the oceanic mesoscale, typically,

$$L = 1{,}000\,\text{km} \qquad U = 10^{-2}\,\text{m/s}. \tag{5.160}$$

We recall that the same values as in (5.160) have been already considered in (3.188), and thus, the beta-plane approximation can be applied at this scale also.

Based on (5.160), we obtain

$$\beta = \frac{\beta_0 L^2}{U} = O(10^3) \tag{5.161}$$

and

$$\varepsilon = \frac{U}{f_0 L} = O(10^{-4}) \tag{5.162}$$

Setting, for shortness, in this section

$$\mathsf{b} := \beta\,\varepsilon = O(0.1) \tag{5.163}$$

Eq. (5.162) implies

$$\varepsilon = O(\mathsf{b}^4) \tag{5.164}$$

We assume the advective timescale, and hence, we get $\varepsilon = \varepsilon_T$, whence

$$\varepsilon_T = O(\mathsf{b}^4) \tag{5.165}$$

Moreover, we take

$$\delta = O(\mathsf{b}^3) \tag{5.166}$$

This choice will be explained below. To summarize, up to now we have expressed all the parameters ε_T, ε, $\beta\,\varepsilon$ and δ which appear into the momentum equations (2.612), (2.613) and (2.616) as integer powers of b.

What about the thermodynamic equation (2.628)? By using (5.163), this equation can be written equivalently as

$$\mathsf{w} = \frac{\mathsf{b}}{\beta\,\mathsf{S}}\frac{\mathsf{D}\rho}{\mathsf{Dt}} \tag{5.167}$$

and the order of magnitude of the quantity $\beta\,\mathsf{S}$ must be estimated at the scale under investigation. We already know, from (5.161), the typical value of β; then we recall (see (2.625)) that

$$\mathsf{S} = \left(\frac{N_s H}{f_0 L}\right)^2$$

where the standard buoyancy frequency is $N_s = O(10^{-3}\,\mathrm{s}^{-1})$ and, according to (5.160), $f_0 L = O(100\,\mathrm{m/s})$.

Since, at the oceanic basin scale, the fluid motion extends from the sea floor up to the free surface, the depth H of the motion is the same as the full ocean depth, whose planetary mean is about 3,800 m. By using the above values in (2.625), we obtain $\mathsf{S} = O(10^{-3})$, and thus,

$$\beta\,\mathsf{S} = O(1) \tag{5.168}$$

Note also that $H \approx 3{,}800\,\mathrm{m}$ and (5.160) are consistent with (5.166).

Governing Equations and Vorticity Dynamics

In terms of the ordering parameter (5.163), the governing equations are

$$\frac{\partial \mathsf{u}}{\partial \mathsf{x}} + \frac{\partial \mathsf{v}}{\partial \mathsf{y}} + \frac{\partial \mathsf{w}}{\partial \mathsf{z}} = 0 \tag{5.169}$$

$$\mathsf{b}^4 \frac{\partial \mathsf{u}}{\partial \mathsf{t}} + \mathsf{b}^4 \mathbf{u} \cdot \nabla' \mathsf{u} - (1 + \mathsf{b}\mathsf{y})\,\mathsf{v} + \mathsf{b}^3 \mathsf{w} = -\frac{\partial \mathsf{p}}{\partial \mathsf{x}} \tag{5.170}$$

$$\mathsf{b}^4 \frac{\partial \mathsf{v}}{\partial \mathsf{t}} + \mathsf{b}^4 \mathbf{u} \cdot \nabla' \mathsf{v} + (1 + \mathsf{b}\mathsf{y})\,\mathsf{u} = -\frac{\partial \mathsf{p}}{\partial \mathsf{y}} \tag{5.171}$$

$$\mathsf{b}^{10} \frac{\partial \mathsf{w}}{\partial \mathsf{t}} + \mathsf{b}^{10} \mathbf{u} \cdot \nabla' \mathsf{w} - \mathsf{b}^3 \mathsf{u} = -\frac{1}{\rho_s}\frac{\partial}{\partial \mathsf{z}}(\rho_s \mathsf{p}) - \rho' \tag{5.172}$$

$$\mathsf{w} = \frac{\mathsf{b}}{\beta\,\mathsf{S}} \frac{\mathsf{D}\rho'}{\mathsf{Dt}} \tag{5.173}$$

Substitution of the expansions of the velocity, pressure and density fields in powers of b, that is,

$$\mathbf{u}(\mathbf{x}, \mathsf{t}; \mathsf{b}) = \mathbf{u}_0(\mathbf{x}, \mathsf{t}) + \mathsf{b}\,\mathbf{u}_1(\mathbf{x}, \mathsf{t}) + O(\mathsf{b}^2)$$

$$\mathsf{p}(\mathbf{x}, \mathsf{t}; \mathsf{b}) = \mathsf{p}_0(\mathbf{x}, \mathsf{t}) + \mathsf{b}\,\mathsf{p}_1(\mathbf{x}, \mathsf{t}) + O(\mathsf{b}^2)$$

$$\rho(\mathbf{x}, \mathsf{t}; \mathsf{b}) = \rho_0(\mathbf{x}, \mathsf{t}) + \mathsf{b}\,\rho_1(\mathbf{x}, \mathsf{t}) + O(\mathsf{b}^2)$$

into (5.169)–(5.173) together with the approximation

$$\frac{1}{\rho_s}\frac{\partial}{\partial \mathsf{z}}(\rho_s \mathsf{p}) + \rho' \approx \frac{\partial \mathsf{p}}{\partial \mathsf{z}} + \rho'$$

already used in the mesoscale dynamics, results in the leading-order equations

$$\frac{\partial \mathsf{u}_0}{\partial \mathsf{x}} + \frac{\partial \mathsf{v}_0}{\partial \mathsf{y}} + \frac{\partial \mathsf{w}_0}{\partial \mathsf{z}} = 0 \tag{5.174}$$

$$v_0 = \frac{\partial p_0}{\partial x} \tag{5.175}$$

$$u_0 = -\frac{\partial p_0}{\partial y} \tag{5.176}$$

$$\frac{\partial p_0}{\partial z} + \rho_0' = 0 \tag{5.177}$$

$$w_0 = 0 \tag{5.178}$$

The first-order equations take the form

$$\frac{\partial u_1}{\partial x} + \frac{\partial v_1}{\partial y} + \frac{\partial w_1}{\partial z} = 0 \tag{5.179}$$

$$v_1 + y v_0 = \frac{\partial p_1}{\partial x} \tag{5.180}$$

$$u_1 + y u_0 = -\frac{\partial p_1}{\partial y} \tag{5.181}$$

and, owing to (5.168),

$$w_1 = \frac{1}{\beta S} \frac{D_0 \rho_0'}{Dt} \tag{5.182}$$

Note that only Eq. (5.182) is non-linear while, unlike the mesoscale ocean dynamics, local acceleration does not appear in the first-order momentum equations, due to the smallness of ε and ε_T with respect to b. To obtain an evolution equation at the geostrophic level of approximation, we preliminarily eliminate the pressure terms from (5.180) and (5.181), to obtain the equation

$$\frac{\partial u_1}{\partial x} + \frac{\partial v_1}{\partial y} + v_0 = 0 \tag{5.183}$$

Then, from (5.179) and (5.183), the relationship

$$v_0 = \frac{\partial w_1}{\partial z} \tag{5.184}$$

is established. Moreover, substitution of (5.177) into (5.182) yields

$$w_1 = -\frac{1}{\beta S} \frac{D_0}{Dt} \frac{\partial p_0}{\partial z} \tag{5.185}$$

Finally, the evolution equation for the sole geostrophic fields follows from (5.184) and (5.185) by eliminating w_1 in favour of p_0 and v_0, whence

$$\frac{\partial}{\partial z}\left(-\frac{1}{\beta S} \frac{D_0}{Dt} \frac{\partial p_0}{\partial z}\right) = v_0 \tag{5.186}$$

Owing to the identity

$$\frac{\partial}{\partial \mathsf{z}}\left(\frac{1}{\beta \mathsf{S}}\frac{\mathsf{D}_0}{\mathsf{Dt}}\frac{\partial \mathsf{p}_0}{\partial \mathsf{z}}\right) = \frac{\mathsf{D}_0}{\mathsf{Dt}}\frac{\partial}{\partial \mathsf{z}}\left(\frac{1}{\beta \mathsf{S}}\frac{\partial \mathsf{p}_0}{\partial \mathsf{z}}\right)$$

Eq. (5.186) is equivalent to

$$\frac{\mathsf{D}_0}{\mathsf{Dt}}\left[\frac{\partial}{\partial \mathsf{z}}\left(\frac{1}{\beta \mathsf{S}}\frac{\partial \mathsf{p}_0}{\partial \mathsf{z}}\right) + \mathsf{y}\right] = 0 \tag{5.187}$$

which is the desired evolution equation of the perturbation pressure at the geostrophic level of approximation. In terms of the stream function $\psi := \mathsf{p}_0$, and by using the Jacobian determinant to express the advective term of D_0/Dt, Eq. (5.187) can be restated as

$$\frac{\partial}{\partial \mathsf{t}}\frac{\partial}{\partial \mathsf{z}}\left(\frac{1}{\beta \mathsf{S}}\frac{\partial \psi}{\partial \mathsf{z}}\right) + \mathscr{J}\left(\psi, \frac{\partial}{\partial \mathsf{z}}\left(\frac{1}{\beta \mathsf{S}}\frac{\partial \psi}{\partial \mathsf{z}}\right) + \mathsf{y}\right) = 0 \tag{5.188}$$

The formal difference between (5.188) and the related mesoscale vorticity equation (5.42) lies in the absence, in (5.188), of relative vorticity. To explain this fact, consider the dimensional advection of total vorticity

$$\mathbf{u}\cdot\nabla[\zeta + f(y)] = \mathbf{u}\cdot\nabla\zeta + \beta_0 v \tag{5.189}$$

In general, $\mathbf{u}\cdot\nabla\zeta = O(U^2/L^2)$ and $\beta_0 v = O(\beta_0 U)$; so, the order of magnitude of the first term at the r.h.s. of (5.189) relative to the second term is $(U^2/L^2)/(\beta_0 U) = \beta^{-1}$. At the mesoscale (see (5.6)), we have $\beta = O(1)$; thus, both relative and planetary vorticities are advected; at the basin scale (see (5.161)), we have $\beta \gg O(1)$, and only planetary vorticity undergoes advection. Thus, $\partial v/\partial x - \partial u/\partial y$ can be reasonably neglected with respect to $\beta_0 \mathsf{y}$ at the basin-scale dynamics.

Ertel's Theorem for Inertial Motions at the Oceanic Basin Scale

Also the conservation statement (5.187), which has been derived under the assumption of a density-conserving flow, can be derived from Ertel's theorem in the well-known form

$$\frac{D}{Dt}\left(\frac{\zeta + f_0 + \beta_0 y}{\rho}\frac{\partial \rho}{\partial z}\right) = 0 \tag{5.190}$$

where, within the beta-plane approximation, $\zeta + f_0 + \beta_0 y = \hat{\mathbf{k}}\cdot(\boldsymbol{\omega} + 2\boldsymbol{\Omega})$. With reference to the 'factors' D/Dt, $\zeta + f_0 + \beta_0 y$ and $\rho^{-1}\partial\rho/\partial z$ of (5.190), we have

$$\frac{D}{Dt} = \frac{U}{L}\left(\frac{\mathsf{D}_0}{\mathsf{Dt}} + \mathsf{b}\mathbf{u}_1\cdot\nabla' + O(\mathsf{b}^2)\right) \tag{5.191}$$

$$\zeta + f_0 + \beta_0 y = f_0 \left(1 + \varepsilon \zeta' + \frac{\beta_0 L}{f_0} \mathrm{y}\right) \approx f_0 \left(1 + \mathrm{b}^5 \zeta' + \mathrm{by}\right) \approx f_0 \left(1 + \mathrm{by}\right) \tag{5.192}$$

and, recalling (5.64),

$$\frac{1}{\rho}\frac{\partial \rho}{\partial z} = -\frac{F}{H}\left(\mathrm{S} + \varepsilon \frac{\partial^2 \mathrm{p}_0}{\partial \mathrm{z}^2}\right) = -\frac{F}{H}\left(\frac{\beta \mathrm{S}}{\beta} + \frac{\beta \varepsilon}{\beta}\frac{\partial^2 \mathrm{p}_0}{\partial \mathrm{z}^2}\right) = -\frac{F}{\beta H}\left(\beta \mathrm{S} + \mathrm{b}\frac{\partial^2 \mathrm{p}_0}{\partial \mathrm{z}^2}\right) \tag{5.193}$$

On the whole, (5.190) yields, to the zeroth and first order in b,

$$\left(\frac{\mathrm{D}_0}{\mathrm{Dt}} + \mathrm{b}\mathbf{u}_1 \cdot \nabla'\right)\left[(1 + \mathrm{by})\left(\beta \mathrm{S} + \mathrm{b}\frac{\partial^2 \mathrm{p}_0}{\partial \mathrm{z}^2}\right)\right] = 0 \tag{5.194}$$

and as $\mathrm{D}_0 \mathrm{S}/\mathrm{Dt} = 0$, the remaining $O(\mathrm{b})$ terms in (5.194) constitute the equation

$$\frac{\mathrm{D}_0}{\mathrm{Dt}}\left(\beta \mathrm{Sy} + \frac{\partial^2 \mathrm{p}_0}{\partial \mathrm{z}^2}\right) + \beta \mathbf{u}_1 \cdot \nabla' \mathrm{S} = 0 \tag{5.195}$$

After a trivial rearrangement, (5.195) takes the form

$$\frac{1}{\mathrm{S}}\frac{\mathrm{D}_0}{\mathrm{Dt}}\frac{\partial^2 \mathrm{p}_0}{\partial \mathrm{z}^2} + \frac{\beta}{\mathrm{S}}\mathrm{w}_1 \frac{\partial \mathrm{S}}{\partial \mathrm{z}} + \beta \mathrm{v}_0 = 0 \tag{5.196}$$

With the aid of (5.185), Eq. (5.196) becomes

$$\frac{1}{\mathrm{S}}\frac{\mathrm{D}_0}{\mathrm{Dt}}\frac{\partial^2 \mathrm{p}_0}{\partial \mathrm{z}^2} - \frac{1}{\mathrm{S}^2}\frac{\partial \mathrm{S}}{\partial \mathrm{z}}\frac{\mathrm{D}_0}{\mathrm{Dt}}\frac{\partial \mathrm{p}_0}{\partial \mathrm{z}} = -\beta \mathrm{v}_0 \tag{5.197}$$

Writing the l.h.s. of (5.197) in compact form, Eq. (5.197) becomes

$$\frac{\mathrm{D}_0}{\mathrm{Dt}}\frac{\partial}{\partial \mathrm{z}}\left(\frac{1}{\mathrm{S}}\frac{\partial \mathrm{p}_0}{\partial \mathrm{z}}\right) + \beta \mathrm{v}_0 = 0 \tag{5.198}$$

which is nothing but (5.187).

In other words, Eq. (5.187) is the statement of conservation of potential vorticity expressed by Ertel's theorem when it is referred to the oceanic basin scale.

5.1.3 *The Effects of Bathymetry*

Ocean Current over Bathymetry

The aim of this section is to point out the behaviour of quasi-geostrophic ocean currents in the presence of bottom bathymetry. To this purpose, we proceed as follows. First, we show that one can readily find a zonal westward flow at the mesoscale assuming a flat bottom, whereas this is not the case if one looks for a similar flow that points eastward. Second, motivated by the previous analysis, we focus our attention on westward flows at the basin scale and show examples of how a topographic feature alters a zonal current.

To elucidate as simply as possible the effects of topographic forcing, we restrict the analysis to steady motions on the beta plane. Under the hypothesis that a zonal current flows over a flat bottom and impinges on a local bottom relief, the resulting current and transport fields are evaluated and explained in terms of potential-vorticity conservation. To develop this programme, a preliminary analysis is in order. In fact, a zonal current may be, a priori, both eastward and westward, but if we realistically suppose that the ocean basin is included between meridional barriers, we are soon led to recognize that only westward zonal flows are admissible in the steady regime. Below we anticipate this aspect and subsequently solve the main problem.

A Steady Westward Zonal Flow at the Mesoscale

Assuming a flat bottom, we investigate the possible formation of the westward asymptotic zonal current ($u_\infty = -1, v_\infty = 0$) arising in the domain

$$D_W :=]-\infty < x \leq 0] \times [0 \leq y \leq 1] \tag{5.199}$$

from the stream function $\psi_{-\infty} = y - y_0$, where $y_0 \in [0, 1]$ is an arbitrary constant. The inertial nature of the flow suggests to start from the vorticity equation (5.42), which becomes

$$\mathscr{J}\left(\psi, \nabla_H'^2 \psi + \beta y\right) = 0 \tag{5.200}$$

when assuming $\rho = \rho_s(z)$, and hence, $\rho' = -\partial\psi/\partial z = 0$. For convenience, we now introduce the function ϕ such that

$$\psi(x, y) = y - y_0 + \phi(x, y) \tag{5.201}$$

Function ψ in (5.201) must satisfy the no–mass flux condition $\psi = 0$ at $x = 0$ and reach the asymptotic limit $\psi(-\infty, y) = y - y_0$. Therefore, we require

$$\phi(0, y) = -y + y_0 \tag{5.202}$$

and

$$\lim_{x \to -\infty} \phi(x, y) = 0 \qquad \forall y \in [0, 1] \tag{5.203}$$

Substitution of (5.201) into (5.200) yields $\mathscr{J}\left(y + \phi, \nabla_H'^2 \phi + \beta y\right) = 0$, that is to say,

$$\nabla_H'^2 \phi + \beta y = F(y + \phi) \tag{5.204}$$

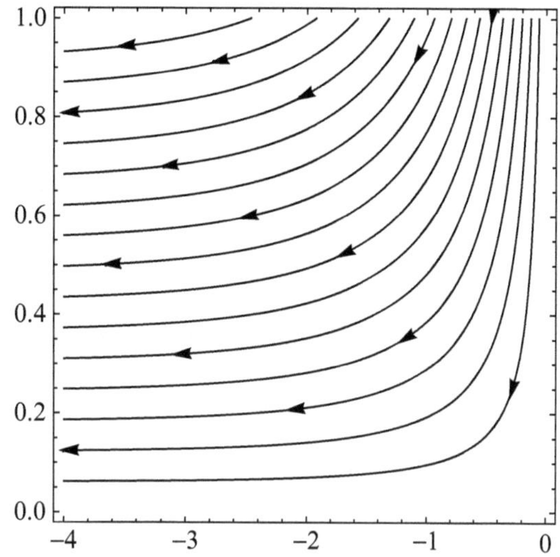

Fig. 5.2 Streamlines and flow directions in the formation region of the asymptotic flow $\psi_{-\infty} = \mathrm{y}$

where F is a function to be determined. To achieve this, we note that the asymptotic version of (5.204) can be evaluated by means of (5.203) to give $\beta\,\mathrm{y} = F(\mathrm{y})$, whence

$$F(\psi) = \beta\,\psi \tag{5.205}$$

Therefore, based on (5.205), Eq. (5.204) becomes

$$\nabla_{\mathrm{H}}^{\prime 2}\phi - \beta\,\phi = 0 \tag{5.206}$$

Problems (5.202), (5.203) and (5.206) have the solution

$$\phi = -(\mathrm{y} - \mathrm{y}_0)\,\exp(\sqrt{\beta}\,\mathrm{x}) \tag{5.207}$$

so finally, from (5.201) and (5.207), we conclude that

$$\psi(\mathrm{x}, \mathrm{y}) = (\mathrm{y} - \mathrm{y}_0)\,[1 - \exp(\sqrt{\beta}\,\mathrm{x})] \tag{5.208}$$

Figure 5.2 shows streamlines and flow directions, in the formation region of the asymptotic flow described by $\psi_{-\infty} = \mathrm{y}$, according to (5.208) with $\mathrm{y}_0 = 0$ and $\beta = 1$. This region represents the south-eastern corner ($\mathrm{x} < 0$, $\mathrm{y} > 0$) of a hypothetical basin of the beta plane. The wind-stress curl is assumed to be negligibly weak in the whole formation region; thus, in this limit, circulation is inertial. Note that, far enough from the meridional wall $\mathrm{x} = 0$, the flow looks actually zonal and with zero relative vorticity (indeed, $\partial \mathrm{u}/\partial \mathrm{y} = 0$). Moreover, the $O(1)$ constant thickness of the layer implies that the non-dimensional transport has the same form as the stream function (5.208).

To summarize, in the region (5.199), the formation of an asymptotic westward zonal current, represented by $\psi_{-\infty} = \mathrm{y} - \mathrm{y}_0$ and governed by (5.200), takes actually place. Equation (5.208) represents the behaviour of the flow field only locally, in the sense that both the meridional northward flow in the region $\mathrm{y} < \mathrm{y}_0$ and the meridional southward flow in the region $\mathrm{y} > \mathrm{y}_0$ of (5.208) should be imagined as branches of a wider flow field fully included in a suitable beta plane.

Search for an Eastward Zonal Flow at the Mesoscale

We now look for a solution, analogous to (5.208), of Eq. (5.200) in the domain

$$\mathrm{D_E} := \; [0 \leq \mathrm{x} < \infty[\times [0 \leq \mathrm{y} \leq 1] \tag{5.209}$$

with asymptotic condition

$$\lim_{\mathrm{x} \to +\infty} \psi(\mathrm{x}, \mathrm{y}) = -(\mathrm{y} - \mathrm{y}_0) \qquad \forall \mathrm{y} \in [0, 1] \tag{5.210}$$

To this purpose, we set, instead of (5.201),

$$\psi(\mathrm{x}, \mathrm{y}) = -(\mathrm{y} - \mathrm{y}_0) + \phi(\mathrm{x}, \mathrm{y}) \tag{5.211}$$

where the term $\phi(\mathrm{x}, \mathrm{y})$ has analogous roles as (5.202) and (5.203), that is,

$$\phi(0, \mathrm{y}) = \mathrm{y} - \mathrm{y}_0 \tag{5.212}$$

and

$$\lim_{\mathrm{x} \to +\infty} \phi(\mathrm{x}, \mathrm{y}) = 0 \qquad \forall \mathrm{y} \in [0, 1] \tag{5.213}$$

Moreover, because of (5.211), the vorticity equation (5.200) implies

$$\nabla_{\mathrm{H}}^{\prime 2} \phi + \beta \mathrm{y} = F(-\mathrm{y} + \phi) \tag{5.214}$$

whence, because of (5.213),

$$F(\psi) = -\beta \, \psi \tag{5.215}$$

Therefore, the problem for ϕ is given by the differential equation

$$\nabla_{\mathrm{H}}^{\prime 2} \phi + \beta \, \phi = 0 \tag{5.216}$$

together with (5.212) and (5.213). The integral of (5.216) satisfying (5.212) is

$$\phi = (\mathrm{y} - \mathrm{y}_0) \cos(\sqrt{\beta}\, \mathrm{x}) \tag{5.217}$$

but (5.217) cannot satisfy the asymptotic boundary condition (5.213).

To summarize, in the region (5.209), there is no asymptotic eastward zonal current represented by $\psi_{+\infty} = -(\mathrm{y} - \mathrm{y}_0)$.

Steady Zonal Flows at the Basin Scale

According to the previous analysis, we consider the asymptotic westward current given by $\psi_{-\infty} = \mathrm{y} - \mathrm{y}_0$ impinging on a bottom relief far enough from $\mathrm{x} = 0$ and note that, trivially, $\nabla_{\mathrm{H}}'^2 \psi_{-\infty} = 0$. This fact leads us to refer the main problem to the basin scale, where relative vorticity is everywhere negligibly small, and to adopt as governing equation the steady version of (5.188), that is,

$$\mathscr{J}\left(\psi, \frac{\partial}{\partial \mathrm{z}}\left(\frac{1}{\beta \mathrm{S}} \frac{\partial \psi}{\partial \mathrm{z}}\right) + \mathrm{y}\right) = 0 \tag{5.218}$$

The further hypothesis is that, far enough from $\mathrm{x} = 0$, the fluid layer is included between the flat free surface and a bottom with modulated bathymetry.

We now derive some preliminary kinematic results. Assume that the flow is vertically included in the interval

$$-h(x, y) \leq z \leq 0 \tag{5.219}$$

with

$$h(x, y) = H + \eta(x, y) \tag{5.220}$$

where H is the typical depth of the motion, in terms of which vertical lengths are scaled. From (5.219) and (5.220), we see that bottom bathymetry is described by

$$z = -H - \eta(x, y) \tag{5.221}$$

Thus, dividing by H, the related non-dimensional profile is given by

$$\mathrm{z} = -1 - \frac{\eta}{H} \tag{5.222}$$

where $\mathrm{z} := z/H$. The boundary condition at the depth (5.221) is given by (1.28), that is to say,

$$w = -\mathbf{u} \cdot \nabla \eta \qquad \text{at } z = -H - \eta \tag{5.223}$$

In the framework of the quasi-geostrophic dynamics at the ocean basin scale, the non-dimensional version of (5.223) is inferred from

$$\frac{UH}{L}[\mathrm{b}\mathrm{w}_1' + O(\mathrm{b}^2)] = -\frac{U}{L}[\mathbf{u}_0 + O(\mathrm{b})] \cdot \nabla' \eta \tag{5.224}$$

where $\nabla' := L\nabla$, while b is the ordering parameter (5.165). In order that (5.224) represents a $O(\mathrm{b})$ balance involving both the vertical velocity w_1 and the topographic gradient, the estimate $\eta = O(\mathrm{b}H)$ must hold, that is,

$$\eta = \mathrm{b}H\eta' \tag{5.225}$$

where $\eta' = O(1)$. In this case, to the leading order in b, Eq. (5.224) gives

$$w_1 = -\mathbf{u}_0 \cdot \nabla' \eta'$$

or, equivalently,

$$w_1 = -\mathscr{J}\left(\psi, \eta'\right) \tag{5.226}$$

Recalling (5.222) and (5.225), we see that (5.226) refers to the non-dimensional bottom depth

$$z = -1 - b\eta' \tag{5.227}$$

Assuming $\eta = b^2 H \eta'$ in place of (5.225) yields $w_1 = 0$ because of (5.224), but, in this case, bottom modulation would be too weak to be detected by the quasi-geostrophic system, which would simply "see" a flat bottom. On the other hand, if $\eta = H\eta'$, then (5.224) implies the stronger constraint $\mathbf{u}_0 \cdot \nabla' \eta' = 0$, that is, motion takes place exactly along the isobaths.

To analyze the system dynamics, the governing equation (5.218) must be complemented with vertical and meridional boundary conditions that ensure a unique solution. Vertical boundary conditions derive from the steady case of Eq. (5.185), that is,

$$w_1 = -\frac{1}{\beta S} \mathscr{J}\left(\psi, \frac{\partial \psi}{\partial z}\right) \tag{5.228}$$

Therefore, owing to (5.228), the free-surface condition $w_1(z = 0) = 0$ means

$$\mathscr{J}\left(\psi, \frac{\partial \psi}{\partial z}\right) = 0 \qquad \text{at } z = 0 \tag{5.229}$$

At the bottom, that is, at $z = -1 - b\eta'$, vertical velocity (5.226) matches vertical velocity (5.228), and therefore,

$$\mathscr{J}\left(\psi, \frac{1}{\beta S}\frac{\partial \psi}{\partial z} - \eta'\right) = 0 \qquad \text{at } z = -1 - b\eta' \tag{5.230}$$

Finally, the latitudinal boundary conditions requested by the beta-plane approximation are

$$\frac{\partial \psi}{\partial x} = 0 \qquad \text{at } y = 0 \text{ and at } y = 1 \tag{5.231}$$

as usual.

We now look for a solution described asymptotically by $\psi_{-\infty} = y - y_0$. Equation (5.218) is equivalent to

$$\frac{\partial}{\partial z}\left(\frac{1}{\beta S}\frac{\partial \psi}{\partial z}\right) + y = F(\psi, z) \tag{5.232}$$

where $F(\psi, \mathrm{z})$ is to be determined. Substitution of $\psi_{-\infty} = \mathrm{y} - \mathrm{y}_0$ into (5.232) gives $\mathrm{y} = F(\psi_{-\infty}, \mathrm{z})$, that is, $F(\psi_{-\infty}, \mathrm{z}) = \psi_{-\infty} + \mathrm{y}_0$, whence equation $F(\psi, \mathrm{z}) = \psi + \mathrm{y}_0$ holds true in the entire fluid domain. Thus, Eq. (5.232) becomes

$$\frac{\partial}{\partial \mathrm{z}}\left(\frac{1}{\beta \mathrm{S}}\frac{\partial \psi}{\partial \mathrm{z}}\right) + \mathrm{y} = \psi + \mathrm{y}_0$$

and, under the further simplifying assumption $\beta\,\mathrm{S} = $ constant, the last equation takes the final form

$$\frac{1}{\beta \mathrm{S}}\frac{\partial^2 \psi}{\partial \mathrm{z}^2} + \mathrm{y} - \mathrm{y}_0 = \psi \tag{5.233}$$

Then, consider boundary condition (5.229) written as

$$\frac{\partial \psi}{\partial \mathrm{z}} = G(\psi) \qquad \text{at } \mathrm{z} = 0 \tag{5.234}$$

Substituting $\psi \to \psi_{-\infty}$ into (5.234) gives $(\partial \psi_{-\infty}/\partial \mathrm{z})_{\mathrm{z}=0} = G(\psi_{-\infty})$, but $\partial \psi_{-\infty}/\partial \mathrm{z}$ is zero, which implies that G is also zero, and so (5.234) gives

$$\frac{\partial \psi}{\partial \mathrm{z}} = 0 \qquad \text{at } \mathrm{z} = 0 \tag{5.235}$$

Analogously, once boundary condition (5.230) is written as

$$\frac{1}{\beta \mathrm{S}}\frac{\partial \psi}{\partial \mathrm{z}} - \eta' = Q(\psi) \tag{5.236}$$

and $\psi_{-\infty}$ is substituted into (5.236) in place of ψ, one obtains $-\eta' = Q(\psi_{-\infty})$. But, in the oceanic area where $\psi \approx \psi_{-\infty}$, the bottom is flat; so, $Q(\psi_{-\infty}) = 0$, and thus, $Q(\psi) = 0\ \forall\, \psi$. Hence,

$$\frac{1}{\beta \mathrm{S}}\frac{\partial \psi}{\partial \mathrm{z}} - \eta' = 0 \qquad \text{at } \mathrm{z} = -1 - \mathrm{b}\eta' \tag{5.237}$$

To summarize, the model under investigation is governed by (5.231), (5.233), (5.235) and (5.237). The general integral of (5.233) is

$$\psi = \mathrm{y} - \mathrm{y}_0 + \mathrm{A}(\mathrm{x}, \mathrm{y})\cosh(\sqrt{\beta \mathrm{S}}\,\mathrm{z}) + \mathrm{B}(\mathrm{x}, \mathrm{y})\sinh(\sqrt{\beta \mathrm{S}}\,\mathrm{z}) \tag{5.238}$$

Because of (5.235), we have $\mathrm{B}(\mathrm{x}, \mathrm{y}) = 0$; so, (5.238) simplifies to

$$\psi = \mathrm{y} - \mathrm{y}_0 + \mathrm{A}(\mathrm{x}, \mathrm{y})\cosh(\sqrt{\beta \mathrm{S}}\,\mathrm{z}) \tag{5.239}$$

Substitution of (5.239) into (5.237) allows us to evaluate

$$A(x, y) = -\frac{\sqrt{\beta S}\,\eta'(x, y)}{\sinh\{\sqrt{\beta S}\,[1 + b\,\eta'(x, y)]\}} \tag{5.240}$$

and finally

$$\psi = y - y_0 - \frac{\sqrt{\beta S}\,\eta'(x, y)}{\sinh\{\sqrt{\beta S}\,[1 + b\,\eta'(x, y)]\}} \cosh(\sqrt{\beta S}\,z) \tag{5.241}$$

Solution (5.241) verifies condition (5.231) if

$$\frac{\partial \eta'}{\partial x} = 0 \qquad \text{at } y = 0 \text{ and at } y = 1 \tag{5.242}$$

Condition (5.242) prevents that a topographic profile forces the flow to go out the latitudinal strip in which the beta-plane approximation is valid; in any case, (5.242) is consistent with physically realistic models of bottom bathymetry. In terms of the (non-dimensional) mass transport

$$\mathscr{T}'(x, y) := \int_{-1-b\eta'(x,y)}^{0} \psi(x, y, z)\,dz \tag{5.243}$$

Eq. (5.241) yields

$$\mathscr{T}' = [1 + b\,\eta'(x, y)]\,(y - y_0) - \eta'(x, y) \tag{5.244}$$

and the isolines of the mass transport are obtained by solving

$$\mathscr{T}'(x, y) = \text{constant} \tag{5.245}$$

The transport streamlines inferred from (5.245) are depth independent, thus giving less information than the streamlines evaluated through (5.241) at each depth, but they are representative of the overall behaviour of the fluid columns extending from the free surface down to the bottom.

Topographic Profiles

Evaluation of (5.241) and (5.244) is carried out once the form of $\eta'(x, y)$ is explicitly given. For example, we consider a depth anomaly oriented in the cross-stream direction, and for mathematical convenience, we assume that η' can be factored as

$$\eta'(x, y) = X(x)\,Y(y) \tag{5.246}$$

where $X(x)$ characterizes the morphology, while function $Y(y)$ is almost constant and $O(1)$ along most of the latitudinal extension of the ridge. Specifically, we now model two somehow realistic cases: a ridge and an escarpment; in both cases, we choose

$$\mathsf{Y}(\mathsf{y}) = \mathsf{Y}_0 \left[\mathsf{S_d}(\mathsf{y}-\mathsf{y}_0) - \mathsf{S_d}(-\mathsf{y}_0) - \mathsf{S_d}(\mathsf{y}-1+\mathsf{y}_0)) + \mathsf{S_d}(-1+\mathsf{y}_0))\right] \quad (5.247)$$

where Y_0 and d are positive parameters to be suitably fixed, while $\mathsf{S_d}$ is the *smooth step* function defined as

$$\mathsf{S_d}(\mathsf{y}) = \frac{1}{2}\left(1+\tanh(2\,\mathsf{d}\,\mathsf{y})\right) \quad (5.248)$$

It is readily seen that $\mathsf{S_d}$ has an inflection point at zero, with derivative d. Finally, note that $\mathsf{Y}(0) = \mathsf{Y}(1) = 0$.

Ridge. A ridge centred at the longitude $\mathsf{x}_0 \ll 0$ can be described through the non-dimensional depth anomaly (5.246)–(5.247) with

$$\mathsf{X}(\mathsf{x}) = -\exp\left[-\left(\frac{\mathsf{x}-\mathsf{x}_0}{\lambda'}\right)^2\right] \quad (5.249)$$

where the length λ' quantifies the steepness and width of the profile.

Escarpment. An escarpment centred at the longitude $\mathsf{x}_0 \ll 0$ can be described through the non-dimensional depth anomaly (5.246)–(5.247) with

$$\mathsf{X}(\mathsf{x}) = -\mathsf{S_d}(\mathsf{x}-\mathsf{x}_0) \quad (5.250)$$

where the parameter d is the slope of the escarpment at its mid-depth and x_0 is its inflection point.

Because of (5.247), boundary conditions (5.242) are verified in both cases (5.249) and (5.250).

Figure 5.3 shows the bathymetry of an idealized ridge and of an idealized escarpment. In both cases, Eq. (5.247) has been used with $\mathsf{Y}_0 = 1$, $\mathsf{y}_0 = 1/8$ and $\mathsf{d} = 6$, together with (5.246). The ridge is described by (5.249) with $\mathsf{x}_0 = -11$ and $\lambda' = 1$. The escarpment is described by (5.250), with $\mathsf{x}_0 = -11$ and $\mathsf{d} = 1$. The dynamics arising in connection with these bathymetric features is illustrated in Fig. 5.4.

Potential Vorticity

Inertial circulation at the ocean basin scale conserves, besides fluid density, also potential vorticity according to the equation

$$\frac{D}{Dt}\left(\frac{f_0+\beta_0 y}{\rho}\frac{\partial \rho}{\partial z}\right) = 0 \quad (5.251)$$

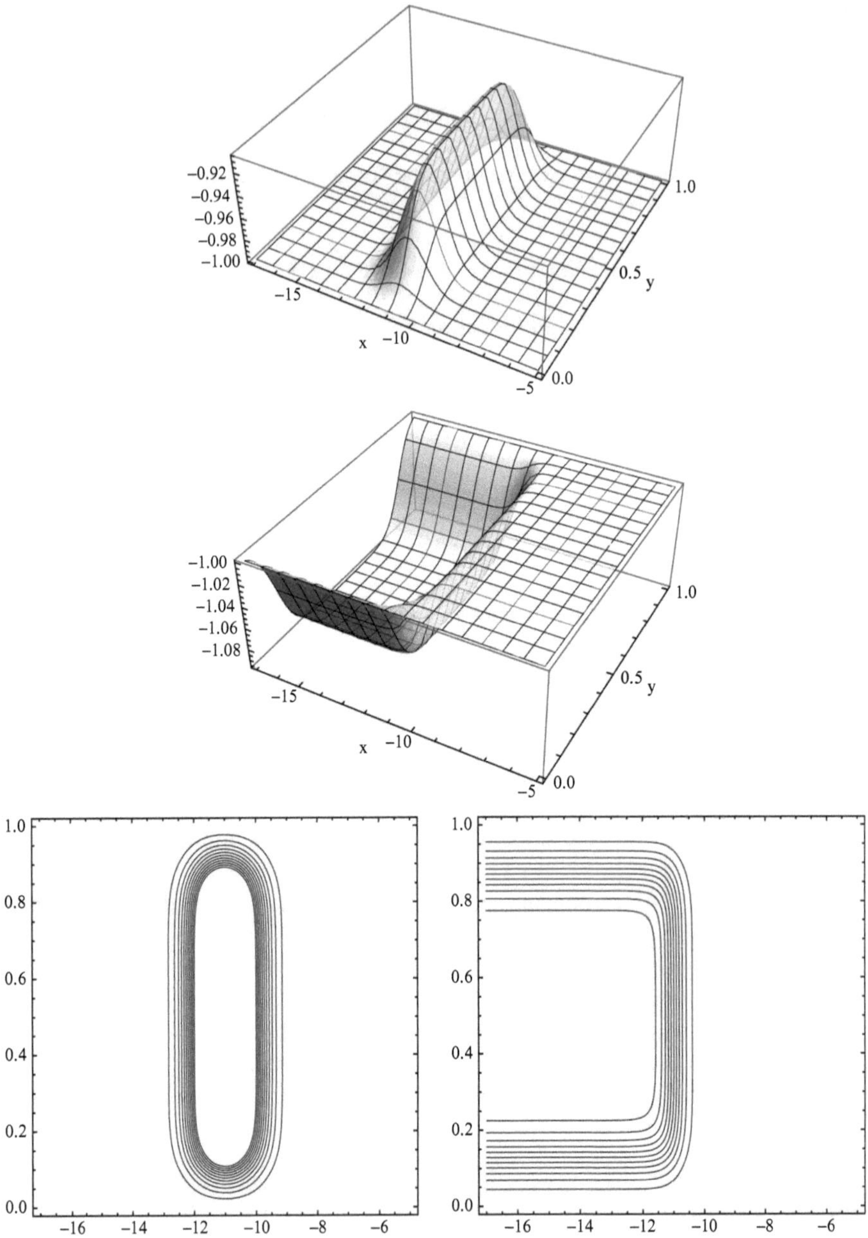

Fig. 5.3 The bathymetry (5.227) of an idealized ridge (*left*) and that of an idealized escarpment (*right*) are shown through 3-D plots and contour plots

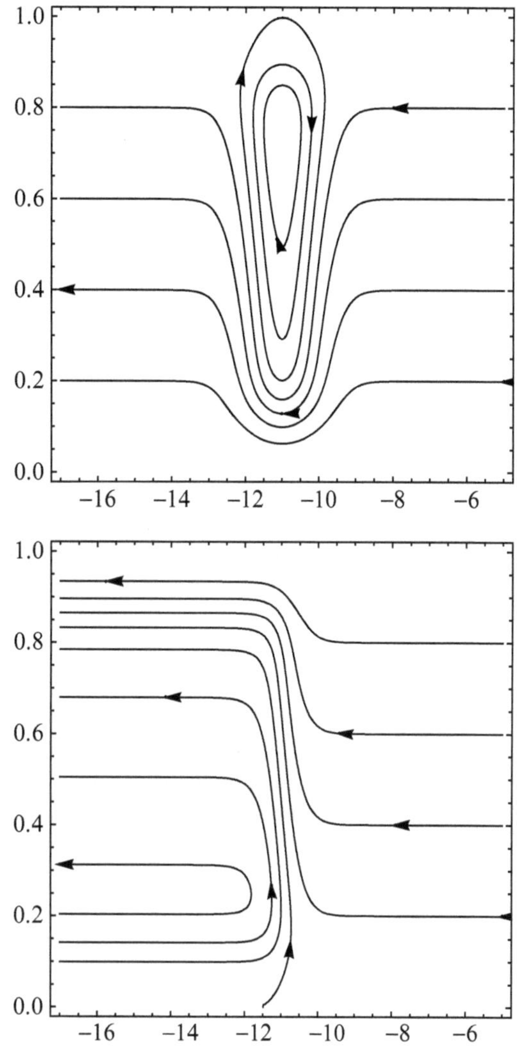

Fig. 5.4 Transport streamlines and transport directions due to a westward flow at basin scale over a bathymetric anomaly. Equation (5.244) with b = 0.1 has been used. *Upper panel*: model ridge (5.249). *Lower panel*: model escarpment (5.250)

that follows from (5.190) and (5.192). Thus, along a streamline, (5.251) implies

$$\frac{f_0 + \beta_0 y}{\rho} \frac{\partial \rho}{\partial z} = C \tag{5.252}$$

where each constant C characterizes a given streamline. Vertical integration of both sides of (5.252) in interval $-H - \eta \leq z \leq 0$ and the subsequent division by $H + \eta$ produces the equation

$$\frac{f_0 + \beta_0 y}{H + \eta} \ln \frac{\rho(z = 0)}{\rho(z = -H - \eta)} = C \tag{5.253}$$

We know that, in general, parcels belonging to the boundary ($z = 0$ and $z = -H - \eta$) never leave it, and in the present model, they also conserve their own density following the motion. Therefore, the factor $\ln[\rho(z=0)/\rho(z=-H-\eta)]$ of (5.253) is a constant; so, (5.253) simplifies to

$$\frac{f_0 + \beta_0 y}{H + \eta} = C \tag{5.254}$$

The non-dimensional version of (5.254) is

$$\frac{1 + \mathsf{b}\mathsf{y}}{1 + \mathsf{b}\eta'} = \mathsf{C} \tag{5.255}$$

Consider now two points of the same streamline, labelled by i and f; then Eq. (5.255) implies

$$\frac{1 + \mathsf{b}\mathsf{y}_\mathrm{i}}{1 + \mathsf{b}\eta'_\mathrm{i}} = \frac{1 + \mathsf{b}\mathsf{y}_\mathrm{f}}{1 + \mathsf{b}\eta'_\mathrm{f}} \tag{5.256}$$

Setting $\mathsf{y}_\mathrm{f} = \mathsf{y}_\mathrm{i} + \Delta\mathsf{y}$ and solving for $\Delta\mathsf{y}$, (5.256) gives

$$\Delta\mathsf{y} = (1 + \mathsf{b}\mathsf{y}_\mathrm{i})\,\frac{\eta'_\mathrm{f} - \eta'_\mathrm{i}}{1 + \mathsf{b}\eta'_\mathrm{i}} \tag{5.257}$$

where $\Delta\mathsf{y}$ is the meridional displacement of the streamline induced by bathymetry and $\eta'_\mathrm{f} - \eta'_\mathrm{i}$ is the difference of the topographic modulation in moving from i to f following the same streamline. Thus, $\eta'_\mathrm{f} > \eta'_\mathrm{i}$ (i.e. a thickening of the fluid layer) implies $\mathsf{y}_\mathrm{f} > \mathsf{y}_\mathrm{i}$ (i.e. a northward displacement of the streamline), while $\eta'_\mathrm{f} < \eta'_\mathrm{i}$ (i.e. a reduction of the fluid layer) implies $\mathsf{y}_\mathrm{f} < \mathsf{y}_\mathrm{i}$ (i.e. a southward displacement of the streamline).

Note that (5.257) can be written as

$$\Delta\mathsf{y} = (\eta'_\mathrm{f} - \eta'_\mathrm{i})\,\frac{1 + (L/R)\cot(\phi_0)\,\mathsf{y}_\mathrm{i}}{1 + (L/R)\cot(\phi_0)\,\eta'_\mathrm{i}} \tag{5.258}$$

where ϕ_0 is the central latitude of the beta plane and R is Earth's mean radius. Although ϕ_0 becomes $-\phi_0$ in passing from the northern hemisphere to the southern hemisphere, we know from (2.609) that $(L/R)\,|\cot(\phi_0)| < O(1)$, and therefore, we see again that $\Delta\mathsf{y}$ has the same sign as $\eta'_\mathrm{f} - \eta'_\mathrm{i}$ in both hemispheres.

Figure 5.4 (upper panel) shows transport streamlines and transport vectors produced by (5.244) for the model ridge (5.249) with $\mathsf{b} = 0.1$, in the proximity of the longitude x_0 where the bathymetric anomaly is located. According to (5.257), a southward displacement of the fluid parcels takes place whenever the ridge increases, mainly with respect to x, while a northward displacement of equal amount follows immediately after, owing to the subsequent decrease of the ridge. Therefore, the transport in the region $\mathsf{x} \ll \mathsf{x}_0$ is the same as that in $\mathsf{x}_0 \ll \mathsf{x} < 0$, at every latitude.

Note that, in spite of the symmetry $\eta'(\mathsf{x},\mathsf{y}) = \eta'(\mathsf{x}, 1-\mathsf{y})$ exhibited by (5.247), this symmetry is broken for $\mathscr{T}'(\mathsf{x},\mathsf{y})$ because of factor y, which is openly not invariant under the transformation $(\mathsf{x},\mathsf{y}) \mapsto (\mathsf{x}, 1-\mathsf{y})$. For $0 < \mathsf{y} < 1$, the shape of the isolines of $\mathscr{T}'(\mathsf{x},\mathsf{y})$ is reminiscent of the graph, in the Cartesian plane (x,y), of equation

$$\mathsf{y} = -\exp\left[-\left(\frac{\mathsf{x}-\mathsf{x}_0}{\lambda'}\right)^2\right] \tag{5.259}$$

The reason is that if b is neglected in (5.244) and Y is approximated by unity, then equation $\mathscr{T}'(\mathsf{x},\mathsf{y}) = \text{constant}$ takes the form (5.259).

Likewise, the model escarpment (5.250) is shown in Fig. 5.4 (lower panel). The main difference between the flows related to the ridge and to the escarpment is due to the fact that, in the escarpment, the increase of the fluid thickness in the westward motion induces a northward displacement of the streamlines in crossing the escarpment which, unlike the ridge, is not compensated by any further displacement. Hence, owing to the conservation of the mass in the latitudinal interval $0 \leq \mathsf{y} \leq 1$, the downstream transport intensifies at higher latitudes.

5.1.4 Forced and Wave-Like Circulation

The Sverdrup Balance

We reconsider the diagnostic equation (5.184), that is,

$$\mathsf{v}_0 = \frac{\partial \mathsf{w}_1}{\partial \mathsf{z}}$$

The dimensional version of (5.184) is obtained assuming

$$\mathsf{v}_0 \approx \frac{v}{U} \qquad \frac{\partial}{\partial \mathsf{z}} = H\frac{\partial}{\partial z} \qquad \mathsf{w}_1 \approx \frac{f_0}{\beta_0 U H} w$$

whence the dimensional equation

$$\beta_0 v = f_0 \frac{\partial w}{\partial z} \tag{5.260}$$

follows. Multiplying this equation by the standard density $\rho_s(z)$ and integrating over the depth H of the geostrophic motion yield

$$\beta_0 \int_0^H \rho_s v \, dz = f_0 \int_0^H \rho_s \frac{\partial w}{\partial z} dz \tag{5.261}$$

where the integral at the l.h.s. of (5.261) expresses the meridional mass transport according to the following equation (5.270). The r.h.s. of (5.261) can be integrated by parts to give

$$f_0 \left[\rho_s w \right]_{z=0}^{z=H} - f_0 \int_0^H w \frac{d\rho_s}{dz} dz \tag{5.262}$$

In the case of a flat bottom, using (3.341), the first term of (5.262) is simply

$$\hat{\mathbf{k}} \cdot \mathrm{rot}\,\boldsymbol{\tau} \tag{5.263}$$

Note that the l.h.s. of (5.261) and (5.263) has the same order of magnitude, that is, $O(10^{-7}\,\mathrm{kg}\cdot\mathrm{m}^{-3}\cdot\mathrm{s}^{-2})$. On the other hand, the order of magnitude of the second term of (5.262) can be evaluated setting

$$f_0 \int_0^H w \frac{d\rho_s}{dz} dz = O\left(\frac{U H^2 \beta_0}{H_\rho} \delta\rho_s \right)$$

where $\delta\rho_s = O(1\,\mathrm{kg}\cdot\mathrm{m}^{-3})$ and $H_\rho = O(4 \times 10^5\,\mathrm{m})$. Hence,

$$\frac{U H^2 \beta_0}{H_\rho} \delta\rho_s = O\left(10^{-12}\,\mathrm{kg}\cdot\mathrm{m}^{-3}\cdot\mathrm{s}^{-2}\right)$$

Therefore, up to a very high degree of approximation, Eq. (5.261) yields

$$\beta_0 \int_0^H \rho_s v\, dz = \hat{\mathbf{k}} \cdot \mathrm{rot}\,\boldsymbol{\tau} \tag{5.264}$$

Although Eq. (5.184) is not an evolution equation, yet Eq. (5.264) may be viewed as a prognostic equation when $\boldsymbol{\tau}$ is time dependent. More precisely, due to the absence of the acceleration terms in the first-order momentum equations (5.180)–(5.181), the meridional component of the mass transport (cf. (5.270)) evolves in phase with the wind forcing.

Equation (5.264) should be compared with the dimensional Sverdrup balance reported in (3.400) and obtained in the framework of the homogeneous models of wind-driven ocean circulation, that is,

$$\beta_0 H \rho_s v = \hat{\mathbf{k}} \cdot \mathrm{rot}\,\boldsymbol{\tau} \tag{5.265}$$

The formal difference lies in the constant density fluid and depth-independent velocity appearing in (5.265), unlike (5.264), which are due to the basic assumptions of the homogeneous models; from this point of view, balance (5.264) is more general than (5.265). More fundamental are the different ways in which (5.264) and (5.265) are derived: while the latter is the consequence of the smallness, with respect to unity, of the ratios $(\delta_I/L)^2$, δ_S/L and $(\delta_M/L)^3$ in the basin interior (see(3.382)), the former comes from the scaling according to (5.160), (5.162) and (5.167) of the governing equations of an inviscid and stratified fluid column extending from the base of the upper Ekman layer down to the sea floor.

The patterns of the mass transport can be derived in a unified way for both (5.264) and (5.265) as follows. Starting from $\rho_s = \rho_s(z)$ and $\mathrm{div}\,\mathbf{u} = 0$, we establish the equation

$$\frac{\partial}{\partial x}(\rho_s u) + \frac{\partial}{\partial y}(\rho_s v) + \frac{\partial}{\partial z}(\rho_s w) - \frac{\mathrm{d}\rho_s}{\mathrm{d}z} w = 0 \qquad (5.266)$$

Since, according to (2.114),

$$\left(\frac{\mathrm{d}\rho_s}{\mathrm{d}z} w\right)\left(\frac{\partial}{\partial z}(\rho_s w)\right)^{-1} = O\left(\frac{H}{H_\rho}\right) = O(10^{-2})$$

Eq. (5.266) can be approximated by

$$\frac{\partial}{\partial x}(\rho_s u) + \frac{\partial}{\partial y}(\rho_s v) + \frac{\partial}{\partial z}(\rho_s w) = 0 \qquad (5.267)$$

where all the terms have the same order of magnitude. Then, integration of (5.267) on the depth of the motion $0 \le z \le H(x,y,t)$ yields

$$\begin{aligned} &\frac{\partial}{\partial x}\int_0^H \rho_s u \,\mathrm{d}z - \frac{\partial H}{\partial x} \rho_s\big|_{z=H} u\big|_{z=H} \\ &+\frac{\partial}{\partial y}\int_0^H \rho_s v \,\mathrm{d}z - \frac{\partial H}{\partial y} \rho_s\big|_{z=H} v\big|_{z=H} l \\ &+\rho_s\big|_{z=H} w\big|_{z=H} = \\ &= 0 \end{aligned}$$

but, recalling that $w\big|_{z=H} = \partial H/\partial t + \mathbf{u}\big|_{z=H} \cdot \nabla H$, this equation simplifies to

$$\frac{\partial}{\partial x}\int_0^H \rho_s u \,\mathrm{d}z + \frac{\partial}{\partial y}\int_0^H \rho_s v \,\mathrm{d}z + \rho_s\big|_{z=H} \frac{\partial H}{\partial t} = 0 \qquad (5.268)$$

The relative order of magnitude of the last term in (5.268), with respect to each of the first two terms of the same equation, turns out to be

$$O\left(\frac{\rho_s U \eta}{L}\frac{L}{\rho_s U H}\right) = O\left(\frac{\eta}{H}\right)$$

where η is the thickness of the oscillating part of the fluid column. Hence,

$$\frac{\eta}{H} \le O\left(\frac{f_0 U L}{g H}\right) = O(10^{-5})$$

This estimate allows us to restate (5.268) as

$$\frac{\partial}{\partial x}\int_0^H \rho_s u\,dz + \frac{\partial}{\partial y}\int_0^H \rho_s v\,dz = 0 \tag{5.269}$$

Equation (5.269) states that the *mass transport*

$$(M_x, M_y) := \left(\int_0^H \rho_s u\,dz, \int_0^H \rho_s v\,dz\right) \tag{5.270}$$

is horizontally non-divergent, that is,

$$\frac{\partial}{\partial x}M_x + \frac{\partial}{\partial y}M_y = 0 \tag{5.271}$$

thus, both the baroclinic balance (5.264) and the barotropic balance (5.265) can be written as

$$\beta_0 M_y = \hat{\mathbf{k}}\cdot \text{rot}\,\boldsymbol{\tau} \tag{5.272}$$

Moreover, because of (5.271), a transport stream function $\mathscr{T} = \mathscr{T}(x, y)$ such that

$$M_x = -\frac{\partial \mathscr{T}}{\partial y} \qquad M_y = \frac{\partial \mathscr{T}}{\partial x} \tag{5.273}$$

can be introduced to satisfy identically (5.271). Owing to the second equation of (5.273), (5.272) becomes

$$\beta_0 \frac{\partial \mathscr{T}}{\partial x} = \hat{\mathbf{k}}\cdot \text{rot}\,\boldsymbol{\tau} \tag{5.274}$$

In accordance with (3.422), the transport stream function is constrained by the boundary condition

$$\mathscr{T}(X_E, y) = 0 \tag{5.275}$$

where X_E is the longitude of the eastern coast of the reference basin. Given $\boldsymbol{\tau} = \boldsymbol{\tau}(x, y)$, problem (5.274), (5.275) yields the unknown stream function $\mathscr{T}$, and then, the desired patterns of the mass transport are represented by the family of curves

$$\mathscr{T}(x, y) = C \tag{5.276}$$

for suitable values of the parameter C. However, in order that a steady mass transport can actually take place, streamlines must be closed. Under such presupposition, which is not satisfied by (5.276) unless some contrived wind stress is considered, the patterns inferred from (5.276) are explained as the outcome of a process model that results only in a local – not global – mass transport field.

We emphasize that the validity of the Sverdrup balance (5.272) has been confirmed also recently, on experimental grounds, by Carl Wunsch (2011) who tells us that

> The goal was to ask whether a particular dynamical relationship, Sverdrup balance, ... would be found to have quantitative skill in describing the observed ocean circulation. The answer is "yes", over a major portion of the observed ocean.

Closure of Transport Streamlines

Like in the homogeneous model, the closure of the streamlines is achieved by adding a small amount of dissipation at the r.h.s.'s of the horizontal momentum equations (5.170) and (5.171), say $\mathsf{b}\,(\delta_{\mathrm{M}}/L)^3\,\nabla_{\mathrm{H}}'^2\mathsf{u}$ and $\mathsf{b}\,(\delta_{\mathrm{M}}/L)^3\,\nabla_{\mathrm{H}}'^2\mathsf{v}$, respectively, where $(\delta_{\mathrm{M}}/L)^3 = \mathsf{E}_{\mathrm{H}}/(2\,\mathsf{b})$. In this way, we have

$$\mathsf{v}_1 + \mathsf{y}\,\mathsf{v}_0 = \frac{\partial \mathsf{p}_1}{\partial \mathsf{x}} - \left(\frac{\delta_{\mathrm{M}}}{L}\right)^3 \nabla_{\mathrm{H}}'^2\,\mathsf{u}_0 \tag{5.277}$$

$$\mathsf{u}_1 + \mathsf{y}\,\mathsf{u}_0 = -\frac{\partial \mathsf{p}_1}{\partial \mathsf{y}} + \left(\frac{\delta_{\mathrm{M}}}{L}\right)^3 \nabla_{\mathrm{H}}'^2\,\mathsf{v}_0 \tag{5.278}$$

in place of (5.180) and (5.181), respectively, while the zero-order equations are left unchanged. Hence, (5.184) takes the more general form

$$\mathsf{v}_0 - \frac{\partial \mathsf{w}_1}{\partial \mathsf{z}} = \left(\frac{\delta_{\mathrm{M}}}{L}\right)^3 \nabla_{\mathrm{H}}'^2 \left(\frac{\partial \mathsf{v}_0}{\partial \mathsf{x}} - \frac{\partial \mathsf{u}_0}{\partial \mathsf{y}}\right)$$

whose dimensional version can be written as

$$\beta_0\,\rho_{\mathrm{s}}\,v - f_0\,\frac{\partial}{\partial z}(\rho_{\mathrm{s}}\,w) = A_{\mathrm{H}}\,\rho_{\mathrm{s}}\,\nabla_{\mathrm{H}}^2 \left(\frac{\partial v}{\partial x} - \frac{\partial u}{\partial y}\right) \tag{5.279}$$

Vertical integration of (5.279) from the (flat) bottom, at $z = 0$, to the depth of the upper Ekman layer, at $z = H$, with the use of positions (5.270) gives

$$\beta_0\,M_y - f_0\left[\rho_{\mathrm{s}}\,w\right]_{z=0}^{z=H} = A_{\mathrm{H}}\,\nabla_{\mathrm{H}}^2 \left(\frac{\partial M_y}{\partial x} - \frac{\partial M_x}{\partial y}\right) \tag{5.280}$$

By resorting to (5.273) together with boundary conditions $w(z=0) = 0$ and (3.341), Eq. (5.280) becomes, in terms of the transport stream function $\mathscr{T}$,

$$\beta_0\,\frac{\partial \mathscr{T}}{\partial x} = \mathbf{k}\cdot \mathrm{rot}\,\boldsymbol{\tau} + A_{\mathrm{H}}\,\nabla_{\mathrm{H}}^4\,\mathscr{T} \tag{5.281}$$

Equation (5.281) has the same form as the dimensional vorticity equation of Munk's model, which can be derived from (3.453) and can be solved exactly along the same lines as at page 234, thus yielding closed transport streamlines. Far from the western boundary layer, Eq. (5.281) simplifies into (5.274), which represents the Sverdrup balance.

Wave-Like Integrals of the Vorticity Equation at the Oceanic Basin Scale

Wave-like integrals of (5.188) can be found, as we ascertain below. Reconsider (5.188) with the simplifying assumption $\mathsf{S} = $ constant, thus written as

$$\frac{\partial}{\partial \mathsf{t}}\frac{\partial^2 \psi}{\partial \mathsf{z}^2} + \mathscr{J}\left(\psi, \frac{\partial^2 \psi}{\partial \mathsf{z}^2}\right) + \beta\,\mathsf{S}\,\frac{\partial \psi}{\partial \mathsf{x}} = 0 \tag{5.282}$$

Substitution of the putative stream function

$$\psi(\mathsf{x}, \mathsf{y}, \mathsf{z}, \mathsf{t}) = \Phi(\mathsf{z})\,\chi(\mathsf{x}, \mathsf{y}, \mathsf{t}) \tag{5.283}$$

into (5.282) gives

$$\frac{\partial \chi}{\partial \mathsf{t}}\frac{\mathrm{d}^2\Phi}{\mathrm{d}\mathsf{z}^2} + \beta\,\mathsf{S}\,\Phi\frac{\partial \chi}{\partial \mathsf{x}} = 0 \tag{5.284}$$

Note that

$$\mathscr{J}\left(\Phi\chi, \frac{\mathrm{d}^2\Phi}{\mathrm{d}\mathsf{z}^2}\chi\right) = \Phi\frac{\mathrm{d}^2\Phi}{\mathrm{d}\mathsf{z}^2}\,\mathscr{J}(\chi, \chi) = 0$$

thus, the Jacobian determinant is not influent in deriving (5.284). Equation (5.284) can be decoupled into an equation for the sole Φ, that is,

$$\frac{\mathrm{d}^2\Phi}{\mathrm{d}\mathsf{z}^2} - \lambda\,\Phi = 0 \tag{5.285}$$

and another equation for the sole χ, that is,

$$\lambda\,\frac{\partial \chi}{\partial \mathsf{t}} + \beta\,\mathsf{S}\,\frac{\partial \chi}{\partial \mathsf{x}} = 0 \tag{5.286}$$

To single out the admissible values of the separation constant λ in (5.285) and (5.286), the vertical boundary conditions in terms of Φ must be determined. Because vertical velocity vanishes at the bottom ($\mathsf{z} = 0$) and at the top ($\mathsf{z} = 1$) of the fluid layer, we have in particular

$$\mathsf{w}_1(\mathsf{z} = 0) = 0 \qquad \mathsf{w}_1(\mathsf{z} = 1) = 0 \tag{5.287}$$

According to (5.185), Eq. (5.287) imply

$$\frac{D_0}{Dt}\frac{\partial p_0}{\partial z}=0 \qquad \text{at } z=0 \text{ and at } z=1 \tag{5.288}$$

Now, position (5.283) is substituted in place of p_0 into (5.288) to give

$$\frac{d\Phi}{dz}\frac{\partial \chi}{\partial t}=0 \qquad \text{at } z=0 \text{ and at } z=1 \tag{5.289}$$

whence the vertical boundary conditions

$$\frac{d\Phi}{dz}=0 \qquad \text{at } z=0 \text{ and at } z=1 \tag{5.290}$$

immediately follow.

Solving problem (5.285), (5.290) allows us to single out $\Phi(z)$ and λ; the result is

$$\Phi(z)=\Phi_m(z)=\cos(m\pi z) \tag{5.291}$$

$$\lambda=\lambda_m=-(m\pi)^2 \tag{5.292}$$

with $m=0,\pm1,\pm2,\ldots$. Given (5.292), Eq. (5.286) becomes for $m\neq 0$

$$\frac{\partial \chi}{\partial t}-\frac{\beta S}{(m\pi)^2}\frac{\partial \chi}{\partial x}=0 \tag{5.293}$$

The general integral of (5.293) is

$$\chi(x,y,t)=\chi_m(x-c_m t,\, y) \tag{5.294}$$

where

$$c_m := -\frac{\beta S}{(m\pi)^2}$$

is the propagation speed and the dependence of χ_m on its arguments $x-c_m t$ and y is arbitrary. In any case, each eigensolution propagates westward, in a non-dispersive way, with its own speed c_m. The total stream function (5.283) is determined by (5.291) and (5.294) to give

$$\psi(x,y,z,t)=\psi_m(x,y,z,t)=\cos(m\pi z)\,\chi_m(x-c_m t,\, y) \tag{5.295}$$

Although (5.295) is an exact solution of (5.282) with (5.290), in general a sum of solutions of the kind (5.295), say $\sum_m N_m\,\psi_m(x,y,z,t)$, is not a solution of (5.282) because of its non-linear term $\mathcal{J}\left(\psi,\, \partial^2\psi/\partial z^2\right)$, which invalidates the superposition principle. This unlike all the linear Rossby waves formerly derived.

The special form

$$\chi_m = A(\mathsf{y}) \cos(\mathsf{kx} - \mathsf{k c}_m \mathsf{t}) \tag{5.296}$$

of (5.294), written as a sinusoid rigidly translating in the zonal direction, that is,

$$\chi_m = A(\mathsf{y}) \cos(\mathsf{kx} - \sigma'_m \mathsf{t}) \tag{5.297}$$

bears some interest. First, the function $A(\mathsf{y})$ of (5.296) may be adapted to confine the solution into a latitudinal strip, for instance, between $\mathsf{y} = 0$ and $\mathsf{y} = 1$, by taking $A(\mathsf{y}) = \sin(\pi \mathsf{y})$. Second, comparison of (5.296) with (5.297) points out the dispersion relation

$$\sigma'_m = -\frac{\mathsf{k}}{(m\pi)^2/(\beta\,\mathsf{S})} \tag{5.298}$$

Relation (5.298) should be compared with the dispersion relation (5.134) of the baroclinic Rossby waves at the oceanic mesoscale, reported below

$$\sigma'_m = -\frac{\mathsf{k}}{\mathsf{k}^2 + \mathsf{n}^2 + (m\pi)^2/\mathsf{S}} \tag{5.299}$$

Because $\beta\,\mathsf{S} = O(1)$ at the basin scale while $\mathsf{S} = O(1)$ at the mesoscale, the dispersion relations (5.298) and (5.299) basically differ each other only in the quantity $\mathsf{k}^2 + \mathsf{n}^2$ of (5.299). At the mesoscale, this quantity is originated from relative vorticity; on the other hand, we already know that, at the basin scale, the advection of relative vorticity is negligibly small with respect to that of planetary vorticity, so (5.299) reduces to (5.298) in the latter case.

Linear Vorticity Dynamics at the Ocean Basin Scale

Following Crisciani and Purini (2012), we shall now derive a linearized version of vorticity dynamics. If one traces back the derivation of (5.188) up to the set of governing equations (5.169)–(5.173), the non-linear term

$$\mathcal{J}\left(\psi, \frac{\partial}{\partial \mathsf{z}}\left(\frac{1}{\beta\,\mathsf{S}}\frac{\partial \psi}{\partial \mathsf{z}}\right)\right)$$

in (5.188) is easily explained as due to the advective term of (5.173), that is, to

$$\frac{\mathsf{b}}{\beta\,\mathsf{S}}\mathbf{u} \cdot \nabla' \rho' \tag{5.300}$$

Thus, linear vorticity dynamics at the ocean basin scale takes place whenever the term (5.300) is significantly smaller than the other terms of the density equation. This happens if the local timescale

$$T_{\mathrm{loc}} = O\left(\frac{1}{\beta_0 L \mathsf{S}}\right) \tag{5.301}$$

is considered, and the order of magnitude

$$\beta \mathsf{S} = O\left(\frac{1}{\mathsf{b}}\right) \tag{5.302}$$

holds in place of $\beta \mathsf{S} = O(1)$, as in (5.168). Hypotheses above imply that T_{loc} is shorter than the advective timescale L/U (and, hence, $\varepsilon < \varepsilon_T$ unlike the non-linear regime); in fact, recalling that $1/\mathsf{b} = O(10)$, from (5.302) we have

$$\beta \mathsf{S} = O\left(\frac{1}{\mathsf{b}}\right) \Longrightarrow \beta \mathsf{S} > 1 \Longrightarrow \frac{\beta_0 L^2 \mathsf{S}}{U} > 1 \Longrightarrow \frac{L}{U} > \frac{1}{\beta_0 L \mathsf{S}} \Longrightarrow \frac{L}{U} > T_{\text{loc}}$$

where (5.301) has been used in the last step. Moreover, estimate (5.301) implies

$$\frac{\varepsilon_T}{\mathsf{bS}} = O(1) \tag{5.303}$$

In fact, we have

$$\frac{\varepsilon_T}{\mathsf{bS}} = O\left(\frac{1}{f_0 T_{\text{loc}} \mathsf{bS}}\right) = O\left(\frac{\beta_0 L \mathsf{S}}{f_0 \mathsf{bS}}\right) = O(1)$$

because of position $\mathsf{b} = \beta_0 L / f_0$. We also note that (5.302) implies

$$\frac{\varepsilon}{\mathsf{bS}} = O(\mathsf{b}) \tag{5.304}$$

since

$$\frac{\varepsilon}{\mathsf{bS}} = \frac{\beta \varepsilon}{\mathsf{b} \beta \mathsf{S}} = O(\beta \varepsilon) = O\left(\frac{\beta_0 L}{f_0}\right) = O(\mathsf{b})$$

Roughly speaking, Eqs. (5.303) and (5.304) make advection negligibly small, while the value of $\beta \mathsf{S} \propto U^{-1}$ becomes significantly larger than $O(1)$. At this point, we use (5.303) and (5.304) to rewrite (2.626) so that parameter b comes into play, thus obtaining

$$\mathsf{w} = \frac{\varepsilon_T}{\mathsf{bS}} \mathsf{b} \frac{\partial \rho'}{\partial \tilde{\mathsf{t}}} + \frac{\varepsilon}{\mathsf{bS}} \mathsf{b} \mathbf{u} \cdot \nabla' \rho' \tag{5.305}$$

where $\tilde{\mathsf{t}} := t / T_{\text{loc}}$. Consider the expansion of each term of (5.305) in powers of b, with (5.303) and (5.304) taken into account, that is to say, after division by b,

$$\mathsf{w}'_1 + O(\mathsf{b}) = \frac{\varepsilon_T}{\mathsf{bS}} \left(\frac{\partial \rho'_0}{\partial \tilde{\mathsf{t}}} + O(\mathsf{b})\right) + \frac{\varepsilon}{\mathsf{bS}} \left(\mathbf{u}_0 \cdot \nabla' \rho'_0 + O(\mathsf{b})\right) \tag{5.306}$$

Because of (5.304), to the leading order in b, Eq. (5.306) yields

$$\mathsf{w}'_1 = \frac{\varepsilon_T}{\mathsf{bS}} \frac{\partial \rho'_0}{\partial \tilde{\mathsf{t}}} \tag{5.307}$$

Equation (5.307) substitutes (5.182) in passing from the non-linear to the linear regime, while all the remaining equations (5.174)–(5.181) are left unaffected, and therefore, the ordering parameter is again b. Then, by proceeding hereafter as in the non-linear regime, the linear quasi-geostrophic potential-vorticity equation

$$\frac{\partial}{\partial \tilde{\mathsf{t}}}\frac{\partial}{\partial \mathsf{z}}\left(\frac{\varepsilon_T}{\mathsf{bS}}\frac{\partial \Psi}{\partial \mathsf{z}}\right)+\frac{\partial \Psi}{\partial \mathsf{x}}=0 \tag{5.308}$$

eventually follows. The dimensional version of (5.308) is

$$\frac{\partial}{\partial t}\frac{\partial}{\partial z}\left(\frac{f_0^2}{N_s^2}\frac{\partial \Psi_*}{\partial z}\right)+\beta_0\frac{\partial \Psi_*}{\partial x}=0 \tag{5.309}$$

where the shorthand notation $\Psi_* := \tilde{p}/(\rho_s f_0)$ has been used in (5.309). The general integral of (5.309) can be evaluated along the same lines as (5.282) in the previous subsection. One can ascertain that the disturbance Ψ_*, anyway fixed at a certain "initial" time, transfers westward with velocity

$$c_m = -\left(\frac{N_s H}{2\pi m f_0}\right)^2 \beta_0 \qquad (m = 1, 2, \ldots)$$

while conserving its shape. The order of magnitude of the maximum velocity is

$$c_1 = O(10^{-2}\,\mathrm{m/s}) \tag{5.310}$$

Thus, in this kind of dynamics, advection (of the order of $10^{-3}\,\mathrm{m/s}$) is about ten times slower than the maximum translatory speed (5.310) of the disturbance.

Remark. In the linearization of the quasi-geostrophic potential-vorticity equation at the ocean basin scale, the ordering parameter b is the same as in the non-linear case. This is somehow different from all the previous cases of linearization, where the change of the ordering parameter $\varepsilon \mapsto \varepsilon_T$ was the key step to eliminate the advective terms of the governing equations at the geostrophic level of approximation. Note, finally, that (5.301) is a generalization, for $\mathsf{S} \neq O(1)$, of the usual estimate $T_{\mathrm{loc}} = (\beta_0 L)^{-1}$ already considered in all the other linear models.

Dimensional Versions of the Quasi-Geostrophic Potential-Vorticity Equations for Density-Conserving Ocean Currents

For the reader's greater convenience and in view of possible further investigations on the subject, we report below the dimensional versions of Eqs. (5.41), (5.57) and (5.188), respectively. In all the cases, the unknown is the field of the perturbation pressure $\tilde{p}(\mathbf{x},t)$ introduced in (2.532), in terms of which the dimensional geostrophic current is

$$(u_0, v_0) = \left(-\frac{\partial}{\partial y}\frac{\tilde{p}}{\rho_s f_0}, \frac{\partial}{\partial x}\frac{\tilde{p}}{\rho_s f_0}\right) \tag{5.311}$$

in accordance with (2.565). Hence, the Lagrangian derivative at the geostrophic level of approximation is given by

$$\begin{aligned}\frac{\mathrm{D}_0}{\mathrm{D}t} &= \frac{\partial}{\partial t} + u_0\frac{\partial}{\partial x} + v_0\frac{\partial}{\partial y} \\ &= \frac{\partial}{\partial t} - \left(\frac{\partial}{\partial y}\frac{\tilde{p}}{\rho_s f_0}\right)\frac{\partial}{\partial x} + \left(\frac{\partial}{\partial x}\frac{\tilde{p}}{\rho_s f_0}\right)\frac{\partial}{\partial y}\end{aligned} \tag{5.312}$$

By using (5.312), Eq. (5.41) yields

$$\frac{\mathrm{D}_0}{\mathrm{D}t}\left[\nabla_{\mathrm{H}}^2\frac{\tilde{p}}{\rho_s f_0} + \frac{\partial}{\partial z}\left(\frac{f_0}{N_s^2}\frac{\partial}{\partial z}\frac{\tilde{p}}{\rho_s}\right) + \beta_0 y\right] = 0 \tag{5.313}$$

Note that, in (5.313), the terms in the square bracket have the physical dimensions of a frequency.

Likewise, the dimensional counterpart of (5.57) is

$$\frac{\partial}{\partial t}\left[\nabla_{\mathrm{H}}^2\frac{\tilde{p}}{\rho_s f_0} + \frac{\partial}{\partial z}\left(\frac{f_0}{N_s^2}\frac{\partial}{\partial z}\frac{\tilde{p}}{\rho_s}\right)\right] + \beta_0 v_0 = 0 \tag{5.314}$$

Moreover, Eq. (5.188) gives

$$\frac{\mathrm{D}_0}{\mathrm{D}t}\left[\frac{\partial}{\partial z}\left(\frac{f_0}{N_s^2}\frac{\partial}{\partial z}\frac{\tilde{p}}{\rho_s}\right) + \beta_0 y\right] = 0 \tag{5.315}$$

In conclusion, we stress that:

- The disappearance of the Jacobian term from (5.313) to (5.314) is basically due to the change of the timescale, from $T_{\mathrm{adv}} = L/U$ to $T_{\mathrm{loc}} = 1/(\beta_0 L)$.
- The disappearance of relative vorticity from (5.313) to (5.315) comes from the increase of the non-dimensional parameter $\beta = \beta_0 L^2/U$ from $O(1)$ to $O(10^3)$ in passing from the mesoscale to the basin scale.

5.2 QG Continuously Stratified Flows in the Atmosphere

In order to show applications of the quasi-geostrophic theory to atmospheric flows, some novelties are preliminarily introduced, that is, a thermal forcing in the proximity of the ground (or at the sea level) and/or a topographic modulation. Moreover, an exponentially decreasing standard density is considered, unlike the constant one used in the oceanographic context. In particular, the quasi-geostrophic potential-vorticity equation includes a special form of thermal vorticity, and in the case of thermal forcing, it becomes inhomogeneous. Moreover, also the no–mass flux condition at the ground is modified accordingly. In both cases, the flow energetics may cause the formation of a vertical energy flux, which allows the propagation of energy towards the open space.

Based on the quasi-geostrophic models so conceived, steady currents, produced by a thermal or topographic perturbation of a zonal one, are analysed. In this context, trapped, non-trapped and singular model solutions are derived.

5.2.1 *Basic Synoptic-Scale Dynamics of the Atmosphere*

The motion of the winds on the synoptic atmospheric scale is characterized by orders of magnitude such that $\varepsilon < O(1)$ and

$$\beta = O(1) \qquad \mathsf{S} = O(1) \qquad \delta = O(\varepsilon^2) \qquad \mathsf{F} = O(\varepsilon) \tag{5.316}$$

Relations (5.316) are verified by the following typical values of the troposphere

$$L = 10^6\,\mathrm{m} \qquad U = 10\,\mathrm{m}\cdot\mathrm{s}^{-1} \qquad H = 10^4\,\mathrm{m} \qquad N_s = 10^{-2}\,\mathrm{s}^{-1} \tag{5.317}$$

The advective timescale $T_{\mathrm{adv}} = L/U$ has the same order of magnitude as the local timescale $T_{\mathrm{loc}} = (\beta_0 L)^{-1}$, whence

$$\varepsilon_{\mathrm{T}} = \varepsilon = 0.1 \tag{5.318}$$

Equations (5.316) and (5.318) show that the Rossby number ε is the ordering parameter of the related non-dimensional equations.

Scaling of the Governing Equations and Vorticity Dynamics

The governing equations (2.597), (2.612), (2.613), (2.616) and (2.633) of frictionless and thermally forced air bodies, written in terms of the ordering parameter, are

$$\frac{\partial \mathsf{u}}{\partial \mathsf{x}} + \frac{\partial \mathsf{v}}{\partial \mathsf{y}} + \frac{1}{\rho_s}\frac{\partial}{\partial \mathsf{z}}(\rho_s \mathsf{w}) = 0 \tag{5.319}$$

$$\varepsilon\frac{\partial \mathsf{u}}{\partial \mathsf{t}} + \varepsilon\mathbf{u}\cdot\nabla'\mathsf{u} - (1+\beta\varepsilon\,\mathsf{y})\mathsf{v} + \varepsilon^2\mathsf{w} = -\frac{\partial \mathsf{p}}{\partial \mathsf{x}} \tag{5.320}$$

$$\varepsilon\,\frac{\partial \mathsf{v}}{\partial \mathsf{t}}+\varepsilon\,\mathbf{u}\cdot\nabla'\mathsf{v}+(1+\beta\,\varepsilon\,\mathsf{y})\,\mathsf{u}=-\frac{\partial \mathsf{p}}{\partial \mathsf{y}} \tag{5.321}$$

$$\varepsilon^5\,\frac{\partial \mathsf{w}}{\partial \mathsf{t}}+\varepsilon^5\,\mathbf{u}\cdot\nabla'\mathsf{w}-\varepsilon^2\,\mathsf{u}=-\frac{1}{\rho_s}\frac{\partial}{\partial \mathsf{z}}(\rho_s\,\mathsf{p})-\rho' \tag{5.322}$$

$$\mathsf{w}+\frac{\varepsilon}{\mathsf{S}}\left(\frac{\mathsf{D}\theta'}{\mathsf{Dt}}-\dot{\mathsf{Q}}\right)=0 \tag{5.323}$$

where

$$\dot{\mathsf{Q}}=\frac{gH}{f_0U^2}\,\frac{\dot{Q}}{c_p\,T_a}$$

has been already met in (2.634). The anomaly of potential temperature θ', appearing in (5.323), is linked to the pressure perturbation p and to the density anomaly ρ' by Eq. (2.559), also reported below:

$$\theta'=\frac{gH\rho_s}{\gamma p_s}\,\mathsf{p}-\rho' \tag{5.324}$$

According to the usual procedure, the expansions

$$\mathbf{u}(\mathbf{x},\mathsf{t};\varepsilon)=\mathbf{u}_0(\mathbf{x},\mathsf{t})+\varepsilon\,\mathbf{u}_1(\mathbf{x},\mathsf{t})+O(\varepsilon^2)$$

$$\mathsf{p}(\mathbf{x},\mathsf{t};\varepsilon)=\mathsf{p}_0(\mathbf{x},\mathsf{t})+\varepsilon\,\mathsf{p}_1(\mathbf{x},\mathsf{t})+O(\varepsilon^2)$$

$$\rho'(\mathbf{x},\mathsf{t};\varepsilon)=\rho_0'(\mathbf{x},\mathsf{t})+\varepsilon\,\rho_1'(\mathbf{x},\mathsf{t})+O(\varepsilon^2)$$

$$\theta'(\mathbf{x},\mathsf{t};\varepsilon)=\theta_0'(\mathbf{x},\mathsf{t})+\varepsilon\,\theta_1'(\mathbf{x},\mathsf{t})+O(\varepsilon^2) \tag{5.325}$$

are substituted into (5.319)–(5.324), whence the resulting equations, to the leading (zeroth) and first order in ε, turn out to be as follows.

Zeroth Order Equations

$$\frac{\partial \mathsf{u}_0}{\partial \mathsf{x}}+\frac{\partial \mathsf{v}_0}{\partial \mathsf{y}}+\frac{1}{\rho_s}\frac{\partial}{\partial \mathsf{z}}(\rho_s\,\mathsf{w}_0)=0 \tag{5.326}$$

$$\mathsf{v}_0=\frac{\partial \mathsf{p}_0}{\partial \mathsf{x}} \tag{5.327}$$

$$\mathsf{u}_0=-\frac{\partial \mathsf{p}_0}{\partial \mathsf{y}} \tag{5.328}$$

$$\frac{1}{\rho_s}\frac{\partial}{\partial z}(\rho_s\,\mathsf{p}_0)+\rho_0'=0 \tag{5.329}$$

$$\mathsf{w}_0=0 \tag{5.330}$$

$$\theta_0'=\frac{gH\rho_s}{\gamma p_s}\,\mathsf{p}_0-\rho_0' \tag{5.331}$$

From (5.331), another equation can be derived by means of which θ_0' is expressed as a function of the sole p_0'. To achieve this, we resort to the dimensional hydrostatic equilibrium condition $\mathrm{d}p_s/\mathrm{d}z+g\,\rho_s=0$ and to (5.329) to write (5.331) as

$$\theta_0'=-\frac{H}{\gamma p_s}\frac{\partial p_s}{\partial z}\,\mathsf{p}_0+\frac{1}{\rho_s}\frac{\partial}{\partial z}(\rho_s\,\mathsf{p}_0)=-H\left(\frac{1}{\gamma p_s}\frac{\mathrm{d}p_s}{\mathrm{d}z}-\frac{1}{\rho_s}\frac{\mathrm{d}\rho_s}{\mathrm{d}z}\right)\mathsf{p}_0+\frac{\partial \mathsf{p}_0}{\partial z} \tag{5.332}$$

On the other hand, according to position (2.95), we have $\theta_s=C_0\,p_s^{1/\gamma}/\rho_s$, and hence,

$$\frac{1}{\theta_s}\frac{\mathrm{d}\theta_s}{\mathrm{d}z}=\frac{1}{\gamma p_s}\frac{\mathrm{d}p_s}{\mathrm{d}z}-\frac{1}{\rho_s}\frac{\mathrm{d}\rho_s}{\mathrm{d}z} \tag{5.333}$$

where the l.h.s. is equal to N_s^2/g by (2.97), so that Eq. (5.333) is equivalent to

$$\frac{N_s^2}{g}=\frac{1}{\gamma p_s}\frac{\mathrm{d}p_s}{\mathrm{d}z}-\frac{1}{\rho_s}\frac{\mathrm{d}\rho_s}{\mathrm{d}z} \tag{5.334}$$

Then, from (5.332) and (5.334), one obtains

$$\theta_0'=-\frac{N_s^2H}{g}\,\mathsf{p}_0+\frac{\partial \mathsf{p}_0}{\partial z} \tag{5.335}$$

Moreover, the identity

$$\frac{N_s^2H}{g}=\frac{f_0^2L^2}{gH}\left(\frac{N_sH}{f_0L}\right)^2=\mathsf{FS}$$

derived from (2.389) and (2.625), allows us to restate (5.335) in the form

$$\theta_0'=-\mathsf{FS}\,\mathsf{p}_0+\frac{\partial \mathsf{p}_0}{\partial z} \tag{5.336}$$

where, according to (5.316), $\mathsf{FS}=O(\varepsilon)$. Therefore, the leading-order balance equation derived from (5.336) is

$$\theta_0'=\frac{\partial \mathsf{p}_0}{\partial z} \tag{5.337}$$

The usefulness of (5.337) lies in the possibility to express, at the geostrophic level of approximation, the anomaly of potential temperature in terms of the perturbation pressure which, unlike potential temperature, is a dynamical field appearing also in the momentum equation.

First-Order Equations

The first-order equations, which are useful in what follows, turn out to be

$$\frac{\partial u_1}{\partial x} + \frac{\partial v_1}{\partial y} + \frac{1}{\rho_s}\frac{\partial}{\partial z}(\rho_s w_1) = 0 \tag{5.338}$$

$$\frac{D_0 u_0}{Dt} - v_1 - \beta y\, v_0 = -\frac{\partial p_1}{\partial x} \tag{5.339}$$

$$\frac{D_0 v_0}{Dt} + u_1 + \beta y\, u_0 = -\frac{\partial p_1}{\partial y} \tag{5.340}$$

$$w_1 = -\frac{1}{S}\left(\frac{D_0}{Dt}\frac{\partial p_0}{\partial z} - \dot{Q}\right) \tag{5.341}$$

Note that (5.341) is the result of the substitution of (5.337) into (5.323) once (5.330) has been taken into account. Cross differentiation of (5.339) and (5.340) yields the vorticity equation

$$\frac{D_0}{Dt}\zeta_0' + \frac{\partial u_1}{\partial x} + \frac{\partial v_1}{\partial y} + \beta\, v_0 = 0 \tag{5.342}$$

where, as usual,

$$\zeta_0' = \frac{\partial v_0}{\partial x} - \frac{\partial u_0}{\partial y}$$

Then, substitution of (5.338) into (5.342) gives

$$\frac{D_0}{Dt}\zeta_0' - \frac{1}{\rho_s}\frac{\partial}{\partial z}(\rho_s w_1) + \beta\, v_0 = 0 \tag{5.343}$$

By eliminating w_1 from (5.343) and (5.341), the vorticity equation at the geostrophic level of approximation

$$\frac{D_0}{Dt}\zeta_0' + \frac{1}{\rho_s}\frac{\partial}{\partial z}\left(\frac{\rho_s}{S}\frac{D_0}{Dt}\frac{\partial p_0}{\partial z} - \frac{\rho_s}{S}\dot{Q}\right) + \beta\, v_0 = 0 \tag{5.344}$$

is achieved. By means of the identity

$$\frac{\partial}{\partial z}\left(\frac{\rho_s}{S}\frac{D_0}{Dt}\frac{\partial p_0}{\partial z}\right) = \frac{D_0}{Dt}\frac{\partial}{\partial z}\left(\frac{\rho_s}{S}\frac{\partial p_0}{\partial z}\right)$$

quite analogous to (5.37), Eq. (5.344) takes the form

$$\frac{\mathrm{D_0}}{\mathrm{Dt}}\left[\zeta_0' + \frac{1}{\rho_s}\frac{\partial}{\partial \mathrm{z}}\left(\frac{\rho_s}{\mathrm{S}}\frac{\partial \mathrm{p_0}}{\partial \mathrm{z}}\right) + \beta \mathrm{y}\right] = \frac{1}{\rho_s}\frac{\partial}{\partial \mathrm{z}}\left(\frac{\rho_s}{\mathrm{S}}\dot{\mathrm{Q}}\right) \tag{5.345}$$

Equation (5.30) will be commented on in the following. Because of (5.327) and (5.328), we have $\zeta_0' = \nabla'^2 \mathrm{p_0}$ and

$$\mathbf{u}_0 \cdot \nabla' = -\frac{\partial \mathrm{p_0}}{\partial \mathrm{y}}\frac{\partial}{\partial \mathrm{x}} + \frac{\partial \mathrm{p_0}}{\partial \mathrm{x}}\frac{\partial}{\partial \mathrm{y}}$$

Thus, (5.345) can be written as

$$\begin{aligned}&\frac{\partial}{\partial \mathrm{t}}\left(\nabla_{\mathrm{H}}'^2 \mathrm{p_0} + \frac{1}{\rho_s}\frac{\partial}{\partial \mathrm{z}}\left(\frac{\rho_s}{\mathrm{S}}\frac{\partial \mathrm{p_0}}{\partial \mathrm{z}}\right)\right)\\ &\quad + \mathscr{J}\left(\mathrm{p_0},\, \nabla_{\mathrm{H}}'^2 \mathrm{p_0} + \frac{1}{\rho_s}\frac{\partial}{\partial \mathrm{z}}\left(\frac{\rho_s}{\mathrm{S}}\frac{\partial \mathrm{p_0}}{\partial \mathrm{z}}\right) + \beta \mathrm{y}\right) = \frac{1}{\rho_s}\frac{\partial}{\partial \mathrm{z}}\left(\frac{\rho_s}{\mathrm{S}}\dot{\mathrm{Q}}\right)\end{aligned} \tag{5.346}$$

Setting, as customary, $\mathrm{p_0} = \psi$, Eq. (5.346) becomes

$$\begin{aligned}&\frac{\partial}{\partial \mathrm{t}}\left(\nabla_{\mathrm{H}}'^2 \psi + \frac{1}{\rho_s}\frac{\partial}{\partial \mathrm{z}}\left(\frac{\rho_s}{\mathrm{S}}\frac{\partial \psi}{\partial \mathrm{z}}\right)\right)\\ &\quad + \mathscr{J}\left(\psi,\, \nabla_{\mathrm{H}}'^2 \psi + \frac{1}{\rho_s}\frac{\partial}{\partial \mathrm{z}}\left(\frac{\rho_s}{\mathrm{S}}\frac{\partial \psi}{\partial \mathrm{z}}\right)\right) + \beta\frac{\partial \psi}{\partial \mathrm{x}} = \frac{1}{\rho_s}\frac{\partial}{\partial \mathrm{z}}\left(\frac{\rho_s}{\mathrm{S}}\dot{\mathrm{Q}}\right)\end{aligned} \tag{5.347}$$

The remarkable resemblance of the unforced version (i.e. for $\dot{\mathrm{Q}} = 0$) of (5.347) with the quasi-geostrophic potential-vorticity equation of the oceanic mesoscale (5.41) shows the common grounds of the large-scale dynamics of the ocean and the atmosphere. However, unlike (5.41), the model governed by (5.347) demands definite hypotheses about the standard density ρ_s.

In the case of an isothermal undisturbed atmosphere (in which $\mathrm{d}T_s/\mathrm{d}z = 0$), the perfect gas law $p_s = R\rho_s T_s$, together with the hydrostatic equation $\mathrm{d}p_s/\mathrm{d}z + g\rho_s = 0$, immediately yields the ordinary differential equation

$$\frac{\mathrm{d}\rho_s}{\mathrm{d}z} + \frac{g}{RT_s}\rho_s = 0$$

Using the non-dimensional coordinate $\mathrm{z} = z/H$, where H is the height of the motion, this equation becomes

$$\frac{\mathrm{d}\rho_s}{\mathrm{dz}} + \frac{gH}{RT_s}\rho_s = 0$$

whose solution is

$$\rho_s(\mathrm{z}) = \rho_{s0}\exp\left(-\frac{H}{H_\rho}\mathrm{z}\right) \tag{5.348}$$

where $H_\rho = RT_s/g$ is the density height scale. In the troposphere, we have $H/H_\rho = O(1)$. Hereafter, we put, in short,

$$\mu' := \frac{H}{H_\rho} \tag{5.349}$$

so (5.348) reads

$$\frac{\rho_s(z)}{\rho_{s0}} = \exp(-\mu' z) \tag{5.350}$$

Assuming the validity of (5.350), the stratification parameter S becomes a $O(1)$ constant, and using (5.349), the thermal vorticity is

$$\frac{1}{\rho_s}\frac{\partial}{\partial z}\left(\frac{\rho_s}{S}\frac{\partial \psi}{\partial z}\right) = \frac{1}{S}\left(\frac{\partial^2 \psi}{\partial z^2} - \mu' \frac{\partial \psi}{\partial z}\right) \tag{5.351}$$

so that (5.347) finally becomes

$$\frac{\partial}{\partial t}\left(\nabla_H'^2 \psi + \frac{1}{S}\left(\frac{\partial^2 \psi}{\partial z^2} - \mu' \frac{\partial \psi}{\partial z}\right)\right)$$
$$+\mathcal{J}\left(\psi, \nabla_H'^2 \psi + \frac{1}{S}\left(\frac{\partial^2 \psi}{\partial z^2} - \mu' \frac{\partial \psi}{\partial z}\right)\right) + \beta \frac{\partial \psi}{\partial x} = \frac{1}{S}\left(\frac{\partial \dot{Q}}{\partial z} - \mu' \dot{Q}\right) \tag{5.352}$$

The no–mass flux boundary condition at the (flat) ground (i.e. $w_1(z=0)=0$), expressed by means of (5.341), gives

$$\frac{D_0}{Dt}\frac{\partial \psi}{\partial z} = \dot{Q} \quad \text{at} \quad z = 0$$

that is to say

$$\frac{\partial}{\partial t}\frac{\partial \psi}{\partial z} + \mathcal{J}\left(\psi, \frac{\partial \psi}{\partial z}\right) = \dot{Q} \quad \text{at} \quad z = 0 \tag{5.353}$$

Since the atmosphere is vertically unbounded, the vertical boundary condition at the top of the atmosphere is less obvious than that at the ground and demands a preliminary discussion about the energetics of the atmosphere.

Available Potential Energy of a Continuously Stratified Adiabatic Atmosphere

To derive the available potential energy for a continuously stratified layer of air, we resort to the simple case of an adiabatic atmosphere; subsequently, this hypothesis will be released.

We already know that, in an adiabatic and compressible atmosphere in which the potential temperature θ is conserved, small air displacements $h(t)$ from the rest state are governed by the dimensional equation (2.85), also reported below:

$$\frac{\mathrm{d}^2 h}{\mathrm{d}t^2} + \frac{g}{\theta}\frac{\partial \theta}{\partial z} h = 0 \tag{5.354}$$

Equation (5.354), multiplied by ρ_s, governs a vertical harmonic oscillator whose restoring force for unit volume is

$$F(h) := \rho_s \frac{g}{\theta}\frac{\partial \theta}{\partial z} h$$

This force can be approximated by using $\theta = \theta_s$, so we may assume

$$F(h) = \rho_s \frac{g}{\theta}\frac{\mathrm{d}\theta_s}{\mathrm{d}z} h \tag{5.355}$$

and the available potential energy of the parcel, relative to the finite displacement δz, is

$$\mathrm{APE(parcel)} = \int_0^{\delta z} F(h)\,\mathrm{d}h = \frac{g\,\rho_s}{2\,\theta_s}\frac{\mathrm{d}\theta_s}{\mathrm{d}z}(\delta z)^2 \tag{5.356}$$

The change of potential temperature $\delta\theta$ due to the displacement δz is, except higher order terms,

$$\delta\theta = \frac{\mathrm{d}\theta_s}{\mathrm{d}z}\,\delta z \tag{5.357}$$

and the substitution of $(\mathrm{d}\theta_s/\mathrm{d}z)^{-1}\,\delta\theta$ in place of δz into (5.356) gives, after a rearrangement,

$$\mathrm{APE(parcel)} = \frac{1}{2} g\,\rho_s\,\theta_s \left(\frac{\mathrm{d}\theta_s}{\mathrm{d}z}\right)^{-1} \left(\frac{\delta\theta}{\theta_s}\right)^2 \tag{5.358}$$

Since, according to (2.560), we have $\theta = (1 + \varepsilon\,\mathsf{F}\,\theta')\,\theta_s \approx (1 + \varepsilon\,\mathsf{F}\,\theta_0')\,\theta_s$ and hence $\delta\theta = \theta - \theta_s = \varepsilon\,\mathsf{F}\,\theta_s\,\theta_0'$, one obtains

$$\left(\frac{\delta\theta}{\theta_s}\right)^2 = (\varepsilon\,\mathsf{F}\,\theta_0')^2 = \left(\frac{f_0\,U\,L}{g H}\right)^2 \theta_0'^2 \tag{5.359}$$

Substitution of (5.359) into (5.358) gives, after little algebra,

$$\mathrm{APE(parcel)} = \frac{\rho_s\,U^2\,\theta_0'^2}{2\,\mathsf{S}} \tag{5.360}$$

where the buoyancy frequency square $N_s^2 = g\,\theta_s^{-1}\,\mathrm{d}\theta_s/\mathrm{d}z$ and the stability parameter $\mathsf{S} = (H\,N_s)^2/(f_0\,L)^2$ have been used to derive (5.360).

By introducing the representative constant value of the standard density, ρ_{s0}, Eq. (5.360) can be usefully recast in the form

$$\mathrm{APE(parcel)} = \rho_{s0} U^2 \frac{\rho_s \theta_0'^2}{2 \rho_{s0} \mathsf{S}} \tag{5.361}$$

In this way, the dimensional part of APE(parcel) is included into the factor $\rho_{s0} U^2$ of (5.361), while the non-dimensional available potential energy APE of the flow in a certain volume V is

$$\mathsf{APE} = \frac{1}{2\rho_{s0}} \int_{\mathsf{V}} \frac{\rho_s \theta_0'^2}{\mathsf{S}} \, d\mathsf{V} \tag{5.362}$$

Recalling (5.337), that is, $\theta_0' = \partial\psi/\partial \mathsf{z}$, Eq. (5.362) takes the form

$$\mathsf{APE} = \frac{1}{2\rho_{s0}} \int_{\mathsf{V}} \frac{\rho_s}{\mathsf{S}} \left(\frac{\partial \psi}{\partial \mathsf{z}}\right)^2 d\mathsf{V} \tag{5.363}$$

In the following, Eq. (5.363) will be derived from the evaluation of the integrated energy equation of the atmosphere starting from the vorticity equation (5.347).

Integrated Energy Equation of a Compressible and Thermally Forced Atmosphere

Consider the vorticity equation (5.347), written in short as

$$\frac{\partial}{\partial \mathsf{t}} \left[\nabla_{\mathrm{H}}'^2 \psi + \frac{1}{\rho_s} \frac{\partial}{\partial \mathsf{z}} \left(\frac{\rho_s}{\mathsf{S}} \frac{\partial \psi}{\partial \mathsf{z}} \right) \right] + \mathscr{J}(\psi, \Pi_0') = \frac{1}{\rho_s} \frac{\partial}{\partial \mathsf{z}} \left(\frac{\rho_s}{\mathsf{S}} \dot{\mathsf{Q}} \right) \tag{5.364}$$

where

$$\Pi_0' := \nabla_{\mathrm{H}}'^2 \psi + \frac{1}{\rho_s} \frac{\partial}{\partial \mathsf{z}} \left(\frac{\rho_s}{\mathsf{S}} \frac{\partial \psi}{\partial \mathsf{z}} \right) + \beta \mathsf{y}$$

The fluid domain taken into account is the unbounded non-dimensional volume

$$\mathsf{V} = \mathsf{D} \times [0 \le \mathsf{z} < \infty[\tag{5.365}$$

where $\mathsf{D} = [0 \le \mathsf{x} < \Lambda'] \times [0 \le \mathsf{y} \le \pi]$ is contained in the beta plane, and hence,

$$\int_{\mathsf{V}} d\mathsf{V} = \int_0^\infty d\mathsf{z} \int_0^{\Lambda'} d\mathsf{x} \int_0^\pi d\mathsf{y} \tag{5.366}$$

The boundary conditions for the stream function are:

- The zonal periodicity

$$\psi(\mathsf{x}, \mathsf{y}, \mathsf{z}, \mathsf{t}) = \psi(\mathsf{x} + \Lambda', \mathsf{y}, \mathsf{z}, \mathsf{t}) \tag{5.367}$$

for a suitable wavelength

- The no–mass flux across the zonal boundaries, that is,

$$\psi = 0 \quad \text{at } y = 0 \text{ and at } y = \pi \tag{5.368}$$

which assures the validity of the beta-plane approximation

The following identities, which hold because of (5.367) and (5.368),

$$\int_D \nabla'_H \cdot \left(\psi \nabla'_H \frac{\partial \psi}{\partial t} \right) dx\,dy = 0 \tag{5.369}$$

$$\int_D \mathscr{J}\left(\psi^2, \Pi'_0\right) dx\,dy = 0 \tag{5.370}$$

$$\int_D \mathscr{J}\left(\psi^2, \frac{\partial \psi}{\partial z}\right) dx\,dy = 0 \tag{5.371}$$

will be useful in a little while. They are proved in the Appendix at the end of this section.

After these preliminaries, multiplication of (5.364) by the non-dimensional factor $-(\rho_s/\rho_{s0})\,\psi$ and the subsequent integration of each product over D with the aid of (5.369) and (5.370) yield, after some trivial rearrangement and recalling (Appendix A, p. 372),

$$\begin{aligned} &\frac{\rho_s}{2\rho_{s0}} \int_D \frac{\partial}{\partial t} |\nabla'_H \psi|^2 \,dx\,dy \\ &+ \frac{1}{\rho_{s0}} \int_D \left[-\frac{\partial}{\partial z}\left(\frac{\rho_s}{S} \psi \frac{\partial}{\partial t} \frac{\partial \psi}{\partial z} \right) + \frac{\rho_s}{S} \frac{\partial \psi}{\partial z} \frac{\partial}{\partial t} \frac{\partial \psi}{\partial z} \right] dx\,dy \\ &= -\frac{1}{\rho_{s0}} \int_D \psi \frac{\partial}{\partial z}\left(\frac{\rho_s}{S} \dot{Q} \right) dx\,dy \end{aligned} \tag{5.372}$$

Equation (5.372) can be written as

$$\begin{aligned} &\frac{\rho_s}{2\rho_{s0}} \int_D \frac{\partial}{\partial t} \left[|\nabla'_H \psi|^2 + \frac{1}{S} \left(\frac{\partial \psi}{\partial z} \right)^2 \right] dx\,dy \\ &+ \frac{1}{\rho_{s0}} \frac{\partial}{\partial z} \left\{ \rho_s \int_D \left[\psi w_1 + \frac{1}{2S} \mathscr{J}\left(\psi^2, \frac{\partial \psi}{\partial z} \right) - \psi \frac{\dot{Q}}{S} \right] dx\,dy \right\} \\ &= -\frac{1}{\rho_{s0}} \int_D \psi \frac{\partial}{\partial z}\left(\frac{\rho_s}{S} \dot{Q} \right) dx\,dy \end{aligned} \tag{5.373}$$

and hence, because of (5.371), Eq. (5.373) simplifies into

$$
\begin{aligned}
&\frac{\rho_s}{2\rho_{s0}} \int_D \frac{\partial}{\partial t} \left[|\nabla'_H \psi|^2 + \frac{1}{S} \left(\frac{\partial \psi}{\partial z} \right)^2 \right] dx dy \\
&+ \frac{1}{\rho_{s0}} \frac{\partial}{\partial z} \rho_s \int_D \left(\psi w_1 - \psi \frac{\dot{Q}}{S} \right) dx dy \\
&= -\frac{1}{\rho_{s0}} \int_D \psi \frac{\partial}{\partial z} \left(\frac{\rho_s}{S} \dot{Q} \right) dx dy
\end{aligned}
\tag{5.374}
$$

Then, vertical integration of (5.374) gives

$$
\frac{d}{dt} \int_V \mathscr{E} \, dV + [F_3]_{z=0}^{z=\infty} = \frac{1}{\rho_{s0}} \int_V \frac{\partial \psi}{\partial z} \frac{\rho_s}{S} \dot{Q} \, dV \tag{5.375}
$$

where

$$
F_3 = \int_D \frac{\rho_s}{\rho_{s0}} \psi w_1 \, dx dy \tag{5.376}
$$

is the non-dimensional vertical energy flux (the subscript "3" is reminiscent of the vertical component of the flux), and

$$
\mathscr{E} = \mathscr{E}_{kin} + ape \tag{5.377}
$$

is the mechanical energy density, with

$$
\mathscr{E}_{kin} = \frac{\rho_s}{2\rho_{s0}} |\nabla'_H \psi|^2 \tag{5.378}
$$

the kinetic energy density and

$$
ape = \frac{\rho_s}{2\rho_{s0} S} \left(\frac{\partial \psi}{\partial z} \right)^2 = \frac{\rho_s \theta_0'^2}{2\rho_{s0} S} \tag{5.379}
$$

the non-dimensional available potential energy density, which is consistent with (5.363). Because of (5.376), the second term at the l.h.s. of (5.375) represents the vertical energy flux at great height.

The physical ground of (5.376) is clarified in the remark reported below. In the case of (5.350), Eqs. (5.377) and (5.376) take the form

$$
\mathscr{E} = \frac{1}{2} \exp(-\mu' z) \left[|\nabla'_H \psi|^2 + \frac{1}{S} \left(\frac{\partial \psi}{\partial z} \right)^2 \right] \tag{5.380}
$$

and

$$
F_3 = \int_D \exp(-\mu' z) \, \psi w_1 \, dx dy \tag{5.381}
$$

respectively.

Remark: The Vertical Energy Flux

To explain the physical meaning of (5.376), we start from Eq. (2.25), which expresses the total rate of work done by the pressure force in the volume δV:

$$\delta \dot{W} = \mathrm{div}\,(p\mathbf{u})\,\delta V \tag{5.382}$$

We have inverted the sign of (2.25) because here we evaluate the work done by the system, rather than that on the system as in Sect. 2.1.2. Assuming that (i) p and $\mathbf{u}$ are Λ-periodic functions of x, (ii) the meridional component of velocity $\mathbf{u}$ fulfils the boundary condition $v = 0$ at $\mathrm{y} = 0$ and at $\mathrm{y} = L\pi$ and (iii) pressure p coincides with pressure perturbation $\tilde{p}$, we integrate (5.382) over the volume V, which is the dimensional counterpart of V given in (5.366). The result is

$$\dot{W} = \left[\int_D \tilde{p}\, w \,\mathrm{d}x\,\mathrm{d}y\right]_{z=0}^{z=\infty} \tag{5.383}$$

because

$$\mathrm{div}\,(p\mathbf{u}) = \mathrm{div}_{\mathrm{H}}(\tilde{p}\mathbf{u}_{\mathrm{H}}) + \frac{\partial}{\partial z}(\tilde{p}w)$$

and the integral of the first term in the r.h.s. of the previous equation vanishes by virtue of assumptions (i) and (ii) .

Moreover, it is known that, at the geostrophic level of approximation, we have $\tilde{p} = f_0\,\rho_{\mathrm{s}}\,U\,L\,\psi$ and

$$w = \frac{U\,H}{L}\,\varepsilon\,\mathrm{w}_1 = \left(\frac{U}{L}\right)^2 \frac{H}{f_0}\,\mathrm{w}_1$$

Thus, in terms of ψ and w_1, Eq. (5.383) gives

$$\dot{W} = \left[\rho_{\mathrm{s}0}\,U^3\,H\,L\int_{\mathrm{D}} \frac{\rho_{\mathrm{s}}(\mathrm{z})}{\rho_{\mathrm{s}0}}\,\psi\,\mathrm{w}_1\,\mathrm{d}x\,\mathrm{d}y\right]_{\mathrm{z}=0}^{\mathrm{z}=\infty}$$

that is, by using (5.376),

$$\dot{W} = \rho_{\mathrm{s}0}\,U^3\,H\,L\,[\mathrm{F}_3]_{\mathrm{z}=0}^{\mathrm{z}=\infty} \tag{5.384}$$

In (5.384), the factor $\rho_{\mathrm{s}0}\,U^3\,H\,L$ represents the order of magnitude of the dimensional total rate of work done by the pressure force in the volume V, while

$$[\mathrm{F}_3]_{\mathrm{z}=0}^{\mathrm{z}=\infty} = \left[\int_{\mathrm{D}} \exp(-\mu'\mathrm{z})\,\psi\,\mathrm{w}_1\,\mathrm{d}x\,\mathrm{d}y\right]_{\mathrm{z}=0}^{\mathrm{z}=\infty} \tag{5.385}$$

is the second term of (5.375) under assumption (5.350).

The Vertical Boundary Conditions at the Top of the Atmosphere

Based on the principle that the energy of every finite portion of fluid be finite, the asymptotic behaviour of every model solution ψ is requested to satisfy the relationship

$$\sup_{0 \leq z < \infty} \mathscr{E}(\psi) < \infty \tag{5.386}$$

where, under the hypothesis of an isothermal undisturbed atmosphere, $\mathscr{E}$ is given by (5.380).

In the case in which (5.386), although verified, may be inconclusive to obtain a unique model solution, the nature of the forcing must be taken into account. In fact, owing to the fact that the atmosphere is heated or cooled from below, the source or sink of the thermal forcing lies in the proximity of the ground; so, the related vertical energy flux must be upward, that is,

$$\mathsf{F}_3 > 0 \tag{5.387}$$

In this case, once (5.386) is verified, condition (5.387) applies.

For practical computations of the l.h.s. of (5.386), it is useful to point out that if

$$\mathscr{E}_k = \frac{1}{2}\exp(-\mu' z)\left[\left|\nabla'_{\mathrm{H}}\psi_k\right|^2 + \frac{1}{S}\left(\frac{\partial \psi_k}{\partial z}\right)^2\right] \qquad (k = 1, 2)$$

and

$$\mathscr{E}_{12} = \frac{1}{2}\exp(-\mu' z)\left\{\left|\nabla'_{\mathrm{H}}(\psi_1 + \psi_2)\right|^2 + \frac{1}{S}\left[\frac{\partial}{\partial z}(\psi_1 + \psi_2)\right]^2\right\}$$

then

$$\mathscr{E}_{12} \leq \left(\sqrt{\mathscr{E}_1} + \sqrt{\mathscr{E}_2}\right)^2 \tag{5.388}$$

Hence, if $\mathscr{E}_1$ and $\mathscr{E}_2$ separately verify (5.386), also $\mathscr{E}_{12}$ verifies the same relationship.

5.2.2 *Thermally Forced Stationary Waves*

The response of the atmosphere to a hypothetical meridional gradient of temperature, from the equator to a pole, would determine a zonally symmetric flow. In the real case, the contrast in temperature between the ocean and the continental land induces also a zonal component of this gradient; thus, the resulting thermal forcing of the atmosphere $\dot{\mathsf{Q}}$ is actually a function of all the coordinates:

$$\dot{\mathsf{Q}} = \dot{\mathsf{Q}}(\mathsf{x}, \mathsf{y}, \mathsf{z}) \tag{5.389}$$

To deal with an analytically solvable model, the thermal forcing (5.389) is assumed to have the special form

$$\dot{Q}(x, y, z) = A \exp(-\alpha' z) \cos(kx) \sin(y) \tag{5.390}$$

where the constant $A > 0$ represents the strength of the forcing, the factor $\exp(-\alpha' z)$ with $\alpha' > 0$ takes into account the extinction of $\dot{Q}$ with height, the factor $\cos(kx)$ is a crude representation of the alternance of $\dot{Q}$ in the transition from ocean to land and vice versa (which is the origin of the above cited zonal component of the temperature gradient) and, finally, $\sin(y)$ confines $\dot{Q}$ in the latitudinal band $0 \leq y \leq \pi$.

Actually, the latter assumption has not a physical ground but is requested in order to develop the steady version of model (5.352), (5.390) within the beta-plane approximation and consistently with (5.368). Without loss of generality, we take

$$k > 0 \tag{5.391}$$

Therefore, by using (5.390), the governing equation inferred from (5.352) is

$$\begin{aligned} &\mathscr{J}\left(\psi, \nabla_H'^2 \psi + \frac{1}{S}\left(\frac{\partial^2 \psi}{\partial z^2} - \mu' \frac{\partial \psi}{\partial z}\right)\right) + \beta \frac{\partial \psi}{\partial x} \\ &\quad = -A \frac{\alpha' + \mu'}{S} \exp(-\alpha' z) \cos(kx) \sin(y) \end{aligned} \tag{5.392}$$

We stress that, in the absence of $\dot{Q}$, that is, if $A = 0$, the unperturbed stream function

$$\psi^{(0)}(y) = -u_0 y \tag{5.393}$$

with u_0 constant for mathematical simplicity, is an exact integral of (5.392). Note that conditions (5.353) with $\dot{Q} = 0$ and (5.386) are satisfied by (5.393). Given (5.393) and A small enough, an asymptotic expansion of the kind

$$\psi(x, y, z; A) = \psi^{(0)}(y) + A\, \psi^{(1)}(x, y, z) + O(A^2) \tag{5.394}$$

is now introduced as a putative solution of the model. Hence, substitution of (5.394) into (5.392) yields, to the leading order in A, the vorticity equation

$$\mathscr{J}\left(\psi^{(0)}, \nabla_H'^2 \psi^{(0)} + \frac{1}{S}\left(\frac{\partial^2 \psi^{(0)}}{\partial z^2} - \mu' \frac{\partial \psi^{(0)}}{\partial z}\right)\right) + \beta \frac{\partial \psi^{(0)}}{\partial x} = 0 \tag{5.395}$$

which, owing to (5.393), is a trivial identity. The first-order equation takes the form

$$u_0 \frac{\partial}{\partial x}\left[\nabla_H'^2 \psi^{(1)} + \frac{1}{S}\left(\frac{\partial^2 \psi^{(1)}}{\partial z^2} - \mu' \frac{\partial \psi^{(1)}}{\partial z}\right)\right] + \beta \frac{\partial \psi^{(1)}}{\partial x}$$
$$= -\frac{\alpha' + \mu'}{S} \exp(-\alpha' z) \cos(kx) \sin(y) \tag{5.396}$$

or, more conveniently,

$$u_0 \frac{\partial}{\partial x}\left[\nabla_H'^2 \psi^{(1)} + \frac{1}{S}\left(\frac{\partial^2 \psi^{(1)}}{\partial z^2} - \mu' \frac{\partial \psi^{(1)}}{\partial z}\right)\right] + \beta \frac{\partial \psi^{(1)}}{\partial x}$$
$$= -\frac{\alpha' + \mu'}{S} \mathrm{Re}\left[\exp(-\alpha' z + ikx)\right] \sin(y) \tag{5.397}$$

Using (5.388) and (5.394), we get

$$\mathscr{E}(\psi) = \left(\sqrt{\mathscr{E}(\psi^{(0)})} + A\sqrt{\mathscr{E}(\psi^{(1)})}\right)^2 \tag{5.398}$$

Moreover,

$$\sup_{0 \le z < \infty} \mathscr{E}(\psi^{(0)}) = \sup_{0 \le z < \infty} \frac{1}{2} \exp(-\mu' z)\, u_0^2 < \infty \tag{5.399}$$

and therefore, the energy density $\mathscr{E}(\psi)$ converges if and only if $\mathscr{E}(\psi^{(1)})$ converges. Thus, we now focus our attention on the behaviour of the first-order term $\psi^{(1)}$. Because of the linearity of (5.397), the trial function

$$\psi^{(1)} = \mathrm{Re}[\phi(z) \exp(ikx)] \sin(y) \tag{5.400}$$

is substituted into (5.397) to give, after little algebra, the ordinary equation (primes denote z-derivatives)

$$\phi''(z) - \mu' \phi'(z) + S\left[\frac{\beta}{u_0} - (1 + k^2)\right] \phi(z) = i \frac{\alpha' + \mu'}{u_0 k} \exp(-\alpha' z) \tag{5.401}$$

for the sole unknown of the problem, that is, $\phi(z)$. The boundary condition (5.353) here becomes

$$\mathscr{J}\left(\psi_0 + A\psi^{(1)} + O(A^2),\ A\frac{\partial \psi^{(1)}}{\partial z} + O(A^2)\right) = A\cos(kx)\sin(y) \tag{5.402}$$

and, to the leading order in A, (5.402) gives

$$u_0 \frac{\partial^2 \psi^{(1)}}{\partial x \partial z} = \cos(kx)\sin(y)$$

whence we get the boundary condition for the sole $\phi(z)$:

$$\mathrm{Re}[\phi'(0)] = 0 \qquad \mathrm{Im}[\phi'(0)] = -\frac{1}{k\,u_0}$$

that is to say,

$$\phi'(0) = -\,\mathrm{Im}\,\frac{1}{u_0\,k} \tag{5.403}$$

The shorthand notations

$$\left(\frac{\mu'}{2}\right)^2 - S\left[\frac{\beta}{u_0} - (1+k^2)\right] =: m^2 \tag{5.404}$$

$$\alpha'^2 + \alpha'\,\mu' + \left(\frac{\mu'}{2}\right)^2 - m^2 =: \Delta'^2 \tag{5.405}$$

where m may be either real or imaginary, will be adopted. Suppose, now, to have singled out a definite integral, say $\phi_{\#}$, of (5.401). Consider, then,

$$\psi_{\#} := \mathrm{Re}[\phi_{\#}(z)\exp(i\,k\,x)]\,\sin(y) \tag{5.406}$$

according to (5.400). Substitution of (5.406) into (5.380) yields

$$\mathscr{E} = \frac{1}{2}\exp(-\mu' z)\,\left|\mathrm{Re}[\phi_{\#}(z)]\,\nabla'_{\mathrm{H}}[\cos(kx)\,\sin(y)] - \mathrm{Im}[\phi_{\#}(z)]\,\nabla'_{\mathrm{H}}[\sin(kx)\,\sin(y)]\right|^2$$
$$+\frac{1}{2S}\exp(-\mu' z)\left|\frac{d\,\mathrm{Re}[\phi_{\#}(z)]}{dz}\cos(kx)\,\sin(y) - \frac{d\,\mathrm{Im}[\phi_{\#}(z)]}{dz}\sin(kx)\,\sin(y)\right|^2 \tag{5.407}$$

Because of (5.388), condition (5.386) is verified if

$$\sup_{0\le z<\infty}\ \exp(-\mu' z)\,\mathrm{Re}^2[\phi_{\#}(z)] < \infty \tag{5.408}$$

$$\sup_{0\le z<\infty}\ \exp(-\mu' z)\,\mathrm{Im}^2[\phi_{\#}(z)] < \infty \tag{5.409}$$

$$\sup_{0\le z<\infty}\ \exp(-\mu' z)\left(\frac{d\,\mathrm{Re}[\phi_{\#}(z)]}{dz}\right)^2 < \infty \tag{5.410}$$

$$\sup_{0\le z<\infty}\ \exp(-\mu' z)\left(\frac{d\,\mathrm{Im}[\phi_{\#}(z)]}{dz}\right)^2 < \infty \tag{5.411}$$

Actually, (5.401) has constant coefficients; so, the dependence of $\phi_{\#}$ on z is exponential in any case, and therefore, conditions (5.410) and (5.411) are a priori assured whenever conditions (5.408) and (5.409) hold true. Thus, (5.410) and (5.411) will not be involved in the following analysis. At this point, the set of linearly independent integrals of (5.401) must be singled out. It is constituted by a particular integral $\tilde{\phi}(z)$ of the full (inhomogeneous) differential equation and by two linearly independent integrals $\phi_{\pm}(z)$ of the related homogeneous equation

$$\phi''(z) - \mu'\phi'(z) + S\left[\frac{\beta}{u_0} - (1+k^2)\right]\phi(z) = 0 \tag{5.412}$$

Provided that $\Delta' \neq 0$, an integral $\tilde{\phi}(z)$ of (5.401), written using position (5.405), is

$$\tilde{\phi}(z) = i\,\frac{\alpha' + \mu'}{\Delta' u_0 k}\exp(-\alpha' z) \tag{5.413}$$

(The case $\Delta' = 0$ will be investigated in the following.) Openly, both (5.408) and (5.409) are verified.

Two linearly independent integrals of (5.412), $\phi_+(z)$ and $\phi_-(z)$, written using position (5.404), are

$$\phi_+(z) = \exp\left[\left(\frac{\mu'}{2} + m\right)z\right] \tag{5.414}$$

$$\phi_-(z) = \exp\left[\left(\frac{\mu'}{2} - m\right)z\right] \tag{5.415}$$

and condition (5.386), in case together with (5.387), is applied to (5.414) and (5.415). The analysis is reported below.

Case 1: $m^2 > 0$. Considering (5.414), we get

$$\sup_{0\leq z<\infty} \exp(-\mu' z)\,[\phi_+(z)]^2 = \sup_{0\leq z<\infty} \exp(2|m|z) = \infty \tag{5.416}$$

and hence, condition (5.386) is violated. On the other hand, (5.415) implies

$$\sup_{0\leq z<\infty} \exp(-\mu' z)\,[\phi_-(z)]^2 = \sup_{0\leq z<\infty} \exp(-2|m|z) = 1 \tag{5.417}$$

and here, condition (5.386) holds. Therefore, with reference to (5.401), we have

$$\phi(z) = \tilde{\phi}(z) + C\phi_-(z) \tag{5.418}$$

where the constant is determined by the boundary condition (5.403), which yields

$$C = -i\,\frac{1}{\Delta' u_0 k}\left(\frac{\mu'}{2} + |m|\right) \tag{5.419}$$

At this point, the complete model solution, up to the first order in A, can be inferred from (5.394), (5.413), (5.415), (5.418) and (5.419):

$$\psi = -u_0 y + \frac{A}{\Delta' u_0 k} \left\{ \left(\frac{\mu'}{2} + |m| \right) \exp\left[\left(\frac{\mu'}{2} - |m| \right) z \right] - (\alpha' + \mu') \exp(-\alpha' z) \right\} \times \sin(kx) \sin(y) \tag{5.420}$$

In order to model the wind pattern at the ground, we fix the parameters in (5.420) as follows:

$$u_0 = \frac{4}{5} \quad A = 0.3 \quad \Delta = \frac{5}{4} \quad k = 1 \quad \mu' = 1 \quad |m| = 1 \quad \alpha' = 1$$

The exponential factors decrease with height, and hence, the meandering is damped to zero. In this limit, the wind blows along the circles of latitude. The situation at the ground is depicted in the top left panel of Fig. 5.5. Moreover, the horizontally averaged energy density (5.380), when computed with with $S = 1$, goes to zero with increasing height, and the wave is trapped to the ground, as shown in the bottom right panel of the same figure.

Case 2: $m^2 < 0$. In this case, with reference to (5.414) and (5.415), we have

$$\mathrm{Re}\,\phi_+(z) = \mathrm{Re}\,\phi_-(z) = \exp\left(\frac{\mu' z}{2}\right) \cos(|m| z) \tag{5.421}$$

$$\mathrm{Im}\,\phi_\pm(z) = \pm \exp\left(\frac{\mu' z}{2}\right) \sin(|m| z) \tag{5.422}$$

and, therefore,

$$\sup_{0 \le z < \infty} \exp(-\mu' z)\, \mathrm{Re}^2 \phi_\pm(z) = 1 \tag{5.423}$$

$$\sup_{0 \le z < \infty} \exp(-\mu' z)\, \mathrm{Im}^2 \phi_\pm(z) = 1 \tag{5.424}$$

Relationships (5.423) and (5.424) show that both ϕ_+ and ϕ_- pass the energy test (5.386); so, (5.400) can be temporarily written as

$$\psi^{(1)} = \mathrm{Re}\{[B_+ \phi_+(z) + B_- \phi_-(z)] \exp(ikx)\} \sin(y) \tag{5.425}$$

where the coefficients B_+ and B_- are still undetermined. Equation (5.425) is then used to evaluate the vertical energy flux (5.381), conventionally averaged on the horizontal fluid domain D, that is,

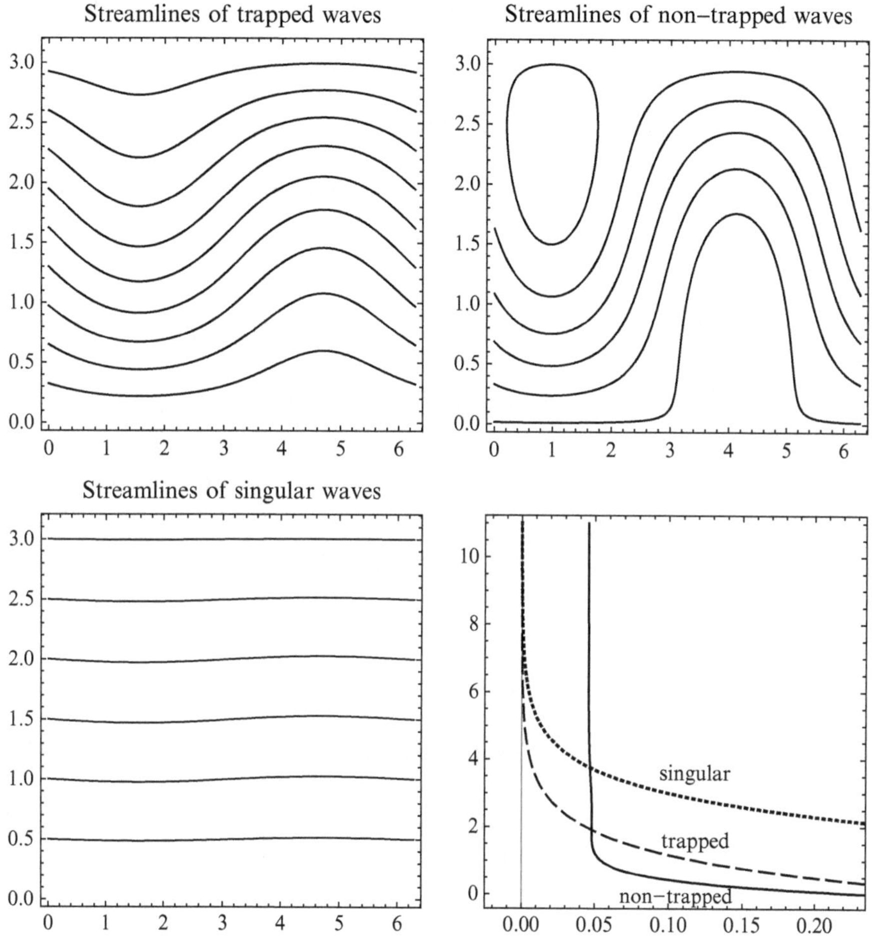

Fig. 5.5 Wind patterns at ground level (z = 0) in a one-period domain of atmospheric waves: trapped (*top left*), non-trapped (*top right*), singular (*bottom left*), together with the vertical structure of horizontally averaged energy density (5.380) (*bottom right*)

$$\langle F_3 \rangle_D = \frac{k}{2\pi^2} \exp(-\mu' z) \int_0^{2\pi/k} \left[\int_0^{\pi} w_1(x, y, z)\, \psi^{(1)}(x, y, z)\, dy \right] dx \tag{5.426}$$

Far from the ground, we have $\dot{Q} \approx 0$; so, the steady version of (5.341) gives, up to the first order in A,

$$w_1 = -A \frac{u_0}{S} \frac{\partial^2 \psi^{(1)}}{\partial x \partial z} \tag{5.427}$$

and (5.427) can be used to evaluate (5.426) for z large enough. We stress that hypothesis $m^2 < 0$ implies, through (5.404),

$$u_0 > 0 \tag{5.428}$$

and inequality (5.428) must be taken into account in determining the sign of (5.426). The integrand of (5.426) may be evaluated with the aid of the expansion

$$w_1 \psi^{(1)} = -A \frac{u_0}{4S} \left\{ \frac{\partial^2}{\partial x \partial z} [\phi(z) \exp(ikx) + \phi^*(z) \exp(-ikx)] \right\}$$
$$\times [\phi(z) \exp(ikx) + \phi^*(z) \exp(-ikx)] \sin^2(y) \tag{5.429}$$

where

$$\phi(z) = B_+ \exp\left[\left(\frac{\mu'}{2} + i|m|\right) z\right] + B_- \exp\left[\left(\frac{\mu'}{2} - i|m|\right) z\right] \tag{5.430}$$

and an asterisk denotes complex conjugation. From (5.429), we have

$$w_1 \psi^{(1)} = -iA \frac{k u_0}{4S} [\phi \phi' \exp(2ikx) + \phi' \phi^* - \phi \phi^{*\prime} - \phi^* \phi^{*\prime} \exp(-2ikx)] \sin^2(y) \tag{5.431}$$

and recalling that $\int_0^{2\pi/k} \exp(\pm 2ikx)\, dx = 0$ while $\int_0^{\pi} \sin^2(y)\, dy = \pi/2$, Eq. (5.426) gives

$$\langle F_3 \rangle_D = -iA \frac{k u_0}{8S} \exp(-\mu' z) (\phi' \phi^* - \phi \phi^{*\prime})$$
$$= A \frac{k u_0}{4S} \exp(-\mu' z) \operatorname{Im}(\phi' \phi^*) \tag{5.432}$$

Therefore, because of (5.391) and (5.428), Eq. (5.432) shows that the averaged energy flux (5.426) is upward if and only if

$$\operatorname{Im}(\phi' \phi^*) > 0 \tag{5.433}$$

The product of

$$\phi' = B_+ \left(\frac{\mu'}{2} + i|m|\right) \exp\left[\left(\frac{\mu'}{2} + i|m|\right) z\right]$$
$$+ B_- \left(\frac{\mu'}{2} - i|m|\right) \exp\left[\left(\frac{\mu'}{2} - i|m|\right) z\right]$$

and

$$\phi^* = B_+^* \exp\left[\left(\frac{\mu'}{2} - i|m|\right) z\right] + B_-^* \exp\left[\left(\frac{\mu'}{2} + i|m|\right) z\right]$$

is

$$\begin{aligned}\phi'\,\phi^* &= \frac{\mu'}{2}\,(|\mathsf{B}_+|^2 + |\mathsf{B}_-|^2)\,\exp(\mu'\,\mathsf{z})\\ &+ 2\,\mathrm{Re}\left\{\mathsf{B}_+\,\mathsf{B}_-^*\,\left(\frac{\mu'}{2} + \mathrm{i}\,|\mathsf{m}|\right)\exp\left[2\left(\frac{\mu'}{2} + \mathrm{i}\,|\mathsf{m}|\right)\mathsf{z}\right]\right\}\\ &+ \mathrm{i}\,|\mathsf{m}|\,(|\mathsf{B}_+|^2 - |\mathsf{B}_-|^2)\,\exp(\mu'\,\mathsf{z})\end{aligned}$$

and therefore,

$$\mathrm{Im}(\phi'\,\phi^*) = |\mathsf{m}|\,(|\mathsf{B}_+|^2 - |\mathsf{B}_-|^2)\,\exp(\mu'\,\mathsf{z}) \tag{5.434}$$

By using (5.434), the average flux (5.432) results to be

$$\langle \mathsf{F}_3\rangle_{\mathsf{D}} = \mathsf{A}\,\frac{\mathsf{k}\,\mathsf{u}_0\,|\mathsf{m}|}{4\,\mathsf{S}}\,(|\mathsf{B}_+|^2 - |\mathsf{B}_-|^2)$$

and hence, in order that (5.387) be verified, only $\phi_+(\mathsf{z})$ turns out to be an admissible integral. Therefore, the total solution is

$$\phi(\mathsf{z}) = \tilde{\phi}(\mathsf{z}) + \mathsf{B}_+\,\phi_+(\mathsf{z}) \tag{5.435}$$

where $\tilde{\phi}(\mathsf{z})$ is given by (5.413), $\phi_+(\mathsf{z}) = \exp[(\mu'/2 + \mathrm{i}\,|\mathsf{m}|)\,\mathsf{z}]$, and B_+ is selected by resorting to (5.403) applied to (5.435), thus obtaining

$$\mathsf{B}_+ = -\,\frac{|\mathsf{m}| + \mathrm{i}\,\mu'/2}{\Delta'\,\mathsf{u}_0\,\mathsf{k}} \tag{5.436}$$

To summarize, in the present context, we have

$$\begin{aligned}\psi^{(1)} &= \mathrm{Re}[\phi(\mathsf{z})\,\exp(\mathrm{i}\,\mathsf{k}\,\mathsf{x})]\,\sin(\mathsf{y})\\ &= \mathrm{Re}\left\{\mathrm{i}\,\frac{\alpha' + \mu'}{\Delta'\,\mathsf{u}_0\,\mathsf{k}}\,\exp(-\alpha'\,\mathsf{z} + \mathrm{i}\,\mathsf{k}\,\mathsf{x})\right.\\ &\quad\left. -\,\frac{|\mathsf{m}| + \mathrm{i}\,\mu'/2}{\Delta'\,\mathsf{u}_0\,\mathsf{k}}\,\exp\left[\left(\frac{\mu'}{2} + \mathrm{i}\,|\mathsf{m}|\right)\mathsf{z} + \mathrm{i}\,\mathsf{k}\,\mathsf{x}\right]\right\}\sin(\mathsf{y})\end{aligned} \tag{5.437}$$

and up to the first order in A, the model solution for $m^2 < 0$ is

$$\begin{aligned}\psi = -\mathsf{u}_0\,\mathsf{y} + \frac{\mathsf{A}\,\sin(\mathsf{y})}{\Delta'\,\mathsf{u}_0\,\mathsf{k}}\,&[(\mu'/2)\,\exp(\mu'\,\mathsf{z}/2)\,\sin(|\mathsf{m}|\,\mathsf{z} + \mathsf{k}\,\mathsf{x})\\ &- |\mathsf{m}|\,\exp(\mu'\,\mathsf{z}/2)\,\cos(|\mathsf{m}|\,\mathsf{z} + \mathsf{k}\,\mathsf{x})\\ &- (\alpha' + \mu')\,\exp(-\alpha'\,\mathsf{z})\,\sin(\mathsf{k}\,\mathsf{x})]\end{aligned} \tag{5.438}$$

In order to model the wind pattern at the ground, we fix the parameters in (5.438) as follows:

$$u_0 = \frac{4}{13} \quad A = 0.3 \quad \Delta = \frac{13}{4} \quad k = 1 \quad \mu' = 1 \quad |m| = 1 \quad \alpha' = 1$$

Only the last exponential factor decreases with height, while the first two terms describe unbounded steady behaviour. Unlike solution (5.420), no zonal wind may arise at any height. The situation at the ground is depicted in the top right panel of Fig. 5.5. Moreover, the horizontally averaged energy density (5.380), when computed with $S = 1$, goes to a constant positive value with increasing height, as shown in the bottom right panel of the same figure; so, the wave is no longer trapped to the ground.

Comments About Trapped and Non-trapped Solutions

Apart from the numerical examples reported above to illustrate some special cases of (5.420) and (5.438), we recall that, quite in general, a trapped solution presupposes $m^2 > 0$ while a non-trapped solution presupposes $m^2 < 0$. The link between the sign of the zonal wind velocity u_0 and that of m^2 follows from (5.404), which can be restated in the form

$$m^2 = \beta S \left(\frac{1}{u_0^c} - \frac{1}{u_0} \right) \tag{5.439}$$

with the critical velocity

$$u_0^c := \frac{\beta S}{(\mu'/2)^2 + S(1+k^2)} > 0$$

Relationship (5.439) shows that if $u_0 < 0$ or $u_0 > u_0^c$ (i.e. if the zonal wind is easterly or westerly and strong enough), then $m^2 > 0$ and the solution is (5.420). On the other hand, if $0 < u_0 < u_0^c$ (i.e. if the zonal wind is westerly but weak enough), then $m^2 < 0$ and the solution is (5.438). It is known that radiating waves of the latter kind occur mainly in spring and autumn.

The Singular Solution

As anticipated in deriving the particular integral (5.413), we now take into account assumption $\Delta' = 0$. The latter implies, through positions (5.404) and (5.405), the special form

$$\phi''(z) - \mu' \phi'(z) - \alpha' (\alpha' + \mu') \phi(z) = i \frac{\alpha' + \mu'}{u_0 k} \exp(-\alpha' z) \tag{5.440}$$

of (5.401). In the case of (5.440), a particular integral $\tilde{\phi}(z)$ may be sought proportional to $z\exp(-\alpha' z)$, thus yielding

$$\tilde{\phi}(z) = -i\,\frac{\alpha' + \mu'}{u_0\,k\,(2\,\alpha' + \mu')}\,z\exp(-\alpha' z) \tag{5.441}$$

On the other hand, two linearly independent integrals, say $\phi_{\pm}$, of the homogeneous equation associated to (5.440) are

$$\phi_{+}(z) = \exp[(\alpha' + \mu')\,z] \tag{5.442}$$

$$\phi_{-}(z) = \exp(-\alpha' z) \tag{5.443}$$

Since

$$\sup_{0\le z<\infty}\ \exp(-\mu' z)\,\phi_{+}^{2}(z) = \infty$$

while

$$\sup_{0\le z<\infty}\ \exp(-\mu' z)\,\phi_{-}^{2}(z) = 1$$

only (5.443) is a physically admissible integral. Thus,

$$\phi(z) = \tilde{\phi}(z) + C\,\phi_{-}(z) \tag{5.444}$$

where the constant C is singled out by applying boundary condition (5.403)–(5.444). One finds $C = i/[u_0\,k\,(2\,\alpha' + \mu')]$, whence

$$\phi(z) = \frac{i}{u_0\,k\,(2\,\alpha' + \mu')}\,[1 - (\alpha' + \mu')\,z]\exp(-\alpha' z) \tag{5.445}$$

Then, substitution of (5.445) into (5.400) gives

$$\psi^{(1)}(z) = \frac{1}{u_0\,k\,(2\,\alpha' + \mu')}\,[(\alpha' + \mu')\,z - 1]\exp(-\alpha' z)\,\sin(kx)\,\sin(y) \tag{5.446}$$

and up to the first order in A, the model solution is

$$\psi = -u_0\,y + \frac{A}{u_0\,k\,(2\,\alpha' + \mu')}\,[(\alpha' + \mu')\,z - 1]\exp(-\alpha' z)\,\sin(kx)\,\sin(y) \tag{5.447}$$

In order to model the wind pattern at the ground in the singular case (i.e. for $\Delta' = 0$), we fix the parameters in (5.447) as follows:

$$u_0 = -1 \quad A = 0.3 \quad k = 1 \quad \mu' = 2 \quad \alpha' = 1$$

The exponential factor decreases with height, and hence, the meandering is damped to zero. In this limit, the wind blows along the circles of latitude like for trapped waves (5.420). Note that the sign of the factor $(\alpha' + \mu')z - 1$ appearing in (5.447 changes with height: for instance, it is negative at $z = 0$ and positive at $z = 1$, and hence, the corresponding meanderings are opposite in phase.

The situation at the ground is depicted in the bottom left panel of Fig. 5.5. Moreover, the horizontally averaged energy density (5.380), when computed with $S = 1$, goes to zero with increasing height, and the wave is trapped to the ground, as shown in the bottom right panel of the same figure.

Remark on the Energy Source of Eq. (5.375)

By using (5.350), (5.390) and (5.394), the energy source of Eq. (5.375), that is,

$$\frac{1}{\rho_{s0}} \int_V \frac{\partial \psi}{\partial z} \frac{\rho_s}{S} \dot{Q} \, dV$$

takes the form

$$\frac{A}{S} \int_V \frac{\partial \psi^{(1)}}{\partial z} \exp[-(\alpha' + \mu')z] \cos(kx) \sin(y) \, dV + O(A^2) \tag{5.448}$$

By inspecting each of the model solutions (5.420), (5.438) and (5.447), one may easily verify that, in any case,

$$\int_0^\infty \frac{\partial \psi^{(1)}}{\partial z} \exp[-(\alpha' + \mu')z] \, dz < \infty$$

and therefore, also the volume integral of (5.448) converges.

5.2.3 *The Effects of Topography*

Wind over Slightly Wavy Terrain

Model Description

A slightly wavy terrain constitutes an idealized topography that induces a disturbance superimposed to a basically zonal flow, that is, a flow which would be exactly zonal on a flat ground. Just the longitudinal wavy modulation of the terrain gives raise to the meridional component of the wind field. Once a suitably simple representation of the topography is fixed and the zonal component of the wind is established, the main problem is to determine the meridional component of the wind field. In the present section, this problem is solved in the framework of the quasi-geostrophic dynamics.

Thus, the governing equation of our model is the steady and unforced version of (5.352), reported below:

$$\mathscr{J}\left(\psi,\ \nabla_{\mathrm{H}}^{\prime 2}\psi+\frac{1}{\mathsf{S}}\left(\frac{\partial^2\psi}{\partial \mathsf{z}^2}-\mu'\frac{\partial\psi}{\partial \mathsf{z}}\right)\right)+\beta\frac{\partial\psi}{\partial \mathsf{x}}=0 \tag{5.449}$$

Equation (5.449) is trivially satisfied by the unperturbed stream function

$$\psi^{(0)}(\mathsf{y})=-\mathsf{u}_0\,\mathsf{y} \tag{5.450}$$

where u_0 is a constant, for mathematical simplicity. We assume (5.450) as the stream function of the zonal wind when the terrain is flat. The question is how (5.450) modifies in the presence of a wavy topography. To answer, we start once again from (1.28) and consider the dimensional topographic disturbance

$$b(x,y)=\varepsilon\,\mathsf{A}H\,\mathsf{b}\left(\frac{x}{L},\frac{y}{L}\right) \tag{5.451}$$

where ε is the Rossby number, while the positive constant A is a "small" amplitude in the sense that

$$O(\varepsilon)<\mathsf{A}<1 \tag{5.452}$$

and, finally, $\mathsf{b}(\mathsf{x},\mathsf{y})$ is the non-dimensional counterpart of $b(x,y)$. Substitution of (5.451) into (1.28), with the use of the well-known expansions

$$w=U\frac{H}{L}\left(\varepsilon\,\mathsf{w}_1+O(\varepsilon^2)\right) \qquad \mathbf{u}=U\left(\mathbf{u}_0+O(\varepsilon)\right)$$

yields the relationship

$$\mathsf{w}_1=\mathsf{A}\,\mathscr{J}\left(\psi,\ \mathsf{b}\right) \tag{5.453}$$

which is referred to the height of the ground. Note that if $\mathsf{A}=O(\varepsilon)$, then $\mathsf{w}_1=0$ while $\mathsf{w}_2\neq 0$, and the model would describe a quasi-geostrophic wind flowing over a negligibly weak spatial modulation. On the other hand, if $\mathsf{A}<1$, owing to the dependence of (5.453) on A, this parameter can be used to expand in powers of it the total stream function, whose first term is (5.450). Indeed, the following expansion of ψ is postulated:

$$\psi(\mathsf{x},\mathsf{y},\mathsf{z},\mathsf{A})=-\mathsf{u}_0\,\mathsf{y}+\mathsf{A}\,\psi^{(1)}(\mathsf{x},\mathsf{y},\mathsf{z})+O(\mathsf{A}^2) \tag{5.454}$$

In (5.454) and in the quasi-geostrophic perspective, $\psi^{(1)}(\mathsf{x},\mathsf{y},\mathsf{z})$ is the basic unknown of the model. Substitution of (5.454) into (5.449) gives the first-order equation

$$u_0 \frac{\partial}{\partial x}\left[\nabla_H'^2 \psi^{(1)} + \frac{1}{S}\left(\frac{\partial^2 \psi^{(1)}}{\partial z^2} - \mu' \frac{\partial \psi^{(1)}}{\partial z}\right)\right] + \beta \frac{\partial \psi^{(1)}}{\partial x} = 0 \tag{5.455}$$

which is nothing but the homogeneous version of (5.397). The model solution is determined by the general integral of (5.455), by the lateral boundary conditions, by the boundary condition at the top of the atmosphere (5.386), if necessary together with (5.387), and by the boundary condition at the ground. The behaviour of $\psi^{(1)}$ at the ground is inferred from (5.453) and from the steady version of (5.341) without the thermal forcing, that is to say,

$$w_1 = -\frac{1}{S}\mathscr{J}\left(\psi, \frac{\partial \psi}{\partial z}\right) \tag{5.456}$$

which holds at every height. In fact, the leading-order equation obtained by substituting (5.454) into (5.453) states that, at the ground,

$$w_1 = A u_0 \frac{\partial b}{\partial x} \tag{5.457}$$

while the analogous procedure applied to (5.456) yields

$$w_1 = -\frac{A}{S} u_0 \frac{\partial}{\partial x}\frac{\partial \psi^{(1)}}{\partial z} \tag{5.458}$$

Then, elimination of the vertical velocity from (5.457) and (5.458) gives

$$-\frac{1}{S}\frac{\partial^2 \psi^{(1)}}{\partial x \partial z} = \frac{\partial b}{\partial x} \tag{5.459}$$

Equation (5.459) describes the desired boundary condition for $\psi^{(1)}$ at the ground. About the form of $b(x, y)$, hereafter we consider the wavy topography

$$b(x, y) = \cos(kx)\sin(y) \tag{5.460}$$

whence (5.459) takes the form

$$\frac{\partial^2 \psi^{(1)}}{\partial x \partial z} = kS \sin(kx)\sin(y) \tag{5.461}$$

Consistently with (5.460), the stream function is assumed to be periodic in x of period $2\pi/k$ and included into the interval $[0 \le y \le \pi]$ of the beta plane; so $\psi(y = 0) = \psi(y = \pi) = 0$, and moreover, the vertical coordinate z spans the half-space $[\varepsilon A b \le z \le +\infty[\approx [0 \le z \le +\infty[$.

Model Solutions

Because of the linearity of the vorticity equation (5.455) and of the above lateral boundary conditions, we set

$$\psi^{(1)} = \mathrm{Re}[\phi(z)\exp(ikx)\sin(y)] \qquad (k>0) \tag{5.462}$$

Substitution of (5.462) into (5.455) produces the ordinary differential equation with constant coefficients

$$\phi''(z) - \mu'\phi'(z) + S\left[\frac{\beta}{u_0} - (k^2+1)\right]\phi'(z) = 0 \tag{5.463}$$

for the sole $\phi(z)$, where the prime symbol ($'$) over ϕ means differentiation with respect to z. Note that (5.463) is the same as (5.412). For mathematical simplicity, the actual vertical coordinate of the ground is approximated by $z = 0$, and the boundary condition of $\phi(z)$ at the ground is obtained by substituting (5.462) into (5.461), whence

$$\mathrm{Re}[\phi'(0)] = -S \qquad \mathrm{Im}[\phi'(0)] = 0$$

that is to say,

$$\phi'(0) = -S \tag{5.464}$$

The general integral of (5.463) can be conveniently written by introducing the notation

$$m^2 := \left(\frac{\mu'}{2}\right)^2 - S\left[\frac{\beta}{u_0} - (k^2+1)\right] \tag{5.465}$$

and, then, by separately considering the three cases $m^2 > 0$, $m = 0$ and $m^2 < 0$.

Case 1: $m^2 > 0$. In this case, the general integral of (5.463) is

$$\phi(z) = C_1 \exp\left[\left(\frac{\mu'}{2} + |m|\right)z\right] + C_2 \exp\left[\left(\frac{\mu'}{2} - |m|\right)z\right] \tag{5.466}$$

If we substitute, separately, $C_1 \exp\left[\left(\frac{\mu'}{2} + |m|\right)z\right]$ and $C_2 \exp\left[\left(\frac{\mu'}{2} - |m|\right)z\right]$ for $\phi_{\#}$ into (5.408), we ascertain that

$$\sup_{0\le z<\infty}\left\{\exp(-\mu' z)\, C_1^2 \exp\left[(\mu' + 2|m|)\,z\right]\right\} = \infty \tag{5.467}$$

$$\sup_{0\le z<\infty}\left\{\exp(-\mu' z)\, C_2^2 \exp\left[(\mu' - 2|m|)\,z\right]\right\} = 0 \tag{5.468}$$

Equation 5.467 rules out the first term at the r.h.s. of (5.466), which simplifies to

$$\phi(z) = C_2 \exp\left[\left(\frac{\mu'}{2} - |m|\right) z\right] \tag{5.469}$$

Moreover, in this case,

$$\sup_{0 \le z < \infty} \left[\exp(-\mu' z) \left(\frac{\partial \phi}{\partial z}\right)^2\right] < \infty$$

and therefore, the boundary condition at the top of the atmosphere (5.386) is verified. Then, substitution of (5.469) into (5.464) singles out the coefficient C_2, whence

$$\phi(z) = -\frac{S}{(\mu'/2) - |m|} \exp\left[\left(\frac{\mu'}{2} - |m|\right) z\right] \tag{5.470}$$

Finally, substitution of (5.470) into (5.462) gives

$$\psi^{(1)}(x, y, z) = -\frac{S}{(\mu'/2) - |m|} \exp\left[\left(\frac{\mu'}{2} - |m|\right) z\right] \cos(kx) \sin(y) \tag{5.471}$$

where

$$|m| = \sqrt{\left(\frac{\mu'}{2}\right)^2 - S\left[\frac{\beta}{u_0} - (k^2 + 1)\right]}$$

Case 2: $m = 0$. In this case, the general integral of (5.463) is

$$\phi(z) = (C_1 + C_2 z) \exp\left(\frac{\mu' z}{2}\right) \tag{5.472}$$

We have

$$\sup_{0 \le z < \infty} \{\exp(-\mu' z) \exp(\mu' z)\} < \infty$$

$$\sup_{0 \le z < \infty} \{\exp(-\mu' z) z \exp(\mu' z)\} = \infty$$

and therefore, the boundary condition at the top of the atmosphere (5.386) demands $C_2 = 0$. Accordingly, we have

$$\sup_{0 \le z < \infty} \left\{\exp(-\mu' z) \left(\frac{\partial \phi}{\partial z}\right)^2\right\} < \infty$$

and hence, relationship (5.386) is verified. Thus, (5.472) reduces to

$$\phi(z) = C_1 \exp\left(\frac{\mu' z}{2}\right) \tag{5.473}$$

Substitution of (5.473) into boundary condition (5.464) gives $C_1 = -2S/\mu'$, whence

$$\phi(z) = -2\frac{S}{\mu'} \exp\left(\frac{\mu' z}{2}\right) \tag{5.474}$$

and by using (5.474), we obtain

$$\psi^{(1)}(x, y, z) = -2\frac{S}{\mu'} \exp\left(\frac{\mu' z}{2}\right) \cos(kx) \sin(y) \tag{5.475}$$

which is nothing but (5.471) with $m = 0$.

Case 3: $m^2 < 0$ The general integral of (5.463) is

$$\phi(z) = C_1 \exp\left[\left(\frac{\mu'}{2} + i|m|\right) z\right] + C_2 \exp\left[\left(\frac{\mu'}{2} - i|m|\right) z\right] \tag{5.476}$$

In this case, if we substitute, separately, $C_1 \exp[(\mu'/2 + i|m|)\, z]$ and $C_2 \exp[(\mu'/2 - i|m|)\, z]$ for $\phi_\#$ into (5.408) and (5.409), we ascertain that

$$\sup_{0 \le z < \infty} \left[\exp(-\mu' z)\, C_1^2 \exp(\mu' z) \cos(2|m|z)\right] < \infty \tag{5.477}$$

$$\sup_{0 \le z < \infty} \left[\exp(-\mu' z)\, C_1^2 \exp(\mu' z) \sin(2|m|z)\right] < \infty \tag{5.478}$$

$$\sup_{0 \le z < \infty} \left[\exp(-\mu' z)\, C_2^2 \exp(\mu' z) \cos(2|m|z)\right] < \infty \tag{5.479}$$

$$\sup_{0 \le z < \infty} \left[-\exp(-\mu' z)\, C_2^2 \exp(\mu' z) \sin(2|m|z)\right] < \infty \tag{5.480}$$

whatever the constants C_1 and C_2 may be.

Therefore, to determine C_1 and C_2 in the present circumstance, one must resort to the supplementary condition (5.387) and evaluate explicitly the r.h.s. of (5.385). We note that $m^2 < 0$ implies

$$\left(\frac{\mu'}{2}\right)^2 < S\left[\frac{\beta}{u_0} - (k^2 + 1)\right]$$

and, in turn,

$$u_0 > 0 \tag{5.481}$$

that is, only westerly winds fit this case. The computation of the averaged vertical energy flux $\langle F_3 \rangle_D$ follows exactly the same lines as (5.426), with C_1 in place of B_+ and C_2 in place of B_-, to give

$$\langle F_3 \rangle_D = \frac{k A u_0 |m|}{4S} \left(|C_1|^2 - |C_2|^2 \right) \tag{5.482}$$

According to (5.481), (5.482) shows that the flux is upward if $C_2 = 0$, and thus, (5.476) simplifies to

$$\phi(z) = C_1 \exp\left[\left(\frac{\mu'}{2} + i|m| \right) z \right] \tag{5.483}$$

The factor C_1 is determined by using boundary condition (5.464), thus yielding

$$\phi(z) = -S \frac{\mu'/2 - i|m|}{(\mu'/2)^2 + |m|^2} \exp\left[\left(\frac{\mu'}{2} + i|m| \right) z \right] \tag{5.484}$$

Unlike (5.470) and (5.474), now ϕ is a complex quantity; so, position (5.462) here means

$$\psi^{(1)} = \mathrm{Re}[\phi(z) \exp(ikx) \sin(y)] = \{\mathrm{Re}[\phi(z)] \cos(kx) - \mathrm{Im}[\phi(z)] \sin(kx)\} \sin(y)$$

and therefore, substitution of (5.484) into (5.462) yields, after little algebra,

$$\psi^{(1)} = -S \frac{\exp(\mu' z/2) \sin(y)}{(\mu'/2)^2 + |m|^2} \left\{ \left[\frac{\mu'}{2} \cos(|m| z) + |m| \sin(|m| z) \right] \cos(kx) \right.$$
$$\left. - \left[\frac{\mu'}{2} \sin(|m| z) - |m| \cos(|m| z) \right] \sin(kx) \right\} \tag{5.485}$$

where

$$|m| = \sqrt{S \left[\frac{\beta}{u_0} - (k^2 + 1) \right] - \left(\frac{\mu'}{2} \right)^2}$$

In a more concise way, solution (5.485) can be written as

$$\psi^{(1)} = -S \frac{\exp(\mu' z/2) \sin(y)}{\sqrt{(\mu'/2)^2 + |m|^2}} \sin(\theta + |m| z + kx) \tag{5.486}$$

where $\tan\theta = \mu'/(2|m|)$.

Trapped and Non-trapped Waves

We have found that, to the first order in A, the model solutions have the form

$$\psi = -\mathsf{u}_0\,\mathsf{y} + \mathsf{A}\,\psi^{(1)}(\mathsf{x},\mathsf{y},\mathsf{z}) \tag{5.487}$$

where $\psi^{(1)}(\mathsf{x},\mathsf{y},\mathsf{z})$ is given by (5.471) if $\mathsf{m}^2 > 0$, by (5.475) if $\mathsf{m} = 0$ and by (5.485) or (5.486) if $\mathsf{m}^2 < 0$. By using (5.487) to evaluate the total mechanical energy density (5.380) as a function of height only, we easily find that

$$\lim_{\mathsf{z}\to+\infty} \mathscr{E}(\mathsf{z}) \begin{cases} = 0 & \text{if } \mathsf{m}^2 > 0 \\ \neq 0 & \text{if } \mathsf{m}^2 > 0 \end{cases}$$

Thus, if $\psi^{(1)}(\mathsf{x},\mathsf{y},\mathsf{z})$ is given by (5.471), the total stream function represents a trapped wave, while if $\psi^{(1)}(\mathsf{x},\mathsf{y},\mathsf{z})$ is given by (5.475) or by (5.485), the total stream function describes a non-trapped wave. In any case, we obtain

$$\lim_{\mathsf{z}\to+\infty} \mathscr{E}(\mathsf{z}) < \infty$$

Numerical Values of Model Parameters

Each solution of the kind (5.487) depends on the parameters β, S, μ', k, A, u_0, and m. In the rest of this section, we take

$$\beta = \mathsf{S} = \mu' = \mathsf{k} = 1 \qquad \mathsf{A} = 0.3 \tag{5.488}$$

while u_0 and m are linked by relationship (5.465), which can be restated as

$$\mathsf{m}^2 = \beta\,\mathsf{S}\left(\frac{1}{\mathsf{u}_0^{\mathsf{c}}} - \frac{1}{\mathsf{u}_0}\right) \tag{5.489}$$

where

$$\mathsf{u}_0^{\mathsf{c}} := \frac{\beta\,\mathsf{S}}{(\mu'/2)^2 + \mathsf{S}(1+\mathsf{k}^2)} \tag{5.490}$$

is the *critical velocity* and, according to (5.488), $\mathsf{u}_0^{\mathsf{c}} = 4/9$. From (5.489), we infer the following alternatives:

$$\mathsf{m}^2 > 0 \Longleftrightarrow \mathsf{u}_0 < 0 \text{ or } \mathsf{u}_0 > \mathsf{u}_0^{\mathsf{c}} \tag{5.491}$$

$$\mathsf{m}^2 = 0 \Longleftrightarrow \mathsf{u}_0 = \mathsf{u}_0^{\mathsf{c}} \tag{5.492}$$

$$\mathsf{m}^2 < 0 \Longleftrightarrow 0 < \mathsf{u}_0 < \mathsf{u}_0^{\mathsf{c}} \tag{5.493}$$

To describe some features of the above derived model solutions, besides (5.488) we take

$$u_0 = -1 \quad \Longrightarrow \quad m^2 = 13/4 \tag{5.494}$$

$$u_0 = 1 \quad \Longrightarrow \quad m^2 = 5/4 \tag{5.495}$$

$$u_0 = 4/9 \quad \Longrightarrow \quad m^2 = 0 \tag{5.496}$$

$$u_0 = 2/9 \quad \Longrightarrow \quad m^2 = -9/4 \tag{5.497}$$

consistently with (5.491)–(5.493).

Density Anomaly

At the geostrophic level of approximation, density anomaly ρ_0' is evaluated from (5.329) with standard density given by (5.350). Thus, we have

$$\rho_0' = \mu' \psi - \frac{\partial \psi}{\partial z} \tag{5.498}$$

where ψ is approximated, to the first order in A, by (5.487). To elucidate the structure of ρ_0', the coordinate y is fixed at the mid-latitude $y = \pi/2$, and for a given set of heights (say, $z = z_n$), the function

$$x \mapsto \rho_0'(x, y = \pi/2, z_n) \tag{5.499}$$

is depicted in Fig. 5.6. We have considered two solutions, that is, (5.471) with (5.494) and (5.485) with (5.497).

Note that, although the density anomaly ρ_0' diverges asymptotically as $\exp(\mu' z/2)$ because of (5.486), total density ρ converges exponentially to zero for $z \to \infty$ since (2.554) implies $\rho \propto \exp(-\mu' z) + r\exp(-\mu' z/2)$.

Transport Stream Function

The non-dimensional transport stream function of the wind is defined as

$$\mathscr{T}'(x, y) := \int_0^{+\infty} \exp(-\mu' z)\, \psi(x, y, z)\, dz \tag{5.500}$$

where ψ is given within approximation (5.487). Hence, we have

$$\mathscr{T}'(x, y) = -\frac{u_0 y}{\mu'} + A \int_0^{+\infty} \exp(-\mu' z)\, \psi^{(1)}(x, y, z)\, dz \tag{5.501}$$

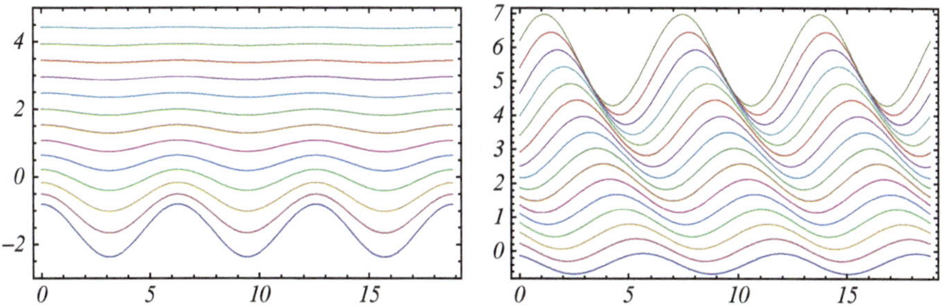

Fig. 5.6 Graphs of density anomaly (5.499) versus longitude at different heights. To enhance readability, graphs have been translated upward proportionally to the corresponding heights. *Left*: trapped waves obtained from (5.499) using (5.471). *Right*: non-trapped waves obtained from (5.499) using (5.486)

Consider now, separately, the first-order terms (5.471), (5.475) and (5.485) to compute the transport stream function.

Substitution of (5.471) into (5.501) yields

$$\mathscr{T}'(x, y) = -\frac{u_0 y}{\mu'} - \frac{AS}{(\mu'/2)^2 - |m|^2}\cos(kx)\sin(y) \tag{5.502}$$

where

$$|m| = \sqrt{\left(\frac{\mu'}{2}\right)^2 - S\left[\frac{\beta}{u_0} - (k^2+1)\right]}$$

Substitution of (5.475) into (5.501) gives

$$\mathscr{T}'(x, y) = -\frac{u_0 y}{\mu'} - \frac{4AS}{\mu'^2}\cos(kx)\sin(y) \tag{5.503}$$

that is, (5.502) with $m = 0$.

Substitution of (5.486) into (5.501) produces the transport stream function

$$\mathscr{T}'(x, y) = -\frac{u_0 y}{\mu'} - \frac{AS}{(\mu'/2)^2 + |m|^2}\cos(kx)\sin(y) \tag{5.504}$$

where

$$|m| = \sqrt{S\left[\frac{\beta}{u_0} - (k^2+1)\right] - \left(\frac{\mu'}{2}\right)^2}$$

At this point, a set of transport streamlines can be inferred from (5.502), (5.503) and (5.504) by using the values reported in (5.488) and (5.494)–(5.497). Thus, (5.502) becomes

$$\mathscr{T}'(x, y) = y + 0.1\cos(kx)\sin(y) \tag{5.505}$$

by using (5.494) and

$$\mathscr{T}'(\mathsf{x},\mathsf{y}) = -\mathsf{y} + 0.3\cos(\mathsf{k}\mathsf{x})\sin(\mathsf{y}) \tag{5.506}$$

by using (5.495). In the same way, (5.503) becomes

$$\mathscr{T}'(\mathsf{x},\mathsf{y}) = -\frac{4}{9}\mathsf{y} - 1.2\cos(\mathsf{k}\mathsf{x})\sin(\mathsf{y}) \tag{5.507}$$

and, finally, (5.504) takes the form

$$\mathscr{T}'(\mathsf{x},\mathsf{y}) = -\frac{2}{9}\mathsf{y} - 0.12\cos(\mathsf{k}\mathsf{x})\sin(\mathsf{y}) \tag{5.508}$$

The transport streamlines obtained from (5.505), (5.506), (5.507) and (5.508) are shown in Fig. 5.7 superimposed to contour lines of topography (5.460).

Ertel's Theorem in the Framework of a Quasi-Geostrophic Adiabatic Atmosphere

The conservation statement established by (5.345), with $\dot{\mathsf{Q}} = 0$, is obtained under the assumption that potential temperature θ is conserved, namely,

$$\frac{\mathrm{D}\theta}{\mathrm{D}t} = 0 \tag{5.509}$$

Thus, the hypotheses of Ertel's theorem are satisfied, and (5.345) with $\dot{\mathsf{Q}} = 0$ is expected to be a special version of this theorem. We shall now see that this is indeed the case.

Starting from (5.509) and assuming that $\partial\theta/\partial z$ is much greater than $\partial\theta/\partial x$ and $\partial\theta/\partial y$, we consider Ertel's theorem in the form (2.507) with $q = \theta$ and $\rho \approx \rho_s$:

$$\frac{\mathrm{D}}{\mathrm{D}t}\left[\frac{\zeta + f}{\rho_s}\frac{\partial\theta}{\partial z}\right] = 0 \tag{5.510}$$

which relies upon the approximation $\nabla\theta \approx (\partial\theta/\partial z)\hat{\mathbf{k}}$. To establish the non-dimensional version of (5.510), we use the expansion

$$\frac{\mathrm{D}}{\mathrm{D}t} = \frac{U}{L}\left[\frac{\mathrm{D}_0}{\mathrm{Dt}} + \varepsilon\mathbf{u}_1\cdot\nabla' + O(\varepsilon^2)\right] \tag{5.511}$$

and set

$$\zeta + f(y) = f_0\left[1 + \varepsilon(\nabla'^2\mathsf{p}_0 + \beta\,\mathsf{y})\right] \tag{5.512}$$

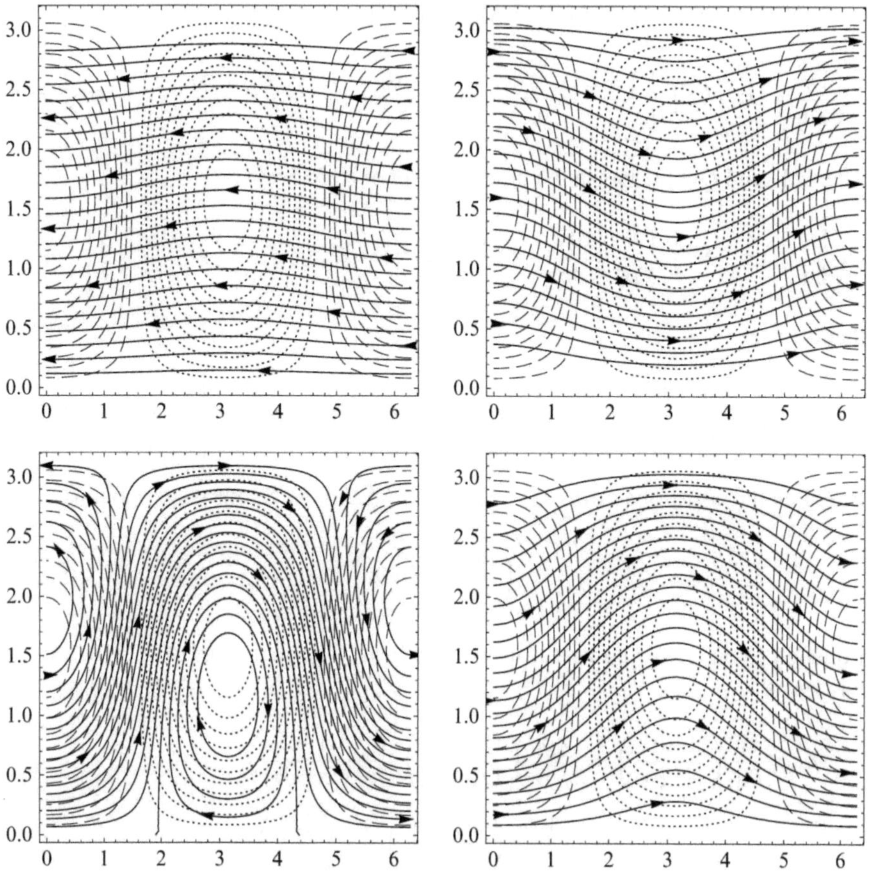

Fig. 5.7 Transport streamlines and topography. *Dashed* (*dotted*) *lines* correspond to positive (negative) topographic anomalies. *Top left*: Transport, related to a trapped wave, represented by (5.505). *Top right*: Transport, related to a trapped wave, represented by (5.506). *Bottom left*: Transport (5.507), computed using the critical velocity (5.490). *Bottom right*: Transport, related to a non-trapped wave, represented by (5.508)

while $\partial\theta/\partial z$ can be manipulated as follows. From $\theta = \theta_s\,(1+\varepsilon\,\mathsf{F}\,\theta_0')$ and (5.337), we have

$$\theta = \theta_s\left(1+\varepsilon\,\mathsf{F}\,\frac{\partial \mathsf{p}_0}{\partial \mathsf{z}}\right) \tag{5.513}$$

whence

$$\frac{\partial\theta}{\partial z} \approx \frac{\mathrm{d}\theta_s}{\mathrm{dz}} + \varepsilon\,\mathsf{F}\,\frac{\partial}{\partial z}\left(\theta_s\,\frac{\partial \mathsf{p}_0}{\partial \mathsf{z}}\right) = \frac{1}{H}\,\frac{\mathrm{d}\theta_s}{\mathrm{dz}} + \frac{\varepsilon\,\mathsf{F}}{H}\,\frac{\partial}{\partial \mathsf{z}}\left(\theta_s\,\frac{\partial \mathsf{p}_0}{\partial \mathsf{z}}\right) \tag{5.514}$$

On the whole, using (5.511), (5.512) and (5.514), (5.510) yields

$$\left(\frac{\mathrm{D}_0}{\mathrm{Dt}} + \varepsilon \mathbf{u}_1 \cdot \nabla'\right) \left\{ \frac{1+\varepsilon(\nabla'^2 \mathrm{p}_0 + \beta\,\mathrm{y})}{\rho_s} \left[\frac{\mathrm{d}\theta_s}{\mathrm{dz}} + \varepsilon\,\mathsf{F} \frac{\partial}{\partial \mathrm{z}}\left(\theta_s \frac{\partial \mathrm{p}_0}{\partial \mathrm{z}}\right)\right]\right\} = 0 \quad (5.515)$$

The $O(1)$ and $O(\varepsilon)$ terms in the brace bracket of (5.515) give

$$\left(\frac{\mathrm{D}_0}{\mathrm{Dt}} + \varepsilon \mathbf{u}_1 \cdot \nabla'\right) \left\{ \frac{1}{\rho_s} \left[\frac{\mathrm{d}\theta_s}{\mathrm{dz}} + \varepsilon \left(\frac{\mathrm{d}\theta_s}{\mathrm{dz}} (\nabla'^2 \mathrm{p}_0 + \beta\,\mathrm{y}) + \mathsf{F} \frac{\partial}{\partial \mathrm{z}}\left(\theta_s \frac{\partial \mathrm{p}_0}{\partial \mathrm{z}}\right)\right)\right]\right\} = 0$$

that is to say,

$$\mathrm{w}_1\,\rho_s \frac{\partial}{\partial \mathrm{z}}\left(\frac{1}{\rho_s}\frac{\mathrm{d}\theta_s}{\mathrm{dz}}\right) + \frac{\mathrm{d}\theta_s}{\mathrm{dz}}\frac{\mathrm{D}_0}{\mathrm{Dt}}(\nabla'^2 \mathrm{p}_0 + \beta\,\mathrm{y}) + \mathsf{F}\frac{\mathrm{D}_0}{\mathrm{Dt}}\frac{\partial}{\partial \mathrm{z}}\left(\theta_s \frac{\partial \mathrm{p}_0}{\partial \mathrm{z}}\right) = 0 \quad (5.516)$$

Based on the equations $g\,\mathrm{d}\theta_s/\mathrm{dz} = N_s^2\,\theta_s$ and $N_s^2 H = g\,\mathsf{F}\,\mathsf{S}$, one easily obtains

$$\left(\frac{\mathrm{d}\theta_s}{\mathrm{dz}} = \mathsf{S}\,\mathsf{F}\,\theta_s\right) \quad (5.517)$$

Equation (5.517) will be useful in what follows. The first term of (5.516) can be written as

$$\begin{aligned}
\mathrm{w}_1\,\rho_s \frac{\partial}{\partial \mathrm{z}}\left(\frac{1}{\rho_s}\frac{\mathrm{d}\theta_s}{\mathrm{dz}}\right) &= -\frac{\rho_s}{\mathsf{S}}\left(\frac{\mathrm{D}_0}{\mathrm{Dt}}\frac{\partial \mathrm{p}_0}{\partial \mathrm{z}}\right)\frac{\partial}{\partial \mathrm{z}}\left(\frac{1}{\rho_s}\frac{\mathrm{d}\theta_s}{\mathrm{dz}}\right) \\
&= -\frac{\mathsf{F}\rho_s}{\mathsf{S}}\left(\frac{\mathrm{D}_0}{\mathrm{Dt}}\frac{\partial \mathrm{p}_0}{\partial \mathrm{z}}\right)\frac{\partial}{\partial \mathrm{z}}\left(\frac{\mathsf{S}\,\theta_s}{\rho_s}\right) \\
&= -\frac{\mathsf{F}\rho_s}{\mathsf{S}}\left(\frac{\mathrm{D}_0}{\mathrm{Dt}}\frac{\partial \mathrm{p}_0}{\partial \mathrm{z}}\right)\left[\theta_s \frac{\partial}{\partial \mathrm{z}}\left(\frac{\mathsf{S}}{\rho_s}\right) + \frac{\mathsf{S}}{\rho_s}\frac{\mathrm{d}\theta_s}{\mathrm{dz}}\right] \quad (5.518)
\end{aligned}$$

The second term of (5.516) becomes, using again (5.517),

$$\frac{\mathrm{d}\theta_s}{\mathrm{dz}}\frac{\mathrm{D}_0}{\mathrm{Dt}}(\nabla'^2 \mathrm{p}_0 + \beta\,\mathrm{y}) = \mathsf{F}\,\mathsf{S}\,\theta_s \frac{\mathrm{D}_0}{\mathrm{Dt}}(\nabla'^2 \mathrm{p}_0 + \beta\,\mathrm{y}) \quad (5.519)$$

The third term of (5.516) yields

$$\mathsf{F}\frac{\mathrm{D}_0}{\mathrm{Dt}}\frac{\partial}{\partial \mathrm{z}}\left(\theta_s \frac{\partial \mathrm{p}_0}{\partial \mathrm{z}}\right) = \mathsf{F}\,\theta_s \frac{\mathrm{D}_0}{\mathrm{Dt}}\frac{\partial^2 \mathrm{p}_0}{\partial \mathrm{z}^2} + \mathsf{F}\frac{\mathrm{d}\theta_s}{\mathrm{dz}}\frac{\mathrm{D}_0}{\mathrm{Dt}}\frac{\partial \mathrm{p}_0}{\partial \mathrm{z}} \quad (5.520)$$

Adding together the r.h.s.'s of (5.518), (5.519) and (5.520), each being divided by $\mathsf{F}\,\mathsf{S}\,\theta_s$, one obtains

$$\frac{\mathrm{D}_0}{\mathrm{Dt}}(\nabla'^2 \mathrm{p}_0 + \beta\,\mathrm{y}) - \frac{\rho_s}{\mathsf{S}^2}\left(\frac{\mathrm{D}_0}{\mathrm{Dt}}\frac{\partial \mathrm{p}_0}{\partial \mathrm{z}}\right)\left(\frac{\partial}{\partial \mathrm{z}}\frac{\mathsf{S}}{\rho_s}\right) + \frac{1}{\mathsf{S}}\frac{\mathrm{D}_0}{\mathrm{Dt}}\frac{\partial^2 \mathrm{p}_0}{\partial \mathrm{z}^2} = 0$$

that is to say,

$$\frac{\mathsf{D}_0}{\mathsf{Dt}}\left[\nabla'^2 \mathsf{p}_0 + \beta\,\mathsf{y} - \frac{\rho_\mathrm{s}}{\mathsf{S}^2}\left(\frac{\partial}{\partial \mathsf{z}}\frac{\mathsf{S}}{\rho_\mathrm{s}}\right)\frac{\partial \mathsf{p}_0}{\partial \mathsf{z}} + \frac{1}{\mathsf{S}}\frac{\partial^2 \mathsf{p}_0}{\partial \mathsf{z}^2}\right] = 0 \tag{5.521}$$

Then, identity

$$-\frac{\rho_\mathrm{s}}{\mathsf{S}^2}\left(\frac{\partial}{\partial \mathsf{z}}\frac{\mathsf{S}}{\rho_\mathrm{s}}\right)\frac{\partial \mathsf{p}_0}{\partial \mathsf{z}} + \frac{1}{\mathsf{S}}\frac{\partial^2 \mathsf{p}_0}{\partial \mathsf{z}^2} = \frac{1}{\rho_\mathrm{s}}\frac{\partial}{\partial \mathsf{z}}\left(\frac{\rho_\mathrm{s}}{\mathsf{S}}\frac{\partial \mathsf{p}_0}{\partial \mathsf{z}}\right)$$

applied to (5.521) leads to its final form

$$\frac{\mathsf{D}_0}{\mathsf{Dt}}\left[\nabla'^2 \mathsf{p}_0 + \beta\,\mathsf{y} + \frac{1}{\rho_\mathrm{s}}\frac{\partial}{\partial \mathsf{z}}\left(\frac{\rho_\mathrm{s}}{\mathsf{S}}\frac{\partial \mathsf{p}_0}{\partial \mathsf{z}}\right)\right] = 0 \tag{5.522}$$

which is nothing but (5.345) with $\dot{\mathsf{Q}} = 0$. Just like in the ocean case, the vorticity equation (5.347), with $\dot{\mathsf{Q}} = 0$, establishes the conservation of potential vorticity $\Pi = (\boldsymbol{\omega}_\mathrm{a} \cdot \nabla\theta)/\rho$ within the geostrophic and hydrostatic approximations.

Appendix: Proof of Identities (5.369), (5.370) and (5.371)

According to Green's identities, the l.h.s. of (5.369) is equivalent to

$$\oint_{\partial \mathsf{D}} \psi \frac{\partial^2 \psi}{\partial \mathsf{x}\partial \mathsf{t}}\,\mathsf{dy} - \oint_{\partial \mathsf{D}} \psi \frac{\partial^2 \psi}{\partial \mathsf{y}\partial \mathsf{t}}\,\mathsf{dx} \tag{5.523}$$

Consider now the first term of (5.523):

$$\oint_{\partial \mathsf{D}} \psi \frac{\partial^2 \psi}{\partial \mathsf{x}\partial \mathsf{t}}\,\mathsf{dy} = \int_\pi^0 \psi(0,\mathsf{y},\mathsf{z})\left[\frac{\partial^2 \psi}{\partial \mathsf{x}\partial \mathsf{t}}\right]_{\mathsf{x}=0}\mathsf{dy} + \int_0^\pi \psi(\Lambda',\mathsf{y},\mathsf{z})\left[\frac{\partial^2 \psi}{\partial \mathsf{x}\partial \mathsf{t}}\right]_{\mathsf{x}=\Lambda'}\mathsf{dy}$$

$$= \int_0^\pi \left\{\psi(\Lambda',\mathsf{y},\mathsf{z})\left[\frac{\partial^2 \psi}{\partial \mathsf{x}\partial \mathsf{t}}\right]_{\mathsf{x}=\Lambda'} - \psi(0,\mathsf{y},\mathsf{z})\left[\frac{\partial^2 \psi}{\partial \mathsf{x}\partial \mathsf{t}}\right]_{\mathsf{x}=0}\right\}\mathsf{dy}$$

where, because of (5.367), both ψ and $\partial^2\psi/\partial \mathsf{x}\partial \mathsf{t}$ are Λ'-periodic with respect to x; so, the quantity in brace brackets is identically zero, whence

$$\oint_{\partial \mathsf{D}} \psi \frac{\partial^2 \psi}{\partial \mathsf{x}\partial \mathsf{t}}\,\mathsf{dy} = 0 \tag{5.524}$$

Consider then the second term of (5.523):

$$\oint_{\partial \mathsf{D}} \psi \frac{\partial^2 \psi}{\partial \mathsf{y}\partial \mathsf{t}}\,\mathsf{dx} = \int_0^{\Lambda'} \psi(\mathsf{x},0,\mathsf{z})\left[\frac{\partial^2 \psi}{\partial \mathsf{y}\partial \mathsf{t}}\right]_{\mathsf{y}=0}\mathsf{dx} + \int_{\Lambda'}^0 \psi(\mathsf{x},\pi,\mathsf{z})\left[\frac{\partial^2 \psi}{\partial \mathsf{y}\partial \mathsf{t}}\right]_{\mathsf{y}=\pi}\mathsf{dx} \tag{5.525}$$

Because of (5.368), the integrands of each integral of (5.525) are zero, and hence,

$$\oint_{\partial D} \psi \frac{\partial^2 \psi}{\partial y \partial t} dx = 0 \tag{5.526}$$

Equations 5.524 and 5.526 prove (5.369).

To prove identity (5.370), we evaluate, by resorting again to Green's identities, the expression

$$\begin{aligned}\int_D \mathscr{J}(\psi^2, \Pi_0')\, dx dy &= \int_D \left[\frac{\partial}{\partial x}\left(\psi^2 \frac{\partial \Pi_0'}{\partial y}\right) - \frac{\partial}{\partial y}\left(\psi^2 \frac{\partial \Pi_0'}{\partial x}\right)\right] dx dy \\ &= \int_{\partial D} \psi^2 \frac{\partial \Pi_0'}{\partial y} dy + \int_{\partial D} \psi^2 \frac{\partial \Pi_0'}{\partial x} dx\end{aligned}$$

In particular, the first term of the r.h.s. is

$$\begin{aligned}\oint_{\partial D} \psi^2 \frac{\partial \Pi_0'}{\partial y} dy &= \int_\pi^0 \psi^2(0, y, z) \left[\frac{\partial \Pi_0'}{\partial y}\right]_{x=0} dy + \int_0^\pi \psi^2(\Lambda', y, z) \left[\frac{\partial \Pi_0'}{\partial y}\right]_{x=\Lambda'} dy = \\ &= \int_0^\pi \left\{\psi^2(\Lambda', y, z) \left[\frac{\partial \Pi_0'}{\partial y}\right]_{x=\Lambda'} - \psi^2(0, y, z) \left[\frac{\partial \Pi_0'}{\partial y}\right]_{x=0}\right\} dy\end{aligned}$$

but, because of (5.367), both ψ and $\partial \Pi_0'/\partial y$ are Λ'-periodic with respect to x; so, the quantity in brace brackets is identically zero, whence

$$\oint_{\partial D} \psi^2 \frac{\partial \Pi_0'}{\partial y} dy = 0 \tag{5.527}$$

Moreover,

$$\oint_{\partial D} \psi^2 \frac{\partial \Pi_0'}{\partial x} dx = \int_0^{\Lambda'} \psi^2(x, 0, z) \left[\frac{\partial \Pi_0'}{\partial x}\right]_{y=0} dx + \int_{\Lambda'}^0 \psi^2(x, \pi, z) \left[\frac{\partial \Pi_0'}{\partial x}\right]_{y=\pi} dx \tag{5.528}$$

Because of (5.368), the integrands of each integral of (5.528) are zero, and hence,

$$\oint_{\partial D} \psi^2 \frac{\partial \Pi_0'}{\partial x} dx = 0 \tag{5.529}$$

Finally, (5.527) and (5.529) imply identity (5.370).

The proof of (5.371) follows the same lines as that of identity (5.370).

Dimensional Version of the Quasi-Geostrophic, Potential-Vorticity Equation for a Thermally Forced Atmosphere

We report below the dimensional version of (5.347), the unknown being the field of the perturbation pressure $\tilde{p}(\mathbf{x}, t)$, just like in the ocean case, while the Lagrangian derivative at the geostrophic level of approximation, D_0/Dt, is the same at that defined in (5.312), so that one obtains

$$\frac{\mathrm{D}_0}{\mathrm{D}t}\left[\nabla_{\mathrm{H}}^2 \frac{\tilde{p}}{\rho_{\mathrm{s}} f_0} + \frac{1}{\rho_{\mathrm{s}}}\frac{\partial}{\partial z}\left(\rho_{\mathrm{s}} \frac{f_0}{N_{\mathrm{s}}^2}\frac{\partial}{\partial z}\frac{\tilde{p}}{\rho_{\mathrm{s}}}\right) + \beta_0 y\right] = \frac{1}{\rho_{\mathrm{s}}}\frac{\partial}{\partial z}\left(\frac{\rho_{\mathrm{s}} f_0 g}{N_{\mathrm{s}}^2}\frac{\dot{Q}}{c_p T_{\mathrm{a}}}\right) \quad (5.530)$$

Note that the terms within the square bracket have the physical dimensions of a frequency, while the r.h.s. has the physical dimensions of the square of a frequency. If potential temperature is conserved, then atmosphere is adiabatic (recall (2.41)) and, hence, the r.h.s. of (5.530) is zero.

Exercises

1. Integrate (5.128) with (5.123) starting from the putative solution

$$\Psi = \frac{1}{2}\,\mathrm{Re}\left\{\exp\left[\mathrm{i}\left(\mathsf{k}\mathsf{x}+\mathsf{k}\mathsf{y}+\mathsf{l}\mathsf{z}-\sigma'\tilde{t}\right)\right]+\exp\left[\mathrm{i}\left(\mathsf{k}\mathsf{x}+\mathsf{k}\mathsf{y}-\mathsf{l}\mathsf{z}-\sigma'\tilde{t}\right)\right]\right\}$$

   Verify that the actual solution is the same as (5.124) with (5.133). Why?

2. Reformulate the lake model under the hypothesis of a circular basin and a stream function of the kind $\psi = \phi(z)\,R(r)$ in place of (5.86), where $r \in [0,\,1]$ is the radial coordinate.

3. Verify that (5.188), together with boundary conditions (5.148) and (5.149), implies the conservation of available potential energy, that is to say,

$$\frac{\mathrm{d}}{\mathrm{dt}}\int_{\mathsf{V}}\left(\frac{1}{\beta\mathsf{S}}\frac{\partial\psi}{\partial\mathsf{z}}\right)\mathrm{d}\mathsf{V} = 0$$

4. Explain the physical grounds of the closed loops of the transport reported in the upper panel of Fig. 5.4.

5. Consider the wind-stress curl

$$\hat{\mathbf{k}}\cdot\mathrm{rot}\,\boldsymbol{\tau} = \frac{\tau_0}{L^2}\,(2x-L)\,\sin\left(\pi\frac{y}{L}\right)$$

   over the fluid domain $D := [0 \le x \le L] \times [0 \le y \le L]$. Evaluate the Sverdrup transport $\mathscr{T}(x,y) = \int_L^x \hat{\mathbf{k}}\cdot\mathrm{rot}\,\boldsymbol{\tau}(\xi,y)\,\mathrm{d}\xi$, and verify that in the present case

$\mathscr{T}(x, y) = 0 \ \forall (x, y) \in D$. Moreover, draw the transport streamlines $\mathscr{T}(x, y) =$ constant, and verify that they are closed. What is the physical reason why this "bizarre" model does not need of any boundary layer?

Bibliographical Note

The quasi-geostrophic dynamics of continuously stratified flows is a central argument in many books of Geophysical Fluid Dynamics, physical oceanography and Meteorology, such as Cushman-Roisin (1994), Gill (1982), Holton (1979), Kamenkovich et al. (1986), Pedlosky (1987), Salby (1996) and Vallis (2006). Here, thermally and topographically forced stationary waves in the atmosphere are expounded following the same lines as in Charney (1973) and Pedlosky (1987) but disregarding for simplicity the production of resonant flows. The effects of thermal forcing are also considered by Vallis (2006), in the quasi-geostrophic framework.

Appendix A
Mathematical Formulas Useful in GFD

For the reader's convenience, we collect in this Appendix some definitions, theorems, identities and inequalities that are useful in GFD but are not easily found all together in a single standard reference book.

Basic Notation

Let $\hat{\mathbf{i}}$, $\hat{\mathbf{j}}$, $\hat{\mathbf{k}}$ be the unit vectors associated to the axes of a right-handed orthogonal Cartesian reference frame. They are related through the constraints

$$\hat{\mathbf{i}} \times \hat{\mathbf{j}} = \hat{\mathbf{k}} \qquad \hat{\mathbf{j}} \times \hat{\mathbf{k}} = \hat{\mathbf{i}} \qquad \hat{\mathbf{k}} \times \hat{\mathbf{i}} = \hat{\mathbf{j}} \tag{A.1}$$

Conditions (A.1) may be written more compactly as

$$\hat{\mathbf{e}}_i \times \hat{\mathbf{e}}_j = \varepsilon_{ijk} \hat{\mathbf{e}}_k$$

where $\hat{\mathbf{e}}_1$, $\hat{\mathbf{e}}_2$, $\hat{\mathbf{e}}_3$ coincide with $\hat{\mathbf{i}}$, $\hat{\mathbf{j}}$, $\hat{\mathbf{k}}$, respectively, and ε_{ijk} is Levi-Civita symbol (Einstein's summation rule is understood).

Let x, y, z be the corresponding coordinates of a point. In GFD, the Cartesian plane (x, y) often represents the f-plane or the β-plane (see p. 88).

Classical Theorems Relative to the Cartesian Plane

We present here the 2-D version of some classical theorems of vector calculus, usually stated for 3-D space. The following formulation is more directly applicable to GFD.

F. Cavallini and F. Crisciani, *Quasi-Geostrophic Theory of Oceans and Atmosphere*, Atmospheric and Oceanographic Sciences Library 45, DOI 10.1007/978-94-007-4691-6,

Notation

Let:

D a bounded and simply connected domain of the plane (x, y)

∂D the boundary of D

$\hat{\mathbf{k}}$ the unit vector normal to D chosen to denote the "outward" direction

$\hat{\mathbf{n}}$ the unit vector in the plane (x, y) locally normal to ∂D and pointing outwards

$\hat{\mathbf{t}}$ the unit vector in the plane (x, y) locally tangent to ∂D and pointing counterclockwise

s counterclockwise curvilinear abscissa, whose differential ds is given by $\mathrm{d}s = \sqrt{x'(\tau)^2 + y'(\tau)^2}\,\mathrm{d}\tau$ with $\{x(\tau), y(\tau)\}$ a parametric representation of the curve ∂D

The three vectors $\hat{\mathbf{k}}, \hat{\mathbf{n}}, \hat{\mathbf{t}}$ constitute a right-handed triple, which means

$$\hat{\mathbf{t}} = \hat{\mathbf{k}} \times \hat{\mathbf{n}} \qquad \hat{\mathbf{k}} = \hat{\mathbf{n}} \times \hat{\mathbf{t}} \qquad \hat{\mathbf{n}} = \hat{\mathbf{t}} \times \hat{\mathbf{k}} \tag{A.2}$$

Moreover, we recall the following identities

$$\mathbf{a} \cdot (\mathbf{b} \times \mathbf{c}) = \mathbf{b} \cdot (\mathbf{c} \times \mathbf{a}) = \mathbf{c} \cdot (\mathbf{a} \times \mathbf{b}) \tag{A.3}$$

which hold for any 3-D vectors **a**, **b**, **c**.

Green's Theorem

For any differentiable scalar function $\psi = \psi(x, y)$, the following identities hold:

$$\begin{aligned} \int_D \frac{\partial \psi}{\partial x}\,\mathrm{d}x\,\mathrm{d}y &= \oint_{\partial D} \psi\,\mathrm{d}y \\ \int_D \frac{\partial \psi}{\partial y}\,\mathrm{d}x\,\mathrm{d}y &= -\oint_{\partial D} \psi\,\mathrm{d}x \end{aligned} \tag{A.4}$$

where the line integrals are to be computed counterclockwise.

2-D Gauss' Theorem

According to (Gauss) *divergence theorem*, a differentiable vector field $\mathbf{a} = \mathbf{a}(x, y)$ defined on D satisfies the identity

$$\int_D \nabla \cdot \mathbf{a} \mathrm{d}x \mathrm{d}y = \oint_{\partial D} \hat{\mathbf{n}} \cdot \mathbf{a} \mathrm{d}s \qquad \left(\nabla := \hat{\mathbf{i}} \frac{\partial}{\partial x} + \hat{\mathbf{j}} \frac{\partial}{\partial y}\right) \tag{A.5}$$

Stokes' Theorem

A 3-D differentiable vector field $\mathbf{a} = \mathbf{a}(x, y)$ defined on D satisfies the identity

$$\int_D \hat{\mathbf{k}} \cdot \nabla \times \mathbf{a} \mathrm{d}x \mathrm{d}y = \oint_{\partial D} \mathbf{a} \cdot \hat{\mathbf{t}} \mathrm{d}s \tag{A.6}$$

where ∇ denotes the 3-D gradient and

$$\hat{\mathbf{t}} \mathrm{d}s = \hat{\mathbf{i}} \mathrm{d}x + \hat{\mathbf{j}} \mathrm{d}y \tag{A.7}$$

whence $\mathrm{d}s^2 = \mathrm{d}x^2 + \mathrm{d}y^2$.

Stokes' theorem (A.6) implies that, if C is a closed curve in the plane (x, y) and $\phi(x, y)$ is a differentiable function, then

$$\oint_C \nabla \phi \cdot \hat{\mathbf{t}} \mathrm{d}s = 0$$

Indeed, denoting by σ the plane region internal to C, identity (A.6) yields

$$\oint_C \nabla \phi \cdot \hat{\mathbf{t}} \mathrm{d}s = \int_\sigma \hat{\mathbf{k}} \cdot \nabla \times \nabla \phi \, \mathrm{d}x \mathrm{d}y$$

which is zero because $\nabla \times \nabla = 0$.

Application of 3-D Divergence Theorem to Surface Forces

Let $V = V(t)$ denote a time-dependent material volume whose boundary $S = S(t)$ is a closed surface. The contact force $\mathbf{f}$ acting on V through S is given by

$$\mathbf{f} = \int_S \mathbf{T} \hat{\mathbf{n}} \mathrm{d}S \tag{A.8}$$

where $\mathbf{T}$ is the *stress tensor* and $\hat{\mathbf{n}}$ is the exterior unit vector normal to S.

For example, the stress may be due to a pressure field p through the relationship

$$\mathbf{T} = -p\mathbf{I} \tag{A.9}$$

where $\mathbf{I}$ is the identity 3×3 matrix. In component notation, Eq. (A.9) reads

$$T_{ij} = -p\,\delta_{ij}$$

with δ_{ij} the Kronecker delta.

We now want to compute force (A.8) under assumption (A.9) by using the 3-D *divergence theorem*, which states that

$$\int_V \nabla \cdot \mathbf{a}\,\mathrm{d}V = \int_S \hat{\mathbf{n}} \cdot \mathbf{a}\,\mathrm{d}S \tag{A.10}$$

for any smooth vector field $\mathbf{a}$. First, recall that the identity matrix $\mathbf{I}$ may be written as

$$\mathbf{I} = \sum_{i=1}^{3} \hat{\mathbf{e}}_i \otimes \hat{\mathbf{e}}_i \tag{A.11}$$

where symbol $\otimes$ denotes the *tensor product* of two vectors, which is defined through the identity $(\mathbf{a} \otimes \mathbf{b})\mathbf{x} = (\mathbf{b} \cdot \mathbf{x})\,\mathbf{a}$, $\forall \mathbf{x}$. Second, substitute (A.11) into (A.9) and then the result in (A.8) to get

$$\mathbf{f} = -\int_S p \left(\sum_{i=1}^{3} \hat{\mathbf{e}}_i \otimes \hat{\mathbf{e}}_i \right) \hat{\mathbf{n}}\,\mathrm{d}S = -\hat{\mathbf{e}}_i \sum_{i=1}^{3} \int_S \hat{\mathbf{n}} \cdot (p\hat{\mathbf{e}}_i)\,\mathrm{d}S$$

whence, using (A.10) and the gradient operator $\nabla = \sum_{k=1}^{3} \hat{\mathbf{e}}_k\,\partial/\partial x_k$,

$$\mathbf{f} = -\hat{\mathbf{e}}_i \sum_{i=1}^{3} \int_V \nabla \cdot (p\hat{\mathbf{e}}_i)\,\mathrm{d}V = -\hat{\mathbf{e}}_i \sum_{i=1}^{3} \int_V \frac{\partial p}{\partial x_i}\,\mathrm{d}V$$

since $\hat{\mathbf{e}}_i \cdot \hat{\mathbf{e}}_j = \delta_{ij}$ and, finally,

$$\mathbf{f} = -\int_V \nabla p\,\mathrm{d}V \tag{A.12}$$

which coincides with the r.h.s. of Eq. (1.38).

From the physical point of view, Eq. (A.8) expresses the force $\mathbf{f}$ exerted on the closed surface $S(t)$ that forms the boundary of the material volume $V(t)$. Equation (A.12) shows that this force is related to the inhomogeneity of pressure p within volume V.

Normal and Tangent to a Plane Curve Implicitly Represented

If we represent ∂D implicitly as

$$B(x,y) = 0 \tag{A.13}$$

then the normal vector $\hat{\mathbf{n}}$ is given by

$$\hat{\mathbf{n}} = \frac{1}{\|\nabla B\|}\nabla B$$

and the tangent vector $\hat{\mathbf{t}}$ is given by

$$\hat{\mathbf{t}} \overset{(A.2)}{=} \frac{1}{\|\nabla B\|}\hat{\mathbf{k}} \times \nabla B = \frac{1}{\|\nabla B\|}\left(-\frac{\partial B}{\partial y}\hat{\mathbf{i}} + \frac{\partial B}{\partial x}\hat{\mathbf{j}}\right) \tag{A.14}$$

For example, if ∂D is a unit circle centred at the origin, then the simplest choice for B is $B(x,y) = x^2 + y^2 - 1$, and hence,

$$\hat{\mathbf{n}} = \hat{\mathbf{x}} = \frac{1}{\sqrt{x^2+y^2}}(x\hat{\mathbf{i}} + y\hat{\mathbf{j}})$$

$$\hat{\mathbf{t}} = \hat{\mathbf{k}} \times \hat{\mathbf{x}} = \frac{1}{\sqrt{x^2+y^2}}(-y\hat{\mathbf{i}} + x\hat{\mathbf{j}})$$

where $\hat{\mathbf{x}} := \mathbf{x}/\|\mathbf{x}\|$ with $\mathbf{x} := x\hat{\mathbf{i}} + y\hat{\mathbf{j}}$ and $\|\mathbf{x}\| := \sqrt{x^2+y^2}$.

Current and Stream Function

At large scale, both winds and marine currents are nearly horizontal, and the corresponding fluids are almost incompressible. Therefore, winds and currents can be expressed mathematically as vector fields of the kind

$$\mathbf{u} = u\hat{\mathbf{i}} + v\hat{\mathbf{j}}$$

that satisfy the condition

$$\operatorname{div}\mathbf{u} = 0 \tag{A.15}$$

In this case, vector $\mathbf{u}$ may be expressed through the *stream function* ψ as

$$\mathbf{u} = \hat{\mathbf{k}} \times \nabla\psi \tag{A.16}$$

or equivalently, using components, as

$$(u,v) = \left(-\frac{\partial\psi}{\partial y}, \frac{\partial\psi}{\partial x}\right) \tag{A.17}$$

Using the vector identity

$$\mathrm{div}\,(\mathbf{a}\times\mathbf{b}) = \mathbf{b}\cdot\mathrm{rot}\,\mathbf{a} - \mathbf{a}\cdot\mathrm{rot}\,\mathbf{b}$$

one verifies that current (A.16) fulfils condition (A.15):

$$\mathrm{div}\,\mathbf{u} = \mathrm{div}\,(\hat{\mathbf{k}}\times\nabla\psi) = \nabla\psi\cdot\mathrm{rot}\,\hat{\mathbf{k}} - \hat{\mathbf{k}}\cdot\mathrm{rot}\,\nabla\psi = 0$$

By applying $\hat{\mathbf{k}}\times$ to (A.16), we may invert it to get

$$\nabla\psi = \mathbf{u}\times\hat{\mathbf{k}}$$

It is evident that (A.16) is invariant with respect to the transformation $\psi \mapsto \psi + C$, where C is any constant. This fact is consistent with the physical meaning of ψ, which is a pressure and hence a quantity that enters the equations of motions only through its gradient.

No–Mass Flux Boundary Condition

The *no–mass flux* boundary condition

$$\hat{\mathbf{n}}\cdot\mathbf{u} = 0 \qquad \text{on } \partial D \tag{A.18}$$

expresses, in mathematical terms, the physical requirement that mass be conserved within the domain D. This interpretation is an obvious consequence of divergence theorem (A.10) and continuity equation (1.17) under the assumption of density conservation following the motion.

Using (A.16), condition (A.18) can be expressed in terms of the stream function as

$$\hat{\mathbf{n}}\cdot(\hat{\mathbf{k}}\times\nabla\psi) = 0 \qquad \text{on } \partial D$$

that is, resorting to (A.2) and (A.3),

$$\hat{\mathbf{t}}\cdot\nabla\psi = 0 \qquad \text{on } \partial D \tag{A.19}$$

Equation (A.19) shows that the directional derivative $\hat{\mathbf{t}}\cdot\nabla\psi = \partial\psi/\partial s$ is zero on the boundary. Hence, assuming D simply connected, we can write

$$\psi = K \qquad \text{on } \partial D$$

where K is a constant. Moreover, since ψ is defined up to an additive constant, we can choose $K = 0$, and hence, the no–mass flux boundary condition finally takes the simple form

$$\psi = 0 \qquad \text{on } \partial D \tag{A.20}$$

Jacobian Determinant

Definition

The *Jacobian determinant* (or, simply, *Jacobian*) of functions $\psi(x, y)$ and $q(x, y)$ is defined as

$$\mathscr{J}(\psi, q) := \det[\nabla_{\mathrm{H}}\psi \quad \nabla_{\mathrm{H}}q] = \det\begin{bmatrix} \dfrac{\partial\psi}{\partial x} & \dfrac{\partial q}{\partial x} \\ \dfrac{\partial\psi}{\partial y} & \dfrac{\partial q}{\partial y} \end{bmatrix} =$$

$$= \frac{\partial\psi}{\partial x}\frac{\partial q}{\partial y} - \frac{\partial\psi}{\partial y}\frac{\partial q}{\partial x} = \hat{\mathbf{k}}\cdot(\nabla\psi \times \nabla q) \tag{A.21}$$

where $\nabla_{\mathrm{H}}\psi$ and $\nabla_{\mathrm{H}}q$ are column vectors with $\nabla_{\mathrm{H}} := (\partial/\partial x, \partial/\partial y)$.

The motivation for this definition lies in the fact that the advection of a scalar q by a current $\mathbf{u}$ is described by $\mathbf{u}\cdot\nabla q$, and assuming (A.16), this means

$$\mathbf{u}\cdot\nabla q \overset{(A.16)}{=} (\hat{\mathbf{k}}\times\nabla\psi)\cdot\nabla q \overset{(A.3)}{=} \hat{\mathbf{k}}\cdot(\nabla\psi\times\nabla q) \overset{(A.21)}{=} \mathscr{J}(\psi, q) \tag{A.22}$$

Basic Identities

The Jacobian determinant verifies several useful identities. Among these, the following are immediate consequences of the definition (A.21) and of the usual differentiation rules:

$$\mathscr{J}(\psi, \psi) = 0 \tag{A.23}$$

$$\mathscr{J}(\psi, C) = 0 \tag{A.24}$$

$$\mathscr{J}(\psi, q) = -\mathscr{J}(q, \psi) \tag{A.25}$$

$$\mathscr{J}(C\psi, q) = C\,\mathscr{J}(\psi, q) \tag{A.26}$$

$$\mathscr{J}(\psi_1 + \psi_2, q) = \mathscr{J}(\psi_1, q) + \mathscr{J}(\psi_2, q) \tag{A.27}$$

$$\mathcal{J}(\psi_1\,\psi_2,\,q) = \psi_1\,\mathcal{J}(\psi_2,\,q) + \psi_2\,\mathcal{J}(\psi_1,\,q) \tag{A.28}$$

$$\mathcal{J}(\psi^\alpha,\,q) = \alpha\,\psi^{\alpha-1}\,\mathcal{J}(\psi,\,q) \tag{A.29}$$

$$\mathcal{J}(\psi,\,q) = \frac{\partial}{\partial x}\left(\psi\frac{\partial q}{\partial y}\right) - \frac{\partial}{\partial y}\left(\psi\frac{\partial q}{\partial x}\right) \tag{A.30}$$

$$\mathcal{J}(\psi,\,q) = \frac{\partial}{\partial y}\left(q\frac{\partial \psi}{\partial x}\right) - \frac{\partial}{\partial x}\left(q\frac{\partial \psi}{\partial y}\right) \tag{A.31}$$

where C is a constant and α is a real number. Identities (A.25)–(A.27) together imply the bilinearity of the Jacobian.

Moreover, the Jacobian determinant $\mathcal{J}(\psi,\,q)$ can be expressed in two ways as a function of $\psi\nabla q$:

$$\mathcal{J}(\psi,\,q) = \hat{\mathbf{k}}\cdot\mathrm{rot}\,(\psi\nabla q) \tag{A.32}$$

$$\mathcal{J}(\psi,\,q) = \mathrm{div}\,(\psi\nabla q\times\hat{\mathbf{k}}) \tag{A.33}$$

Indeed, applying identity

$$\nabla\psi\times\nabla q = \mathrm{rot}\,(\psi\nabla q)$$

to definition (A.21) yields (A.32) and hence (A.33) because of (A.3).

Using components, both (A.32) and (A.33) yield (A.30).

Functional Dependence

This topic will be presented here in a short heuristic way. For a thorough treatment, see Newns (1967).

Roughly speaking, functions $\psi(x,y)$ and $q(x,y)$ are *functionally dependent* if there exists a function $F(\psi,q)$ such that

$$F[\psi(x,y),\,q(x,y)] = 0 \qquad \forall x,y \tag{A.34}$$

Condition (A.34) is equivalent to

$$\mathcal{J}(\psi,\,q) = 0 \tag{A.35}$$

To justify this equivalence, we first show that (A.34) implies (A.35). Condition (A.34) implies $\nabla F = 0$, that is,

$$\begin{cases} \dfrac{\partial F}{\partial \psi}\dfrac{\partial \psi}{\partial x} + \dfrac{\partial F}{\partial q}\dfrac{\partial q}{\partial x} = 0 \\[2ex] \dfrac{\partial F}{\partial \psi}\dfrac{\partial \psi}{\partial y} + \dfrac{\partial F}{\partial q}\dfrac{\partial q}{\partial y} = 0 \end{cases} \tag{A.36}$$

The linear system (A.36) in the unknowns $\partial F/\partial\psi$ and $\partial F/\partial q$ has nonzero solutions (if and) only if

$$\det\begin{bmatrix}\dfrac{\partial\psi}{\partial x} & \dfrac{\partial q}{\partial x}\\[2mm] \dfrac{\partial\psi}{\partial y} & \dfrac{\partial q}{\partial y}\end{bmatrix}=0$$

which implies (A.35).

Conversely, (A.35) implies (A.34). By definition, (A.35) implies that $\nabla_{\mathrm{H}}\psi$ and $\nabla_{\mathrm{H}}q$ are proportional, and hence, ψ and q have the same isolines. Let C be a curve that crosses once all the isolines, and let s be the curvilinear abscissa on C. On C, functions ψ and q depend on s only:

$$\psi=\Psi(s) \qquad q=Q(s)$$

whence

$$s=\Psi^{-1}(\psi)=Q^{-1}(q)$$

and therefore,

$$\psi=\Psi[Q^{-1}(q)]$$

yields (A.34) with $F(\psi,q)=\psi-\Psi[Q^{-1}(q)]$.

Example

If C is the axis of abscissas and

$$\psi=2x+3y \qquad q=\exp(2x+3y)$$

then

$$Q(s)=\exp(2s) \qquad Q^{-1}(q)=\frac{1}{2}\log(q) \qquad \Psi(s)=2s$$

and therefore,

$$\psi=2\frac{1}{2}\log(q)=\log(q)$$

that is, $q=\exp(\psi)$ as it was a priori evident in this simple case.

Boundary as an Isoline of Stream Function

The boundary ∂D is an isoline of the stream function ψ. Indeed, using (A.19) and (A.14), we obtain $\mathscr{J}(B,\ \psi)=0$. By definition, this implies that $\nabla_{\mathrm{H}}B$ and $\nabla_{\mathrm{H}}\psi$ are proportional, and hence, B and ψ have the same isolines. Since ∂D is an isoline of B [namely, $B(x,\ y)=0$ by (A.13)], we conclude that ∂D is an isoline of ψ as well.

An Integral Property of the Jacobian

Let ∂R denote a simple closed curve on which a given stream function ψ takes the constant value ψ_0, and let R be the bounded region whose frontier is ∂R. The following identity holds:

$$\int_R \mathscr{J}(H(\psi), q)\,\mathrm{d}x\mathrm{d}y = 0 \tag{A.37}$$

for any functions H and q.

Indeed, the l.h.s of (A.37) can be rewritten as

$$\begin{aligned}\int_R \mathscr{J}(H(\psi), q)\,\mathrm{d}x\mathrm{d}y &\overset{(A.33)}{=} \int_R \mathrm{div}\,[H(\psi)\nabla q \times \hat{\mathbf{k}}]\,\mathrm{d}x\mathrm{d}y = \\ &\overset{(A.5)}{=} H(\psi_0) \oint_{\partial R} \hat{\mathbf{n}} \cdot (\nabla q \times \hat{\mathbf{k}})\,\mathrm{d}s = \\ &\overset{(A.3)\,(A.2)}{=} H(\psi_0) \oint_{\partial R} \hat{\mathbf{t}} \cdot \nabla q\,\mathrm{d}s = \\ &\overset{(A.6)}{=} H(\psi_0) \int_R \hat{\mathbf{k}} \cdot (\nabla \times \nabla q)\,\mathrm{d}x\mathrm{d}y = \\ &= 0\end{aligned}$$

Jacobian in Polar Coordinates

In polar coordinates (r, θ), the gradient may be expressed as

$$\nabla = \hat{\mathbf{r}}\frac{\partial}{\partial r} + \hat{\boldsymbol{\theta}}\frac{1}{r}\frac{\partial}{\partial \theta}$$

where $\hat{\mathbf{r}}$ and $\hat{\boldsymbol{\theta}}$ are unit vectors associated to coordinates r and θ, respectively. Accordingly, the Jacobian is given by

$$\begin{aligned}\mathscr{J}(\psi, q) &\overset{(A.21)}{=} \hat{\mathbf{k}} \cdot \left[\left(\hat{\mathbf{r}}\frac{\partial \psi}{\partial r}\right) + \left(\hat{\boldsymbol{\theta}}\frac{1}{r}\frac{\partial \psi}{\partial \theta}\right)\right] \times \left[\left(\hat{\mathbf{r}}\frac{\partial q}{\partial r}\right) + \left(\hat{\boldsymbol{\theta}}\frac{1}{r}\frac{\partial q}{\partial \theta}\right)\right] = \\ &= \frac{1}{r}\left(\frac{\partial \psi}{\partial r}\frac{\partial q}{\partial \theta} - \frac{\partial q}{\partial r}\frac{\partial \psi}{\partial \theta}\right)\end{aligned} \tag{A.38}$$

because $\hat{\mathbf{k}} \cdot (\hat{\mathbf{r}} \times \hat{\boldsymbol{\theta}}) = 1$ since the sequence $(\hat{\mathbf{k}}, \hat{\mathbf{r}}, \hat{\boldsymbol{\theta}})$ is orthonormal and right-handed.

Looking at the r.h.s. of (A.38), one immediately realizes that rotational symmetry (i.e. $\partial/\partial\theta = 0$) implies that the Jacobian is zero.

The Average Current Is Zero

Let R be as in (A.37); for example, we may take $R = D$ as noted before. The average current over R is zero.

Indeed, denoting by $m(R)$ the area of R, the average current components are:

$$\langle u \rangle := \frac{1}{m(R)} \int_R u \, \mathrm{d}x \mathrm{d}y \overset{(A.17)}{=} -\frac{1}{m(R)} \int_R \frac{\partial \psi}{\partial y} \mathrm{d}x \mathrm{d}y \tag{A.39}$$

$$\langle v \rangle := \frac{1}{m(R)} \int_R v \, \mathrm{d}x \mathrm{d}y \overset{(A.17)}{=} \frac{1}{m(R)} \int_R \frac{\partial \psi}{\partial x} \mathrm{d}x \mathrm{d}y \tag{A.40}$$

Moreover, the integral in the r.h.s. of (A.40) is given by

$$\begin{aligned} \int_R \frac{\partial \psi}{\partial x} \mathrm{d}x \mathrm{d}y &\overset{(A.4)}{=} \oint_{\partial R} \psi \, \mathrm{d}y = \psi_0 \oint_{\partial R} \mathrm{d}y = \psi_0 \oint_{\partial R} \nabla y \cdot (\hat{\mathbf{i}} \mathrm{d}x + \hat{\mathbf{j}} \mathrm{d}y) = \\ &\overset{(A.7)}{=} \psi_0 \oint_{\partial R} (\hat{\mathbf{t}} \cdot \nabla y) \, \mathrm{d}s \overset{(A.6)}{=} \psi_0 \int_R \hat{\mathbf{k}} \cdot (\nabla \times \nabla y) \, \mathrm{d}x \mathrm{d}y = \\ &= 0 \end{aligned} \tag{A.41}$$

Likewise,

$$\int_R \frac{\partial \psi}{\partial y} \mathrm{d}x \mathrm{d}y = 0 \tag{A.42}$$

Finally, substituting (A.41) and (A.42) in (A.39) and (A.40), we conclude that the average current is zero.

Integral Inequalities

Wirtinger-Like Inequalities

Let $\phi(x, y)$ be a differentiable function defined over the square domain $D = [a \leq x \leq b] \times [a \leq y \leq b]$ and such that

$$\phi = 0 \qquad \text{on } \partial D \tag{A.43}$$

The following inequalities are immediate consequences of Wirtinger's inequality (Ito (1987)):

$$\frac{\pi^2}{(b-a)^2} \int_a^b \phi^2(x, y) \, \mathrm{d}x \leq \int_a^b \left(\frac{\partial \phi}{\partial x} \right)^2 \mathrm{d}x \qquad \forall y \tag{A.44}$$

$$\frac{\pi^2}{(b-a)^2}\int_a^b \phi^2(x,y)\,\mathrm{d}y \le \int_a^b \left(\frac{\partial\phi}{\partial y}\right)^2 \mathrm{d}y \qquad \forall x \tag{A.45}$$

In turn, inequalities (A.44) and (A.45) imply

$$\frac{\pi^2}{(b-a)^2}\int_D \phi^2(x,y)\,\mathrm{d}x\,\mathrm{d}y \le \int_D \left(\frac{\partial\phi}{\partial x}\right)^2 \mathrm{d}x\,\mathrm{d}y \qquad \forall y \tag{A.46}$$

and

$$\frac{\pi^2}{(b-a)^2}\int_D \phi^2(x,y)\,\mathrm{d}x\,\mathrm{d}y \le \int_D \left(\frac{\partial\phi}{\partial y}\right)^2 \mathrm{d}x\,\mathrm{d}y \qquad \forall x \tag{A.47}$$

respectively.

By adding (A.46) and (A.47), we finally obtain

$$\frac{2\pi^2}{(b-a)^2}\int_D \phi^2\,\mathrm{d}x\,\mathrm{d}y \le \int_D |\nabla\phi|^2\,\mathrm{d}x\,\mathrm{d}y \qquad \forall x \tag{A.48}$$

An Inequality Relating Gradient and Laplacian

Consider the integral

$$I := \int_D \phi\,\nabla^2\phi\,\mathrm{d}x\,\mathrm{d}y \tag{A.49}$$

Integrating by parts (A.49), we get $I = \oint_{\partial D} \phi\,\hat{\mathbf{n}}\cdot\nabla\phi\,\mathrm{d}s - \int_D |\nabla\phi|^2\,\mathrm{d}x\,\mathrm{d}y$, and hence,

$$I = -\int_D |\nabla\phi|^2\,\mathrm{d}x\,\mathrm{d}y \tag{A.50}$$

because of (A.43).

On the other hand, by applying Schwarz inequality in the form

$$\left(\int_D f g\,\mathrm{d}x\,\mathrm{d}y\right)^2 \le \left(\int_D f^2\,\mathrm{d}x\,\mathrm{d}y\right)\left(\int_D g^2\,\mathrm{d}x\,\mathrm{d}y\right)$$

to the r.h.s. of (A.49), we get

$$I^2 \le \left(\int_D \phi^2\,\mathrm{d}x\,\mathrm{d}y\right)\left(\int_D (\nabla^2\phi)^2\,\mathrm{d}x\,\mathrm{d}y\right) \tag{A.51}$$

and hence, because of (A.50),

$$\left(\int_D |\nabla\phi|^2\,\mathrm{d}x\,\mathrm{d}y\right)^2 \le \left(\int_D \phi^2\,\mathrm{d}x\,\mathrm{d}y\right)\left(\int_D (\nabla^2\phi)^2\,\mathrm{d}x\,\mathrm{d}y\right) \tag{A.52}$$

Using (A.48) and simplifying, (A.52) takes the final form

$$\frac{2\pi^2}{(b-a)^2}\int_D |\nabla\phi|^2\,\mathrm{d}x\,\mathrm{d}y \leq \int_D (\nabla^2\phi)^2\,\mathrm{d}x\,\mathrm{d}y \tag{A.53}$$

An Application of (A.53)

Inequality (A.53) is useful, for instance, to derive an integral inequality that allows one to prove the decay of kinetic energy

$$K(t) = \frac{1}{2}\int_D |\nabla\psi|^2\,\mathrm{d}x\,\mathrm{d}y$$

in a system subjected to turbulent dissipation and no–mass flux boundary condition.

Indeed, in this case, the kinetic energy fulfils an equation of the kind

$$\frac{\mathrm{d}}{\mathrm{d}t}K(t) = -\varepsilon_1\,K(t) - \varepsilon_2\int_D (\nabla^2\psi)^2\,\mathrm{d}x\,\mathrm{d}y \tag{A.54}$$

with $\varepsilon_1 > 0$ and $\varepsilon_2 > 0$ [cf. (3.417)]. Applying (A.53) to the last term of (2.501), we obtain the differential inequality

$$\frac{\mathrm{d}}{\mathrm{d}t}K(t) \leq -\delta\,K(t) \tag{A.55}$$

with

$$\delta := \varepsilon_1 + \frac{4\pi^2}{(b-a)^2} > 0$$

which is equivalent to

$$\frac{\mathrm{d}}{\mathrm{d}t}[\exp(\delta\,t)\,K(t)] \leq 0 \tag{A.56}$$

By integrating (A.56) over the time interval $[0,t]$, one obtains

$$K(t) \leq \exp(-\delta\,t)\,K(0) \tag{A.57}$$

Inequality (A.57) shows that kinetic energy, whose time evolution is governed by Eq. (A.54), decays in time with (at least) exponential rate.

References

Apel JR (1987) Principles of ocean physics. Academic, London, p 631
Batchelor GK (1967) An introduction to fluid dynamics. Cambridge University Press, London, p 615
Boning CW (1986) On the influence of frictional parameterization in wind-driven ocean circulation models. Dyn Atmos Ocean 10:63–92
Brown RA (1991) Fluid mechanics of the atmosphere. Academic, San Diego, p 486
Bryan K (1963) A numerical investigation of a non-linear model of a wind-driven ocean. J Atmos Sci 20:594–606
Carnevale GF, Cavallini F, Crisciani F (2001) Dynamic boundary conditions revisited. J Phys Oceanogr 31(8): 2489–2497
Charney JG (1955) The generation of oceanic currents by wind. J Mar Res 14:477–498
Charney JG (1973) Planetary fluid dynamics. In: Morel P (ed) Dynamic meteorology. Reidel, Dordrecht, pp 97–352
Crisciani F, Cavallini F (2007) An inner additional condition for solving Munk-like ocean circulation models. Il Nuovo Cimento 30C(N.4): 347–364
Crisciani F, Özgökmen TM (2001) Symmetry properties of barotropic bottom-dissipated single-gyre systems in inertial regime. Dyn Atmos Ocean 33(2): 135–156
Crisciani F, Purini R (1997) Boundary condition of the Sverdrup solution from flow energetics. J Phys Oceanogr 27(N.2): 357–361
Crisciani F, Purini R (2011) A note on long Rossby waves on a quiescent abyss. Eur Phys J Plus 126:36
Crisciani F, Purini R (2012) On the linearization of the quasi-geostrophic potential vorticity equation at the ocean basin-scale. Eur Phys J Plus 127:6
Crisciani F, Cavallini F, Mosetti R (2007) On the additional boundary condition of wind- driven ocean models on the eastern coast. Il Nuovo Cimento 30C(N.2): 219–228
Csanady GT (1982) Circulation in the coastal ocean. Reidel, Dordrecht/Boston/London, p 279
Cushman-Roisin B (1994) Introduction to geophysical fluid dynamics. Prentice Hall, Englewood Cliffs, p 320
Dijkstra HA (2005) Nonlinear physical oceanography. Springer, Dordrecht/Berlin/Heidelberg/New York, p 532
Ertel H (1942a) Ein neuer hydrodynamischer Wirbesatz. Meteorol Z (Braunschweig) 59(N.9): 277–281
Ertel H (1942b) Ein neuer hydrodynamischer Erhaltungssatz. Die Naturwiss (Berlin) 30(N.36): 543–544
Ertel H (1942c) Uber hydrodynamische Wirbelsatze. Phys Z (Leipzig) 43:526–529

F. Cavallini and F. Crisciani, *Quasi-Geostrophic Theory of Oceans and Atmosphere*, Atmospheric and Oceanographic Sciences Library 45, DOI 10.1007/978-94-007-4691-6,

Euler L. (1757) Principes generaux du mouvement des fluides. Mémoires de l'académie des sciences de Berlin 11:274–315
Fletcher CAJ (1988) Computational techniques for fluid dynamics, vol II. Springer, Berlin, p 484
Fofonoff NP (1954) Steady flow in a frictionless homogeneous ocean. J Mar Res 13:254–262
Gill AE (1982) Atmosphere-ocean dynamics. Academic/Harcourt Brace, San Diego, p 662
Hellerman S, Rosenstein M (1983) Normal monthly wind-stress over the world ocean with error estimates. J Phys Oceanogr 13:1093–1104
Hendershott MC (1987) Single layer models of the general circulation. In: Abarbanel DI, Young WR (eds) General circulation of the ocean. Springer, Berlin/Heiderlberg/New York, p 291
Hide R (1978) General introduction: dynamics of rotating fluids. In: Roberts PH, Soward AM (ed) Rotating fluids in geophysics. Academic, London/New York, p 551
Holton JR (1979) An introduction to dynamic meteorology. Academic, New York/San Francisco/London, p 391
Ito K, Mathematical Society of Japan (ed) (1987), Encyclopedic dictionary of mathematics, 2nd edn. MIT Press, Cambridge
Kamenkovich VM, Koshlyakov MN, Monon AS (1986) Synoptic eddies in the ocean. Reidel, Dordrecht/Boston, p 433
Krauss K (1973) Methods and results in theoretical oceanography – Part 1: dynamics of the homogeneous and the quasihomogeneous ocean. Gebruder (u con dieresi)Borntraeger, Berlin/Stuttgart
LeBlond PH, Mysak LA (1978) Waves in the ocean. Elsevier, Amsterdam/Oxford/New York, p 602
Lorenz E (1955) Available potential energy and the maintenance of the general circulation. Tellus 7:157–167
Massel SR (1999) Fluid mechanics for marine ecologists. Springer, Berlin/Heidelberg, p 566
Müller P (1995) Ertel's potential vorticity theorem in physical oceanography. Rev Geophys 33: 67–97
Munk WH (1950) On the wind-driven ocean circulation. J Meteor 7:79–93
Newns WF (1967) Functional dependence. Am Math Monthly 74:911–920
Pedlosky J (1965) A note on the western intensification of the oceanic circulation. J Marine Res 23:207–209
Pedlosky J (1987) Geophysical fluid dynamics. Springer, New York, p710
Pedlosky J (1996) Ocean circulation theory. Springer, Berlin, p 453
Pedlosky J (2003) Waves in the ocean and atmosphere. Springer, Berlin, p 260
Peixoto JP, Oort AH (1992) Physics of climate. Springer, New York, p 520
Pond S, Pickard GL (1983) Introductory dynamical oceanography. Pergamon, Oxford, p 329
Proudman J. (1953) Dynamical oceanography. Methuen, London/Wiley, New York, p 409
Rhines PB, Young WR (1982) A theory of the wind-driven circulation. I. Mid-ocean gyres. J Mar Res 40(suppl): 559–596
Rossby CG, Collaborators (1939) Relation between variations in the intensity of the zonal circulation of the atmosphere and the displacement of the semipermanent centers of action. J Mar Res 2:38–55
Salby ML (1996) Fundamentals of atmospheric physics. Academic, New York/SanFrancisco/London, p 627
Salmon R (1998) Lectures on geophysical fluid dynamics. Oxford University Press, New York/Oxford, p 378
Stommel H (1948) The westward intensification of wind-driven ocean currents. Trans Am.Geophys Union 29:202–206
Stull RB (1988) An introduction to boundary layer meteorology. Kluwer, Dordrecht/Boston/London, p 666
Sverdrup HU (1947) Wind-driven currents in a baroclinic ocean; with application to the equatorial currents of the eastern Pacific. Proc Nat Acad Sci 33:318–326
Vallis GK (2006) Atmospheric and oceanic fluid dynamics. Cambridge University Press, Cambridge, p745

Veronis G (1966) Wind-driven ocean circulation – Part 2. Numerical solutions of the non-linear problem. Deep-Sea Res 13:31–55

Wunsch C (2011) The decadal mean ocean circulation and Sverdrup balance. J Mar Res 69: 417–434

Young WR (1987) Baroclinic theories of the wind driven circulation. In: Abarbanel DI, Young WR (ed) General circulaton of the ocean. Springer, Berlin/Heidelberg/New York, p 291

Zeidler E (1985) Nonlinear functional analysis and its applications III, Springer, New York

Index

F. Cavallini and F. Crisciani, *Quasi-Geostrophic Theory of Oceans and Atmosphere*, Atmospheric and Oceanographic Sciences Library 45, DOI 10.1007/978-94-007-4691-6,

MIX
Papier aus verantwortungsvollen Quellen
Paper from responsible sources
FSC® C105338

If you have any concerns about our products,
you can contact us on
ProductSafety@springernature.com

In case Publisher is established outside the EU,
the EU authorized representative is:
Springer Nature Customer Service Center GmbH
Europaplatz 3, 69115 Heidelberg, Germany

Printed by Libri Plureos GmbH
in Hamburg, Germany